AF712617

CONSTRUCTION ESTIMATING

second edition

Featuring
Sage Timberline
Estimating Software

sage

AMERICAN TECHNICAL PUBLISHERS
ORLAND PARK, ILLINOIS 60467-5756

Leonard P. Toenjes

Construction Estimating contains procedures commonly practiced in industry and the trade. Specific procedures vary with each task and must be performed by a qualified person. For maximum safety, always refer to specific manufacturer recommendations, insurance regulations, specific job site and plant procedures, applicable federal, state, and local regulations, and any authority having jurisdiction. The material contained is intended to be an educational resource for the user. American Technical Publishers, Inc. assumes no responsibility or liability in connection with this material or its use by any individual or organization.

American Technical Publishers, Inc., Editorial Staff

Editor in Chief:
Jonathan F. Gosse
Vice President—Production:
Peter A. Zurlis
Art Manager:
James M. Clarke
Multimedia Manager:
Carl R. Hansen
Technical Editor:
Richard A. Scheffrahn
Copy Editor:
Dane K. Hamann
Cover Design:
Melanie G. Doornbos
Illustration/Layout:
Melanie G. Doornbos
Jennifer M. Hines
CD-ROM Development:
Robert E. Stickley
Daniel Kundrat
Amanda N. Sidorowicz
Nicole S. Polak

Microsoft, Windows, Windows Vista, PowerPoint, and Internet Explorer are either registered trademarks or trademarks of Microsoft Corporation in the United States and/or other countries. Adobe, Acrobat, and Reader are registered trademarks of Adobe Systems Incorporated in the United States and/ or other countries. Intel is a registered trademark of Intel Corporation in the United States and/or other countries. Firefox is a registered trademark of the Mozilla Foundation. Autodesk Buzzsaw and Autodesk Revit are registered trademarks of Autodesk, Inc. ConcensusDocs is a trademark of Associated General Contractors of America Corp. Ditch Witch is a registered trademark of The Charles Machine Works, Inc. e-builder is a trademark of E-Builder, Inc. MasterFormat is a trademark of The Construction Specifications Institute, Inc. NEC is a registered trademark of National Fire Protection Association. UL is a registered trademark of Underwriters Laboratories Inc. Quick Quiz, Quick Quizzes, and Master Math are either registered trademarks or trademarks of American Technical Publishers, Inc.

© 2012 by American Technical Publishers, Inc.
All rights reserved

2 3 4 5 6 7 8 9 – 12 – 9 8 7 6 5 4 3 2 1

Printed in the United States of America

ISBN 978-0-8269-0545-1

This book is printed on recycled paper.

Acknowledgments

The author and publisher are grateful to the following companies, organizations, and individuals for providing software, photographs, information, and technical assistance.

American Hardwood Export Council
Andersen Windows, Inc.
APA—The Engineered Wood Association
Autodesk, Inc.
Baldor Electric Co.
Baldwin
California Redwood Association
Carlisle
Case, LLC
The Charles Machine Works, Inc., Manufacturer of Ditch Witch® Products
Classic Post & Beam
Cleaver-Brooks
ELE International, Inc.
Gates & Sons, Inc.
General Electric Company
Hilti, Inc.
iBeam Systems, Inc.
JCB Inc.
JLG Industries, Inc.
John Deere Construction & Forestry Equipment
Multiquip Inc.
National Wood Flooring Association
Pella Corporation
Portland Cement Association
Sage
Southern Forest Products Association
Symons Corporation
TrusSteel Div. of Alpine Engineering Products, Inc.

Department of Construction Science and Management
Clemson University
Gregg R. Corley, P.E.

Department of Building Science
Auburn University
Roger Killingsworth
Associate Professor

Sage
Vicki Roberge
Aaron Jost

Contents

Contents

Interactive CD-ROM Contents

Introduction

Construction Estimating is a comprehensive introduction to estimating practices used in the construction industry. The textbook presents estimating practices and methods that features content applicable to ledger sheet, electronic spreadsheet, and estimating software use. Applicable CSI MasterFormat™ divisions are referenced in each chapter of the textbook.

Construction Estimating covers how to estimate and provides a guide to sources of information on prints and specifications. The chapters provide a review of construction materials and methods prior to the introduction of estimating practices in order to provide an intuitive transition for students. This organization builds a solid foundation for understanding and developing accurate estimates.

Each chapter of *Construction Estimating* includes several features to enhance learning—Key Concepts, Key Terms, Web Links, Review Questions, and Activities. Key Concepts presented at the beginning of each chapter are designed to focus on important estimating concepts presented in each chapter. Key Terms at the end of each chapter identify terms that are critical to understanding the topics presented. Web Links identify websites of trade associations and organizations where additional information on material, labor, or equipment can be obtained. Review Questions can be used to evaluate student comprehension of chapter content. Activities provide opportunities for students to prepare estimates using ledger sheets, electronic spreadsheets, or estimating software through the use of prints, specifications, and cost data.

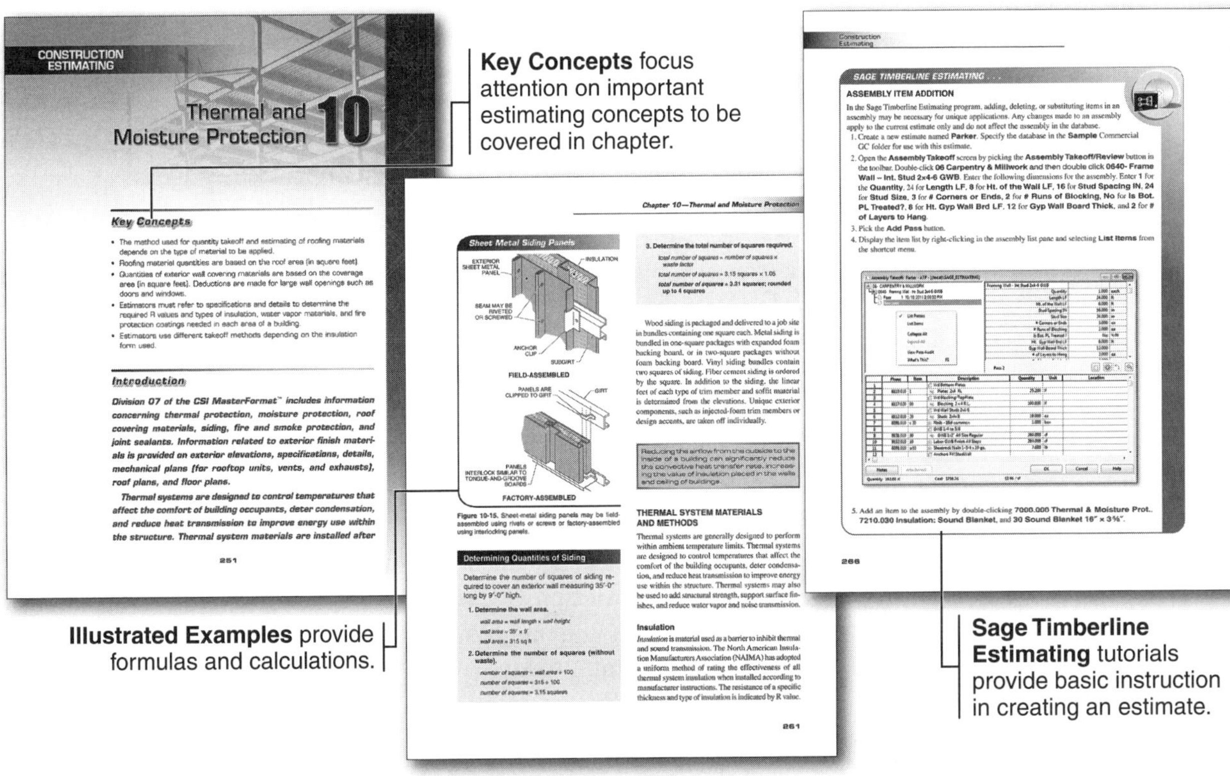

Interactive CD-ROM Features

The *Construction Estimating* Interactive CD-ROM provided with the textbook is designed to enhance and supplement the content in the textbook.

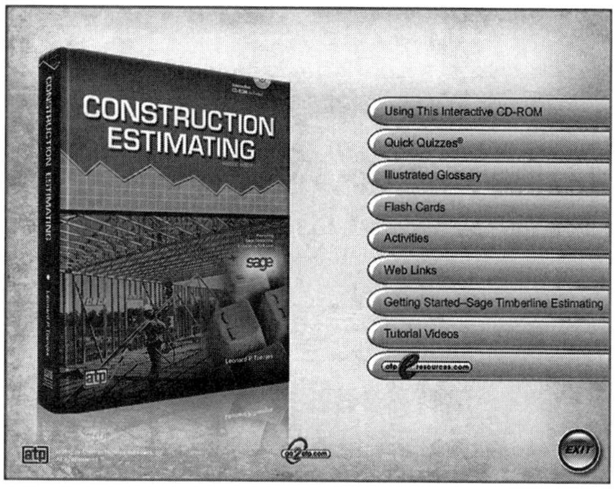

- Quick Quizzes® with 10 questions per chapter to reinforce fundamental concepts
- Illustrated Glossary of industry terms with links to illustrations and video clips
- Flash Cards for review of common industry terms and definitions
- Activities include prints, specifications, and spreadsheets and provide opportunities to prepare estimates using Microsoft® Excel® or Sage Timberline Estimating
- Web Links with links to online reference material that support continued learning
- Getting Started—Sage Timberline Estimating provides information for installation of the Sage Timberline Estimating demonstration release and instructions for updating the General Contractor's database
- Tutorial Videos provide interactive hands-on tutorials covering fundamental Sage Timberline Estimating procedures
- Access to ATPeResources.com, which provides a comprehensive array of instructional resources

The Sage Timberline Estimating DVD provided with the textbook includes a 180-day demonstration release of the Extended Edition (Version 11.2). Sample MasterFormat™ databases are provided with the software that can be used to complete activities presented in the textbook. Installation information for the Sage Timberline Estimating software is included on the *Construction Estimating* CD-ROM and can be accessed by picking the Getting Started—Sage Timberline Estimating section. When used in conjunction with the textbook, the *Construction Estimating* CD-ROM and Sage Timberline Estimating software provide an exceptional estimating training package.

Construction Estimating is one of many high-quality training products available from American Technical Publishers. To obtain information about related training products, visit the American Technical Publishers website at www.go2atp.com.

The Publisher

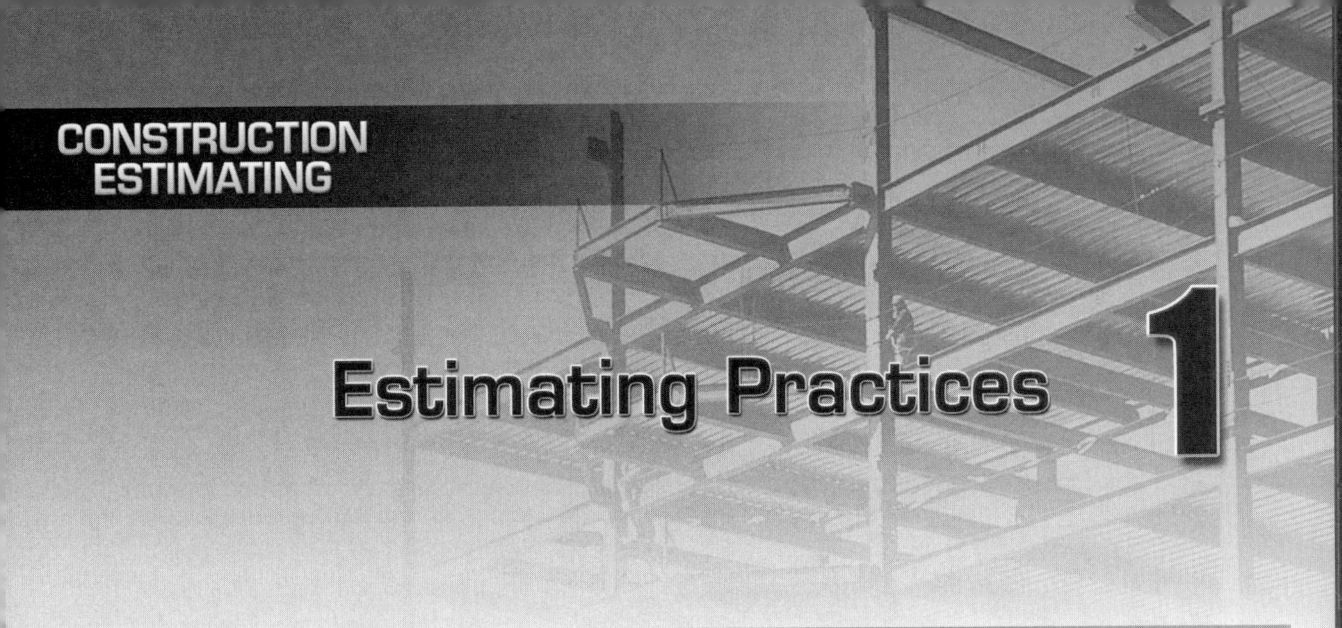

Estimating Practices 1

Key Concepts

- Levels of interim estimates include order of magnitude estimates, conceptual estimates, design development estimates, and construction document estimates.
- Levels of final estimates include detail takeoff estimates and crew-based estimates.
- The estimating process begins with a review of all construction contract documents including detail drawings, and specification books including the general specifications, addenda, and agreement forms.
- Estimators are involved in almost every phase of the planning and construction process.
- Design features may be changed based on estimator input in value engineering.
- Estimators use a vast library of standardized information in the pricing and bidding of materials, equipment, and labor.

Introduction

To complete the takeoff and estimating processes successfully, good estimating practices should be employed. By using a systematic approach to putting an estimate together, estimators ensure accuracy, comprehensiveness, consistency, and timeliness for each project. An estimator must work with the most current and complete set of plans and specifications, including all addenda, when working on a design-build or design-bid-build project. An estimator must have a complete understanding of the concepts and requirements of the owner and design team when developing a conceptual or preliminary estimate.

ESTIMATING PRACTICES

An *estimating practice* is a practice used to integrate all parts of the estimating process in a cohesive, consistent, and reliable manner to ensure an accurate final bid. Interim estimates are prepared as the design process proceeds towards completion. Final estimates are prepared when all construction documents, including detail drawings and specifications, are completed.

> The purpose of this document is to present the Schematic Design Phase for the first Academic Building at the Spokane Higher Education Park at Riverpoint. The 100,000 square foot facility will be part of the future Academic Campus which calls for the construction of a projected 377,000 gross square feet of space to accommodate the Spokane programs of Washington State University and Eastern Washington University. The development of the Higher Education Park beyond that date will be accommodated on the remainder of the 48-acre site.
>
> The location for this first academic building is directly southwest across Riverpoint Blvd. from Spokane Intercollegiate Research and Technology Institute (SIRTI),

Conceptual estimates require a minimum amount of information such as building function, size, and location.

Estimating Levels

Various levels of estimates may be required for a project at different stages of the project. Owners typically prepare a series of initial estimates to determine project feasibility under various financial scenarios and design options. As a project moves from concept to design to completion of construction documents and detail drawings, various interim estimates may be required to ensure that the project is within the price range and desired function of the owner and the capability of the contractors and subcontractors attempting to bid on the project. The final owner determination for pricing of the entire construction project as planned is completed when all owner decisions are finalized and construction documents and detail drawings for the project are ready to be distributed to contractors to begin the bidding process.

Levels of interim estimates include order of magnitude estimates, conceptual estimates, design development estimates, and construction document estimates. The levels of final estimates include detail takeoff estimates and crew-based estimates. The estimating practices used depend on company practice, the project being estimated, and/or estimator preference.

Order of Magnitude Estimates. Order of magnitude estimates allow for the general scope of the entire construction project to be defined. It is most economical for the purchaser of a construction project to balance available finances with needed project functionality early in the planning process. Order of magnitude estimates are used to develop various basic project budgets based on a variety of project options and require the least amount of construction documentation to complete. In some instances, an order of magnitude estimate may indicate that it is not feasible to move forward with a particular desired project based on a lack of financial capacity for the desired structure.

Information developed by the project owner related to the project type, fundamental design, site location, and quality of construction is required to prepare an order of magnitude estimate. An order of magnitude estimate should be created by an owner when the initial concept and design is in the very early stages or approximately 5% complete. For example, an owner may desire that an extremely high level of sustainable construction practices and materials be included in a building. Order of magnitude estimates can be developed with various options and pricing to allow the owner to make realistic determinations about options such as sustainable roofing materials or reuse of surface water in the context of available financing. The estimate options are compared to the project budget to determine which project design and components are most appropriate for the available financial resources.

A well-designed order of magnitude estimate can save a project from significant construction problems in the future. Full consideration and definition of design and construction options during the initial stages of a project minimize financial issues, clarify future design options for the architectural and engineering team, minimize change orders during the construction process, and help ensure the owner receives the final product desired.

Conceptual Estimates. Conceptual estimates, also referred to as schematic or model estimates, are based on the building function or the functional area of the building. Conceptual estimates can be prepared for all types of construction delivery systems but are more common for design-build projects. Each building function or functional area is priced based on company historical records of costs per square foot according to the function of a similar building. Because of this, in most cases, conceptual estimates are based on incomplete construction documents.

Construction documents for conceptual estimates may include general information related to a building site, the foundation and structural systems, and the general electrical and mechanical requirements. Where possible, conceptual estimates are priced by individual building systems, such as exterior walls, roofs, or structural members. These types of conceptual estimates require a certain amount of general quantity takeoff but are not as detailed as detail takeoff or crew-based estimates.

Additional detailed cost information is included in conceptual estimates that may again change some elements of building functionality, size, or design. For example, a conceptual estimate may indicate that substitutions of various types of curtain wall systems with various window or glass options in certain portions of a structure may be more economical for the owner and still serve the desired functionality. Conceptual estimates of this type allow the owner to make more informed construction decisions, again minimizing future change orders or disagreements about building performance.

Conceptual estimates in their most basic form require only minimal information related to the building, such as function, size, height, location, and construction system used and should be created when the design is 5% to 40% complete.

Design Development Estimates. Design development estimates require completion of the majority of the construction documents and confirm the overall project costs as the design nears completion. During the development of a design development estimate, considerations are again given to specific construction options. In good construction practices, this would be the final and most appropriate point to make any significant design or function changes. A minimum of 25% of the construction documents should be completed before beginning to complete a design development estimate.

However, a design development estimate is best created by the owner when construction documents are 40% to 50% complete. In some instances, design development estimates for certain portions of the project could be completed after construction has started. When these design development estimates are utilized later in the project, contract price adjustments must be taken into consideration.

Construction Document Estimates. Construction document estimates are prepared when the construction documents, including detail drawings and specifications, are more than 90% complete. Construction document estimates are the most accurate cost estimates prepared prior to the final bid, but they require a great deal of effort to produce. Owners commonly compare the costs determined at this level of estimating with bids received from various contractors and subcontractors. Variances in pricing between owner estimates and contractor estimates are then discussed and reviewed before the project is awarded to the contractor.

Detail Takeoff Estimates. A *detail takeoff estimate* is an estimate in which each individual construction component is accounted for, or taken off. For example, a detail takeoff estimate for a section of concrete sidewalk may include individual entries and calculations for the cubic yards of concrete, forming materials, subsurface preparation, excavation labor, excavation equipment costs, concrete placement labor costs, concrete finishing labor costs, form removal labor, and final grading. **See Figure 1-1.** When developing an estimate for a project, detail takeoff estimates should only be prepared for work done by the contractor; subcontractors should prepare detail takeoff estimates for work they will be performing.

Portland Cement Association

Figure 1-1. In detail takeoff estimating, each individual construction component is accounted for, or taken off.

Crew-Based Estimates. A *crew-based estimate* is an estimate in which material quantities, equipment, and labor costs are included in a single calculation based on a specific quantity of construction put in place. For example, the material quantities, equipment, and labor costs for a concrete element of building might be based on cubic yards of concrete.

Care must be taken when using crew-based estimates to ensure that all labor and equipment costs are current and included in the estimate. In all instances, the amount and quality of equipment available will impact productivity and cost. For equipment-intensive operations, such as earthmoving, special attention is paid to the type of equipment provided for a particular operation and the costs for operating that equipment, since that is the primary expense for this crew-based operation. For labor-intensive operations requiring a large amount of handwork, more attention must be paid to the quality and productivity of the labor force.

Estimators must work closely with project managers and other field staff to develop accurate crew-based estimates for various construction operations. Many variables in equipment technology and condition, labor productivity and training, and specific job-site conditions must be considered when preparing crew-based estimates. For cost effectiveness, the variables must be monitored, reviewed, and kept current in the contractor database to ensure crew-based estimates match the actual job-site requirements.

For crew-based estimates, material quantities, equipment, and labor costs are included in a single calculation. For example, a single takeoff entry would be entered in an estimate for a section of concrete sidewalk. The estimate would be based on the number of square feet of sidewalk to be put in place and a factor that represents all labor, materials, and equipment costs for the operation.

Labor costs for this particular operation would include excavation, forming, concrete placement, and finishing. Materials would include any required subsurface treatment, forming materials, concrete, and any finishing treatments. Equipment could include various concrete transporting devices, screeding equipment, power trowels, or other equipment based on the size of the concrete placement. Company historical data of total costs per unit of construction is commonly factored into crew-based estimates.

ESTIMATING PROCESS

Estimating is the computation of construction costs of a project. The estimating process begins with a review of all construction contract documents, including detail drawings, and specification books, including the general specifications, addenda, and agreement forms. **See Figure 1-2.** An *addendum* is a change to the originally issued construction contract documents. Construction contact documents are reviewed to determine if a particular construction project contains items that can be built, installed, or provided by a particular construction company or supplier. The review also assists the estimator in developing a conceptual idea of the scope of the project and the approximate project budget.

Figure 1-2. Printreading skills are the basis of successful estimating.

Bid Determination and Strategy

The estimating and bidding process can be time-consuming and can involve many company and outside resources, generating an initial cost even before entering into an agreement. A decision by management about the sizes and scopes of projects to bid on is important for long-term company success. Prior to entering into the estimating and bidding process, the owner of a contracting, subcontracting, or material-supply firm must take several factors into consideration to help ensure success for a project. These factors include type of construction, company capacity, and market conditions.

Type of Construction. General contractors and subcontractors often specialize in various types of construction projects. For example, a contractor may specialize in health care projects such as hospitals and outpatient care facilities. Unless there are special market conditions or company issues, this contractor would not likely bid on a movie theater or series of cellular phone towers. It is common for contractors to seek projects that fit into their construction experience to take full advantage of their equipment, human resources, and expertise and to maximize their bidding and estimating experiences and resources.

Company Capacity. Construction companies manage their firms by maintaining an appropriate backlog of work. The amount of management staff, labor, and equipment resources must be balanced against the number of projects to be estimated and bid on. Too few projects in the backlog require that the estimating department seek additional projects to help maintain company operations. Too many projects in the backlog can result in a failure to complete the contracted work by the required date. Maintaining the proper backlog plays a role in the number of projects bid and the profit margins that management may apply to a particular project.

Market Conditions. The laws of supply and demand apply to bidding determination and strategy. When there is a large supply of projects with few contractors available to supply the construction services, higher profit margins and estimates can be anticipated. Conversely, when there are many contractors bidding on a limited number of projects, lower profit margins and estimates can be expected. Market conditions apply to the general contractor, subcontractors, and suppliers. An evaluation of the number of contractors or subcontractors bidding on a particular project will provide the company owner and the estimating department with information about the market conditions and assist in evaluating whether resources spent in preparing a bid will be justified.

Prebid Meeting

A *bid* is an offer from a contractor, subcontractor, or supplier to, respectively, perform construction work, provide a construction service, or provide materials or supplies at a stated price. A *prebid meeting* is a conference in which all interested parties in a construction project review the project, ask questions of the architect or owner concerning methods for accomplishing the work, and share information necessary to understand the entire scope of the work. The estimator, architect or owner, project manager, engineers, and contractors are typically in attendance at a prebid meeting.

Estimators need to understand the limits of the contractor's capacity to do the work that is bid. Some work may need to be subcontracted to other contractors.

At a prebid meeting, the estimator meets with owners, and possibly with the future occupants of the structure. A *contractor* is an individual or company responsible for performing construction work in accordance with the terms of a contract. A *general contractor* is the contractor who has overall responsibility for a construction project. A *subcontractor* is a specialty contractor who works under the direct control of the general contractor. General contractor and subcontractor representatives attend the prebid meeting to gain a better understanding of the project and make contacts to obtain future bidding information. In some cases, attendance at a prebid meeting is required in order to submit a bid that meets the owner requirements.

The architect and engineers are also normally in attendance at a prebid meeting to field questions concerning the construction project. The estimator may contact the architect in advance of the prebid meeting to acquire the names of all plan holders and obtain information concerning other potential bidders, including subcontractors and suppliers who may have assisted in compiling the bidding documents. Any problems with the bidding documents that may result from the number of final copies required, alternate proposals required, subcontractor and supplier listings required, or unit pricing requirements may also be discussed prior to the prebid meeting.

Estimators are involved in almost every phase of the planning and construction process. Owners will employ estimators to develop initial cost estimates at the order of magnitude and conceptual levels. Architectural, engineering, and design firms utilize estimators to prepare design development estimates and construction document estimates. Contractors, subcontractors, and suppliers rely on estimators to provide detail takeoff estimates or crew-based estimates to develop costs for inclusion in the bid to the owner.

Bid Proposal

Bids and bid proposals include prices and costs for completion of a construction project. A bid proposal includes pricing, performance criteria, and scheduling and staffing information as needed by the owner for selection of the contractor or contracting team. Bids and bid proposals must be presented to the owner and architect in a format that is clear and comprehensible. Each firm seeking the construction contract submits information including quantity, cost, overall pricing, and any other owner requirements specific to the construction project.

The bid generated by the estimator covers each section of the specifications and includes all items that are estimated and priced. Subcontractors are contacted for bids on any portion of the work that is not taken off and estimated by the estimator. *Quantity takeoff* is the practice of reviewing construction contract documents to determine the quantities of materials to be included in a bid. Bids are requested from subcontractors through personal contacts or written requests. These requests note the work to be performed, the project schedule, and the time at which bids are due.

When preparing an estimate, representatives from a general contractor, subcontractor, or supplier may visit the job site to determine working and building conditions. Such a visit can provide information that may not be apparent in the construction documents. Additional costs that may impact the final bid can be identified during a job site visit.

The estimator begins to take off and price each of the specification sections. For large projects, the quantity takeoff process may be divided among several estimators, each with their own specialty. For example, individual estimators may specialize in electrical, mechanical, or earthwork. Items are entered into columns of the master bid sheet or an estimating program as the estimate progresses. **See Figure 1-3.**

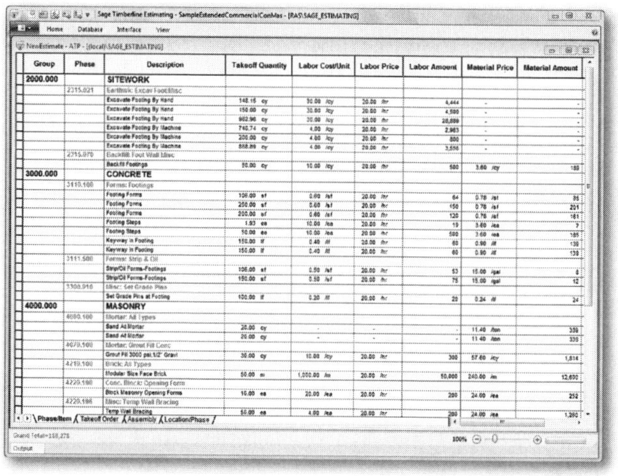

Figure 1-3. Estimating programs or spreadsheets are commonly used to compile bid information.

Labor, material, and equipment costs are typically determined after quantity takeoff is completed. However, labor, material, and equipment cost factors may be part of the quantity takeoff process in a spreadsheet or estimating software. Labor costs are highly variable and must be carefully determined based on worker skill and productivity. Material costs are somewhat easier to determine based on supplier bids; however, future material cost fluctuations and variations must be carefully monitored and taken into account when preparing an estimate. Equipment expenses, including depreciation, maintenance, and insurance costs, must also be taken into account.

Subcontractor Quotations. Some portions of a construction project may be assigned to subcontractors for bidding and performance of the work. The scope of the work to be bid and performed by a subcontractor must be clearly defined. A scope-of-work letter may be provided to subcontractors who are bidding on portions of the work to ensure complete, consistent, and accurate subcontractor bids. **See Figure 1-4.** As subcontractor bids are received, they are entered into the quoted bid column of a ledger sheet or in the proper cell of spreadsheets or estimating software.

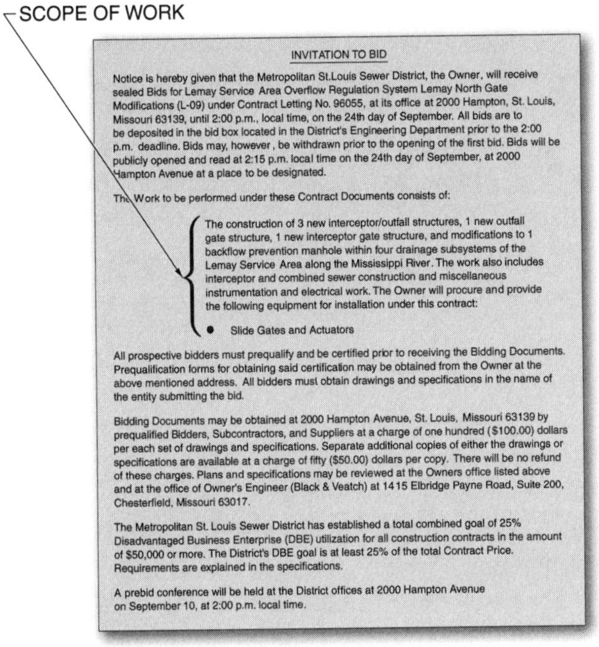

INVITATION TO BID

Notice is hereby given that the Metropolitan St.Louis Sewer District, the Owner, will receive sealed Bids for Lemay Service Area Overflow Regulation System Lemay North Gate Modifications (L-09) under Contract Letting No. 96055, at its office at 2000 Hampton, St. Louis, Missouri 63139, until 2:00 p.m., local time, on the 24th day of September. All bids are to be deposited in the bid box located in the District's Engineering Department prior to the 2:00 p.m. deadline. Bids may, however, be withdrawn prior to the opening of the first bid. Bids will be publicly opened and read at 2:15 p.m. local time on the 24th day of September, at 2000 Hampton Avenue at a place to be designated.

The Work to be performed under these Contract Documents consists of:

The construction of 3 new interceptor/outfall structures, 1 new outfall gate structure, 1 new interceptor gate structure, and modifications to 1 backflow prevention manhole within four drainage subsystems of the Lemay Service Area along the Mississippi River. The work also includes interceptor and combined sewer construction and miscellaneous instrumentation and electrical work. The Owner will procure and provide the following equipment for installation under this contract:

- Slide Gates and Actuators

All prospective bidders must prequalify and be certified prior to receiving the Bidding Documents. Prequalification forms for obtaining said certification may be obtained from the Owner at the above mentioned address. All bidders must obtain drawings and specifications in the name of the entity submitting the bid.

Bidding Documents may be obtained at 2000 Hampton Avenue, St. Louis, Missouri 63139 by prequalified Bidders, Subcontractors, and Suppliers at a charge of one hundred ($100.00) dollars per each set of drawings and specifications. Separate additional copies of either the drawings or specifications are available at a charge of fifty ($50.00) dollars per copy. There will be no refund of these charges. Plans and specifications may be reviewed at the Owners office listed above and at the office of Owner's Engineer (Black & Veatch) at 1415 Elbridge Payne Road, Suite 200, Chesterfield, Missouri 63017.

The Metropolitan St. Louis Sewer District has established a total combined goal of 25% Disadvantaged Business Enterprise (DBE) utilization for all construction contracts in the amount of $50,000 or more. The District's DBE goal is at least 25% of the total Contract Price. Requirements are explained in the specifications.

A prebid conference will be held at the District offices at 2000 Hampton Avenue on September 10, at 2:00 p.m. local time.

Figure 1-4. A scope-of-work letter provides complete information to subcontractors bidding on portions of the work.

A conceptual estimate of subcontractor-supplied work should be generated by the owner or the estimator for the general contractor for comparison with the subcontractor bids. This comparison ensures accuracy and provides verification of the costs, especially when using unfamiliar subcontractors. An estimator for a general contractor may not always use the lowest subcontractor bid. Care is taken to use the lowest qualified bid that addresses the scope of work and can complete the project successfully.

Comprehensive requests for subcontractor and supplier quotations are required. Estimators for general contractors should maintain a current distribution list and e-mail and telephone lists of subcontractors and suppliers. All notifications should include complete information about the project being bid, including the job name, architect, bid date and time, information concerning plan availability, and the name, telephone number, and e-mail address of the estimator. Estimators should make follow-up calls or send e-mails to all notified subcontractors and suppliers to determine if they will be providing a bid for a specific project. Suppliers may be a source of information for the general contractor estimator concerning subcontractors who may be submitting bids for a project.

Systematic Approach

A systematic approach to estimating incorporates standard practices and procedures to ensure an accurate and comprehensive estimate. Successful estimating begins with an organized approach to quantity takeoff. The estimator must fully review and understand all portions of the project pertaining to the estimate being prepared. In addition, the estimator should use a consistent process when preparing all costs and quantities of work with the most current set of detail drawings and specifications, including all addenda.

Accuracy. Accuracy in an estimate includes precise calculations of all labor, materials, and equipment quantities and costs. Mathematical and printreading skills are essential for accurate estimates. Accurate totals are required as items are counted or calculated from the drawings and specifications. Estimators commonly markup drawings and specifications to ensure that all necessary items are counted and that no items are counted twice. **See Figure 1-5.** Accurate interpretation of all print symbols, abbreviations, and dimensions is also required for quantity takeoff and estimating.

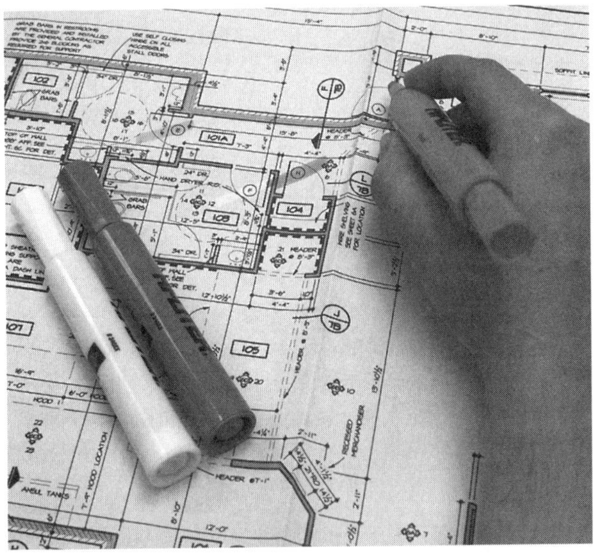

Figure 1-5. Prints are marked by an estimator to indicate that quantities have been taken off.

Comprehensiveness. Comprehensiveness in estimating ensures that all portions of the proposed work are included in the estimate. Care must be taken when reviewing the specifications and drawings as some items may be indicated in the specifications and not on the drawings, or indicated on the drawings and not in the specifications. Marking up both the drawings and specifications as a bid is generated ensures that all items are counted and no items are overlooked.

A complete review of the specifications and addenda for all items that may affect the estimate is required. Items may be described in several portions of the drawings and specifications. Estimators should not rely on a single source of product or material information in a set of drawings and specifications.

Consistency. Consistent estimating practices include the common procedures used by a company or an individual that allow for quick reference to items in the estimate. Patterns and habits should be developed for creating good estimates. While procedures used by individuals vary, standard steps or procedures are followed for reviewing plans and specifications and developing each bid. A common standard procedure for creating an accurate and consistent estimate is as follows:

1. Review the specifications.
2. Review the detail drawings.
3. Set up the bidding sheets or a database (either ledger-based or computer-based).
4. Perform the quantity takeoff for self-performed work.
5. Determine the price of materials, equipment, and labor for self-performed work.
6. Obtain estimates for subcontracted work.
7. Review the final estimate.
8. Double-check for accuracy.

After the initial process is complete, all items are priced and double-checked. Profit, overhead, and taxes are added, and the job is ready for final review and a definitive bid. Estimators should track last-minute items that could cause problems on the final bid day and anticipate possible actions and solutions.

Timeliness. A timely estimating process requires that the most current construction documents, including the detail drawings and specifications, are used for the bidding process. After the initial plans are released, an architect or owner may make changes to the project in the form of addenda.

Addenda may include changes to the specifications and detail drawings. Estimators must monitor the design process and maintain awareness of any addenda up to the time of submission of the final bid. Lack of knowledge of addenda can create a situation where a bid is developed based on outdated project information. An estimator should contact the architect a few days prior to the bid submission to ensure that all addenda have been taken into account in preparing the bid.

Full consideration of the many variables associated with estimating is required from the beginning of the estimating process. Estimators analyze the entire project, paying special attention to unusual or highly specialized items. Work on estimating or obtaining subcontractor or supplier bids should begin early in the process to minimize delays in completion of the final bid.

Estimators should ask questions and obtain timely and complete answers concerning any items in the contract documents that require clarification. Lack of timely information at the time of submitting a bid may put the estimator in the position of making a judgment about the project based on incomplete information. Timely collection of information is required when preparing a complete, accurate bid.

> Contractors that develop a consistent process for creating estimates can rely on using these estimates for bidding on projects. Comparing the construction costs of a completed project to the original estimate, contractors can adjust the estimating methods for future estimates.

VARIABLES

An estimator begins the bidding process with a set of detail drawings (plans), specifications, and addenda. Estimating costs, including preliminary and conceptual costs, design development, budgeting, and the construction bid, can be affected by the completeness of the drawings and specifications.

In addition to the drawings and specifications, an estimator may visit the job site to directly observe existing working conditions. Items such as existing structures, access to and from the job site, trees, shrubs, and surrounding buildings and neighborhoods may affect quantities and costs of the job. The estimator must be aware of labor and material availability, the construction design for the project, and various government requirements in the construction area.

No two projects or estimates are the same. Conditions change involving the weather, job-site conditions, employees, architectural drawings, subcontractors, suppliers, local government agencies, and a variety of other items. Estimators using bidding information from standardized industry information, company historical data, or electronic databases must keep the variability of each job in mind.

Project Location

The location of a project can introduce many variables into the estimating process. Access to the job site can greatly impact material hauling costs. A job site storage area for delivery of materials may or may not be available. Some materials required by the architectural design may not be available without high shipping costs to reach the job site location. Labor markets are always changing based on the levels of employment in the local construction industry and other industries.

Overhead expenses also vary based on job site location. Overhead expenses include taxes, insurance, permits, and other location-specific expenses. A variety of building code requirements, permitting processes, and legislative rules apply to construction projects and are different for various structures depending on the location.

Material Availability and Pricing. Material availability refers to the availability of material items needed for a project. Developments in the designs and properties of new materials provide a wide variety of material options for structural engineers and architects. The exchange of information between engineers, product designers and manufacturers, building owners, architects, and estimators is required to ensure that the appropriate available materials are designed and installed into the structure. Catalogs, web sites, and CD-ROMs that describe the latest developments in construction materials can be a helpful resource. **See Figure 1-6.**

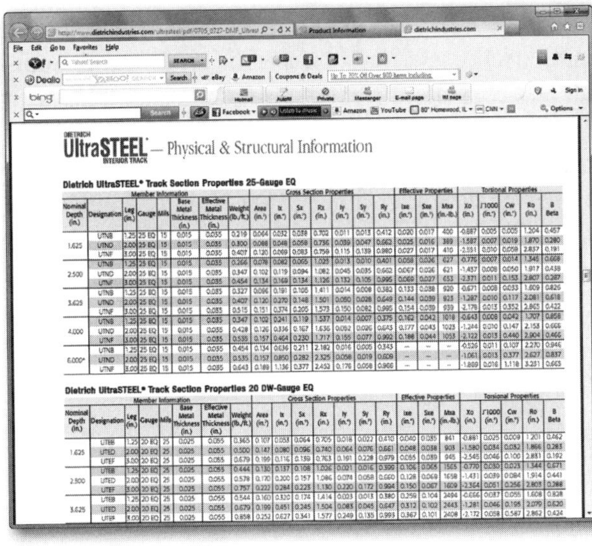

Figure 1-6. Web sites of material and equipment suppliers provide information about the latest developments in construction materials.

Estimators must take material and product availability into account when developing an estimate because availability affects price and job-site scheduling. Use of a scarce material or difficult-to-obtain product in a particular geographic location may result in higher prices.

Potential fluctuations in future costs of materials and products must also be taken into account as national and global markets have a direct impact on the costs, especially for construction projects with a long duration. Lengthy delivery times for scarce materials may also affect the project completion date. Proper planning for material pricing and ordering is essential to avoid costly penalties on projects with financial penalties for time overruns. Material and product prices should be checked at the project location prior to pricing.

Estimators must examine the bid documents to determine if the use of alternative materials or products is allowed. Some owners and architects may require cost breakdowns to be included in the final bid proposal. Material pricing is tracked more carefully and divided into individual categories when cost breakdowns are required. Additional staffing may be required at the time of the final bid preparation if cost breakdowns are required.

Labor Availability and Productivity. Labor is a high-risk item in an estimate. Issues such as labor availability and productivity comprise a large portion of the final estimate and bid.

Labor availability refers to the availability of skilled labor required on a specific project. Estimators should consider the employment and unemployment levels in a construction market and assess the impact the levels may have on labor resources. In some situations, tradesworkers may be temporarily moved from another labor market to increase labor availability for a project, resulting in additional housing and travel costs and premium compensation. The availability of various types of skilled labor in an area must also be assessed when pricing a job and developing a bid.

Labor productivity is another labor component that estimators need to take into consideration when developing an estimate. New construction materials and methods have greatly changed the skills that laborers involved in construction need to have. Automation and prefabrication are playing a large role in the building process. New tools and procedures have increased the skill levels required by tradesworkers on a job site. Labor contracts with particular work rules may also have an impact on labor productivity factors. Trade associations or other contractors may be able to provide information regarding labor productivity in a given market.

While standard labor-cost rates are available from a variety of industry sources, an estimator must dedicate time to becoming familiar with the labor market in the project construction area. The availability and variability of union labor or open shop labor can have a significant effect on labor rates and productivity.

Labor availability and productivity impact labor rates and ultimately job scheduling. The construction industry commonly faces variability of labor supply, productivity, and costs in various locations. Estimators should be aware of job-site labor conditions and their potential impact on labor costs and project scheduling.

Subcontractor Arrangements. Owners and architects may require a listing of subcontractors and suppliers to be submitted by the general contractor or construction manager at the time of the final bid. Some owners may provide a listing of preapproved subcontractors and suppliers as the only firms that may submit bid proposals. Requirements for utilization of disadvantaged-, minority-, or women-owned businesses may be included in the specifications or local governmental regulations and can also impact the subcontractors on a project. Estimators should check the general conditions and specifications to ensure that only approved subcontractors and suppliers are solicited for bid proposals.

Overhead Expenses. *Overhead* is any business expense that is not chargeable to a particular part of a construction project. Home-office overhead cannot be readily charged to a specific project and includes office employee salaries, office expenses, travel, professional services, and taxes and fees. The sales taxes for materials must be determined to ensure proper material pricing. For example, some materials on construction projects for not-for-profit entities may be exempt from local or state sales taxes depending on local and state legislation. **See Figure 1-7.** Taxes for certain materials or products may be different in different cities, counties, or states.

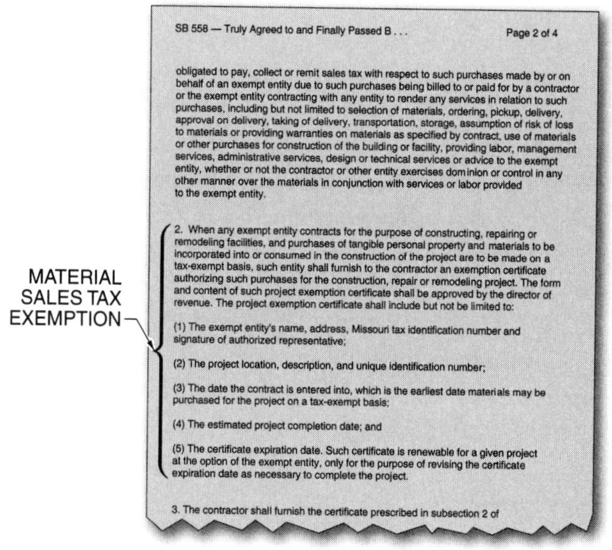

SB 558 — Truly Agreed to and Finally Passed B . . . Page 2 of 4

obligated to pay, collect or remit sales tax with respect to such purchases made by or on behalf of an exempt entity due to such purchases being billed to or paid for by a contractor or the exempt entity contracting with any entity to render any services in relation to such purchases, including but not limited to selection of materials, ordering, pickup, delivery, approval on delivery, taking of delivery, transportation, storage, assumption of risk of loss to materials or providing warranties on materials as specified by contract, use of materials or other purchases for construction of the building or facility, providing labor, management services, administrative services, design or technical services or advice to the exempt entity, whether or not the contractor or other entity exercises dominion or control in any other manner over the materials in conjunction with services or labor provided to the exempt entity.

2. When any exempt entity contracts for the purpose of constructing, repairing or remodeling facilities, and purchases of tangible personal property and materials to be incorporated into or consumed in the construction of the project are to be made on a tax-exempt basis, such entity shall furnish to the contractor an exemption certificate authorizing such purchases for the construction, repair or remodeling project. The form and content of such project exemption certificate shall be approved by the director of revenue. The project exemption certificate shall include but not be limited to:

(1) The exempt entity's name, address, Missouri tax identification number and signature of authorized representative;

(2) The project location, description, and unique identification number;

(3) The date the contract is entered into, which is the earliest date materials may be purchased for the project on a tax-exempt basis;

(4) The estimated project completion date; and

(5) The certificate expiration date. Such certificate is renewable for a given project at the option of the exempt entity, only for the purpose of revising the certificate expiration date as necessary to complete the project.

3. The contractor shall furnish the certificate prescribed in subsection 2 of

Figure 1-7. Estimators require knowledge of any legislation that may affect product pricing.

Mobilization costs vary based on the location of the construction site and are determined by visiting or viewing the job site. *Mobilization* is the process of moving the required construction equipment and support systems to the current construction site. Mobilization costs include items such as the installation of temporary electrical service, job site office and storage trailer transportation and rental, and the creation and removal of temporary access roads if required.

Job overhead includes all costs associated with a specific project but cannot be assigned to a specific item in the project. Fees for permits must be submitted to each municipality and government agency responsible for building permits and inspections.

To properly determine costs for permitting fees, the estimator must contact all applicable permitting agencies for a particular locality and determine the types of permits required for all types of construction and their individual costs. In some instances, multiple copies of construction documents must be provided to the permitting agency for their review, resulting in additional overhead expenses for printing and delivery.

Insurance rates for a project vary depending on the contractor experience, security of the area surrounding the project, type of project, and insurance industry cost trends. The estimator must carefully assess the insurance costs for each project. Location of a project in a dangerous or unsecured area may require the addition of security patrols, alarms, or security systems.

Project start-up and close-out costs may be included in overhead expenses. Printing multiple sets of prints and specifications for members of the construction team may be required. Installation and maintenance of job site signage and protective barriers may also be part of overhead costs.

When construction is concluded, the final preparation of the building site for usage includes the removal of many of the temporary structures and facilities that were needed only during the construction process. In addition, owners may require a set of as-built construction documents in either printed or electronic format that include all the changes made to the project during construction resulting from change orders or other issues. Owners may also require information for the ongoing maintenance of building systems such as heating, ventilating, and air conditioning (HVAC) or electrical controls. All of these close-out costs are included in overhead expenses.

Building Codes. Individual municipalities and government agencies have adopted various building codes. Many locations throughout the United States have adopted the International Building Code (IBC) for commercial construction and the International Residential Code (IRC) for one- and two-family dwellings. A variety of other codes and regulatory agencies' rules are used for various portions of the building process such as electrical installation and fire protection.

Even though standard building codes exist, local building and zoning commissions may adopt and modify the codes to address the needs of their residents. Building codes and legal requirements may vary from city to city, state to state, and country to country, sometimes in an overlapping manner for various agencies. Estimators must remain up-to-date on the latest codes and legislation to conform to these requirements.

Environmental Factors. Climatic requirements and the protection of the surrounding environment are two variables estimators must make allowances for in the bidding process. Specifications may describe certain climatic requirements that must exist for the placement of concrete, masonry, or other materials that may be affected by extremely hot, cold, wet, or dry conditions. **See Figure 1-8.** Protection from the elements and/or heated enclosures may be required to allow the project to stay on schedule and for building work to continue during inclement weather.

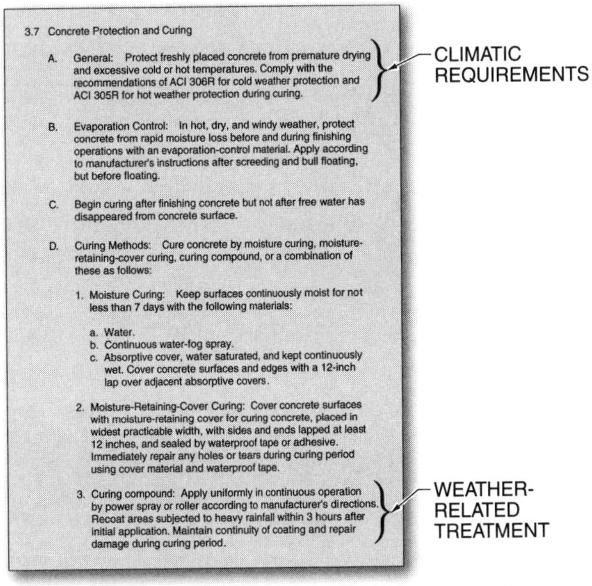
3.7 Concrete Protection and Curing

A. General: Protect freshly placed concrete from premature drying and excessive cold or hot temperatures. Comply with the recommendations of ACI 306R for cold weather protection and ACI 305R for hot weather protection during curing.

B. Evaporation Control: In hot, dry, and windy weather, protect concrete from rapid moisture loss before and during finishing operations with an evaporation-control material. Apply according to manufacturer's instructions after screeding and bull floating, but before floating.

C. Begin curing after finishing concrete but not after free water has disappeared from concrete surface.

D. Curing Methods: Cure concrete by moisture curing, moisture-retaining-cover curing, curing compound, or a combination of these as follows:

1. Moisture Curing: Keep surfaces continuously moist for not less than 7 days with the following materials:
   a. Water.
   b. Continuous water-fog spray.
   c. Absorptive cover, water saturated, and kept continuously wet. Cover concrete surfaces and edges with a 12-inch lap over adjacent absorptive covers.
2. Moisture-Retaining-Cover Curing: Cover concrete surfaces with moisture-retaining cover for curing concrete, placed in widest practicable width, with sides and ends lapped at least 12 inches, and sealed by waterproof tape or adhesive. Immediately repair any holes or tears during curing period using cover material and waterproof tape.
3. Curing compound: Apply uniformly in continuous operation by power spray or roller according to manufacturer's directions. Recoat areas subjected to heavy rainfall within 3 hours after initial application. Maintain continuity of coating and repair damage during curing period.

Figure 1-8. Estimators must consider climatic requirements that may affect work scheduling and material placement.

Protection of the environment surrounding a construction project may also be noted in the specifications. For example, a silt fence may be required around a job site to retain topsoil so it does not erode into the stormwater drainage system. A job site visit is beneficial for observing the conditions that exist and how they might affect job progress and scheduling. Additional work may be required to protect the environment around the work area. This additional work could affect work schedules.

Governmental Factors. A variety of local, state, and national organizations have developed legislative and industry requirements that affect construction site practices and scheduling. Local building codes or ordinances may contain specific limits on various job-site activities. State codes may allow work to proceed only when certain conditions exist related to environmental factors. For example, many state transportation departments allow certain types of pavement to be placed only within certain temperature limits. Noise ordinances may exist that restrict high decibel levels during certain times of day or night.

Several United States agencies regulate construction site activities including the Occupational Safety and Health Administration (OSHA), the Environmental Protection Agency (EPA), and the Office of Federal Contract Compliance Programs (OFCCP). The *Occupational Safety and Health Administration* (OSHA) is a federal agency, established under the Occupational Safety and Health Act of 1970, that requires employers to provide a safe environment for their employees. OSHA reviews and monitors safety conditions on construction job sites. Estimators must ensure that any costs associated with meeting OSHA regulation requirements are included in the bid.

> Construction cost manuals should be used only as pricing guidelines since the cost information represents average regional costs of a number of contractors and may not correspond to the costs incurred by a particular construction company. The most reliable source of cost information is company historical data.

The *Environmental Protection Agency* (EPA) is a federal government agency established in 1970 to control and reduce pollution in the areas of air, water, solid waste, pesticides, radiation, and toxic substances. The EPA monitors the effects of a project on the environment. Estimators must include costs of environmental protection in bids as mandated by the EPA.

The *Office of Federal Contract Compliance Programs* (OFCCP) is a federal government agency whose mandate is to promote affirmative action and equal employment opportunity on behalf of minorities, women, the physically challenged, and veterans. The OFCCP monitors federally funded projects to ensure that civil rights laws are being followed. Costs incurred that are associated with OFCCP recordkeeping, recruitment, and training must be included in the job overhead of the project being bid.

Value Engineering

In some cases, there may be allowances in the bidding documents for value engineering. *Value engineering* is the process whereby the owner and contractor agree on changes to the original design documents that will result in a comparable finished construction project with some substitutions and cost savings. The allowances defined in the construction contract allow for the contractor to share in a portion of the financial benefits derived from their suggested changes to the project.

The changes are intended to deliver the same results to the owner as presented in the original documents but through a less expensive or more productive method. Value engineering allows architects and estimators to work together and design components of a structure that may be more efficient and economical. It also allows all members of the construction team to share in the financial gains resulting from the agreed-to changes.

Design features may be changed based on estimator input in value engineering. For example, a contractor with experience in building retaining walls may review a retaining wall design on a particular project and recommend that it could be built with precast concrete sections rather than a cast-in-place concrete. The precast sections would be installed more quickly and at a lower cost than the cast-in-place wall, yet provide the same structural benefits. The contractor and architect would meet to review the potential design changes and cost savings related to this value-engineered change in the design.

Value engineering can be utilized prior to beginning construction or during the construction process. After a contractor is selected by an owner, the design and construction teams will meet prior to beginning construction and review the value engineering changes proposed by the contractor. In some instances, additional value engineering recommendations may be made by the contractor during the construction process as additional elements of the project take place.

Labor

One of the highest risk variables in construction project estimating is the cost of construction labor. Labor cost per hour, availability, and skill level may vary greatly depending on the project location, time of year, and work conditions. Estimators must take these items into account when creating labor pricing for various items in an estimate.

Working Conditions and Wage Rates. Working conditions at a job site impact work productivity. Cramped working conditions, limited equipment and supply storage, and difficult delivery situations usually result in lower productivity. For example, a small job site may create congestion of workers and materials, thereby lowering work productivity and increasing labor costs. An extremely large job site may also lower work productivity and increase labor costs because of the extra time necessary to reach various areas of the site. This can apply to both vertical and horizontal distances.

Wage rates refer to costs of worker wages and benefits. Working conditions and wage rates vary significantly from project to project. Factors affecting working conditions and wage rates include project location, material storage space, the trade performing a given construction operation, the existence of a building trade union, and any contract agreements that may exist.

For example, the costs per hour of worker wages and benefits for a tradesworker covered by a union labor agreement are different than tradesworkers who are not covered by such an agreement. Jurisdictional claims by various trade unions must also be taken into account in preparing labor costs. Labor costs for various material installations vary depending on the trade performing the installation and wage and benefit rates.

Training. Labor training includes skill development programs such as vocational education programs, apprenticeships, and journeyman upgrade courses and programs. The quality of work performed by tradesworkers can vary greatly from area to area and from trade to trade. An estimator should contact industry sources in the area of the construction project to determine if significant quality concerns due to untrained construction labor exist. **See Figure 1-9.**

Figure 1-9. Highly trained labor can reduce construction costs and time and improve the quality of the work.

Additional work hours may be required for project completion if training is insufficient to produce quality construction in a minimum amount of time. Additional costs may also be incurred due to rework, repair, and patching of inadequate work to meet the quality standards required by the owner and architect.

Cost Resources. A variety of resources are used to price labor in an estimate. For established companies, company historical records are the most reliable source of labor cost data. By tracking costs on completed projects, companies develop labor costs for various types of construction processes and projects. These labor costs can be utilized for future estimates. For newer companies, standard labor-rate information for various construction markets and various types of construction is available for purchase from industry information sources.

Scheduling

Project scheduling also affects the estimated costs for materials, labor, and overhead. Material costs may be higher or lower at various times of the year or during various market cycles. For example, concrete costs may be higher during winter months in cold climates due to the inclusion of required admixtures in the concrete that inhibit freezing and assist the concrete in developing a quick set. Construction projects that require quick turnaround may demand overtime or shift work for tradesworkers. Overhead may be affected by the time necessary to complete a project, the number of tradesworkers required for completion, and penalties for not completing a project on a specific schedule.

Estimators need to include the cost and time required when working with hazardous materials.

Liquidated damages are penalties assessed against the contractor or subcontractor for failure to complete work within a specified period of time. Estimators should check the specifications for the inclusion of liquidated damages to ensure that the estimate takes the project schedule into consideration.

Bid Schedule and Location. An estimator should consult the general conditions of the specifications to determine the date, time, and location for submission of the final bid. The day of the week, potential holiday schedules, staffing availability on the bid day, and conflicts with other construction projects that may be bidding at the same time should be taken into consideration. Bids due early in the morning require scheduling considerations on the day prior to bidding. The location for the delivery of the bid must be considered in order to leave an adequate amount of time to properly prepare the final bid documents and deliver them to the bid location.

REFERENCE DATA

Estimators use a vast library of standardized information in the pricing and bidding of materials, equipment, and labor. Many private vendors collect market information and publish reference materials. These reference materials include material, labor, and equipment costs that can be combined with the quantity takeoff for pricing.

Associations such as the American Institute of Architects (AIA) and the Construction Specifications Institute (CSI) provide a variety of reference materials for building standards and specifications. Contractors who have been in business for several years often accumulate their own historical reference data. Contractors analyze previously constructed projects to determine costs for labor, materials, equipment, and overhead. Historical or third-party reference data may be available and stored electronically or in print form.

Printed and Electronic References

Many printed and electronic references containing tables, charts, and various cost information are available for use by an estimator. The references are organized to allow an estimator to quickly determine material, equipment, and labor costs in relation to quantity takeoff. **See Figure 1-10.** Component costs, prices based on square, cubic, or linear measure, and standard labor rates may be obtained from reference charts and tables.

Printed References

SWITCH AND RECEPTACLE PLATES					
Material	**Craft @ Hr**	**Unit**	**Material**	**Labor**	**Total**
Combination Decorator and Three Standard Switch Plates					
4 gang brown	L1 @ 0.20	Ea	5.62	6.77	12.39
4 gang ivory	L1 @ 0.20	Ea	5.62	6.77	12.39
4 gang white	L1 @ 0.20	Ea	5.62	6.77	12.39
Semi-Jumbo Switch Plates					
1 gang brown	L1 @ 0.05	Ea	1.15	1.69	2.84
1 gang ivory	L1 @ 0.05	Ea	1.03	1.69	2.72
1 gang white	L1 @ 0.05	Ea	1.15	1.69	2.84
1 gang gray	L1 @ 0.05	Ea	1.15	1.69	2.84
2 gang brown	L1 @ 0.10	Ea	2.55	3.39	5.94
2 gang ivory	L1 @ 0.10	Ea	[illegible]	[illegible]	5.94

PRE-ENGINEERED STEEL BUILDINGS*						
	Craft @ Hr	**Unit**	**Material**	**Labor**	**Equipment**	**Total**
40′ × 100′ (4000 SF)						
14′ eave height	H5 @ 0.074	SF	3.83	3.42	1.02	8.27
16′ eave height	H5 @ 0.081	SF	4.31	3.74	1.11	9.16
20′ eave height	H5 @ 0.093	SF	4.87	4.29	1.28	10.44
60′ × 100′ (6000 SF)						
14′ eave height	H5 @ 0.069	SF	3.72	3.19	0.95	7.86
16′ eave height	H5 @ 0.071	SF	3.79	3.28	0.98	8.05
14′ eave height	H5 @ 0.065	SF	3.10	3.00	0.89	6.99
16′ eave height	H5 @ 0.069	SF	3.28	3.19	0.95	7.42
20′ eave height	H5 @ 0.074	SF	3.65	3.42	1.02	8.09
100′ × 200′ (20,000 SF)						
14′ eave height	H5 @ 0.063	SF	3.02	2.19	0.87	6.80
16′ eave height	H5 @ 0.066	SF	3.16	3.05	0.91	7.12
20′ eave height	H5 @ 0.071	SF	3.40	3.28	0.98	7.66

* 26 gauge colored galvanized steel roof and siding with 4 in 12 (20 lb live load) roof. Cost per square foot of floor area. Costs do no include foundation or floor slab. Add delivery cost to site. Equipment is a 15-ton truck crane and a 2-ton truck.

Figure 1-10. Printed references contain charts that provide information covering material, equipment, and labor costs.

Computers have become common in the estimating process, and most reference tables and databases are available through the Internet or on CD-ROM. Reference tables and databases containing material, equipment, and labor costs can be purchased and integrated into various estimating software packages or spreadsheets. **See Figure 1-11.** Experienced contractors may develop their own electronic databases for costs based on historical data obtained through their job site experiences.

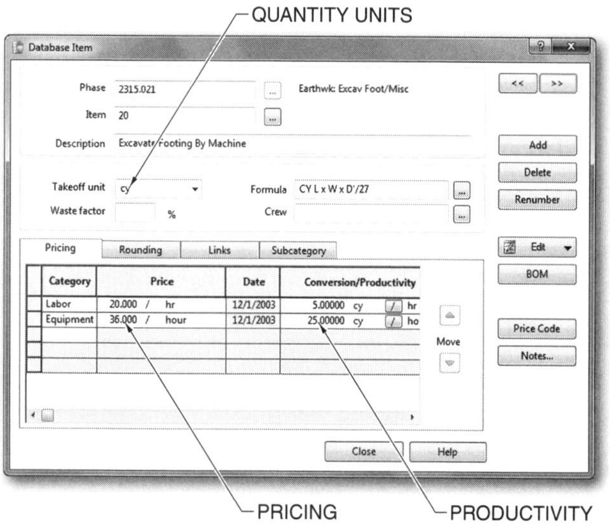

Figure 1-11. Estimating programs use electronic databases of cost information for quantity pricing.

Component Costs. Items that are counted from a set of prints can be priced according to component costs. For example, a structure may contain five Type A windows in the west wall. The estimator can locate Type A windows in the reference material and determine a cost per window including material and installation labor. Component costs are available for a variety of materials such as doors, windows, special fixtures, louvers, and other individual items that are shown on a drawing.

In an electronic format, component costs are located by searching a database for particular words or items. The costs can be copied into a spreadsheet or may be entered automatically in an estimating program that connects the component cost database with the takeoff spreadsheet.

Unit Prices. Unit (crew-based) pricing is the calculation of material, equipment, and labor prices in a single step based on a specific quantity of construction put in place. Labor rates are calculated based on standard material quantities, and judgments are made about the production levels of labor per material quantity unit based on historical data. Items such as concrete flatwork, painting, gypsum board, and flooring are priced according to the square foot or square yard.

For example, a standard concrete driveway estimate is developed by calculating the number of square feet of driveway to be placed and finished. The area of the driveway directly affects material costs for the amount of concrete and reinforcement and the labor costs for placing and finishing the concrete. **See Figure 1-12.** A labor cost for concrete finishing based on the number of square feet to be finished is included in reference tables or databases. Other materials are unit-priced for labor and materials based on linear feet or other unit-price measures.

Portland Cement Association

Figure 1-12. In crew-based pricing, material and labor costs for items like concrete flatwork are calculated based on the square feet of work to be performed.

Electronic databases contain similar information for unit pricing as printed references. When estimating with a computerized system, electronic information concerning unit pricing may be calculated automatically by entering the dimensions and type of material to be bid.

Equipment Rates. Costs for rental, maintenance, and operations of construction equipment are included in various printed and electronic references. Rates vary according to equipment availability, the volume of construction work in an area, and the equipment required. These rates can be entered into databases or spreadsheets where necessary.

Labor Rates. Many government organizations and trade associations provide labor rates for various geographic regions. Estimators use these resources to determine costs per hour for various construction tradesworkers. After an estimator has determined the number of work hours required for a particular unit price of material from reference tables, wage rates can be added to the calculations to determine total labor costs.

For example, a reference table may indicate the amount of time necessary for a carpenter to hang a 3′-0″ metal swinging door in a metal frame is 45 min. Another reference table is accessed to determine the wage and benefit costs for a carpenter per hour. Multiplication of these two numbers provides the labor cost based on these reference tables.

A wide range of labor rate information is available in electronic databases. Various services are available to electronically download wage rates for different tradesworkers in various locations. This adds accuracy to the estimate and may be tied directly to the estimate spreadsheet or estimating program.

Cost Indexes. A *cost index* is a compilation of a number of cost items from various sources in a common reference table. Estimators consult cost indexes to obtain information related to current prices and potential changes in construction-cost items. A cost index factor is assigned to the cost items so an estimator can make necessary adjustments to an estimate to account for potential cost changes. For example, a building may have cost $6.2 million to construct in the early 2000s. When bidding on a new building of similar size and construction, the original cost of $6.2 million is multiplied by a cost index factor to determine the approximate cost of a similar building today.

Cost indexes are divided into a broad range of categories, including construction system, labor, material, equipment, and geographic location. Sources of cost indexes include public agencies, such as the Bureau of Reclamation and the U.S. Department of Commerce, and many private sources specializing in specific cost indexes, such as for industrial buildings, chemical processing plants, reinforced concrete buildings, and other specialty structures.

Assembly Costs. An *assembly cost* is the price of a number of common construction materials combined into a unit assembly, such as a wall constructed of studs, top and bottom plates, headers, and strap bracing. Common construction assemblies may be included in databases of estimating programs. Items such as wall assemblies, floor assemblies and coverings, or ceiling finishes may be taken off and priced as assemblies. **See Figure 1-13.**

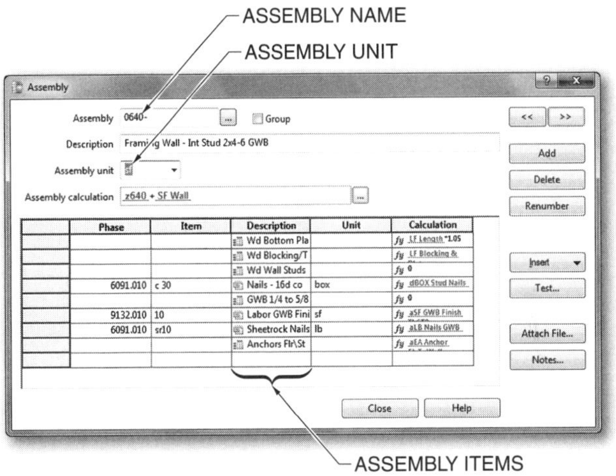

Figure 1-13. Common construction assemblies may be unit-priced in estimating programs.

For example, a construction assembly may consist of 2 × 6 studs spaced 16″ on center (OC) with 2 × 6 top and bottom plates and braced with metal strap bracing. The number of linear feet of this wall type is determined by the estimator. In conjunction with the database, the estimating program will automatically determine the number of studs, plates, amount of blocking and bracing materials, and labor costs related to each item. This greatly simplifies the bidding process for standard assemblies.

SAGE TIMBERLINE ESTIMATING . . .

Implementing an estimating software solution requires an investment in hardware, software, and training. However, the return on investment is quickly realized through more complete, accurate, and timely estimates. Sage Timberline Estimating is the estimating module of Sage Software, Inc. By allowing the software to perform data handling, calculations, and presentation of the results, an estimator can focus more closely on the creative aspects of estimating.

. . . SAGE TIMBERLINE ESTIMATING . . .

Opening the Sage Timberline Estimating Program

1. Pick the **Start** button followed by **Programs** → **Sage** → **Sage Timberline Estimating** → **Sage Timberline Estimating** to launch Sage Timberline Estimating software. As an alternative option, click on the **Sage Timberline Estimating** icon to start the software.

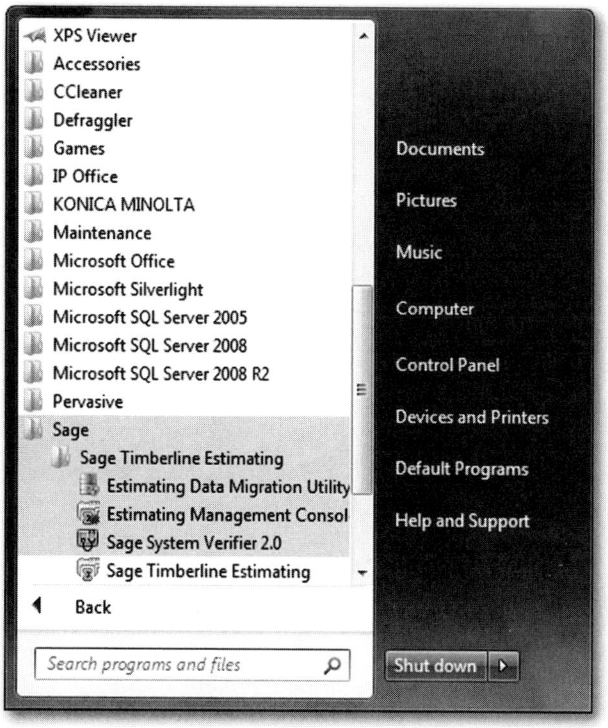

2. When the Sage Timberline Estimating program is launched, an **Activation Error** dialog box appears indicating that the user needs a license to use the software. Pick **Demo** to proceed.

. . . SAGE TIMBERLINE ESTIMATING

3. Next, a **Demonstration Mode** dialog box appears, indicating that 180 days remain for using the Sage Timberline Estimating software. Pick **OK** to proceed.

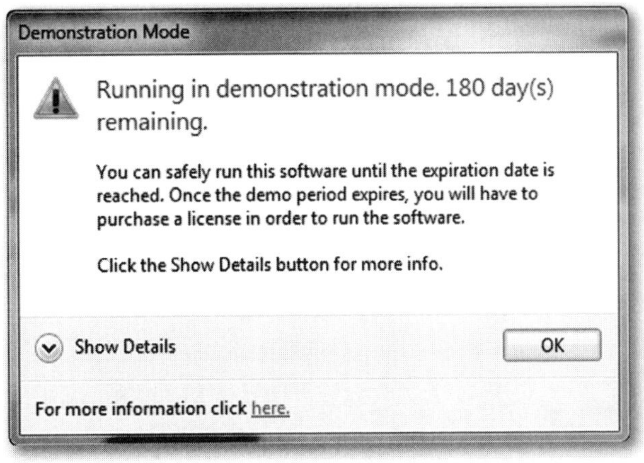

The Sage Timberline Estimating DVD includes several document files provided by Sage Software, Inc. Read these documents to learn more about the use of this software, including access to customer support.

- *What's New* provides an update on this software release. It includes contact information for more information and assistance on the software.
- *Release Notes* provides the general requirements of this software release and directions to access further assistance for the use of the software.
- *Getting Started Guide* provides guidance in creating, organizing, managing, and analyzing estimates. This includes developing basic concepts, creating new estimates, performing takeoffs, working with the spreadsheets, and printing reports.

Use Sage Timberline Estimating Help to get answers to software questions.

Quick Quiz®

Quick Quiz®

Refer to the CD-ROM for the Quick Quiz® questions related to chapter content.

Key Terms

Illustrated Glossary

- *addendum*
- *assembly cost*
- *bid*
- *contractor*
- *cost index*
- *crew-based estimate*
- *detail takeoff estimate*
- *Environmental Protection Agency (EPA)*
- *estimating*
- *estimating practice*
- *general contractor*
- *liquidated damages*
- *mobilization*
- *Occupational Safety and Health Administration (OSHA)*
- *Office of Federal Contract Compliance Programs (OFCCP)*
- *overhead*
- *prebid meeting*
- *subcontractor*
- *quantity takeoff*
- *value engineering*

Web Links

Web Links

American Institute of Architects
www.aia.org

American Society of Professional Estimators
www.aspenational.org

Construction Specifications Institute
www.csinet.org

Environmental Protection Agency
www.epa.gov

International Code Council
www.iccsafe.org

Occupational Safety and Health Administration
www.osha.gov

United States Department of Labor - Office of Federal Contract Compliance Programs
www.dol.gov/ofccp

Estimating Practices 1

Review Questions

Estimating

________ **1.** ___ is the computation of construction costs of a project.

________ **2.** A(n) ___ is a change to the originally issued construction contract documents.

________ **3.** A(n) ___ is an offer to perform construction work at a stated price.

________ **4.** A(n) ___ is a conference in which all interested parties in a construction project review the project, ask questions of the architect or owner concerning methods for accomplishing the work, and share information necessary to understand the entire scope of the work.

________ **5.** ___ is any business expense that is not chargeable to a particular part of a construction project.

________ **6.** A cost ___ is a compilation of a number of cost items from various sources in a common reference table.

________ **7.** An estimating ___ is the system used to integrate all parts of the estimating process in a cohesive, consistent, and reliable manner to ensure an accurate final bid.

________ **8.** ___ is the practice of reviewing construction contract documents to determine quantities of materials that are included in a bid.

________ **9.** ___ is a process in which construction personnel employed by the firm developing the bid, such as project managers, estimators, and engineers, are allowed to recommend changes to the construction contract documents.

________ **10.** ___ estimating is an estimating practice in which material quantities, equipment, and labor costs are included in a single calculation based on a specific quantity of construction put in place.

U.S. Government Agencies

________	**1.** Occupational Safety and Health Administration (OSHA)	**A.** Controls and reduces pollution of air, water, solid waste, pesticides, radiation, and toxic substances
________	**2.** Environmental Protection Agency (EPA)	**B.** Requires employers to provide a safe environment for their employees
________	**3.** Office of Federal Contract Compliance Programs (OFCCP)	**C.** Promotes affirmative action and equal employment opportunity on behalf of minorities, women, the physically challenged, and veterans

Short Answer

1. Identify the four levels of estimating for interim estimates and briefly define each level.

2. Contrast detail takeoff estimating and crew-based estimating.

3. Describe the purpose of a prebid meeting and list the people who are typically involved in the meeting.

4. Identify four variables that may be introduced into the estimating process due to project location.

5. Define "value engineering" and how it may impact a building design and estimate.

6. Cite one advantage of using web-based reference material over printed reference material.

7. Describe overhead expenses and explain how each project contributes to covering those costs.

8. Describe mobilization and how it impacts the project.

Estimating Practices 1

Activities

Activity 1-1—Interpreting Specifications

__________ **1.** Concrete surfaces must be kept continuously moist for not less than ___ days.

__________ **2.** Comply with recommendations of ___ for cold-weather protection during curing.

__________ **3.** The sides and ends of a moisture-retaining cover must be lapped at least ___″.

__________ **4.** Concrete may be cured by ___ curing, moisture-retaining-cover curing, curing compound, or a combination of these.

T F **5.** Continuous water-fog spray may be used for moisture curing.

__________ **6.** The sides and ends of a moisture-retaining cover must be sealed by ___ tape or adhesive.

__________ **7.** Comply with recommendations of ___ for hot-weather protection during curing.

3.7 Concrete Protection and Curing

A. General: Protect freshly placed concrete from premature drying and excessive cold or hot temperatures. Comply with the recommendations of ACI 306R for cold weather protection and ACI 305R for hot weather protection during curing.

B. Evaporation Control: In hot, dry, and windy weather, protect concrete from rapid moisture loss before and during finishing operations with an evaporation-control material. Apply according to manufacturer's instructions after screeding and bull floating, but before floating.

C. Begin curing after finishing concrete but not after free water has disappeared from concrete surface.

D. Curing Methods: Cure concrete by moisture curing, moisture-retaining-cover curing, curing compound, or a combination of these as follows:

1. Moisture Curing: Keep surfaces continuously moist for not less than 7 days with the following materials:
   a. Water.
   b. Continuous water-fog spray.
   c. Absorptive cover, water saturated, and kept continuously wet. Cover concrete surfaces and edges with a 12-inch lap over adjacent absorptive covers.
2. Moisture-Retaining-Cover Curing: Cover concrete surfaces with moisture-retaining cover for curing concrete, placed in widest practicable width, with sides and ends lapped at least 12 inches, and sealed by waterproof tape or adhesive. Immediately repair any holes or tears during curing period using cover material and waterproof tape.
3. Curing Compound: Apply uniformly in continuous operation by power spray or roller according to manufacturer's directions. Recoat areas subjected to heavy rainfall within 3 hours after initial application. Maintain continuity of coating and repair damage during curing period.

Activity 1-2—Interpreting Cost Data

_______________ **1.** The material cost per square foot of floor area for a 40′ × 100′ steel building with a 14′ eave height is $___.

_______________ **2.** The labor cost per square foot of floor area for a 60′ × 100′ steel building with a 16′ eave height is $___.

_______________ **3.** The equipment cost per square foot of floor area for a 40′ × 100′ steel building with a 20′ eave height is $___.

_______________ **4.** The hours required per square foot of floor area for a 60′ × 100′ steel building with a 14′ eave height is ___ hr.

T F **5.** Costs include a foundation or floor slab.

PRE-ENGINEERED STEEL BUILDINGS*						
	Craft @ Hr	Unit	Material	Labor	Equipment	Total
40′ × 100′ (4000 SF)						
14′ eave height	H5 @ 0.074	SF	3.83	3.42	1.02	8.27
16′ eave height	H5 @ 0.081	SF	4.31	3.74	1.11	9.16
20′ eave height	H5 @ 0.093	SF	4.87	4.29	1.28	10.44
60′ × 100′						
14′ eave height	H5 @ 0.069	SF	3.72	3.19	0.95	7.86
16′ eave height	H5 @ 0.071	SF	3.79	3.28	0.98	8.05

* 26 gauge colored galvanized steel roof and siding with 4 in 12 (20 lb live load) roof. Cost per square foot of floor area. Costs do no include foundation or floor slab. Add delivery cost to site. Equipment is a 15-ton truck crane and a 2-ton truck.

Activity 1-3—Takeoff

__________________ **1.** The stud spacing for the assembly is ___″.

__________________ **2.** The phase number for the 2 × 4 RL treated plate is ___.

__________________ **3.** The unit of measure for the 2 × 4 plates is ___.

T F **4.** The assembly is for a 2 × 4 stud wall.

__________________ **5.** The item number for the 2 × 4 RL temporary bracing material is ___.

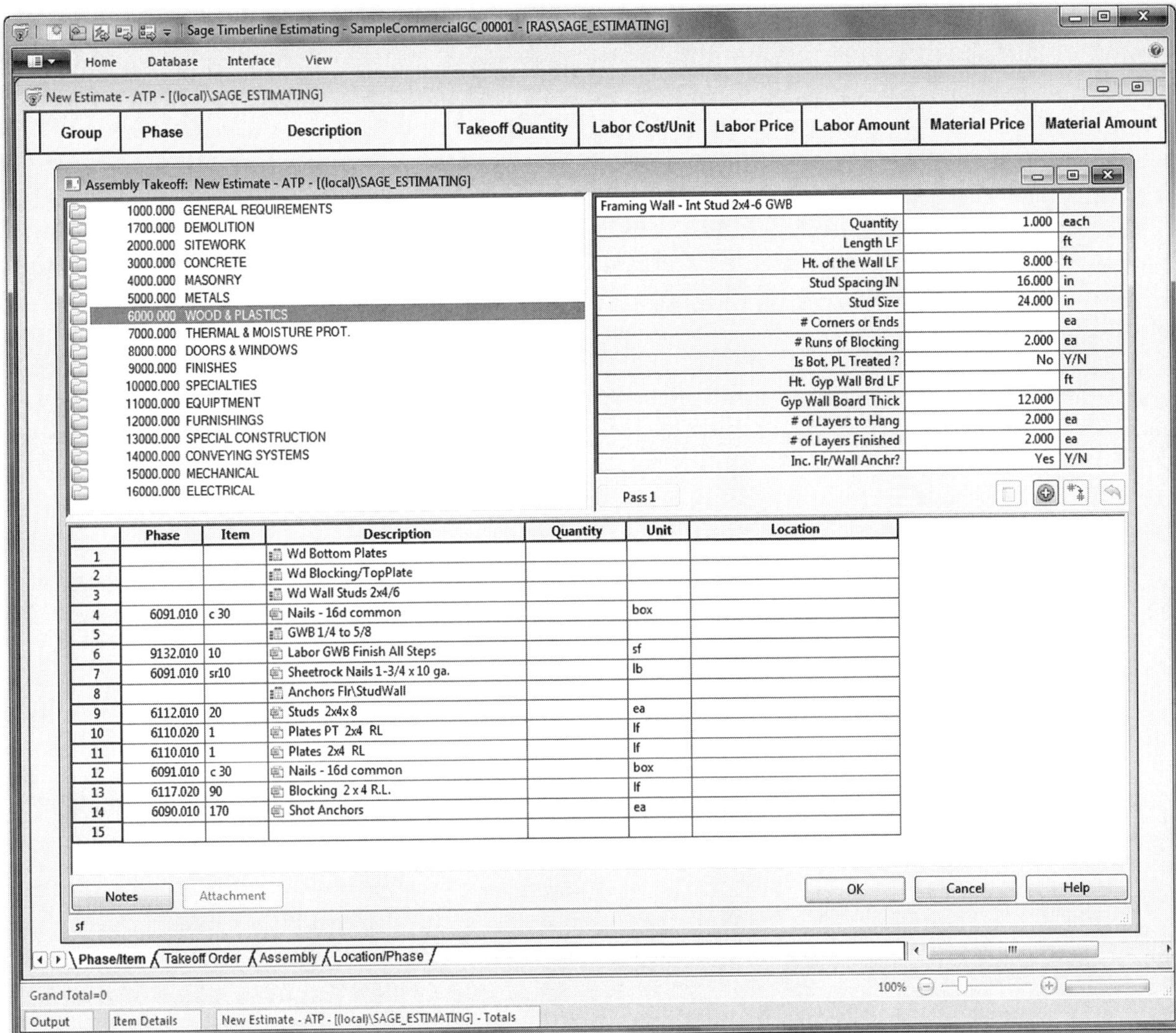

Estimating Methods 2

Key Concepts

- Estimators use a variety of estimating methods for quantity takeoff and pricing of construction projects.
- An estimator must have printreading skills in order to mark and take off the portions of the plans that pertain to the work being bid.
- The MasterFormat™ is a uniform system of numbers and titles for organizing information about construction requirements, products, and activities into a standard sequence.
- One of the biggest advantages of estimating programs is consistency of data within the database.
- Electronic plan markup is becoming more widely used for estimating.
- Building information modeling (BIM) is a methodology that uses coordinated, consistent formats for electronic design, construction, and product information to optimize planning and construction of a project.
- Use of estimating programs allows for the material quantity, labor, equipment, and overhead information to be used by others on the construction project team.

Introduction

An estimating method is an approach used by an estimator to perform project analysis, quantity takeoff, and pricing in a consistent and organized manner. Estimating is performed by either traditional or electronic estimating methods. In traditional estimating, quantities are calculated and entered onto ledger sheets for pricing. In electronic estimating, quantity takeoff and pricing are accomplished using computer spreadsheets and integrated database systems. Advantages of electronic estimating include faster estimates, the ease

with which last-minute changes can be incorporated into the bid, information sharing across a larger estimating team, and assistance in reducing mathematical errors or missed bid items.

ESTIMATING METHODS

An *estimating method* is an approach used by an estimator to perform project analysis, quantity takeoff, and pricing in a consistent and organized manner. Estimators use a variety of estimating methods for quantity takeoff and pricing of construction projects. The estimating process begins with a set of detail drawings and specifications.

The specifications are reviewed to determine the types and amounts of work contained in a particular construction project. With paper documents, the applicable portions of the specifications are highlighted and the drawings are marked for quantity takeoff. With electronic documents, on-screen methods are utilized for both specification and plan quantity review. Material quantities are calculated and entered into ledger sheets or cells of electronic spreadsheets for pricing. All quantities are multiplied by the appropriate unit costs and then totaled. Overhead, profit, and other costs are included to determine a final project price.

John Deere Construction & Forestry Company

Equipment and labor comprise a large share of an estimate.

Legal information is included in the specifications to fully describe the rights and responsibilities of all parties involved in the construction project. Bidding requirements, procurement and contracting information, and general requirements for every aspect of the construction process are included in the contract documents.

General conditions are the written agreements describing building components and various construction procedures included in the specifications for a construction project. The relationships between the various parties to the construction contract are also expressed in the general conditions. The general conditions and specifications include items that affect a project but do not appear on the drawings, such as weather protection requirements, sustainability goals, modification procedures, and bonding and insurance requirements.

Items in the specifications related to the estimate include materials, equipment, labor, overhead, and profit. Material quantities are calculated based on the type of each material required. Equipment used on a construction project includes power equipment such as earthmoving, hauling, and lifting equipment, temporary job-site equipment, and communication equipment. Labor pricing is determined based on the number of labor hours required and the cost of labor per hour. Overhead costs are expenses that are not attributable to any one specific operation. Profit is the monetary benefit realized by a construction firm at the completion of the building process.

Estimators should be attentive to owner requirements for reporting and recordkeeping. Collecting, documenting, and distributing or communicating the necessary construction information in the proper format may require additional time and resources throughout the job. Estimators must take all portions of the general conditions and specifications into account when calculating overall job costs. The architect and owner may include a list of standard bid categories. The estimator can review the listing of bid categories to organize quantity takeoff and pricing.

A decision is made by a contractor to bid on a project if the project includes the type of work that the contractor can perform. For example, a contractor that performs interior finish work will review the standard bid categories to determine if this particular project includes any interior-finish-system installation.

TRADITIONAL ESTIMATING METHODS

The *traditional estimating method* is an estimating method in which material quantities are calculated and entered onto ledger sheets for pricing. The quantities are multiplied by the appropriate unit costs to determine total material costs. The material costs are then added to determine the final material cost. Material quantities may also be used as the multiplier when using unit labor costs to calculate total labor costs.

Plan Markup

Plan markup is the color coding or marking off of items on the actual plan documents during the quantity takeoff process. **See Figure 2-1.** For most construction projects, it is difficult to remember the sections of the project that have been taken off and quantities determined without a visual reminder. An estimator studies and reviews the plans and uses several colored markers to indicate each type of material as it is taken off of the plans.

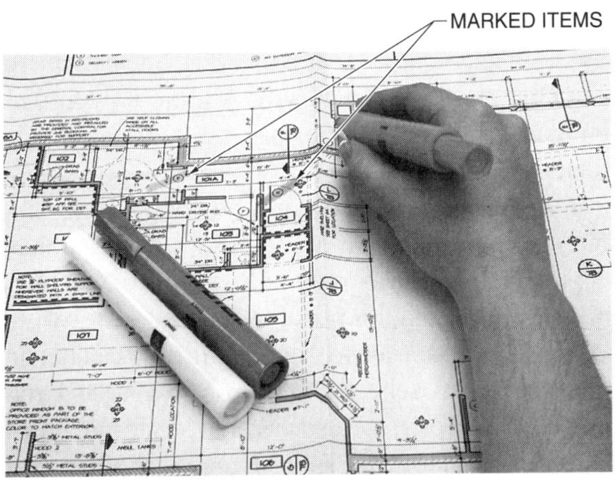

Figure 2-1. Marking up plans ensures all items to be taken off are accounted for by an estimator.

Accuracy. An estimator must have printreading skills in order to mark and take off the portions of the plans that pertain to the work being bid. For example, an estimator working with a set of plans to determine the number of light standards in a parking lot may mark each light standard with a red "X" or a colored number to indicate it has been included in the total calculations. Mistaking drains for light standards would lead to an inaccurate bid.

Comprehensiveness. After plans are marked and quantities listed on a ledger sheet, the estimator reviews all of the specifications and plans a final time to ensure all applicable items have been taken off. The final review should include a review of the entire project, including the general conditions, detail drawings, and portions of the drawings that may not appear to apply to the work being bid. This ensures all items have been included in the plan markup and quantity takeoff.

In addition to materials, a variety of general conditions, permitting requirements, and other overhead costs must be included in a comprehensive bid. Estimators should pay special attention to scheduling of construction on the job and any other items that may affect the overall bid.

Ledger Sheets

A *ledger sheet* is a grid consisting of rows and columns on a sheet of paper into which item descriptions, locations, code numbers, quantities, and costs are entered. As prints and specifications are marked up and quantities determined, the estimator enters these quantities on a ledger sheet. **See Figure 2-2.** Estimators should use standard categories on ledger sheets to ensure consistency and to present the information in a format familiar to others in the construction process. Skill in mathematical calculations reduces the amount of time required for the final tabulation of material quantities. A well-designed ledger sheet can be cross-referenced to other portions of the bid and used in other estimating calculations.

> Ledger sheets organize data for mathematical calculations involved in performing a takeoff, which will directly translate to the development of formulas in electronic spreadsheets.

QUANTITY SHEET

Sheet No. 1

Project: *Airport Site* Date:

Estimator: *RFK* Checked:

No.	Average Elevation ft	Average Cut ft/sq ft	Quantity Cu ft	Average Fill ft/sq ft	Quantity Cu ft	Total Cut Cu ft	Total Fill Cu ft	Difference Cu ft
	0′ grade level = 162′							
1	169.22	7.22	72,200					
2	171.30	9.30	93,000					
3	168.50	6.50	65,000					
4	164.55	2.55	25,500					
5	161.40			0.60	6,000			
6	157.87			4.13	41,300			
31	172.17	10.17	101,700					
32	166.60	4.60	46,000					
33	160.40			1.60	16,000			
34	154.00			8.00	80,000			
35	146.65			15.35	153,500			
36	139.00			23.00	230,000		2,022000	978

Figure 2-2. The traditional estimating method uses ledger sheets to track quantities taken off from prints.

Categories. The general conditions section of the specifications commonly contains a list of the bid categories. Companies that bid on a project must submit their estimate according to a format provided by the architect and owner. **See Figure 2-3.** Individual ledger sheets may be developed for each type of work.

For example, all items pertaining to pavement on a project, including curbs, sidewalks, concrete flatwork, asphalt paving, striping, parking blocks, and other incidental pavement items may be placed on one ledger sheet.

ARTICLE 5 - PROPOSALS AND BIDDING PROCEDURE

A. All proposals shall be submitted without modification or reservation on the proposal form with each space properly filled. Proposals not on the form will be rejected. Proposal forms are available on disk.

State Project No. 30-936-96-0010 (B)
Eastern Reception, Diagnostic and Correctional Center

The unit prices include all labor, overhead and profit, materials, equipment, appliances, etc. to cover the finished work of the several kinds of work called for. Only a single unit price shall be given and it shall apply for either MORE or LESS work.

Earth Excavation: Add/Deduct __________ per cubic yard. Price includes classified earth excavation as defined in Section 02200.

Rock Excavation: Add/Deduct __________ per cubic yard. Price includes removing and hauling to on-site spoil areas classified rock excavation as defined in Section 02200.

General Site Fill: Add/Deduct __________ per cubic yard. Price includes hauling, placing, and compacting of on-site material consisting of suitable soil material, obtained on site as defined in Section 02200.

Structural Fill: Add/Deduct __________ per cubic yard. Price includes hauling, placing, and compacting structural fill as defined in section 02200.

Figure 2-3. The general conditions commonly contain standardized bid categories to provide consistency of bids from all bidding companies.

A numerical code is given to each item. For building construction, these codes are commonly based on the numbered divisions of the MasterFormat™, which is published by the Construction Specifications Institute (CSI). **See Figure 2-4.**

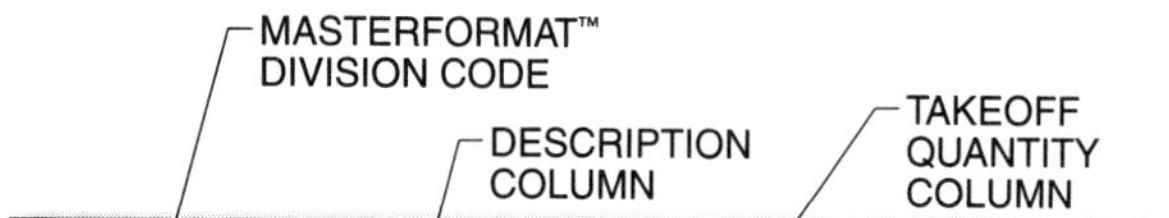

QUANTITY SHEET

Sheet No.

Project: Date:

Estimator: Checked:

No.	Group	Phase	Description	Takeoff Quantity	Labor Cost/Unit	Labor Price	Total Labor	Total Material
	32 10 00		Sitework					
		32 13 00	Paving, Asphalt					
			Asphalt Paving Lab./Equip	sy		20.00/hr		
			H Asphalt Material	cy	–	–		
			E Asphalt Material	cy	–	–		
			F Asphalt Material	cy	–	–		
			Crusher Run Base	cy		20.00/hr		
			Soil Cement Base	cy		20.00/hr		
			Asphalt Primer	sy		20.00/hr		

Figure 2-4. The standard MasterFormat™ division codes help organize the bidding process.

The *MasterFormat*™ is a uniform system of numbers and titles used for organizing information about construction requirements, products, and activities into a standard sequence. The MasterFormat™ code is entered in the left column of a ledger sheet. Internal company codes may be entered in an adjoining column. A description column is created on the ledger sheet indicating the material taken off, such as gravel, concrete, or carpet. A takeoff quantity column is included to indicate the numerical quantity and the units used, such as square feet (sf or sq ft), linear feet (lf), or cubic yards (cy or cu yd). Additional columns include material and labor pricing information and total cost for each item.

For road, bridge, and highway construction, numerical codes are commonly developed by each state department of transportation. **See Figure 2-5.** Bids must be assembled according to these categories, which vary from state to state. Estimators can obtain the codes from each state's department of transportation. The categories make it easier to locate individual items and make necessary changes when project requirements shift due to addenda, value engineering adjustments, or material and labor price changes.

BID ITEMS		
612.0412	Pipe Undedrain Wrapped 12-Inch	LF
612.0415	Pipe Undedrain Wrapped 15-Inch	LF
612.0504	Pipe Undedrain Wrapped and Plowed 4-Inch	LF
612.0506	Pipe Undedrain Wrapped and Plowed 6-Inch	LF
612.0600	Underdrain Trench	LF
612.0700	Drain Tile Exploration	LF
612.0804	Apron Endwalls for Underdrain Reinforeced Concrete 4-Inch	EACH
612.0806	Apron Endwalls for Underdrain Reinforeced Concrete 6-Inch	EACH
614.0100	Cable Guard Fence	LF
614.0103	Anchorages for Cable Guard Fence	EACH
614.0105	Anchorages for Steel Plate Beam Guard	EACH
614.0110	Anchorages for Steel Plate Beam Guard Temporary	EACH
614.0115	Anchorages for Steel Plate Beam Guard Type 2	EACH
614.0150	Anchor Assemblies for Steel Plate Beam Guard	EACH
614.0200	Steel Thrie Beam Structure Approach	LF
614.0250	Steel Thrie Beam Structure Approach Temporary	LF
614.0305	Steel Plate Beam Guard Class A	LF
614.0310	Steel Plate Beam Guard Class B	LF
614.0340	Steel Plate Beam Guard Over Low-Fill Culverts Class A	LF
614.0355	Steel Plate Beam Medium Guard	LF
614.0360	Steel Plate Beam Guard Temporary	LF
614.0370	Steel Plate Beam Guard Energy Absorbing Terminal	EACH
614.0380	Steel Plate Beam Guard Energy Absorbing Terimal Temporary	EACH

Figure 2-5. Standard estimating categories may be mandated by government agencies.

Mathematical Requirements. Estimating requires proficiency with many types of mathematical calculations. Estimators must add quantities and linear totals, multiply quantities by costs per unit and labor costs to determine pricing, and calculate areas and volumes to determine square feet, square yards, and various cubic measurements. Familiarity with common formulas speeds the process of bid preparation and ensures a high level of accuracy.

Cross-References. As an estimate is developed, columns from various ledger sheets may be cross-referenced to each other. For example, there may be a certain type and number of steel reinforcing bars required in concrete foundation walls, sidewalks, and roof decks. Cross-referencing across the various portions of the ledger sheet allows for a total number of reinforcing bars to be calculated and one order to be placed at one price rather than three orders submitted separately, costing additional ordering and shipping fees.

ELECTRONIC ESTIMATING METHODS

The *electronic estimating method* is an estimating method in which quantity takeoff and pricing calculation is accomplished using computer spreadsheets and integrated database systems. Electronic estimating methods have had a tremendous effect on the estimating process. Electronic estimating methods include computerized spreadsheet and estimating programs.

Spreadsheets have streamlined the calculation process by allowing estimators to enter quantities, costs, and standard calculations into individual cells. Estimating programs allow material pricing, labor costs, and quantity calculations to be determined electronically through the use of standard item, crew-based, and assembly information that may be entered manually or retrieved from database history files.

Advantages of electronic estimating methods include faster estimates, the ease with which last-minute changes can be incorporated into the bid, the ability to share estimating components for a large structure with a team of estimators, and assistance in reducing mathematical errors or missed bid items. In addition, electronic estimating methods enable standardization throughout a large construction company; integration with other company departments such as accounting, scheduling, and job costing operations; possible integration with subcontractor and supplier bid information; and flexibility of final bid reporting.

Electronic estimating also allows estimators to easily import and utilize information from previous bids in new bids. Some electronic estimating programs allow easy conversion from English to metric measurements.

A variety of estimating programs are available ranging from simple to complex. Programs range from spreadsheet programs that are developed and customized by an estimator or construction estimating department, to ledger sheet formats, to more sophisticated estimating programs that integrate the quantity takeoff process with design documents and scheduling operations through the use of a digitizer, electronic construction documents, and quantity and pricing calculations.

Standardized Formats

Electronic estimating programs vary from estimator-created spreadsheets to estimating programs that include descriptions, material quantities, labor costs, and totals. Standard row and column formats are common in all electronic estimating programs. The features available depend on the sophistication of the software.

Advanced features include database interactivity, the use of standard assemblies and standardized preset formulas for various estimating activities, a digitizer to calculate quantities, the ability to interface with drawings created on computer-aided drafting (CAD) or building information modeling (BIM) systems or with detail drawings scanned into a computer, and the ability to compile a variety of final reports and forms.

Spreadsheets. A *spreadsheet* is a computer program that uses cells organized into rows and columns to perform various mathematical calculations when formulas and numbers are entered. The foundation of most estimating programs is a computer spreadsheet. **See Figure 2-6.** Spreadsheets are designed with a standardized row and column format for the entry of material quantities and descriptions, labor costs and material prices, and other items required by the estimator. The quality of the final bid is determined by the accuracy of the calculation information entered into spreadsheet cells.

Care must be taken by the estimator to verify that the spreadsheet is performing the proper mathematical calculation for the proper range of cells. Estimators can purchase software or electronic templates with preformatted cell calculations that can be copied into an existing spreadsheet.

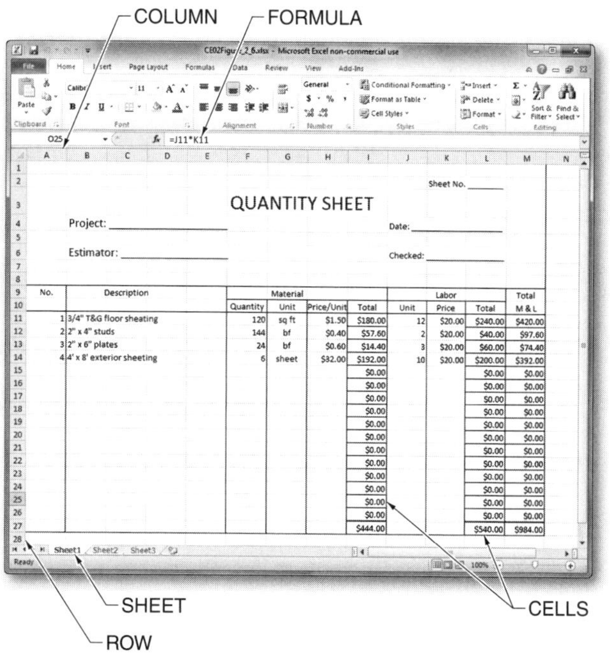

Figure 2-6. A spreadsheet uses cells organized into rows and columns to perform various mathematical calculations when formulas and numbers are entered into them.

Database Transfer. *Database transfer* is the direct transfer by an estimating program of stored information from a database into a spreadsheet. One of the most time-saving features of sophisticated estimating programs is their ability to correlate quantity takeoff amounts with material and labor unit pricing information to quickly calculate costs. The database for the estimating program stores standard costs and formulas for material, equipment, labor, and other items. **See Figure 2-7.**

Item Pricebook

Phase	Item	Description	Labor Price	Material Price	Sub Price
90000		DRYWALL/TRIM/PAINT/.FLOOR			
92000		Lath & Plaster			
	1	Stucco			1.90 /sf
	2	Stucco			2.20 /sf
	3	Stucco			2.40 /sf
	10	Stucco			/sf
92500		Drywall			
	1	Drywall			0.40 /sf
	2	Drywall			5.00 /sf
	3	Drywall			0.86 /sf
	10	Drywall			/sf
	100	Labor to Hang	0.15 /sf		
	110	Labor to Tape & Finish	0.15 /sf		
	401	Drywall Wall ½″		0.50 /sf	
	405	Drywall Ceiling ½″x		0.20 /sf	
	406	Drywall Ceiling ½″		0.50 /sf	
	410	Water-Rock ½″		0.25 /sf	
	500	Drywall Walls ⅝″x		0.25 /sf	
	505	Drywall Ceiling ⅝″x		0.25 /sf	
	510	Water-Rock ⅝″		0.27 /sf	
	1210	Texture by Machine			/sf
	1220	Texture Smooth			/sf

Figure 2-7. Estimating programs use electronic databases containing labor and material costs that interface with computer spreadsheets to facilitate the estimating process.

Proper installation, organization, and maintenance of the database are required to ensure the accuracy of estimating programs. Database codes must be organized to ensure consistency of estimates and ease in finding various materials and work processes.

As each quantity for an item is entered, material unit costs and labor cost information per unit are automatically transferred from the database to the spreadsheet. The estimator can make final adjustments in the spreadsheet calculations without changing or disturbing the database source information.

Estimating programs are commonly available with preloaded databases for estimating highway or building construction projects. The database items are then customized by the estimator for company-specific applications. Preloaded databases include data items such as types and rates for labor, equipment, materials, and overhead.

For transportation projects governed by state departments of transportation, state codes and preliminary quantities may be transferred electronically from Internet-based data-sharing services. State-maintained web sites allow estimators to download material and work codes, descriptions, and preliminary material quantities by logging onto the web site, requesting the appropriate job information, and transferring this information into an estimating program specifically designed to accommodate road and bridge construction.

One of the biggest advantages of computerized estimating programs is consistency of data within the database. Information concerning labor and material pricing and formulas are stored in the database for reuse. This feature produces consistent costs and formulas for all estimators on a large project or for companies with several estimators when managed properly.

Care should be taken by the lead estimator on a project to ensure that the database pricing is both current and applicable to the project being bid. Estimators should make any necessary adjustments to the database pricing for the specific project. Even when using standard information from a database, estimators must be aware that no two projects are exactly alike and modifications to database information may be required to produce an accurate final estimate.

Assemblies. An *assembly* is a collection of items needed to complete a particular unit of work. Standard construction assemblies are entered into the database as a standard material, equipment, and labor quantity and cost per standard unit of measure. For example, a common wall assembly includes cost information for material, equipment, and labor based on the number of linear feet of wall. **See Figure 2-8.** The wall assembly includes quantities and costs for materials, equipment, and labor for structural framing, interior finish, and possibly door and window openings.

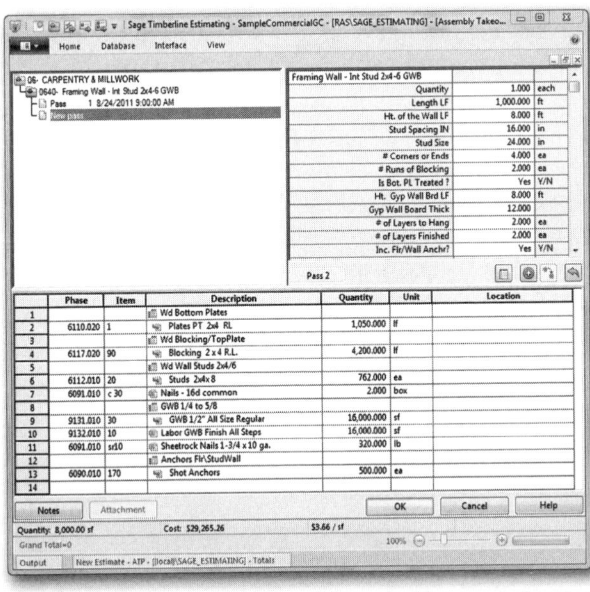

Figure 2-8. Standard assemblies enable an estimator using an estimating program to calculate several items concurrently.

The linear feet of the assembly is determined from the detail drawings and entered into the estimating program spreadsheet. The quantity for each item in the assembly is calculated by the program based on the assembly information obtained from the database. Estimating programs also allow for activity setups that can transfer information to a spreadsheet concerning labor crew costs, materials, equipment, and overhead based on the construction operation, such as pipe laying or paving work.

Final Reporting. Final reporting includes the compilation of all the various information required for members of the construction team. After all the components of a bid are entered and the final cost estimate is determined, some estimating programs allow the final bid to be printed in various formats for suppliers, subcontractors, construction personnel (such as project managers) and for presentation to the owner.

Estimating programs have the flexibility to rearrange information for reports as needed by the various groups without reentry of the information. Other estimating program reporting features include subcontractor comparisons, cost reporting by construction activity, customization of bid formats as required by various owners such as state departments of transportation, and review of company job costs based on historical data.

> Estimating programs may offer a parametric assembly feature that allows the estimator to view an assembly as the dimensions and specifications of the assembly are entered or revised.

Electronic Plan Markup

Electronic plan markup is becoming more widely used for estimating. Electronic plan markup can be performed with a digitizer on paper prints, with an interface for CAD drawings, or using electronic image formats (such as jpg or pdf). A *digitizer* is an electronic input device that converts analog data into digital data. A digitizer is used with an estimating program to input drawing coordinates directly into the program, reducing the need to enter quantities and dimensions using the computer keyboard. The estimating program reads coordinates for various points on the prints and determines linear feet, square feet, and other measurements depending on the drawing scale.

For CAD drawings, a quantity takeoff is performed on the computer screen. **See Figure 2-9.** Care must be taken not to change or manipulate various components of the drawings. Architects often attach specific information codes or tags to individual components of CAD drawings to interface with BIM or estimating programs.

Figure 2-9. Prints can be scanned or CAD drawings can be used with an estimating program to input drawing coordinates directly into the program.

The CAD drawings are transferred into the estimating software with quantity calculations made during the transfer. Some CAD systems allow drawings that are scanned into the computer to be marked with various colors on the computer screen. As with the traditional estimating method, in some instances marking the components as they are being entered into the estimating program is required to avoid duplication and ensure all required items are accurately and completely included in the estimate.

Building Information Modeling. As design and construction information become increasingly available in electronic format, integration of design and construction data becomes more important to the entire construction process. *Building information modeling (BIM)* is a methodology that uses coordinated, consistent formats for electronic design, construction, and product information to optimize planning and construction of a project. Use of a BIM system creates a design and construction process where owners, architects, engineers, estimators, contractors, subcontractors, and suppliers utilize electronic information in a consistent format to identify the best construction practices and avoid costly oversights. The estimator must review the specifications to determine if the owner requires that a specific type of BIM model or software system be included for the project.

A BIM model is developed and presented in an electronic three-dimensional (3-D) format. All members of the construction team on a BIM project provide their specific individual documents in an intelligent BIM format. The BIM model allows for integration of the structural, mechanical, electrical, and all other applicable building systems. The coordination and collaboration of all of the various electronic components from architects, engineers, contractors, and subcontractors creates a complete building model. This model is then utilized to produce views of buildings and building systems for review and modification as needed prior to construction. Modifications can be made in instances where construction conflicts are identified and clash detection occurs. **See Figure 2-10.**

For example, when HVAC and electrical systems are added into the BIM model, a clash detection analysis may indicate that both an air duct and an electrical conduit are intended for the same location. The BIM model will indicate this potential clash. Both the HVAC subcontractor and electrical subcontractor can revisit their documents and make the necessary adjustments prior to construction. In many instances, this will reduce job-site conflicts and the need to redesign various construction elements in the field.

An accurate BIM model will also allow for the prefabrication of larger building components. Where the BIM model indicates all potential clashes have been eliminated, the use of prefabrication can reduce overall labor costs at the jobsite, thereby requiring the estimator to make the necessary adjustments in transportation and rigging expenses and on-site labor requirements. Coordination of this opportunity to use prefabrication must be done with the appropriate members of the building team during development of the estimate. When opportunities for prefabrication savings are determined during the project, value engineering adjustments may be made.

> An accurate and complete BIM model allows a total design review. Design modifications can resolve any identified construction conflicts and detected clashes of systems.

Construction Modeling Tools

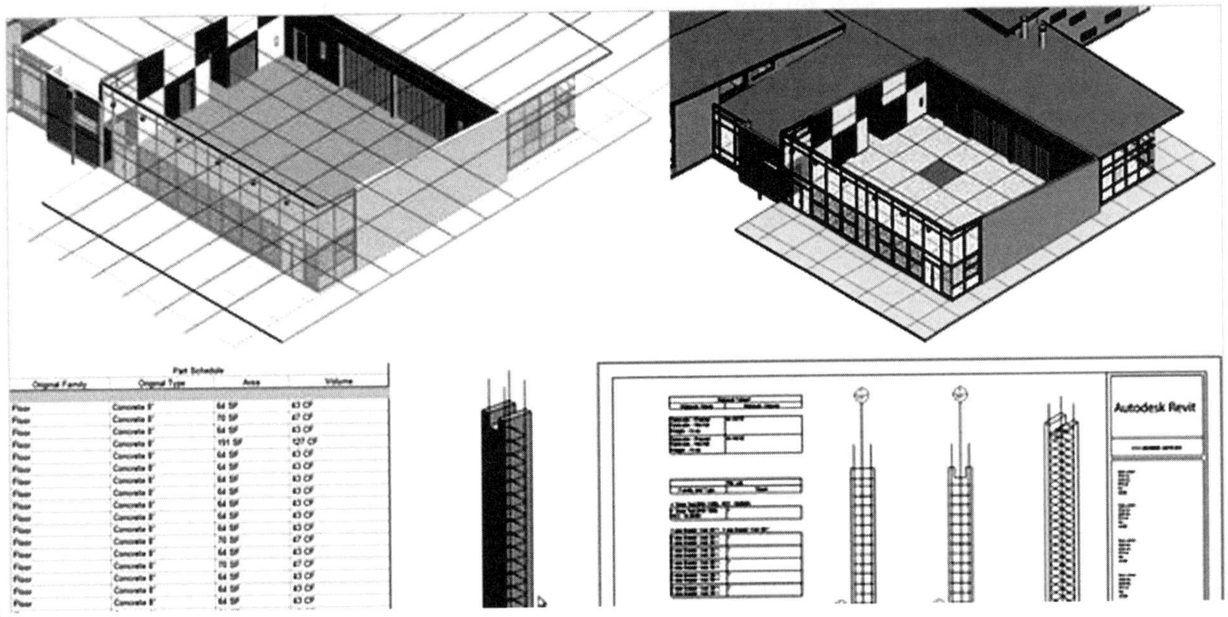

Autodesk, Inc.

Figure 2-10. Three-dimensional models are used to produce buildings and building systems to locate and avoid potential construction conflicts.

BIM models may also provide fourth and fifth "dimensions," with the fourth dimension being quantity takeoff and estimating, and the fifth dimension being project scheduling and planning. When construction projects utilizing BIM are designed, intelligent model objects have data attached to them. In some instances, the object data is collected by the BIM software to generate quantity takeoff information and estimates in conjunction with estimating information databases and/or be used to develop a schedule for the project.

In these instances, some virtual construction of the building can be modeled, allowing the construction team and the estimator to develop various sequences for the construction process. More economical or efficient building scheduling can be determined with this system, impacting the estimate in terms of labor costs.

Estimating Program Internet Capability. Some estimating programs have the ability to link to the Internet and transfer information to the program. Prints can be digitized or scanned into an electronic format, or CAD drawings can be uploaded and made available to estimators through the Internet. Autodesk® Revit® is a commonly utilized program for BIM model development and integration.

Electronic "plan rooms" are provided by a variety of local and national associations and construction information firms. After obtaining a password, members of electronic plan rooms can search through specifications and prints on a computer screen. When projects or items are found that can be bid by a company, the specifications and plans can be electronically retrieved for use with the estimating program.

In some cases, these plans can be marked up on the computer screen to facilitate quantity takeoff. For example, e-builder™ is an Internet-based communication tool designed to enhance the exchange of information and collaboration among construction project participants. In addition, Autodesk® Buzzsaw® is a project management system that assists in the planning, construction, and operation phases of a building. Buzzsaw helps to ensure that accurate information is always available to everyone involved in the creation and management of a construction project.

Integration with Job Site Management Systems

Use of estimating programs allows for the material quantity, labor, equipment, and overhead information to be used by others on the construction project team. Project managers can compare accounting information to the bid information to track job costs throughout a construction project. Other computer programs that assist in construction scheduling can interface with an estimating program to track material quantities and labor use throughout the course of a project. Electronic job tracking systems can be used to enter actual labor hours and identify the hours using cost codes. Radio frequency identification (RFID) or bar scanning technology may be utilized to track various materials as they are delivered and moved at a job site and to track with estimates throughout the construction process.

The compatibility of estimating programs with CAD and BIM programs and scheduling and tracking programs affects the estimating process by enabling the automatic calculation of quantities and costs. Architectural CAD and BIM programs that are compatible with estimating programs enable material quantities and job cost tracking to be calculated from a CAD drawing by the estimating program and entered into the spreadsheet automatically.

Job Cost Analysis. *Job cost analysis* is the study of the final costs of building a project as compared to the original estimate. As a construction project proceeds, the owner, architect, and project managers for various contractors need to ensure that the costs of the project are staying within the estimated amounts.

Integration of estimating programs with job cost analysis programs can indicate areas where savings or cost overruns may have occurred. This integration allows for quick, timely information availability to assist in decisions concerning the remainder of the project. At the completion of the project, the information obtained concerning the initial estimate and the final job cost can be integrated into company historical cost records to improve the accuracy of that database, which will be utilized for future estimates.

Scheduling and Progress Tracking. Construction scheduling includes information concerning material delivery times, work start and completion times, and subcontractor start and completion times. Several computer programs are available for the specific purpose of scheduling construction projects and keeping track of various work phases. **See Figure 2-11.** Integration of estimating programs with scheduling programs allows for ongoing analysis of areas where changes in scheduling or estimated materials and/or labor may be necessary. As with job cost information, this information can be integrated into company historical data for future use.

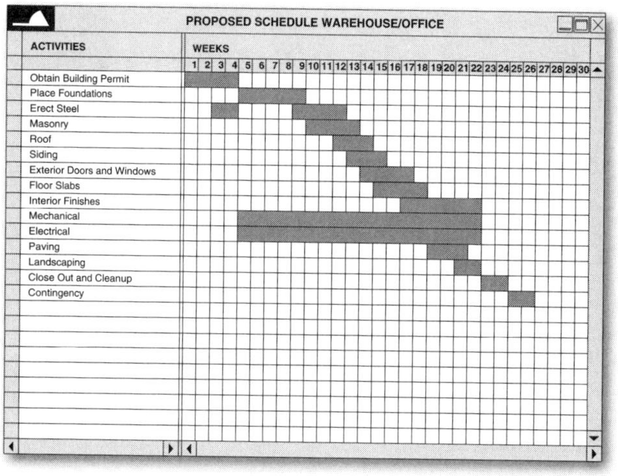

Figure 2-11. A job scheduling program can interface with an estimating program to create an integrated project management system.

CAD and BIM Compatibility. *CAD and BIM compatibility* is the ability of an estimating program to directly transfer quantity takeoff information from CAD-generated drawings or BIM systems. This allows estimating programs to automatically develop quantity takeoffs based on intelligent object codes embedded in the electronic plan documents. As labor unit prices and material costs are added to an integrated database, a single program may be used to design and produce the drawings, as well as develop material and labor quantities and costs.

SAGE TIMBERLINE ESTIMATING . . .

GRAPHICAL USER INTERFACE

The graphical user interface (GUI) for the Sage Timberline Estimating program consists of several common components in a ribbon layout, including dialog boxes, spreadsheets, and databases. A better understanding of these components and their purposes will allow for easier navigation in the Sage Timberline Estimating program and proper preparation of an estimate.

The title bar of the Sage Timberline Estimating program displays the database and estimate names. The database contains all the information required to prepare an estimate, including labor, material, and equipment costs. The Sage Timberline Estimating databases typically consist of a three-tiered hierarchy. Groups are the most general level of the database. Group numbering correlates to the division numbering used in the CSI MasterFormat, such as **Sitework** or **Concrete**. Phases are the next level of the database and allow you to organize items into collections of work-related tasks or material types. Examples of phases include **3110.100 Forms: Footings** or **3310.140 Conc: Footings**. Items are the lowest and most specific level of the database. Items represent specific tasks or materials such as **Footing Forms** or **Footing Conc 3000 psi**.

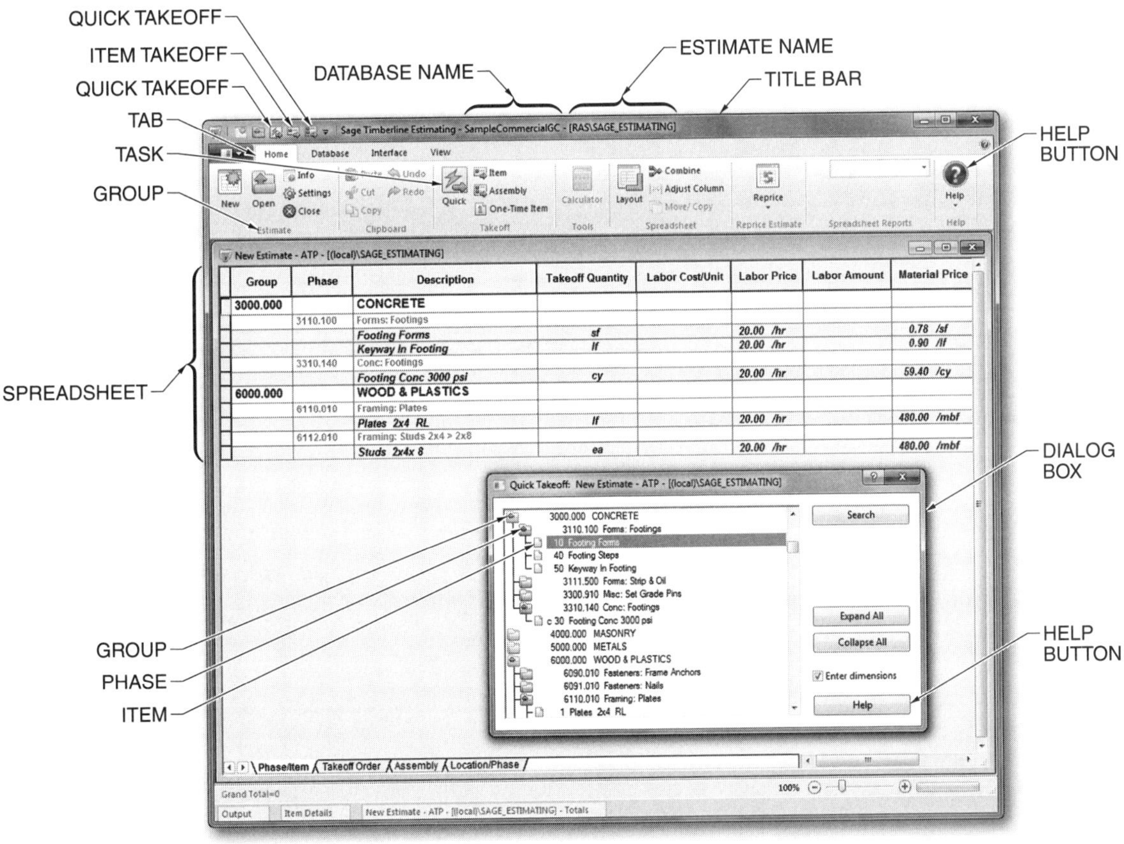

The ribbon layout of the Estimating Management Console contains all of the commands used in the Sage Timberline Estimating program. Picking one of the tab menu headings displays the associated commands. Some commands may be followed by an arrow or ellipses (**...**). A dialog box is displayed when a takeoff menu entry is picked from the Estimating Management Console.

. . . SAGE TIMBERLINE ESTIMATING

The Estimating Management Console contains buttons representing the most frequently used commands and features of the Sage Timberline Estimating program (the console can be customized). The ribbon can be collapsed to save space by double-clicking on any tab. The cursor is hovered over any tab on the collapsed ribbon to view the ribbon contents. Any tab is double-clicked again to expand the ribbon.

The spreadsheet is the actual estimate being developed and represents the items, quantities, and monetary amounts that compose the estimate. Spreadsheets are automatically saved as work is being done. When creating an estimate, the information from the database is connected to the spreadsheet. Sage Timberline Estimating stores estimating information in three places.

- Estimate database: This database stores individual project estimates. Each estimate contains information for a specific project, such as quantities, labor, equipment, and subcontractors.
- Standard database: This database contains a catalog of information for use with estimates, such as a list of items and assemblies, prices, calculation methods, and conversions.
- Address book database: This database stores all contact information for individuals and organizations.

The estimate database stores the estimates. All estimates could be stored in the same estimate database or in more than one estimate database if different types of projects are involved.

Quick Quiz®

Quick Quiz®

Refer to the CD-ROM for the Quick Quiz® questions related to chapter content.

Key Terms

Illustrated Glossary

- *assembly*
- *building information modeling (BIM)*
- *CAD and BIM compatibility*
- *database transfer*
- *digitizer*
- *electronic estimating method*
- *estimating method*
- *general conditions*
- *job cost analysis*
- *ledger sheet*
- *MasterFormat™*
- *plan markup*
- *spreadsheet*
- *traditional estimating method*

Web Links

Web Links

Autodesk® Buzzsaw®
www.autodesk.com/buzzsaw

Construction Specifications Institute
www.csinet.org

e-builder™
www.e-builder.net

CONSTRUCTION ESTIMATING

Estimating Methods 2

Review Questions

Estimating

_______________ **1.** The ___ estimating method is an estimating method in which quantities are calculated and entered into ledger sheets for pricing.

_______________ **2.** Plan ___ is the color coding or marking off of items during the quantity takeoff process.

_______________ **3.** A(n) ___ sheet consists of rows and columns on a sheet of paper into which item descriptions, locations, code numbers, quantities, and costs are manually entered into them.

_______________ **4.** A(n) ___ is a computer program that uses cells organized into rows and columns to perform various mathematical calculations when formulas and numbers are entered into them.

_______________ **5.** The ___ estimating method is an estimating method in which takeoff and pricing is accomplished using computer spreadsheets and integrated database systems.

_______________ **6.** A(n) ___ is a collection of items needed to complete a particular unit of work.

_______________ **7.** A(n) ___ is an electronic input device that converts analog data to digital data.

T F **8.** The estimating process begins with a set of detail drawings and specifications.

_______________ **9.** The ___ is a uniform system of numbers and titles for organizing information about construction requirements, products, and activities into a standard sequence.

_______________ **10.** ___ is the direct transfer of stored information by an estimating program from a database into a spreadsheet.

T F **11.** An estimating method is the approach used by an estimator to perform project analysis, quantity takeoff, and pricing in a consistent and organized manner.

_______________ **12.** Job cost ___ is the study of the final costs of building a project as compared to the original estimate.

_______________ **13.** CAD ___ is the ability of an estimating program to directly transfer quantity takeoff information from CAD-generated drawings.

T F **14.** The foundation of most estimating programs is the bid document.

T F **15.** Consistency of data is a major advantage of estimating programs.

Job Scheduling

Refer to the proposed schedule for the warehouse/office to answer the following questions.

T F 1. Obtaining the building permit is the first activity performed.

T F 2. The roof is completed before concrete is placed for the floor slabs.

______________ 3. Mechanical and ___ work require the most time.

______________ 4. Placing the foundations requires ___ weeks.

______________ 5. Including the contingency, the Warehouse/Office will require ___ weeks to complete.

T F 6. Landscaping is started after paving is completed.

T F 7. No other activities are scheduled during cleanup.

______________ 8. ___ weeks are required to erect the steel.

T F 9. Interior finishes are completed before the electrical work is completed.

T F 10. Masonry and siding are started at the same time.

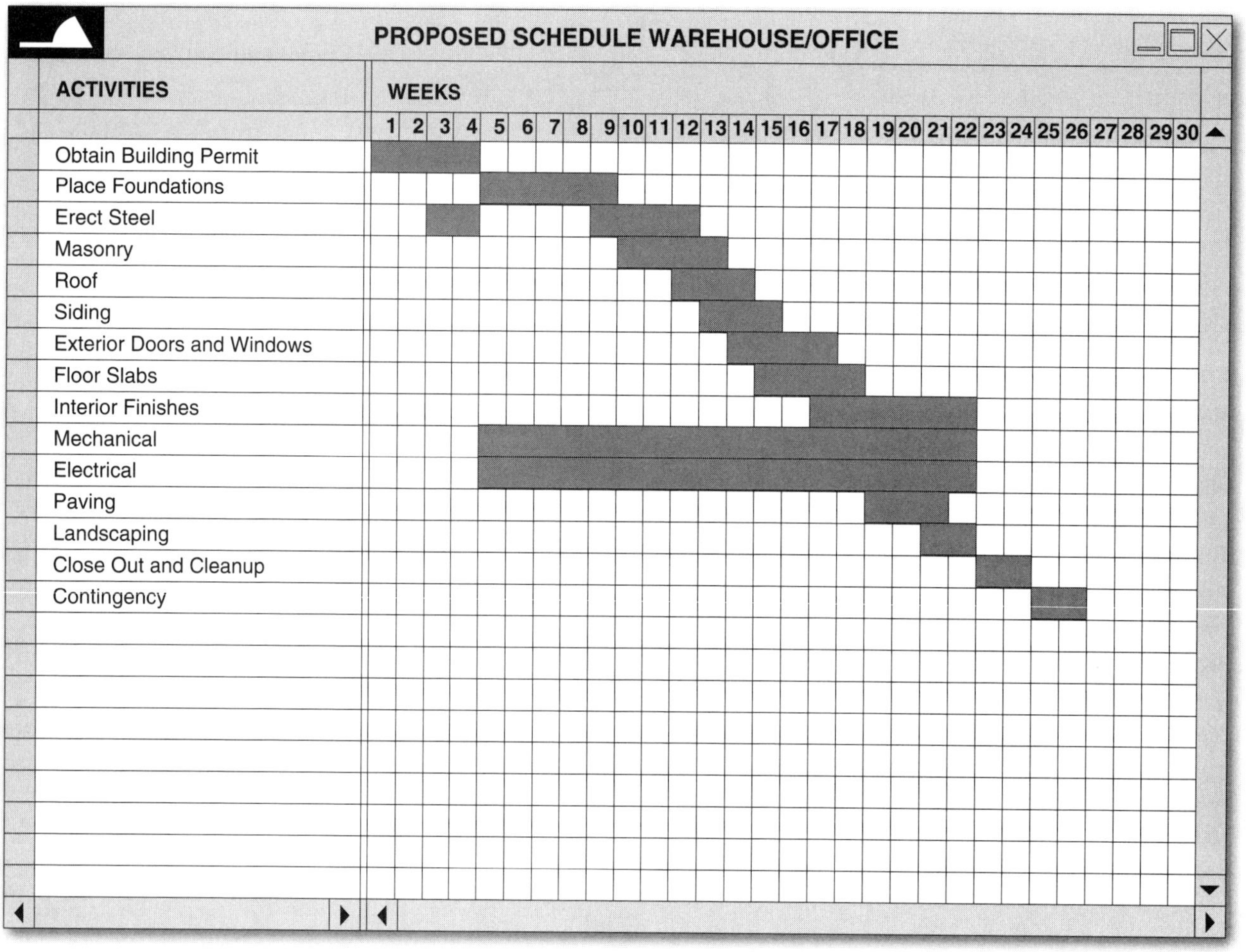

Short Answer

1. Compare and contrast traditional and electronic estimating methods.

2. Cite advantages of using electronic estimating methods.

3. Identify the advantages of interfacing an estimating program with scheduling, tracking, and CAD programs.

4. Discuss the advantages of using building information modeling (BIM).

Estimating Methods 2

Activities

Activity 2-1—Field Conversion

Use the field conversion method to convert the inch values to decimal feet values. Refer to the Inch-Decimal Foot Conversion table in the Appendix.

1. $8\frac{5}{8}'' =$ ___′.
2. $4\frac{3}{8}'' =$ ___′.
3. $5\frac{1}{4}'' =$ ___′.
4. $3\frac{3}{8}'' =$ ___′.
5. $1\frac{1}{8}'' =$ ___′.
6. $\frac{7}{8}'' =$ ___′.
7. $9\frac{5}{8}'' =$ ___′.
8. $7\frac{1}{2}'' =$ ___′.
9. $11\frac{5}{8}'' =$ ___′.
10. $2\frac{3}{4}'' =$ ___′.

Activity 2-2—Unit Conversion

1. The total material unit cost for fifty-two 2 × 4 joist hangers is $___.
2. The total labor unit cost for two hundred sixty-six 2 × 6 joist hangers is $___.
3. The material unit cost for each 4 × 4 joist hanger is $___.
4. The total material and labor cost for a project requiring one hundred twenty-four 2 × 10 joist hangers is $___.
5. The total material unit cost for ninety-six 2 × 14 joist hangers is $___.

COST DATA			
Material	**Unit**	**Material Unit Cost***	**Labor Unit Cost***
Joist hangers heavy-duty 12 gauge, galvanized			
2 × 4	100	63.00	120.00
2 × 6	100	72.00	138.00
2 × 10	100	106.00	138.00
2 × 14	100	176.00	156.00
4 × 4	100	114.00	138.00

* in $

______ **6.** The total material and labor cost for 1025 lf of No. 14 gauge, 2-conductor armored cable is $___.

______ **7.** The total material unit cost for 365 lf of No. 12 gauge, 2-conductor armored cable is $___.

______ **8.** The total labor unit cost of 837 lf of No. 10 gauge, 2-conductor armored cable is $___.

______ **9.** The cost per linear foot for No. 10 gauge, 3-conductor armored cable is $___.

______ **10.** The total material and labor cost for 482 lf of No. 12 gauge, 2-conductor armored cable, including 12% for overload and profit, is $___.

COST DATA			
Material	Unit	Material Unit Cost*	Labor Unit Cost*
Copper solid armored (MC) cable			
No. 14 gauge, 2-conductor	100 lf	55.00	28.80
No. 12 gauge, 2-conductor	100 lf	55.60	32.20
No. 10 gauge, 2-conductor	100 lf	101.10	37.20
No. 10 gauge, 3-conductor	100 lf	136.00	47.40

* in $

______ **11.** The total concrete cost for a 12″ thick by 8′ high by 27′ long concrete wall is $___.

______ **12.** The total labor unit cost (including setting rebar, erecting forms, and placing concrete) for a 12″ thick by 14′ high concrete wall is $___.

______ **13.** The total material, labor, and equipment cost for an 8″ thick by 8′ high by 66′ long concrete wall is $___.

______ **14.** The total equipment cost for an 8″ thick by 14′ high by 42′ long concrete wall is $___.

______ **15.** The total material, labor, and equipment cost for an 8″ thick by 8′ high by 135′ long concrete wall, including 8% for overhead and profit, is $___.

COST DATA				
Material	Unit	Material Unit Cost*	Labor Unit Cost*	Equipment Unit Cost*
Concrete walls, 8″ thick × 8′ high with two mats of No. 5 rebar 12″ OC each way				
Reinforcing Steel	sf	2.25	1.75	0.02
Forms	sf	1.40	3.95	0.06
Concrete	sf	2.69	1.69	0.30
Concrete walls, 8″ thick × 14′ high with two mats of No. 5 rebar 12″ OC each way				
Reinforcing Steel	sf	2.25	2.00	0.10
Forms	sf	1.40	4.74	0.28
Concrete	sf	2.69	2.01	0.44
Concrete walls, 12″ thick × 8′ high with two mats of No. 5 rebar 12″ OC each way				
Reinforcing Steel	sf	2.25	1.75	0.02
Forms	sf	1.40	3.95	0.06
Concrete	sf	4.05	2.59	0.46
Concrete walls, 12″ thick × 14′ high with two mats of No. 5 rebar 12″ OC each way				
Reinforcing Steel	sf	2.25	2.00	0.10
Forms	sf	1.40	4.74	0.28
Concrete	sf	4.05	2.96	0.67

* in $

Specifications and Drawings 3

Key Concepts

- Construction of residential, commercial, and industrial buildings requires a set of construction documents including drawings and specifications. Specifications provide details that cannot be shown on drawings or that require further description.
- Specifications are divided into sections that cover the requirements of each specific type of construction.
- The MasterFormat™ is a uniform system of numbers and titles used for organizing information about construction requirements, products, and activities into a standard sequence. Division 01 of the specifications describes the overall construction work and includes site restrictions and utility sources.

Introduction

Specifications are written supplements to a set of construction documents and drawings that provide additional construction information. The specifications along with the drawings describe the entire construction project and process. The MasterFormat™ is a master list of numbers and titles used for organizing information about construction requirements, products, and activities into a standard sequence. The MasterFormat™ contains 50 divisions that define the broad areas of construction. An estimator must be able to read and interpret architectural and shop drawings.

SPECIFICATIONS

Construction of residential, commercial, and industrial buildings requires a set of construction documents including drawings and specifications. *Specifications* are written supplements to a set of construction documents and drawings that provide additional construction information. Specifications provide details that cannot be shown on drawings or that require further description. Specifications, along with the drawings, describe the specific items and practices included in the construction project. Specifications contain information related to legal issues, construction materials and procedures, and quality control issues of construction.

Specifications are an organized presentation of bidding information, contract requirements, and all phases of the construction process for a particular project. Standard forms, language, and formats included in the specifications provide direction for all parties involved to ensure clear and accurate communication. Individual construction projects and architects require the specifications to be used in a variety of ways. The specifications must be completely reviewed during the estimating process to fully understand the project and to develop an accurate estimate.

Estimators should note any areas where the specifications are in conflict or inconsistent with the drawings. For example, a foundation plan may indicate that a poured-in-place concrete basement slab is to be placed directly on the soil below the slab. The specifications may state that all slabs are to be placed on 4″ of crushed gravel. The estimator or contractor must obtain written clarification from the architect and/or owner when a conflict exists between the drawings and specifications.

An architect is typically responsible for the detailed design and coordination of the construction project. The architect is the central person in a group of design professionals including civil engineers, structural engineers, mechanical engineers, and electrical engineers. The design-bid-build traditional project delivery system is the process of designing the structure, then making the project available for bid, and then building the structure. With this system, meetings are scheduled between the architect and the owner or developer to determine the needs of the owner or developer. After deliberation and conferences, the architect prepares preliminary design ideas and conceptual estimates for the project and submits them to the owner or developer for approval. The owner or developer suggests changes and returns the final requirements to the architect, who then begins work on the detail drawings.

The intended function of the structure, materials used in the building of the structure, and methods by which these materials are placed must be clearly specified. All parties involved in the project must understand the architect's concept of the structure. The architect or developer submits the specifications and drawings to several contractors for bidding. The specifications are the most important document for the contractor at the beginning of the estimating and bidding process.

Specifications are divided into sections that cover the requirements of each specific type of construction. Estimators must be familiar with the parts of the specifications that apply to the type of work performed by their particular construction firm and carefully analyze every section, paragraph, and line. Misreading one word or paragraph may distort the entire meaning and requirements of a particular section of the specifications and can lead to a faulty bid or a dangerous deviation from the drawings. Care must be taken, because in some cases, various sections of the specifications can contain conflicting information.

General Conditions

The General Conditions section of the specifications includes overhead expenses that cover the entire construction project. The construction process involves many people, agencies, and trades. It includes city, county, state, and federal agencies, banks and financing entities, bonding companies, and insurance companies. The concerns of all these parties must be stated in writing so that all parties bidding for the job are aware of the requirements. This gives all the bidding parties a fair and equal chance of arriving at the best and most complete bid.

The general conditions also state the method used for bidding and the selection process for awarding the project. The general conditions indicate whether the project is for a fixed lump-sum bid, cost plus, time and materials, or another project delivery method. When the owner utilizes a project delivery system other than design-bid-build, other considerations must be taken into account when reviewing the specifications. If program managers, construction managers, or other parties are involved in the construction process, information about the various decision-making processes concerning the project can impact the overall job cost. The general conditions may also contain standardized bidding forms for contractors to use to submit their bids.

MasterFormat™

The *Construction Specifications Institute (CSI)* is an organization of individuals and organizations that includes architects, engineers, contractors, specifiers, and suppliers of construction products. The CSI, in conjunction with the American Institute of Architects (AIA), the Associated General Contractors of America (AGC), the Associated Specialty Contractors (ASC), and other industry groups, has developed the *CSI MasterFormat™ for Construction Specifications* and *The Uniform System for Construction Specifications, Data Filing, and Cost Accounting*. These specification standards apply primarily to projects in the United States and Canada. Estimators working on projects developed outside the United States and Canada must review the specifications more carefully.

The *MasterFormat™* is a uniform system of numbers and titles used for organizing information about construction requirements, products, and activities into a standard sequence.

Several years ago, specifications for construction projects were organized by the MasterFormat 16-division structure. **See Figure 3-1.** The current edition of the MasterFormat contains 50 divisions that define broad areas of construction. **See Figure 3-2.** Assigned divisions are numbered 00 to 14; 21 to 23; 25 to 28; 31 to 35; 40 to 45; and 48. Divisions such as 15 to 20 were not assigned to construction areas to allow for future expansion of the MasterFormat without the entire system having to be restructured. Each assigned division is designed to provide complete written information about individual construction requirements for building and material needs.

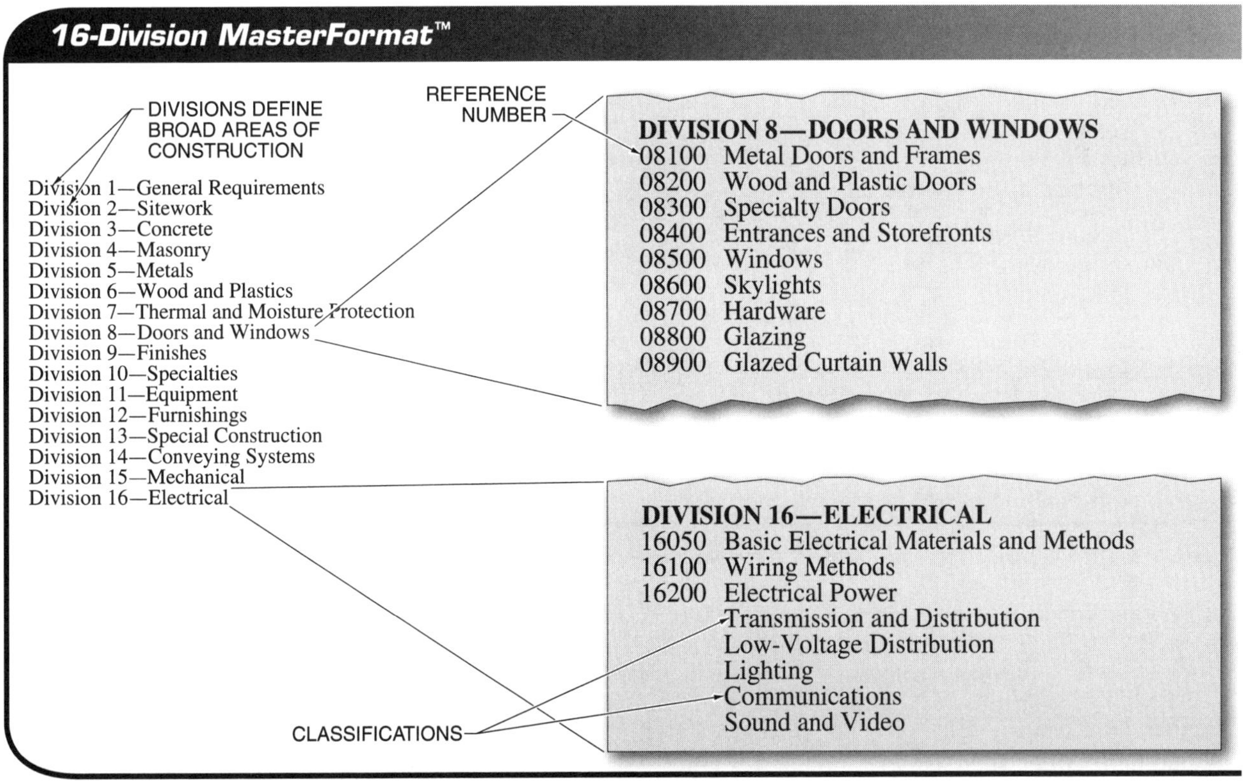

Figure 3-1. The 16-division MasterFormat™ was used by specifiers and estimators prior to the release of the revised MasterFormat™.

50-Division MasterFormat™

Procurement and Contracting Requirements Group
- Division 00—Procurement and Contracting Requirements

Specifications Group

General Requirements Subgroup
- Division 01—General Requirements

Facility Construction Subgroup
- Division 02—Existing Conditions
- Division 03—Concrete
- Division 04—Masonry
- Division 05—Metals
- Division 06—Woods, Plastics, and Composites
- Division 07—Thermal and Moisture Protection
- Division 08—Openings
- Division 09—Finishes
- Division 10—Specialties
- Division 11—Equipment
- Division 12—Furnishings
- Division 13—Special Construction
- Division 14—Conveying Equipment
- Divisions 15 to 19—Unassigned

Facility Services Subgroup
- Division 20—Unassigned
- Division 21—Fire Suppression
- Division 22—Plumbing
- Division 23—Heating, Ventilating, and Air-Conditioning (HVAC)
- Division 24—Unassigned
- Division 25—Integrated Automation
- Division 26—Electrical
- Division 27—Communications
- Division 28—Electronic Safety and Security
- Division 29—Unassigned

Site and Infrastructure Subgroup
- Division 30—Unassigned
- Division 31—Earthwork
- Division 32—Exterior Improvements
- Division 33—Utilities
- Division 34—Transportation
- Division 35—Waterway and Marine Construction
- Divisions 36 to 39—Unassigned

Process Equipment Subgroup
- Division 40—Process Integration
- Division 41—Material Processing and Handling Equipment
- Division 42—Process Heating, Cooling, and Drying Equipment
- Division 43—Process Gas and Liquid Handling, Purification, and Storage Equipment
- Division 44—Pollution Control Equipment
- Division 45—Industry-Specific Manufacturing Equipment
- Divisions 46 to 47—Unassigned
- Division 48—Electrical Power Generation
- Division 49—Unassigned

DIVISIONS DEFINE BROAD AREAS OF CONSTRUCTION

DIVISION 4—MASONRY

04 01 00	Maintenance of Masonry
04 05 00	Common Work Results for Masonry
04 06 00	Schedules for Masonry
04 08 00	Commissioning of Masonry
04 20 00	Unit Masonry
04 40 00	Stone Assemblies
04 50 00	Refractory Masonry
04 60 00	Corrosion-Resistant Masonry
04 70 00	Manufactured Masonry

TITLES DIVIDE DIVISIONS INTO MORE DISCRETE AREAS

DIVISION 26—ELECTRICAL

26 01 00	Operation and Maintenance of Electrical Systems
26 05 00	Common Work Results for Electrical
26 06 00	Schedules for Electrical
26 08 00	Commissioning of Electrical Systems
26 09 00	Instrumentation and Control for Electrical Systems
26 10 00	Medium-Voltage Electrical Distribution
26 20 00	Low-Voltage Electrical Distribution
26 30 00	Facility Electrical Power Generating and Storing Equipment
26 40 00	Electrical and Cathodic Protection
26 50 00	Lighting

Figure 3-2. The 50-division MasterFormat™ provides room for inclusion of new construction materials and techniques.

Each division includes sets of numbered titles. For example, Division 48—Electrical Power Generation has numbered titles such as Operation and Maintenance for Electrical Power Generation (48 01 00), Electrical Power Generation Equipment (48 10 00), and Fuel Cell Electrical Power Generation Equipment (48 19 00). The following discussion of the MasterFormat is based on the 50-division structure, which is currently in use by architects, engineers, specifiers, estimators, and related construction personnel.

Division 00—Procurement and Contracting Requirements. Division 00 of the specifications includes information regarding items that must be submitted by those bidding on a project. Division 00 also includes information about the scope required for various proposals and the procurement meetings that must take place prior to a bid being received. Preliminary project schedules and project budget information are specified in Division 00. Other project information cited in Division 00 includes survey information, photographs related to the project, environmental assessment information, existing building information (for remodeling work), and seismic and geotechnical information.

Division 00 includes various forms that must be submitted by contractors for bidding. These forms are of particular interest to an estimator and contractor since they may include unit prices, allowances, alternate material or substitution forms, quantities forms, wage rate information, and a bid submittal checklist. Documentation of various project requirements pertaining to revisions, addendum, health and safety, project delivery systems, subcontractor requirements, quality control, contract closeout, and other conditions of the contract is also described in Division 00.

Division 01—General Requirements. Division 01 of the specifications describes the overall construction work and includes site restrictions and utility sources. **See Figure 3-3.** Language regarding the potential use and processes for value engineering is also included in Division 01. Coordination of project management and progress documentation requirements is specified, including such items as photographic documentation and periodic work observations. Quality, quality assurance, quality control, and product requirements are also specified in Division 01.

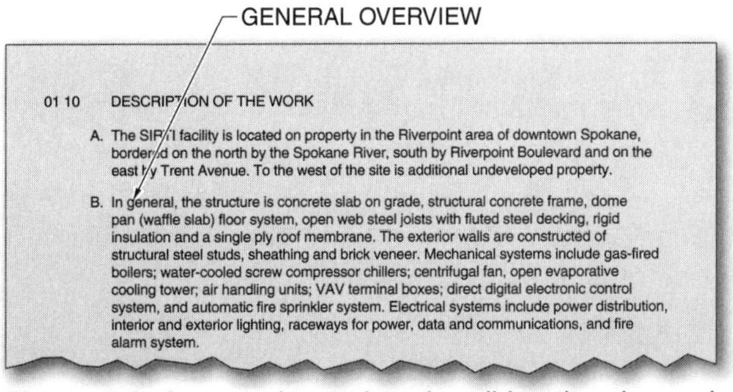

01 10 DESCRIPTION OF THE WORK

A. The SIR/I facility is located on property in the Riverpoint area of downtown Spokane, border/d on the north by the Spokane River, south by Riverpoint Boulevard and on the east b/y Trent Avenue. To the west of the site is additional undeveloped property.

B. In general, the structure is concrete slab on grade, structural concrete frame, dome pan (waffle slab) floor system, open web steel joists with fluted steel decking, rigid insulation and a single ply roof membrane. The exterior walls are constructed of structural steel studs, sheathing and brick veneer. Mechanical systems include gas-fired boilers; water-cooled screw compressor chillers; centrifugal fan, open evaporative cooling tower; air handling units; VAV terminal boxes; direct digital electronic control system, and automatic fire sprinkler system. Electrical systems include power distribution, interior and exterior lighting, raceways for power, data and communications, and fire alarm system.

Figure 3-3. A general overview describing the size and scope of the construction project is included in Division 01 of the specifications.

Reference materials for project abbreviations and definitions are included, as well as information regarding temporary lighting and telecommunications facilities, construction facilities, temporary elevators and scaffolds, barricades, fences, stormwater pollution control, and signage. Project closeout issues related to cleanup, starting and adjusting of equipment, final site survey, training of owner staff, and other transition operations are also described.

Commissioning is the process of reviewing the final performance of building systems to ensure they meet owner and design requirements. Any commissioning or final operation and maintenance requirements are shown in Division 01.

Division 02—Existing Conditions. Division 02 of the specifications includes information concerning existing items, such as underground storage tanks and concrete foundations and piers, as well as such items as any existing thermal or moisture protection. Responsibilities for subsurface exploration, excavation, compaction, and disposal of excavated materials are also part of Division 02. Information related to the potential existence or discovery of hazardous materials at the building site and proper handling and disposal of these hazardous materials is described.

Division 02 details the contractor responsibilities for testing and proper handling of construction-site soil, fill, and backfill materials. Bearing capacities for the subsurface must be achieved and measured according to industry standards, such as American Society for Testing and Materials (ASTM) standards. Information regarding the removal and use of surface soils, soil stabilization, and necessary soil improvements is provided. Specific soil and slope finish information may also be given in Division 02. **See Figure 3-4.** Various types of investigation of subsurface conditions related to seismic, electrical, and magnetic properties are also identified.

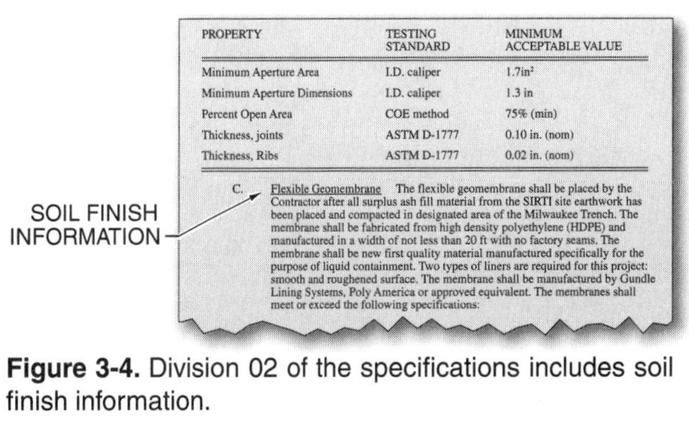

PROPERTY	TESTING STANDARD	MINIMUM ACCEPTABLE VALUE
Minimum Aperture Area	I.D. caliper	1.7in²
Minimum Aperture Dimensions	I.D. caliper	1.3 in
Percent Open Area	COE method	75% (min)
Thickness, joints	ASTM D-1777	0.10 in. (nom)
Thickness, Ribs	ASTM D-1777	0.02 in. (nom)

C. Flexible Geomembrane The flexible geomembrane shall be placed by the Contractor after all surplus ash fill material from the SIRTI site earthwork has been placed and compacted in designated area of the Milwaukee Trench. The membrane shall be fabricated from high density polyethylene (HDPE) and manufactured in a width of not less than 20 ft with no factory seams. The membrane shall be new first quality material manufactured specifically for the purpose of liquid containment. Two types of liners are required for this project: smooth and roughened surface. The membrane shall be manufactured by Gundle Lining Systems, Poly America or approved equivalent. The membranes shall meet or exceed the following specifications:

Figure 3-4. Division 02 of the specifications includes soil finish information.

A significant amount of information provided in Division 02 covers demolition procedures and requirements, removal of construction materials, and site remediation related to chemical and biological contaminants, including such materials as asbestos, lead, polychlorinated biphenyls (PCBs), and mold.

Division 03—Concrete. Division 03 of the specifications contains information concerning concrete, such as maintenance, schedules, materials, placement procedures, curing, finishing, formwork construction and removal, and reinforcing methods. Precast concrete members, which are commonly used in commercial construction, are also described in Division 03.

Concrete material information includes the types of cement, aggregate, admixtures, and quality of water needed for the concrete mixture. Reinforcing steel is described according to ASTM references. Accessories to be placed in concrete, such as chairs, dowels, and anchors, may be described according to specific manufacturer references. **See Figure 3-5.** Form materials such as insulating concrete forms, form-release agents, grout, joint fillers, waterstops, and curing compounds are also described in Division 03.

> While most information regarding concrete is included in Division 03, information regarding specialty projects may be included in other divisions. For example, cast-in-place concrete for tunnel or shaft lining is included in Division 31.

Division 04—Masonry. Division 04 of the specifications addresses masonry construction, including masonry units, mortar, reinforcement, schedules, masonry unit materials, and accessories. Different types of mortar are used in various applications, such as load-bearing masonry walls, non-load-bearing masonry walls, and tuckpointing. Mortar ingredients detailed in Division 04 include types of cement, aggregate, water, bonding agents, coloring, and admixtures. **See Figure 3-6.** An *admixture* is a substance other than water, aggregate, and portland cement that is added to concrete to modify its properties.

Sizes and colors of face brick and concrete masonry units may be described in Division 04 with reference to a specific manufacturer. When stone is supplied, a specific supplier or quarry may be identified to ensure stone quality and uniformity. Other masonry material information in Division 04 covers metal ties and anchors, flashing, and control joints.

Division 05—Metals. Division 05 of the specifications describes metal used on a construction project, including the following:

- structural metal framing members such as columns, beams, and joists
- metal decking for floors, walls, and roofs
- cold-formed metal framing members
- metal fabrications such as stairs, railings, and grates
- decorative metals such as handrails, ladders, and expansion joints

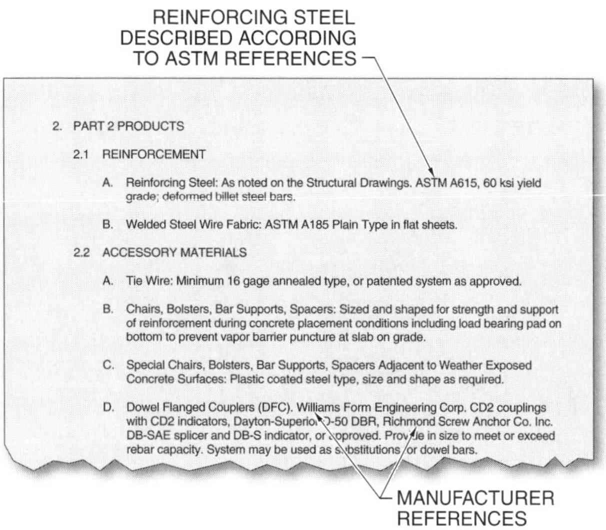

2. PART 2 PRODUCTS

2.1 REINFORCEMENT

A. Reinforcing Steel: As noted on the Structural Drawings. ASTM A615, 60 ksi yield grade; deformed billet steel bars.

B. Welded Steel Wire Fabric: ASTM A185 Plain Type in flat sheets.

2.2 ACCESSORY MATERIALS

A. Tie Wire: Minimum 16 gage annealed type, or patented system as approved.

B. Chairs, Bolsters, Bar Supports, Spacers: Sized and shaped for strength and support of reinforcement during concrete placement conditions including load bearing pad on bottom to prevent vapor barrier puncture at slab on grade.

C. Special Chairs, Bolsters, Bar Supports, Spacers Adjacent to Weather Exposed Concrete Surfaces: Plastic coated steel type, size and shape as required.

D. Dowel Flanged Couplers (DFC). Williams Form Engineering Corp. CD2 couplings with CD2 indicators, Dayton-Superior D-50 DBR, Richmond Screw Anchor Co. Inc. DB-SAE splicer and DB-S indicator, or approved. Provide in size to meet or exceed rebar capacity. System may be used as substitutions for dowel bars.

Portland Cement Association

Figure 3-5. Specifications include specific manufacturer references to define types and qualities of materials and hardware.

Portland Cement Association

PORTLAND CEMENT/LIME MORTARS			
Type/Description	**Portland Cement***	**Hydrated Lime or Lime Putty***	**Sand***
M—Mortar of high compressive strength (minimum 2500 psi) after curing 28 days and with greater durability than some other types; used for masonry below ground and in contact with the earth, such as foundations, retaining walls, and access holes; Type M withstands severe frost action and high lateral loads	1	¼	3
S—Mortar with a fairly high compressive strength (minimum 1800 psi) after curing 28 days; used in reinforced masonry and for standard masonry where maximum flexural strength is required; also used when mortar is the sole bonding agent between facing and backing units	1	½	4¼
N—Mortar with a medium compressive strength (minimum 750 psi) after curing 28 days; used for exposed masonry above ground and where high compressive strength or lateral masonry strengths are required	1	1	6
O—Mortar with a low compressive strength (minimum 350 psi) after curing 28 days; used for general interior walls; may be used for load-bearing walls of solid masonry if axial compressive stress does not exceed 100 psi and wall is not exposed to weathering or freezing	1	2	9

* proportion by volume

Figure 3-6. Division 04 of the specifications addresses components of masonry construction, including types of masonry units, mortar, reinforcement, and accessories.

Most cold-formed metal framing members are included in Title 05 40 00, but supports for gypsum board and plaster are covered in Title 09 22 00. Flashing and sheet metal are described in Title 07 60 00. Metal pipe and conduit specifications are described in Divisions 21, 22, 23, and 26.

Metal material specifications commonly refer to ASTM standards for structural metal shapes, coatings, and connectors such as bolts, nuts, and washers to ensure the correct materials are specified. **See Figure 3-7.** The shape, diameter, type of metal, and pipe schedule are provided in the specifications for railings and other decorative metal.

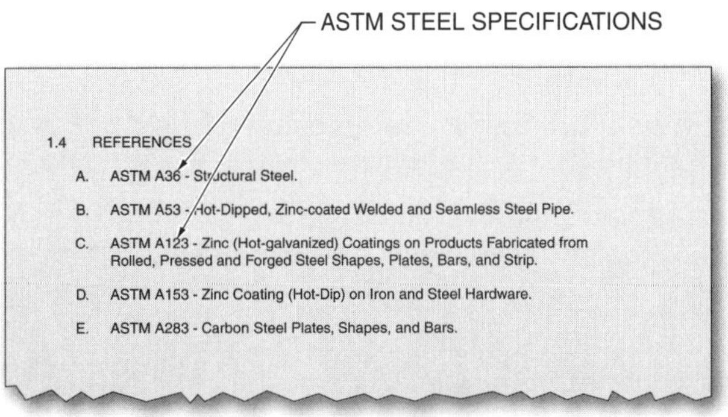

1.4 REFERENCES

A. ASTM A36 - Structural Steel.

B. ASTM A53 - Hot-Dipped, Zinc-coated Welded and Seamless Steel Pipe.

C. ASTM A123 - Zinc (Hot-galvanized) Coatings on Products Fabricated from Rolled, Pressed and Forged Steel Shapes, Plates, Bars, and Strip.

D. ASTM A153 - Zinc Coating (Hot-Dip) on Iron and Steel Hardware.

E. ASTM A283 - Carbon Steel Plates, Shapes, and Bars.

Figure 3-7. Steel material specifications commonly refer to ASTM standards for structural steel shapes, coatings, and connectors.

Division 06—Wood, Plastics, and Composites. Division 06 of the specifications provides information regarding rough carpentry, finish carpentry, architectural woodwork, structural plastics and plastic fabrications, and structural composite materials. Rough carpentry includes wood framing, structural panels, heavy timber construction, treated wood foundations, wood decking, sheathing, and shop-fabricated structural wood members such as laminated veneer lumber (LVL), wood I-joists, and wood trusses.

As with other materials in the specifications, standards from various industry groups are used including the American Forest and Paper Association (AF&PA), the Western Wood Products Association (WWPA), and the American Plywood Association (APA). Finish woodwork quality standards are provided by the Architectural Woodworking Institute (AWI). **See Figure 3-8.**

Related materials described in Division 06 of the specifications include lumber treatments, such as fire retardants and wood preservatives, and fasteners, such as nails, bolts, and lag screws. Finish cabinetry (casework) information includes plastic laminate grades and various hardware. Where manufactured wood casework is installed instead of shop-built cabinets and millwork, the manufactured wood casework is described in Division 12.

> The specifications and drawings must be thoroughly reviewed to ensure an accurate estimate. Specifications normally include specific time constraints for the completion of phases of the project. The contractor is responsible for requirements included in the specifications and drawings.

FINISH REQUIREMENTS RELATED TO AWI STANDARDS

WOODWORK QUALITY STANDARDS

Joint Tolerance*	Premium Grade	Custom Grade	Economy Grade
Maximum gap between exposed components	1/64	1/32	1/16
Maximum length of gap in exposed components	3	5	8
Maximum gap between semi-exposed components	1/32	1/16	1/8
Maximum length of gap in semi-exposed components	6	8	12

* in in.
Note: No gap may occur within 48″ of another gap

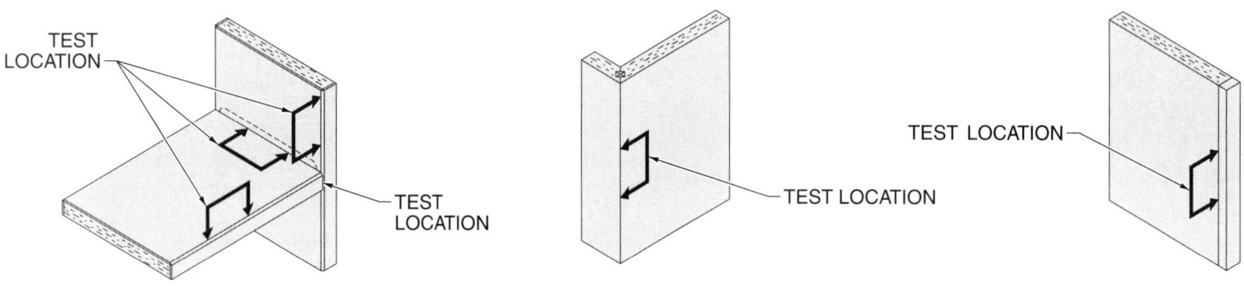

JOINERY TEST LOCATIONS

Figure 3-8. The Architectural Woodworking Institute (AWI) provides woodwork quality standards.

Division 07—Thermal and Moisture Protection. Division 07 of the specifications covers construction materials used to stop moisture movement and penetration and provide thermal insulation and fire protection to a structure. Construction materials included in Division 07 include bituminous products, asphalt roofing, rubberized roofing, mastics, shingles and shakes, roof tiles, waterproof coatings, vapor barriers, water repellents, sheet metal flashing, insulation materials, fireproofing materials, and joint sealants. Exterior insulation and finish systems (EIFS), wall panels, and siding are also included in Division 07.

Due to the specialized nature of many thermal and moisture protection products, Division 07 of the specifications relies heavily on manufacturer names and product numbers. **See Figure 3-9.** Performance-based information is included in the specifications where common materials are used. For example, specifications for expanded polystyrene (EPS) insulation board include physical characteristics, such as required board density, thermal resistance, compressive strength, and performance-based information such as R-factor. Schedules for application of various materials may be included to assist tradesworkers in locating the placement of each material described in the specifications.

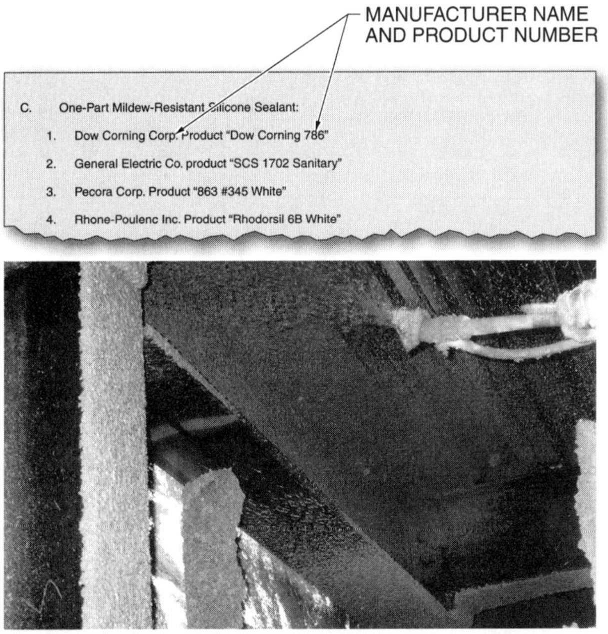

Multiquip Inc.

Figure 3-9. Thermal and moisture protection and fireproofing products in Division 07 of the specifications are designated by manufacturer names and product numbers.

Division 08—Openings. Division 08 of the specifications contains information concerning wall and roof openings, such as doors, windows, and skylights. Division 08 also contains schedules for doors, windows, louvers, vents, and the related hardware. **See Figure 3-10.**

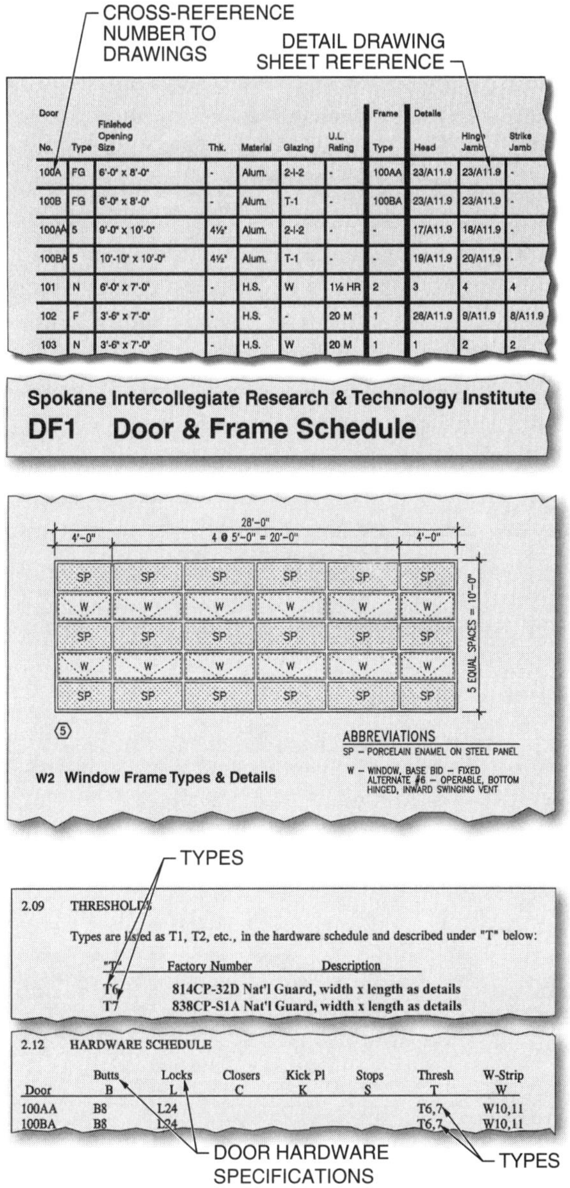

Door No.	Type	Finished Opening Size	Thk.	Material	Glazing	U.L. Rating	Frame Type	Details Head	Details Hinge Jamb	Details Strike Jamb
100A	FG	6'-0" x 8'-0"	-	Alum.	2-I-2	-	100AA	23/A11.9	23/A11.9	-
100B	FG	6'-0" x 8'-0"	-	Alum.	T-1	-	100BA	23/A11.9	23/A11.9	-
100AA	5	9'-0" x 10'-0"	4½"	Alum.	2-I-2	-	-	17/A11.9	18/A11.9	-
100BA	5	10'-10" x 10'-0"	4½"	Alum.	T-1	-	-	19/A11.9	20/A11.9	-
101	N	6'-0" x 7'-0"	-	H.S.	W	1½ HR	2	3	4	4
102	F	3'-6" x 7'-0"	-	H.S.	-	20 M	1	28/A11.9	9/A11.9	8/A11.9
103	N	3'-6" x 7'-0"	-	H.S.	W	20 M	1	1	2	2

"T"	Factory Number	Description
T6	814CP-32D	Nat'l Guard, width x length as details
T7	838CP-S1A	Nat'l Guard, width x length as details

Door	Butts B	Locks L	Closers C	Kick Pl K	Stops S	Thresh T	W-Strip W
100AA	B8	L24				T6,7	W10,11
100BA	B8	L24				T6,7	W10,11

Figure 3-10. Division 08 of the specifications provides door, window, and hardware schedules and detail information that can be cross-referenced to the prints.

Door information is included for swinging wood and metal doors, access doors, overhead doors and grilles, glass doors, and sliding doors. Information concerning door frames, louvers, glass lights, and astragals is also provided. An *astragal* is a one- or two-piece molding used to cover the opening between the stiles of a pair of doors to provide a weathertight seal. A door schedule containing information relating each door to a numbered opening on the architectural drawings is also included. Title 08 40 00 specifically pertains to entrances, storefronts, and curtain-wall construction.

Glazing, weatherstripping, and types of frame finishes comprise a large portion of Division 08. Windows are typically specified by manufacturer product codes. Window performance-based requirements, including deflection, air leakage, thermal performance, and water leakage, are provided in Division 08. Drawings in the specifications may furnish additional details for large or complicated window installations. Division 08 may also include a schedule for placement of various glazing materials, such as glass types and glazing compounds.

A variety of hardware components are described in the specifications with information covering their types and locations. Manufacturer names, designs, sizes, finishes, and functions are described in the hardware specifications. The hardware schedule for doors may include information regarding hinges, locks, door closers, door pulls, push plates, kickplates, stops, bolts, coordinators, thresholds, weatherstripping, and door seals.

Division 09—Finishes. Division 09 of the specifications provides specific information concerning finish materials. The finish materials included in Division 09 include metal lath and plaster, gypsum products, nonstructural metal stud framing and accessories, ceramic floor and wall tile, resilient flooring, carpeting, wood flooring, suspended ceiling systems, special wall and ceiling coverings, and coatings such as stains, varnishes, and exterior and interior paint.

Different areas of a building have different uses and require different floor, wall, and ceiling finishes. Various tile, wood products, gypsum products, plaster, cementitious materials, paint, and special treatments are necessary to accommodate the various usage requirements. Finish materials for each area of a large building may be noted on a room finish schedule. **See Figure 3-11.** Manufacturers of various finish materials are typically identified in the room finish schedule of Division 09.

INTERIOR FINISH SCHEDULE — SUMMIT

FLOORING		FURNISHED BY			INSTALLED BY		
		O	GC	ES	O	GC	ES
CARPET (STANDARD)	"US AXMINSTER" #15766-07-07 "SIERRA"			O		O	
CARPET (ALTERNATE)	"MASLAND" #11418 BURGUNDY FAN			O		O	
PAD/TRANSITION STRIP	HARTEX 3/8" CONTRACT CUSHION w/ #W-136 "TACK" TRANSITIONAL STRIP w/ BROWN VINYL INSERT			O		O	
FIELD FLOOR TILE FT-1	"CROSSVILLE" 8"x8" CROSSGRIP #A790 BURGUNDY SMOKE w/ 6"x8" COVE BASE & MAPEI/DAL-QUALITY #09 GRAY/ LATICRETE #542 PLATINUM GROUT		O			O	
DINING ROOM - ACCENT FT-3 DECORATIVE FLOOR TILE	"CROSSVILLE" 8"x8" CROSSGRIP #A675 STONEHENGE W/ MAPEI/DAL QUALITY #09 GRAY/ LATICRETE #542 PLATINUM GROUT		O			O	
DINING ROOM - ACCENT FT-4 DECORATIVE FLOOR TILE	"CROSSVILLE" 8"x8" CROSSGRIP #A634 TERRA ROSATA w/ MAPEI/DAL - QUALITY #09 GRAY/ LATICRETE #542 PLATINUM GROUT		O			O	

WALLS		FURNISHED BY			INSTALLED BY		
		O	GC	ES	O	GC	ES
CERAMIC WALL TILE C-1/C-3 KITCHEN/RESTROOMS	"DAL-TILE" DAL-MATTE GLAZED WALL TILE #D-735 ALMOND 4 1/4"x4 1/4" WALL TILE w/ MAPEI/DAL-QUALITY #0 WHITE/ LATICRETE #544 BRIGHT WHITE GROUT		O			O	
CERAMIC WALL TILE C-4 KITCHEN/RESTROOM ACCENTS	"DAL-TILE" #DH-57 GRAPE 4 1/4"x4 1/4" WALL TILE w/ MAPEI / DAL-QUALITY #0 WHITE/LATICRETE #544 BRIGHT WHITE GROUT		O			O	
NON-TILE WALL FINISH (FOOD PREP AREA)	4'-0"x9'-0" IVORY "FRP" w/ J-MOLDING & CORNER TRIM TO MATCH (MARLITE OR KAL-WALL)		O			O	
WALLCOVERING WC-B1 (BACKGROUND ZONE)	"ESSEX" SEVERIN #72-10064 (54" WIDE)			O		O	
PAINTED WALL FINISH SP-1 (ALTERNATE BACKGROUND)	"LE FLECK" CUSTOM #FA5056D (HEAVY DENSITY COVERAGE) BASE COAT @ 300 SQ. FT. PER GALLON; FINISH COAT @ 175 SQ. FT. / GAL. (FASHION WALLCOVERING)		O			O	
WALLCOVERING WC-C (CHAIR RAIL)	WENDY'S SAGE CUSTOM BORDER #G2142679 6 1/8" H x 15 YDS PER BOLT (WOLF GORDON)			O		O	
PANELING (HIGH IMPACT ZONE)	"MARLITE" 319 WHITTIER ANIGRE DESIGN 16"x96"x1/4") PLANKS W/ J MOLDING & CORNER TRIM TO MATCH		O			O	

INTERIOR FINISH SCHEDULE — SUMMIT

CEILING		FURNISHED BY			INSTALLED BY		
		O	GC	ES	O	GC	ES
CEILING TILE CT-1 (DINING ROOM)	"CELOTEX" NUTEX NTX154-FB MANILLA 24"x24" REVEAL EDGE, FOIL BACKED CEILING TILE w/ MATCHING GRID SYSTEM, COLOR TO MATCH TILE		O			O	
CEILING TILE CT-2 (KITCHEN)	"CELOTEX" CAPAUL VINYLROCK GYPSUM PANELS WHITE 24"x48" W/ MATCHING GRID SYSTEM OR "USG" #3270-WHITE w/ "DX" GRID SYSTEM; GRID COLOR TO MATCH CEILING TILES		O			O	
DRYWALL (RESTROOMS)	SMOOTH PAINTED DRYWALL / FINISH P-3			O			O

SEATING		FURNISHED BY			INSTALLED BY		
		O	GC	ES	O	GC	ES
TABLES 20"x24" (STANDARD)	"SWAN" TABLE TOPS WT-2024 ONE PIECE SOLID SURFACE COLOR: TAHITI SAND			O			O
TABLES 24"x42" (STANDARD QUAD BOOTH ONLY)	"SWAN" TABLE TOPS BT-2442 ONE PIECE SOLID SURFACE COLOR: TAHITI SAND			O			O
FREESTANDING TABLE BASES	FALCON #203-22 (20"x24" TOPS):BROWN EPOXY FALCON #204-2230 (24"x42" TOPS): BROWN EPOXY			O			O
CANTILEVERED TABLE BASES (QUAD BOOTH)	PREFINISHED WALL MOUNTED CANTILEVERED TABLE BASE			O			O
CHAIRS	SHELBY WILLIAMS #WY-97 BENTWOOD SIDE CHAIR IN NATURAL FRAME UPHOLSTER UP-1			O			O
BOOTH SEATING	AVIS FURNITURE COMPANY VINYL UPHOLSTERED BOOTH WITH REMOVABLE SEAT AND BACK OF 4" HIGH DENSITY COMMERCIAL FOAM. FRAME FINISHED IN PL-1 W/ PL3 ACCENT, OAK TRIM FINISHED IN S-1, UPHOLSTER IN UP-1, AND BOOTH BASE IN V-5			O			O
UPHOLSTERY UP-1 (STANDARD)	"NAUGAHYDE" #62 - PORT			O			O
UPHOLSTERY UP-1 (ALTERNATE)	"SPRADLING" #8533 PORT			O			O

Figure 3-11. Room finish schedules in Division 09 of the specifications contain room finish information.

Division 10—Specialties. Division 10 of the specifications contains a list of the specialty items that may be part of a construction project. Examples of specialty items include visual display surfaces, signage, toilet compartments and accessories, fireplaces, fire extinguishers, flagpoles, awnings, shelving, laundry accessories, and access flooring. **See Figure 3-12.** In a manner similar to other divisions of the specifications, extensive reliance is placed on manufacturer names and products. Specialty items are items commonly not bid by a general contractor, but more likely provided by subcontractors and specialty contractors.

Figure 3-12. Division 10 of the specifications includes specialized items such as signage and fire extinguishers.

Division 11—Equipment. Special equipment may be built into or installed in some structures. Loading docks, pedestrian control devices, projection screens, dishwashers, ovens, water-treatment equipment, and a variety of industrial equipment may be installed during construction.

Industry-specific commercial equipment such as library, office, food service, bank, healthcare, detention, and athletic equipment is also included in Division 11. The equipment is part of the initial construction bid. The general contractor and subcontractors are responsible for obtaining and installing the special equipment prior to completion of the project according to the information provided in Division 11. Equipment specifications rely heavily on specific manufacturer model numbers.

Division 12—Furnishings. Division 12 of the specifications provides information concerning manufactured metal, wood, and plastic cabinetry and countertops, including materials, finishes, hardware, fabrication, and installation. **See Figure 3-13.** Division 12 also includes information about installed seating such as for a theater or classroom, as well as window blinds, curtains, entrance mats and grilles, and specialized furniture, such as for banks, health care facilities, and laboratories.

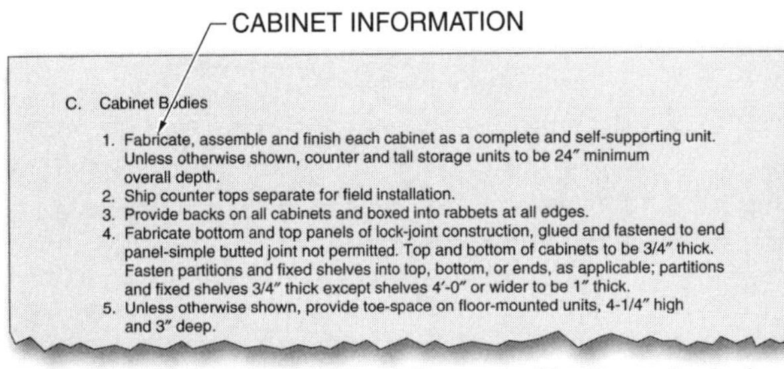

C. Cabinet Bodies

1. Fabricate, assemble and finish each cabinet as a complete and self-supporting unit. Unless otherwise shown, counter and tall storage units to be 24″ minimum overall depth.
2. Ship counter tops separate for field installation.
3. Provide backs on all cabinets and boxed into rabbets at all edges.
4. Fabricate bottom and top panels of lock-joint construction, glued and fastened to end panel-simple butted joint not permitted. Top and bottom of cabinets to be 3/4″ thick. Fasten partitions and fixed shelves into top, bottom, or ends, as applicable; partitions and fixed shelves 3/4″ thick except shelves 4′-0″ or wider to be 1″ thick.
5. Unless otherwise shown, provide toe-space on floor-mounted units, 4-1/4″ high and 3″ deep.

Figure 3-13. Division 12 of the specifications includes cabinet, countertop, installed seating, and window treatment information.

Division 13—Special Construction. Division 13 of the specifications provides information concerning highly specialized construction projects including swimming pools, amusement park equipment, controlled environment rooms, and building automation systems that control such items as lighting, fire protection, and heating and cooling equipment. Information from the manufacturer for each product or material is required.

Division 14—Conveying Equipment. Division 14 of the specifications describes the various conveying systems used in a building. Equipment for moving people, materials, and equipment is installed in most large construction projects. Elevators are essential in multistory office buildings. Elevators and escalators are often installed in multistory retail stores to move customers and goods from floor to floor. Pneumatic tubes are commonly installed in banks to shuttle containers from a drive-up window to the teller area. Conveyors and monorail systems are common in large manufacturing and power plants to move materials and products from one area to another.

Elevators are the most common conveying system defined in Division 14. Elevator specifications include the rated net capacity, rated speed, travel distance, number of stops, car size, and other specifications. **See Figure 3-14.**

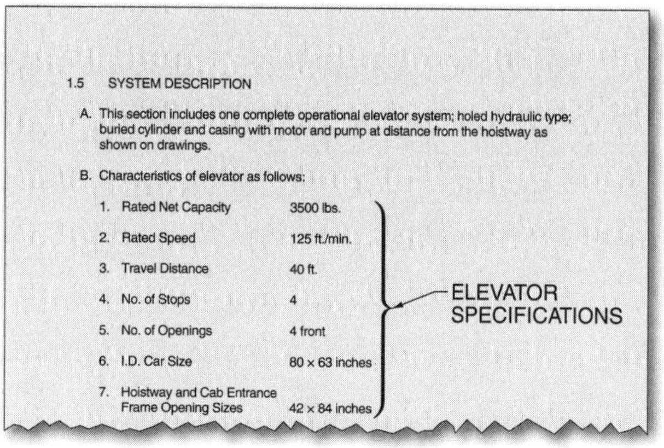
1.5 SYSTEM DESCRIPTION

A. This section includes one complete operational elevator system; holed hydraulic type; buried cylinder and casing with motor and pump at distance from the hoistway as shown on drawings.

B. Characteristics of elevator as follows:

1. Rated Net Capacity 3500 lbs.
2. Rated Speed 125 ft./min.
3. Travel Distance 40 ft.
4. No. of Stops 4
5. No. of Openings 4 front
6. I.D. Car Size 80 x 63 inches
7. Hoistway and Cab Entrance Frame Opening Sizes 42 x 84 inches

Figure 3-14. Division 14 of the specifications provides information concerning conveying systems such as elevators.

A variety of government and industry organizations such as the American National Standards Institute (ANSI), the National Fire Protection Association (NFPA), and Underwriters Laboratories Inc. (UL®) publish regulations and standards that apply to elevator construction and operation. In addition, the Americans with Disabilities Act (ADA), published by the U.S. Department of Justice, provides guidelines for accommodating physically challenged individuals.

Information regarding plumbing systems is included in Division 22 of the specifications.

A variety of conveying systems are designed for commercial and industrial use. Typically, overhead lifts and monorail systems are electrically controlled. Overhead lifts and monorail systems are described in the specifications according to their lifting capacity, speed of lift and movement, and height of lift required. Conveying systems move items as varied as cans, coal, luggage, and automobile components. Conveyor type, speed, and capacity are part of Division 14 specifications. Scaffold requirements are also included in this section of the specifications for any scaffold that will become a permanent portion of the building.

Division 21—Fire Suppression. Division 21 is divided into four main areas—water-based fire-suppression systems, fire-extinguishing systems, fire pumps, and fire-suppression water storage. Carbon-dioxide, clean-agent, wet-chemical, dry-chemical and wet-pipe systems are covered in Division 21. The specifications describe piping, sprinkler heads, check valves, tanks, pumps, and fire-department connections as applicable.

Agencies that regulate fire-protection equipment and publish standards include the NFPA, the International Association of Plumbing and Mechanical Officials (IAPMO), the International Code Council (ICC), the Occupational Safety and Health Administration (OSHA), and local fire authorities. All fire-protection systems are designed and installed in accordance with the applicable standards from these agencies and all applicable codes.

Division 22—Plumbing. Plumbing information in Division 22 of the specifications includes connections to available water supplies, pipe and/or tubing, pumps, valves, meters, gauges, and hydrants. A plumbing fixture schedule is typically included in this portion of the specifications.

Wastewater disposal and treatment information is provided for waste pipes and stormwater drains. Pipes and valves are specified by type, size, material, and ability to withstand a certain amount of pressure, typically expressed in pounds per square inch (psi). **See Figure 3-15.** Pipe and fittings for gas, compressed air, and vacuum systems are also described in Division 22. Tanks such as water heaters and storage tanks are specified in Division 22. Fixtures such as sinks, water closets, bathtubs, drinking fountains, lavatories, urinals, bidets, laundry tubs, and faucets are also specified in Division 22.

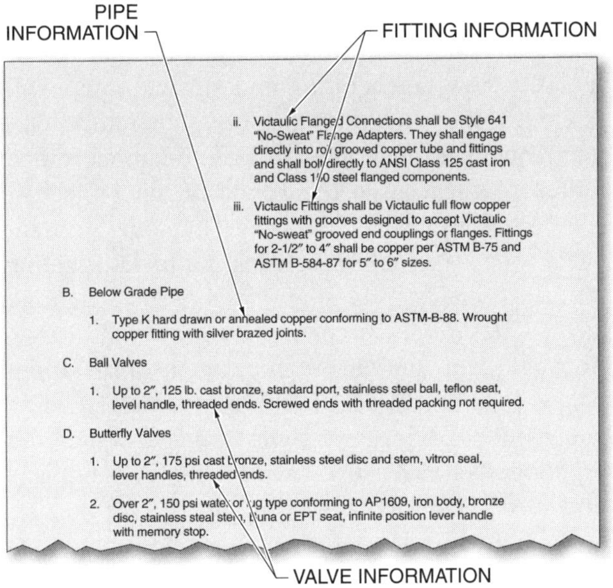

ii. Victaulic Flanged Connections shall be Style 641 "No-Sweat" Flange Adapters. They shall engage directly into roll grooved copper tube and fittings and shall bolt directly to ANSI Class 125 cast iron and Class 150 steel flanged components.

iii. Victaulic Fittings shall be Victaulic full flow copper fittings with grooves designed to accept Victaulic "No-sweat" grooved end couplings or flanges. Fittings for 2-1/2″ to 4″ shall be copper per ASTM B-75 and ASTM B-584-87 for 5″ to 6″ sizes.

B. Below Grade Pipe

1. Type K hard drawn or annealed copper conforming to ASTM-B-88. Wrought copper fitting with silver brazed joints.

C. Ball Valves

1. Up to 2″, 125 lb. cast bronze, standard port, stainless steel ball, teflon seat, level handle, threaded ends. Screwed ends with threaded packing not required.

D. Butterfly Valves

1. Up to 2″, 175 psi cast bronze, stainless steel disc and stem, vitron seal, lever handles, threaded ends.

2. Over 2″, 150 psi wafer or lug type conforming to AP1609, iron body, bronze disc, stainless steal stem, Buna or EPT seat, infinite position lever handle with memory stop.

Figure 3-15. Plumbing specifications provide pipe, valve, and fitting information.

Figure 3-16. Ductwork carries heated or cooled air from a forced-air HVAC system through a building.

Division 23—Heating, Ventilating, and Air-Conditioning (HVAC). Division 23 of the specifications describes various mechanical systems, such as heating, ventilating, and air-conditioning equipment and its related fixtures, pipes, instrumentation, and controls. Architects, mechanical engineers, electrical engineers, estimators, and various general contractors and subcontractors must coordinate their work to ensure proper specification and operation of mechanical systems. Basic material and installation information for pipe, pipe hangers and supports, duct and pipe connectors, insulation, gauges, flow control and measurement devices, electric motors and starters, and fuel tanks is provided. Controls installed on HVAC systems are typically also defined in Division 23.

HVAC systems generate heat from sources such as boilers, geothermal, solar, or natural gas heaters, provide cooling with systems containing refrigerants, chillers, and compressors, and distribute air throughout a building with various air handlers, motors, fans, and duct systems. **See Figure 3-16.** In hot and cold water systems, specifications are provided for pipe, circulating pumps, and heating and cooling transfer equipment. Small heating and cooling systems, such as unit heaters and air conditioners for special conditions, are also specified in Division 23. Air-handling applications such as filtering, removal, and exhausting of fumes and smoke is also described.

Estimators must take care when evaluating subcontractor bids related to Divisions 21, 22, and 23. Plumbing and mechanical subcontractors may have overlapping areas or uncovered areas of responsibility, depending on their assessment of the specifications and the scope of their bids. Estimators must carefully assemble and completely analyze the scope of subcontractor bids in these divisions to ensure that the entire scope of work is covered without redundancies that may result in either an incomplete bid or excessive costs.

Division 25—Integrated Automation. To increase efficiency and environmental performance, many new buildings have control systems that are used to remotely monitor and control other building systems, such as fire-suppression, plumbing, electrical, HVAC, and conveying systems. Division 25 covers integrated automation systems and includes information on computer networking equipment such as servers, routers, hubs, modems, control panels, and related wiring.

Information on computer software for integrated automation systems is also included in Division 25. Since the terminal devices and instrumentation for most integrated automation systems requires seamless integration with other building systems, terminal devices and instrumentation may be included to prevent conflict between the requirements of the building systems and the integrated automation systems.

Division 26—Electrical. Division 26 of the specifications provides information regarding wiring, equipment, and finish of electrical systems. Five titles comprise the majority of Division 26:

- Medium-Voltage Electrical Distribution (26 10 00)
- Low-Voltage Electrical Transmission (26 20 00)
- Facility Electrical Power Generating and Storing Equipment (26 30 00)
- Electrical and Cathodic Protection (26 40 00)
- Lighting (26 50 00)

Medium-voltage electrical distribution systems include substations, transformers, circuit protection, and switchgear that are used to distribute medium-voltage (2400 V to 69 kV) electrical power from the facility service point to the point of delivery. Low-voltage electrical transmission systems include overhead power systems, transformers, switchgear, panelboards, power distribution units, controllers, wiring, and circuit protection devices that are used to distribute low-voltage electrical power from the point of power transformation to the point of use.

Typical voltages included in Title 26 20 00 are 120 V, 208 V, 230 V, 240 V, 277 V, 460 V, and 480 V. Facility power-generating and storing equipment includes photovoltaic collectors, generators, battery equipment, power filters and conditioners, and transfer switches. Electrical and cathodic protection systems provide protection against lightning and transient voltages. Lighting includes luminaries (lighting fixtures), ballasts, dimming controls, and lighting accessories for fluorescent, high-intensity discharge, incandescent, mercury vapor, neon, and sodium vapor lighting. In addition, information regarding emergency lighting, exit signs, and special-purpose lighting, such as theatrical and healthcare lighting, is included in Division 26.

Electrical wiring information in Division 26 includes conductors and cables, raceways and boxes, and wiring devices and connections. Wire size, insulation, and installation information is typically described. **See Figure 3-17.** The specifications contain quality assurance requirements and approved manufacturer names and product numbers for each of these areas.

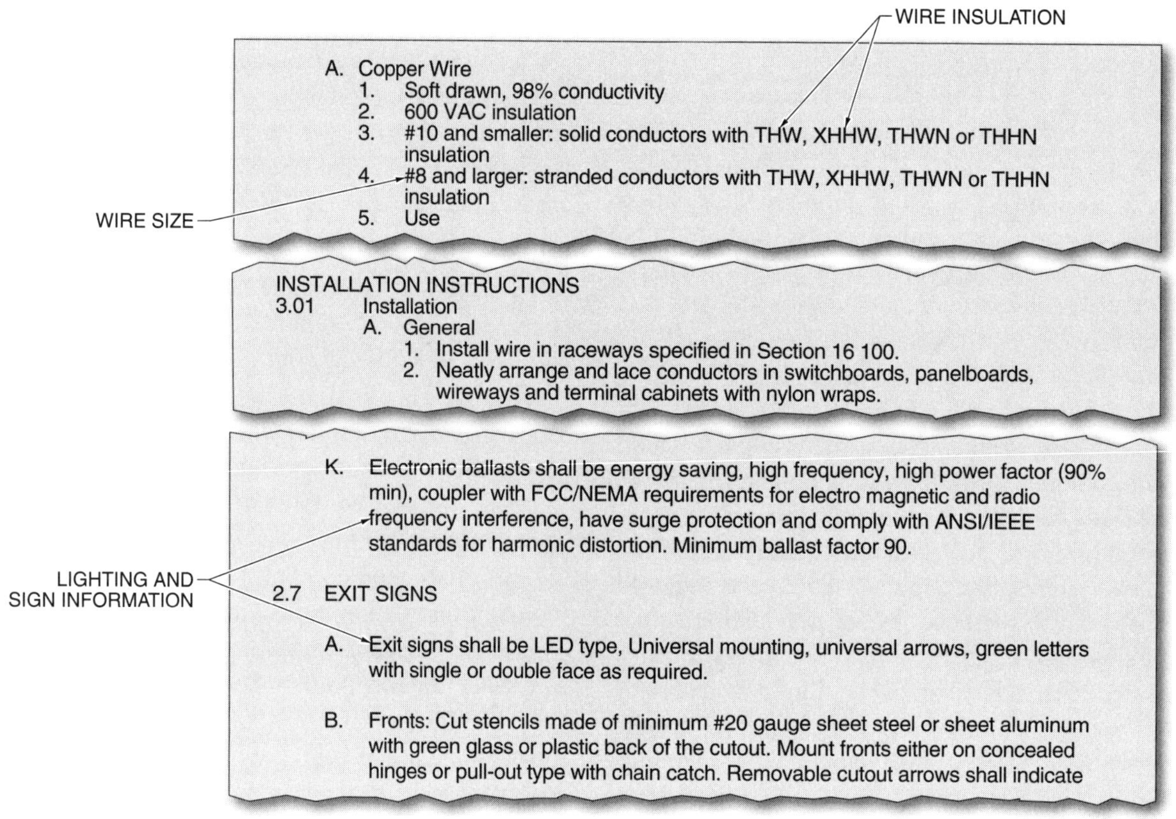

Figure 3-17. Division 26 of the specifications includes lighting, sign, and wire information.

The National Electrical Code® (NEC®), which is updated every three years, is commonly cited in the specifications for electrical systems. For example, subsection 700.9(D)(2) requires that equipment for feeder circuits for occupancies greater than 1000 people, or for buildings above 75′ with assembly, educational, residential, detention, correctional, business, or mercantile occupancies, be located in spaces that are fully protected by approved automatic fire-suppression systems or in spaces with a 1-hr fire resistance rating. **See Figure 3-18.**

Division 27—Communications. Division 27 of the specifications describes communications systems including data, voice, and audio-video equipment. Information related to equipment-room fittings, cables, fiber optics, antennas, media converters, and data communications computer hardware and software requirements is included.

Division 28—Electronic Safety and Security. Division 28 contains information related to electronic safety and security systems, including alarm systems, surveillance and monitoring systems, access control and intrusion detection systems, and X-ray screening. Electronic fire alarms, smoke detectors, and various radiation and hazardous material detectors are also described in Division 28.

Division 31—Earthwork. Earthwork to be done on a building site, including clearing, earth moving, soil treatment, erosion control, and excavation, is included in Division 31 of the specifications. **See Figure 3-19.** Special foundations and load-bearing structures, such as piles and caissons, are also covered in this area of the specifications. Tunneling and mining operations, such as boring, and equipment such as linings and casings are also detailed.

Case, LLC

Figure 3-19. Earth moving is described in Division 31 of the specifications.

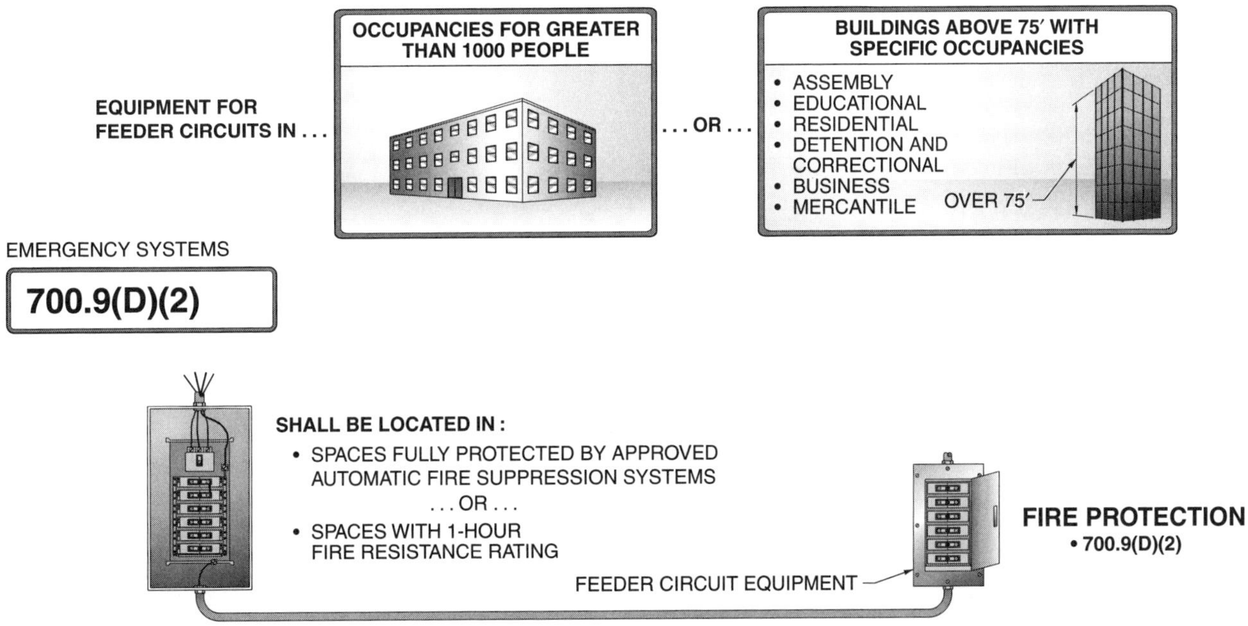

Figure 3-18. Many electrical specifications rely on information from the National Electrical Code® (NEC®).

Division 32—Exterior Improvements. Division 32 includes information on a variety of materials and methods for improving the appearance and use of a building exterior, ranging from paving to landscaping and irrigation. Paving materials and standards for their use and installation are described in Division 32, including the weather conditions under which paving may or may not be installed and pavement slope and smoothness requirements. Curbs and gutters, athletic surfaces, fences, gates, retaining walls, and sound barriers are also covered in Division 32.

Landscaping information, such as soil mixtures, trees and shrubs, grasses, finish grading, fertilizers, and mulches, is given in Division 32. Installation, protection, and maintenance procedures for plant materials are also given. Piping and connection information is included where irrigation systems are installed.

Division 33—Utilities. The utility services required for a building are described in Division 33 including water, sanitary sewage, storm drainage, fuel distribution, electrical, and communications utilities. The utilities referred to in Division 33 are provided by public service agencies, municipalities, or other public/private partnerships. Division 33 does not include facility-level utilities and related services as described in other divisions, such as Division 22 for plumbing and Division 26 for electrical. Connection and instrumentation requirements for the utilities are cited.

Information regarding septic fields, culverts, ponds, reservoirs, and wastewater systems is included in Division 33. Title 33 70 00, Electrical Utilities, provides information related to electrical power distribution through substations, transformers, switchgear, circuit protection devices, utility towers and poles, and wiring.

Division 34—Transportation. Specialized requirements for roadways, railways, airfields, and bridges are included in the specifications for Division 34. Power system requirements for electrically powered lifts, trains, or airfield equipment is also described. For bridge construction projects, information regarding machinery for operating bridges, as well as specialty items such as vibration dampers or ice shields, is provided.

Division 34 also includes information regarding vehicle and passenger fare collection systems.

Division 35—Waterway and Marine Construction. Construction of dams, levees, gates, valves, dredging, seawalls, revetments, and breakwaters is included in Division 35. **See Figure 3-20.** Specialized shoreline protection systems, such as piling, riprap, and concrete masonry unit walls, are described. Division 35 specifications also pertain to floating construction projects such as offshore platforms, buoys, marine bollards, and underwater construction.

Portland Cement Association

Figure 3-20. Waterway construction, such a dam construction, is specified in Division 35.

Division 40—Process Integration. A variety of industrial processes are used in the manufacture of commercial and consumer goods. These industrial processes commonly use various gases, liquids, or solid fuels and materials. The fuels and materials must be conveyed throughout a building with process piping. Division 40 of the specifications describes gas and vapor, liquid, and solid and mixed material process piping. Electrically, hydraulically, and pneumatically operated control systems are included in Division 40, as well as process control hardware and software.

Division 41—Material Processing and Handling Equipment. Specifications for construction of systems to handle bulk materials, piece materials, manufacturing equipment, and containers are included in Division 41. Specialized equipment related to the type of material being moved, including various conveyor systems, cranes, hoists, and mobile plant equipment, is included in this portion of the specifications. Information related to manufacturing lines and equipment, machining equipment, and finishing equipment is provided in Division 41.

Division 42—Process Heating, Cooling, and Drying Equipment. The specifications for Division 42 describe temperature-control equipment used in industrial processes. Process heating equipment includes low- and high-pressure boilers; electric, fuel-fired, and solar heaters; heat exchangers and furnaces; and ovens. Process cooling equipment includes cooling towers, chillers, condensers, evaporators, and humidifiers. Process drying equipment includes equipment used to remove moisture from gases and solid/bulk materials that are used in industrial processes.

Division 43—Process Gas and Liquid Handling, Purification, and Storage Equipment. Gases and liquids used for industrial processes must be properly handled, purified, and stored. Division 43 covers process gas and liquid handling equipment, including fans, pumps, blowers, compressors, filters, tanks, and related equipment such as agitators, emulsifiers, and mixers.

Prior to use in their intended processes, gases and liquids may need to be filtered or purified so the equipment can perform efficiently. Filters, decarbonators, deionization units, scrubbers, and degasifiers are used to filter or purify gases or liquids used in industrial processes. Pressure and atmospheric tanks and vessels used to store process gases and liquids are also covered in Division 43.

Division 44—Pollution Control Equipment. The specifications for Division 44 provide information regarding equipment used for control of air pollution, noise pollution, water pollution, and solid waste materials. The Clean Air Act and Clean Water Act, developed by the Environmental Protection Agency (EPA), specify air and water quality and emissions limitations, respectively. Additional local environmental requirements may also be taken into consideration when specifying the equipment required in Division 44.

Air pollution control equipment includes cyclonic separators, dust collectors, filters, precipitators, and air pollution scrubbers. Noise pollution control equipment includes physical barriers, silencers, and electronic frequency cancellers. Water pollution equipment includes water-treatment equipment used to collect water for treatment and distribution. Mechanical, chemical, thermal, and other water-treatment and filtration equipment is used to minimize water pollution and is covered in Division 44. Solid waste control equipment, such as balers, liquid extractors, compactors, and waste transfer equipment, is also included in Division 44.

Division 45—Industry-Specific Manufacturing Equipment. Specifications for Division 45 include various industry-specific equipment used for oil and gas extraction, mining, and printing, and equipment used for food and beverage, tobacco, textile, leather, wood, paper, computer, electronics, and metal manufacturing. Division 45 also covers manufacturing equipment for many other industries.

Division 48—Electrical Power Generation. Specifications for Division 48 include information regarding electrical power-generation equipment utilizing fossil fuel, nuclear fuel, hydroelectric energy, solar energy, wind energy, geothermal energy, electromechanical energy, and fuel cell equipment. Power plant boilers, condensers, steam and gas turbines, and generators are described in Division 48 specifications.

Departments of Transportation

Specialized numerical and description classification systems are used for road, highway, bridge, and other heavy construction performed by each state department of transportation. Each state uses a classification code developed internally by their state department of transportation. **See Figure 3-21.** Proposals for bids are commonly distributed for each category listed. Some general contractual conditions that should be considered by the estimator are similar to building construction and include the type of project delivery system and the potential for value engineering.

Portland Cement Association

MISSOURI DEPARTMENT OF TRANSPORTATION CONSTRUCTION CLASSIFICATION CODES	
CODE	**DESCRIPTION**
Division 200—Clearing and Grubbing	
201	Clearing and Grubbing
202	Removals
202.10	Removal of Bridges
202.20	Removal of Improvements
203	Roadway and Drainage Excavation, Embankment, and Compaction
204	Embankment Control
204.10	Embankment Control Stakes
204.20	Settlement Gauges
204.30	Pore Pressure Measurement Devices
205	Overhaul
206	Excavation for Structures
207	Linear Grading
208	Interception Ditch
209	Subgrade Preparation
210	Subgrade Compaction
211	Subgrade Scarifying
212	Subgrading and Shouldering
213	Shaping Shoulders
214	Water
215	Shaping Slopes
Division 300—Bases and Aggregate Surfaces	
301	Plant Mix Bituminous Base Course
302	Stabilized Permeable Base
303	Rock Base
304	Aggregate Base Course

Figure 3-21. Each state department of transportation uses its own construction codes.

A preliminary quantity may be provided along with each material or work type code. These quantities indicate the preliminary estimating calculations made by department of transportation engineers for a specific project. Estimators must use these codes to submit a bid to a department of transportation in the required format. Some electronic estimating programs designed specifically for heavy/highway construction can be preloaded with the codes for a particular state or series of states.

ARCHITECTURAL PRACTICES

An estimator must be able to read and interpret architectural and shop drawings. Familiarity with various lines, drawing scales, symbols, and abbreviations is required to develop an accurate bid. **See Appendix.**

All individuals involved in the construction process require a common source of information that is legally and functionally reliable. All members of the building team must read and interpret all sections of the construction documents, including the prints, with skill and accuracy to ensure the project is built in accordance with applicable building codes, the needs of the owner, and the design of the architect. The detail drawings that make up a large set of prints are separated into different divisions to help individuals locate information easily. The use of symbols and abbreviations and the ability to interpret and interrelate portions of the prints across the divisions are necessary skills in the construction and estimating processes.

One of the most necessary and difficult skills to develop when creating an estimate from a large set of prints is the ability to visualize an entire project and the relationship of all components from a set of prints and specifications. In addition to an overall view, it is often necessary to obtain information from several different drawings and the specifications to fully understand a single building element. The ability to efficiently obtain and locate information from a variety of sources and combine this information into a common understanding is the most important requirement quantity takeoff for a large building project. The integration of many building components into a single model can be performed with the proper use of computer technology and building information modeling (BIM) techniques.

Print Divisions

Prints are separated into different divisions to help individuals find information. Print divisions are denoted by a capital letter shown in the title block. For example, architectural prints are denoted by a capital "A" followed by a sheet number. In a similar manner, structural prints are denoted by an "S", mechanical prints by an "M", electrical prints by an "E", and civil prints by a "C". **See Figure 3-22.**

Architects may use other divisions depending on the nature of the building project. For example, site mechanical plans may be identified by the designation MPE, denoting mechanical, plumbing, and electrical on one drawing. Sheet numbering begins with the number 1 within each division. For example, the architectural prints may run from page A1 to page A65. Print divisions may be subdivided when several pages apply to the same elements. For example, the structural prints begin with page S1, and subdivisions may be given as S1.1, S1.2, and S1.3.

Architectural. Architectural prints include general building information, floor plans, elevation views, section drawings, and detail drawings. Architectural prints are commonly the largest division of prints for a construction project.

Structural. Structural prints provide information about sizes, styles, and placement for foundations, beams, columns, joists, braces and other framing and load-bearing members. Framing and load-bearing members may be built of wood, masonry, reinforced concrete, or structural steel.

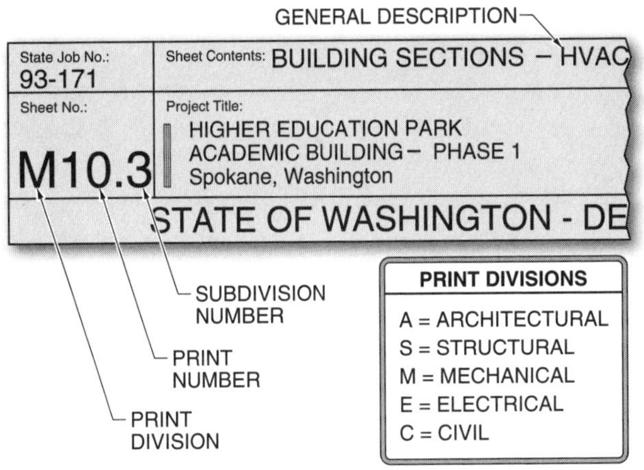

Figure 3-22. Prints are separated into divisions for ease of information retrieval.

Mechanical. Mechanical prints include plumbing, heating, ventilating, and air conditioning information including ductwork, piping, and equipment placement and sizes. Information about piping for fire-suppression systems may also be part of the mechanical prints or may be provided in a separate division.

Electrical. Electrical prints indicate the capacity and placement of power plant systems, luminaries, cable trays, conduit and panel schedules, finish fixtures, wiring, switches, and any other electrical installations. Electrical prints commonly contain numerical and alphabetic codes referring to electrical controls and fixtures referenced in schedules in the specifications.

Civil. Civil prints include overall site layout, grading, elevations, and topographical information. Other information includes site drainage, paving designs, and parking layout. Landscaping may also be a part of the civil prints or provided separately.

SAGE TIMBERLINE ESTIMATING . . .

CREATING AN ESTIMATE AND STORING ESTIMATE INFORMATION

With Sage Timberline Estimating software, a new estimate is created by specifying a folder and file name for the estimate. Estimating information is stored by inputting the information in the **Estimate Information** dialog box.

. . . SAGE TIMBERLINE ESTIMATING . . .

Creating an Estimate

1. Pick the **New Estimate** button from the Home Tab to display the **New Estimate** dialog box.

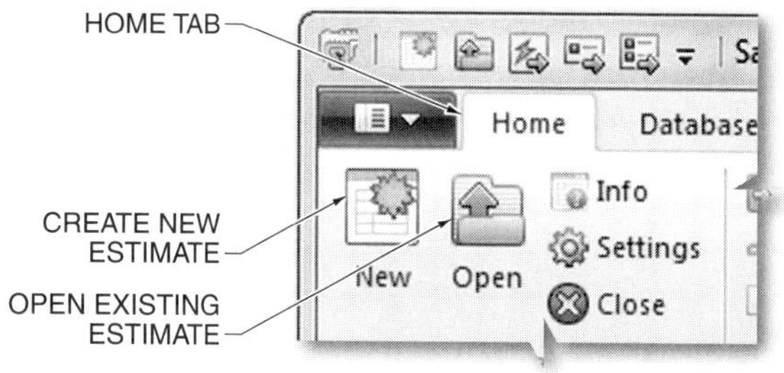

2. Enter a file name (**Stanton Industries**) for the new estimate at the **Enter a name for the new estimate** tab. Continue with selecting a storage place for the estimate at the **Choose the estitmate database where this estimate will be stored** tab, select the database (**SampleCommercialGC**) for the database at the **Choose a database to associate with the new estimate** tab, and select the branch for the database at the **Choose or browse to the branch where you want your estimate displayed** tab.

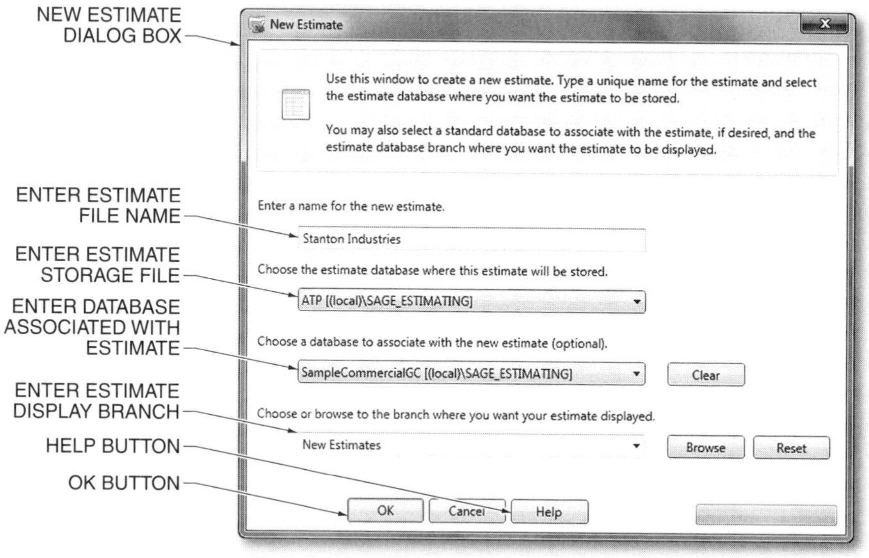

3. Pick the **Open** button from the Database Tab to display the **Database** dialog box. This displays the existing databases that can be associated with the estimate. If needed, a new a database can be created to suit the needs of an estimate.

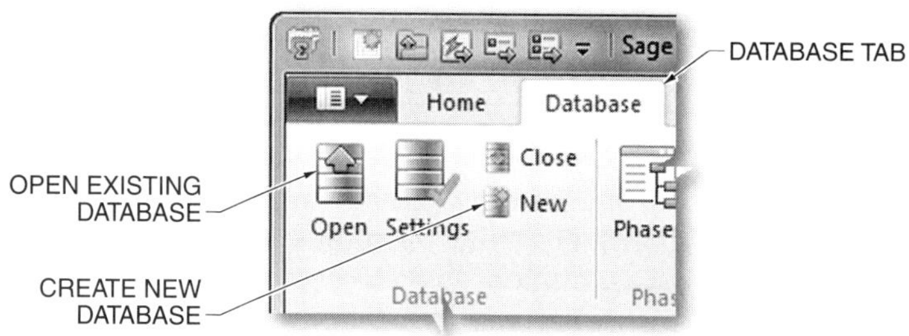

. . . SAGE TIMBERLINE ESTIMATING

4. Review the available databases that can be associated with the estimate.

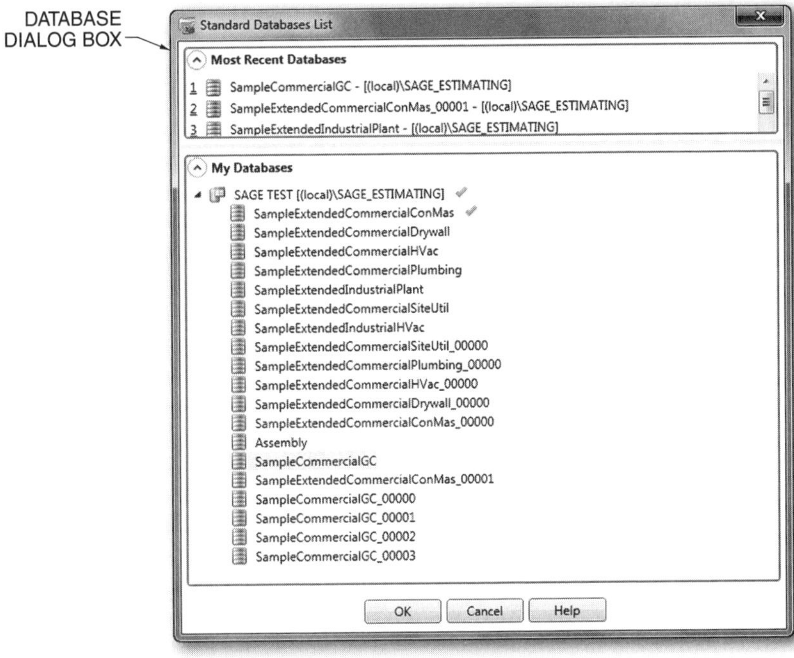

5. Once the New Estimate dialog box is complete, choose the **OK** button to close the **New Estimate** dialog box and open the **Estimate Information** dialog box.

Storing Estimate Information

The **Estimate Information** dialog box consists of six tabs through which the estimator can enter general information about the project. The first tab—**Main**—contains the only required field: **Project name**. The project name will appear in the heading of every printed page of the estimate report and will also appear on the report cover page. All other fields are optional. Pick the tab headings to explore the contents of the other tabs. Close the **Estimate Information** dialog box and the spreadsheet will be displayed. If general estimate information needs to be updated, select **Estimate Information** from the **Home** tab.

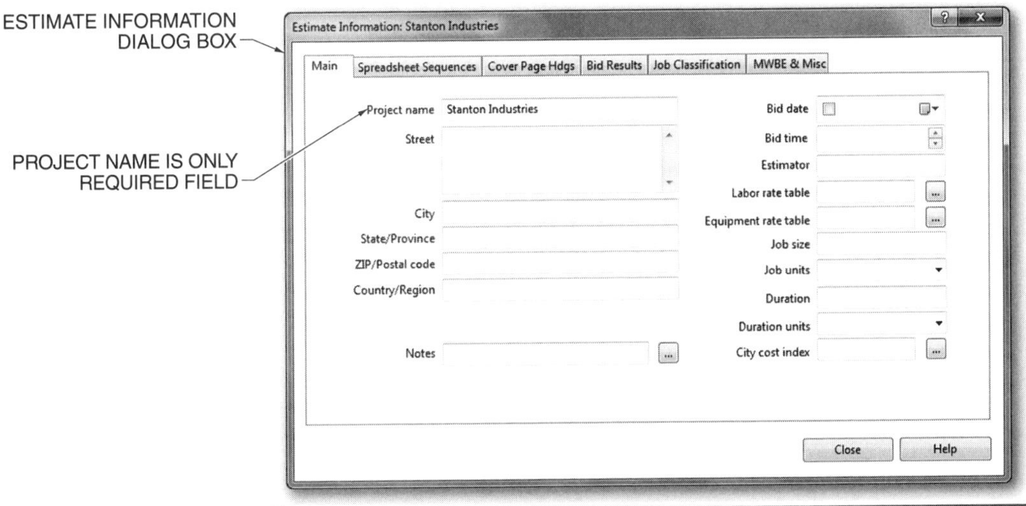

Quick Quiz®

Quick Quiz®

Refer to the CD-ROM for the Quick Quiz® questions related to chapter content.

Key Terms

Illustrated Glossary

- *admixture*
- *astragal*
- *commissioning*
- *Construction Specifications Institute (CSI)*
- *MasterFormat™*
- *specifications*

Web Links

Web Links

Americans with Disabilities Act
www.ada.gov

American Forest and Paper Association
www.afandpa.org

American Institute of Architects
www.aia.org

American National Standards Institute
www.ansi.org

APA – The Engineered Wood Association
www.apawood.org

ASTM International
www.astm.org

Architectural Woodwork Institute
www.awinet.org

Associated General Contractors of America
www.agc.org

Associated Specialty Contractors
www.assoc-spec-con.org

National Fire Protection Association
www.nfpa.org

Underwriters Laboratories Inc.
www.ul.com

Western Wood Products Association
www.wwpa.org

Specifications and Drawings 3

Review Questions

Estimating

______________ **1.** ___ are written supplements to a set of construction documents and drawings that provide additional construction information.

______________ **2.** The current MasterFormat™ contains ___ divisions.

T F **3.** Division 01 covers general requirements.

______________ **4.** Division 02 of the specifications provides information on ___.
- A. existing site conditions
- B. masonry
- C. concrete
- D. metals

______________ **5.** Information on admixtures for masonry mortar is included in Division ___.
- A. 03
- B. 04
- C. 08
- D. 10

______________ **6.** Data and voice communications are described in Division ___.

______________ **7.** Division ___ of the specifications provides information on concrete.

T F **8.** Division 07 covers woods, plastics, and composites.

______________ **9.** Division 11 of the specifications covers ___.

______________ **10.** Specifications in Division ___ address doors and windows.

______________ **11.** Division 23 specifications cover ___.
- A. sheet metals
- B. furnishings
- C. masonry
- D. air-conditioning units

T F **12.** Conveying systems are not covered in the MasterFormat™.

T F **13.** A different numerical and description classification system is used for road, highway, bridge, and other heavy construction performed by each state department of transportation.

______ **14.** Prints are separated into different divisions, denoting their type by ___.

A. sequential numbers
B. capital letters
C. icons
D. none of the above

Material Symbols

Refer to the Symbols tables in the Appendix.

______ **1.** Wood panel
______ **2.** Plywood
______ **3.** Glass
______ **4.** Concrete
______ **5.** Gravel
______ **6.** Rock
______ **7.** Earth
______ **8.** Cut stone
______ **9.** Brick
______ **10.** Rubble
______ **11.** Steel
______ **12.** Small-scale metal
______ **13.** Blocking
______ **14.** Rough wood member
______ **15.** Lath and plaster
______ **16.** Acoustical tile
______ **17.** Ceramic tile
______ **18.** Structural clay tile
______ **19.** Blanket insulation
______ **20.** Concrete block

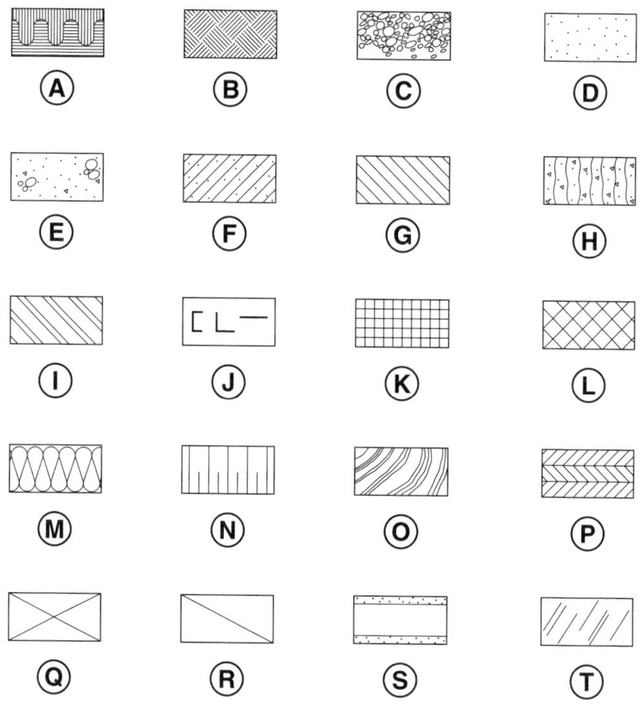

Electrical Lines

__________ 1. Three wires

__________ 2. Exposed wiring

__________ 3. Home run

Plumbing Lines

__________ 1. Waste

__________ 2. Vent

__________ 3. Cold water

__________ 4. Hot water

Printreading

__________ 1. Earth

__________ 2. Gravel

__________ 3. Concrete

__________ 4. Rough wood member

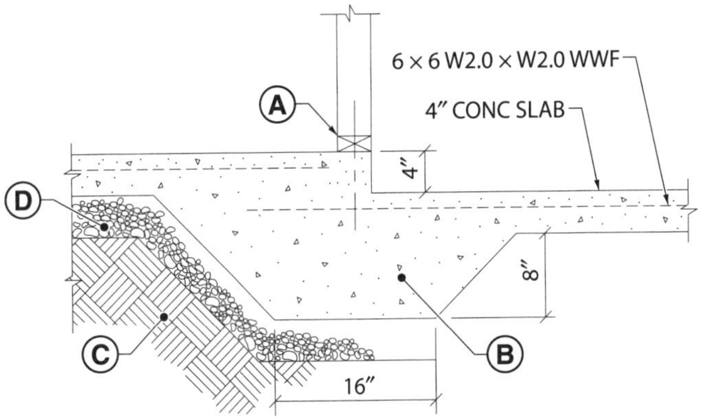

SECTION A
SCALE 3/4″ = 1′-0″

Short Answer

1. Discuss the importance of using the specifications when bidding on a construction project.

2. Other than the difference in number between the 16-division and 50-division MasterFormat structure, discuss the differences and advantages gained through the use of the 50-division structure.

Specifications and Drawings 3

Activities

Activity 3-1—Ledger Sheet Activity

Refer to Print 3-1 and Quantity Sheet No. 3-1. Take off the square feet of floor area for TV Seminar Room 320, Project Lab 321, and Project Lab 323. Calculate the total area (in square feet) of the three rooms.

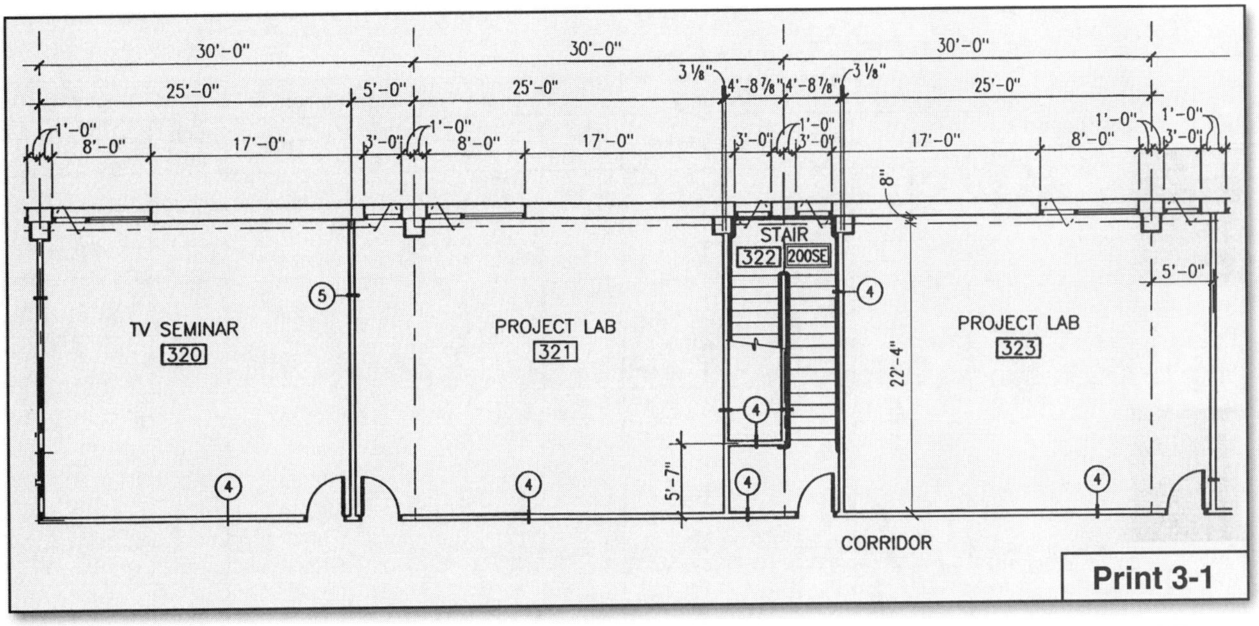

Print 3-1

QUANTITY SHEET

Project: ____________ Sheet No. 3-1
Estimator: ____________ Date: ____________
Checked: ____________

No.	Description	Dimensions L	Dimensions W			Unit		Unit		Unit		Unit
	TV seminar room 320											
	Project lab 321											
	Project lab 323											
	Total area (sq ft)											

Activity 3-2—Spreadsheet Activity

Refer to Print 3-2 and Quantity Sheet No. 3-2 on the CD-ROM. Determine the area (in square feet) of Microcomputer Classroom 326. Convert all dimensions to decimal foot equivalents when calculating the area.

Activity 3-3—Material Costs Activity

Refer to Print 3-3 (following page) and the Estimate Summary Sheet No. 3-3. Determine the material costs of the equipment for the dining room and entrance area as noted on Estimate Summary Sheet 3-3.

ESTIMATE SUMMARY SHEET

Project: Wendy's seating and entrance
Estimator: ____________

Sheet No. 3-3
Date: ____________
Checked: ____________

No.	Description	Dimensions			Quantity		Material		Equipment		Total	
						Unit	Unit Cost	Total	Unit Cost	Total	Unit Cost	Total
R01							$ 480					
R05							$ 205					
R06							$ 125					
R07							$ 515					
R08							$ 310					
R10							$ 415					
R11							$2150					

EQUIPMENT SCHEDULE

MARK	EQUIPMENT NUMBER	QUANTITY	DESCRIPTION OF EQUIPMENT	REMARKS	FURNISHED BY	INSTALLED BY
A01		2	DRINK TOWER 23 x 23		O	O
A03		1	BAG N BOX RACK		O	O
A04		1	BULK CO2 TANK		O	O
A05		2	ICED TEA DISPENSER		ES	ES
A06		1	COFFEE BREWER		ES	ES
A07		1	COFFEE WARMER		ES	ES
B01		120YD	CARPET		ES	GC
B02		1	CARPET/TILE TRANSITION STRIP		ES	GC
B06		1	SILK PLANT PACKAGE		ES	GC

EQUIPMENT SCHEDULE

MARK	EQUIPMENT NUMBER	QUANTITY	DESCRIPTION OF EQUIPMENT	REMARKS	FURNISHED BY	INSTALLED BY
N16		2	OFFICE WALL CABINET		ES	ES
N17		1	PIGEON HOLE CABINET		ES	ES
N18		1	OFFICE OVERHEAD SHELF		ES	ES
Q01		1	REMOTE CONDENSING UNIT (ROOF)		ES	ES
Q02		1	ICE MAKER		ES	ES
Q04		1	FREEZER		ES	ES
Q06		2	UC REFRIDGERATOR		ES	ES
Q08		2	FROSTY MACHINE		ES	ES
R01		2	SINGLE TRASH UNIT		ES	ES
R03		1	CONDIMENT STAND		ES	ES
R05		77	CHAIRS		ES	ES
R06		31	21x24 TABLETOPS		ES	ES
R07		31	21x24 TABLE BASES		ES	ES
R08		8	26x42 TABLE TOP		ES	ES
R10		1	BENCH SEAT (102")		ES	ES
R11		2	QUAD BOOTH		ES	ES
R15		1	SERPENTINE RAIL	@ 36" HIGH	O	GC
R16		3	HIGH CHAIRS		ES	ES
R17		2	BOOSTER CHAIRS		ES	ES

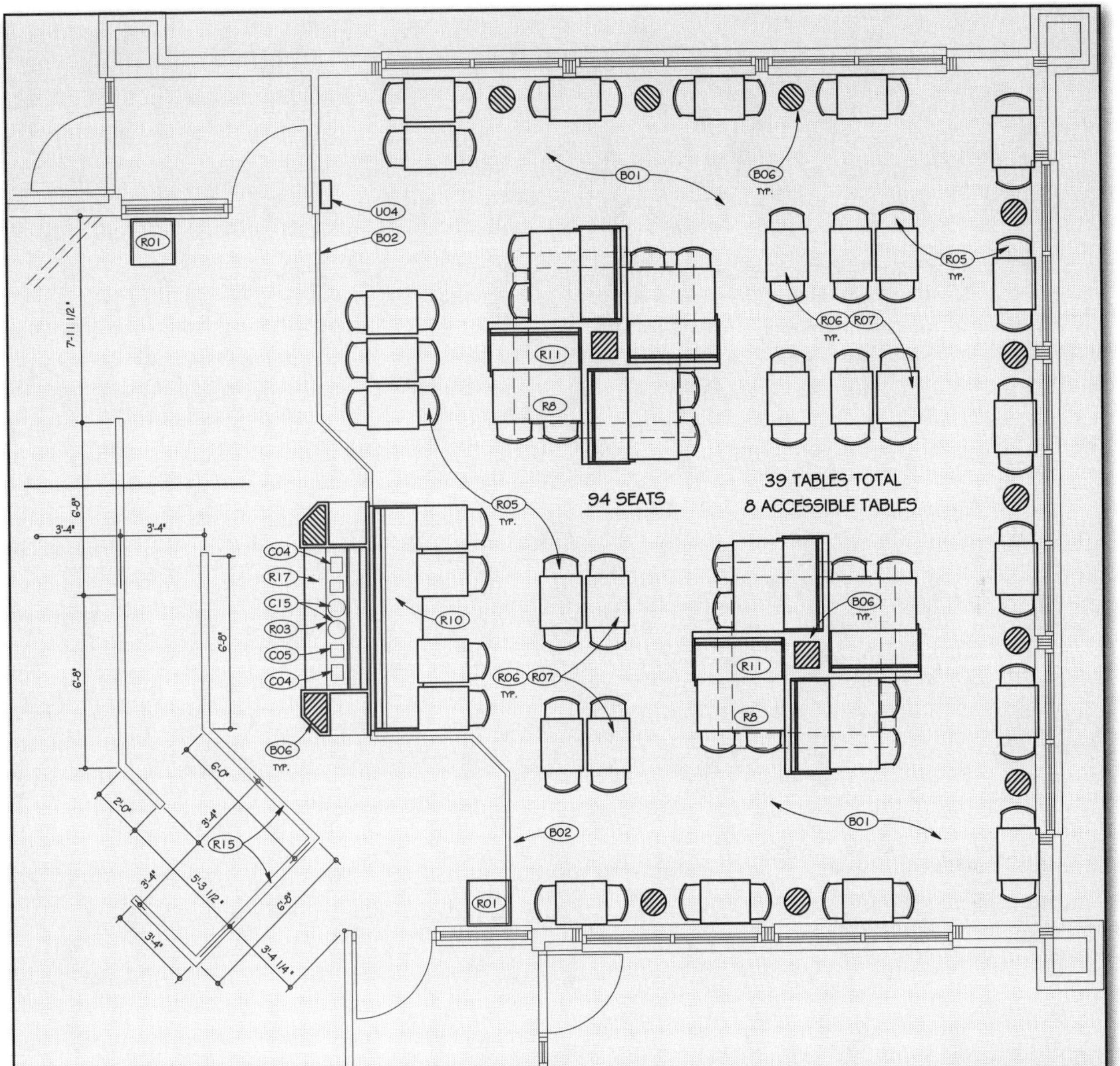

Print 3-3

Activity 3-4—Sage Timberline Estimating Activity

Create a new estimate and name it Activity 3-4. Input *Roberts Industries* for the client name and *Johnson Group* for the architect in the **Notes** of the **Estimate Information** Dialog Box. Pick the **Quick Takeoff** button and take off a job by double-clicking **1000.00 General Requirements**, double-clicking **1520.030 Temp: Office Supplies** and double-clicking **10 Job Sign**. Close the **Quick Takeoff** dialog box. Enter *1* for the **Takeoff Quantity**. Print a standard estimate report by selecting **Reports** and then selecting **Standard Estimate**. Pick **Print** to print the report.

Contracting Requirements and General Requirements 4

Key Concepts

- The business of construction is becoming more complex in terms of project delivery systems, business procurement processes, and general requirements.
- Owners, architects, engineers, program managers, construction managers, contractors, subcontractors, suppliers, and a host of other construction service providers must have clearly defined roles throughout the construction process.
- As owners search for project-delivery systems that meet their specific needs, clarification of individual roles and duties is essential for an efficient construction process.
- Estimators should understand the various entities involved in the contracting and general requirements to ensure that all fees and costs are included in the final cost estimate.

Introduction

Review of the contracting requirements by the estimator is critical to making important decisions related to company workload, schedules, and a variety of other business issues. Construction estimating items include material, labor, equipment, overhead, and profit.

Material quantities are calculated based on the type of each material required. Labor pricing is determined based on the number of labor hours required and the cost of labor per hour. Equipment used on a construction project includes power equipment such as earthmoving, hauling, and lifting equipment; temporary job-site equipment; and communication

equipment. Overhead costs are expenses that are not attributable to any one specific operation. Profit is the monetary benefit realized by a construction firm at the completion of the building process.

PROCUREMENT AND CONTRACTING REQUIREMENTS

When determining whether to undertake a construction project, a construction firm begins by reviewing the contracting requirements. Review of the contracting requirements, which are part of the construction documents, is critical to making important decisions related to company workload, schedules, and a variety of other business issues.

Solicitation and Procurement Information

Information related to the structure of the contracting requirements for a project is included in the Procurement and Contracting Requirements section of the construction documents. Instructions to bidders related to the scope of the bids and the relationships of the owner, primary contractor, and subcontractors are provided. Other entities may be mentioned in the instructions to bidders and the scope of the project, where applicable. These entities may include a program manager, owner's representative, construction manager, special inspector, or commissioning agent.

With complex construction projects, it is critical for estimators that all members of the construction team, including the contractor and all subcontractors, understand the scope of the work on which they are bidding. Inclusion or exclusion of items can create costly overlapping of bids or underbidding of items that may be missed in all bids.

Many items that affect the project but do not appear on the prints are included in the procurement and contracting requirements. Construction contracts are available in many forms, ranging from standard documents to job-specific contracts. Estimators must take all portions of the specifications into account when calculating overall job costs for items that may otherwise be overlooked in the final price of a project.

Construction Contracts. A variety of construction contract formats are used throughout the construction industry. The three basic categories are standard contracts, owner- or contractor-developed contracts, and job-specific contracts. Standard contracts are available from sources such as the American Institute of Architects or ConcensusDOCS™. Both of these organizations provide an assortment of standard construction contracts covering various project-procurement models and contracting methods. Standard contracts are available in hard copy or electronic versions.

Owner- or contractor-developed contracts are developed for an owner, general contractor, or construction manager for the majority of the projects they bid. The legal counsel for these entities develops contract documents that fit the business model of the individual owner or contractor. Finally, job-specific contracts are documents customized for a specific job or project. In each of these contract models, construction company owners, estimators, subcontractors, and suppliers should carefully review all aspects of the construction contract prior to beginning any portion of the construction process, including bidding.

Schedules. Schedules related to the entire project, project phases, and milestones during the project are included in the procurement and contracting requirements. When determining whether to bid on a project, estimators must coordinate schedules with existing company projects and company capacities at any point in relation to management capability, available workforce and equipment, and any bonding and insurance requirements.

Site Conditions. The condition of the existing building site also has an impact on the decision of whether to proceed with a project. Procurement information specifies information about the existing site, including surveys, soil conditions, hazardous materials, any existing materials or utilities, and all geophysical and geotechnical information. **See Figure 4-1.**

ELE International, Inc.

Figure 4-1. Geotechnical and geophysical information, such as soil bearing capacity, is specified in the Procurement and Contracting Requirements.

Forms. Procurement forms, including various bid forms that must be submitted to the owner, are included in the Procurement and Contracting Requirements. Forms related to various project-delivery systems, such as construction management, cost-plus fee, unit price, or design-build are included. Other forms related to allowances, bonding, certificate of insurance, substitutions, minority business enterprises (MBEs), work plan schedules, and project closeout forms are also included.

General Conditions. General conditions is a common term used to describe the written agreements for components and various construction procedures included in the Procurement and Contracting Requirements and the General Requirements sections of the specifications. These two sections specify the rights and responsibilities of all parties involved in a construction project and the relationships between the various parties.

General Requirements

Items in the General Requirements that are taken into consideration during the preliminary estimate include price and payment procedures, potential substitution of materials or products and the necessary procedures to follow, administrative requirements that the contractors and subcontractors must meet, and documentation and submittal procedures. In addition, issues related to applicable building codes, legislation, sustainable construction requirements, and permits are included in the General Requirements. Estimators must also consider the quality control issues for manufacturers, suppliers, and contractors.

Temporary facilities and utilities provided at the job site are included in the estimate. Estimators coordinate this process with the project team and project managers to ensure that all required temporary utility services, field offices, temporary material-handling and transportation equipment, barriers, and signage are included. Mobilization costs for deploying and removing these temporary services must be included by the estimator. In addition, environmental issues such as air quality, dust containment, noise reduction, and erosion control are noted as part of the final project estimate. Environmental requirements must be met in accordance with EPA and local environmental regulations.

Progress Reports. A variety of progress reports for the owner and architect may be required. **See Figure 4-2.** Costs associated with generating progress reports include time for field personnel to generate written and photographic information, information tracking and recording, and additional recordkeeping. Making accurate progress reports requires the services of a person knowledgeable about the times required for each phase of the work and each trade.

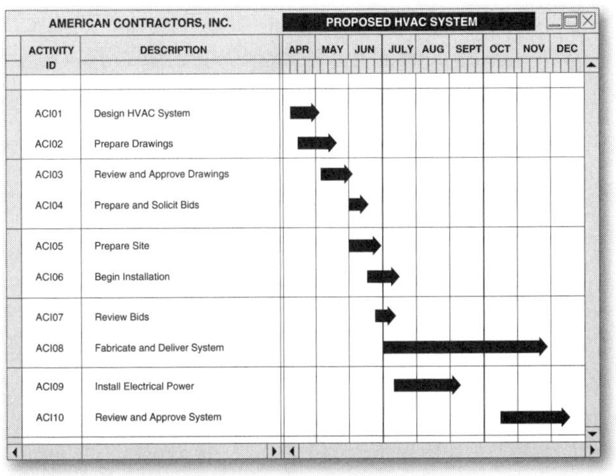

Figure 4-2. A contractor may be required to submit regular progress reports, thereby creating additional overhead costs.

Project Closeout. Issues that must be addressed at the conclusion of a construction project are described in the General Requirements. Final cleanup, final construction closeout procedures, a final site survey, warranties, disposition of project record documents, and final performance requirements are included. In some instances, owners may require as-built drawings or other operational documentation be provided at the conclusion of construction to allow for ongoing maintenance and repairs of the facility.

Commissioning is the process of reviewing the final performance of building systems to ensure they meet owner and design requirements. Third-party commissioning firms may be engaged for complex construction projects to ensure the operation of all building systems are in accordance with the design specifications.

Temporary Job-Site Requirements

The need for job-site construction equipment and temporary construction materials is determined by the project team. Temporary job-site requirements for project construction include items such as temporary utilities, field offices, access roads, signage, materials for temporary barriers or enclosures, heat and weather protection, and plumbing. Many of these expenses are categorized as mobilization costs.

Temporary Utilities. Electric, natural gas, and other utilities are necessary on a job site. Electrical power is provided for the field office, power tool operation, lighting, and security. Natural gas may be necessary to heat a structure during construction. Costs for installation, use, and removal of temporary utilities must be included in the estimate.

Field Offices. Large job sites require one or more modular office trailers or offices to provide on-site recordkeeping and project management facilities. A field office contains basic office equipment, space for review of the prints and specifications, telephone and facsimile (fax) equipment, and computer equipment. The cost of modular office trailers is allocated to several jobs depending on company policies for length of use of each trailer. Job-site trailers are commonly rented or leased similar to other job-site construction equipment. When owned by the company, costs associated with office trailers are depreciated in a manner similar to construction equipment.

Access Roads. An *access road* is a temporary road installed by a general contractor to facilitate transportation of materials, equipment, and tradesworkers onto and out of a job site. An access road may be necessary depending on the job-site location. Access road costs may include building and maintaining the road and removal of the road at the completion of construction. Access roads are commonly formed by digging a shallow path and filling it with crushed stone.

Municipalities also may specify requirements related to job-site access. Municipalities may require a contractor to clean equipment as it exits the job site or provide street cleaning adjacent to the job site to minimize construction dirt and other debris in the area around the site. Where cleaning requirements exist, labor and equipment costs must be included in the estimate.

Signage. Construction projects are often marked with a sign indicating the general contractor, architect, owner, subcontractors, and project name. **See Figure 4-3.** Signage costs include sign design, erection, painting, maintenance throughout the life of the project, and removal at project completion. Signage costs are included in the temporary job-site equipment portion of the estimate.

Figure 4-3. Costs for such items as signs should be reviewed by the estimator and included in the overall estimate.

Barriers and Enclosures. Temporary barriers and enclosures, such as fencing and protective walkways, may be required for a job site. Temporary fencing may be required at various locations on a job site for security, safety, or limiting public access to the site. An estimator should include costs from a surveyor to establish the proper fence lines. Information concerning temporary fencing is included in section 01 56 26 of the specifications.

For example, the specifications may state that the entire construction site must be enclosed by a fence extending 7′-0″ above the respective grades as indicated on the drawings attached to the specifications. The contractor also includes the cost of maintaining and providing protective fences for any existing trees.

Existing fencing at the site may be used to enclose the job site if allowed in the specifications. If the fence is built along the curb, free access from the street to hydrants, fire and police alarm boxes, and light standards must be maintained. Fencing provisions are also made to allow for truck, tradesworker, and equipment and material passage to and from the job site. Temporary fences are constructed and maintained during construction.

Other temporary protection may be required for various environmental conditions such as dust control and noise abatement. When working in areas that adjoin ongoing business operations, such as working in a hospital, provisions must be made to contain construction dust using practices such as temporary walls with plastic sheeting. In such instances, additional provisions may be necessary to maintain a safe and healthy environment in work areas and adjoining areas with minimal airflow between the areas.

Costs for materials and labor for installation, maintenance and removal of barriers, air-moving equipment, and other necessary equipment must be included in the estimate.

Heat and Weather Protection. Temporary heat includes costs such as renting, owning, and operating heating units, fuel costs, and possible ventilation during construction. **See Figure 4-4.** Heating costs vary depending on the construction project, location, weather conditions, schedule, and materials being installed. Projects such as commercial buildings with masonry, plaster, or other temperature-sensitive materials may require additional heating equipment at certain points in the project schedule. Projects in cold climates may also require additional heating equipment. Future fuel costs must be considered when estimating heating costs.

COLD-WEATHER REQUIREMENTS

M. Cold-Weather Placement: Comply with provisions of ACI 306R and as follows. Protect concrete work from physical damage or reduced strength that could be caused by frost, freezing actions, or low temperatures. Obtain approval of the engineer for concrete placement for ambient temperatures of 32°F and falling.

1. When air temperature has fallen to or is expected to fall below 40°F (4°C), uniformly heat water and aggregates before mixing to obtain a concrete mixture temperature of not less than 50°F (10°C) and not more than 80°F (10°C) at point of placement.
2. Do not use frozen materials or materials containing ice or snow.
3. Do not use calcium chloride, salt, or other materials containing antifreeze agents or chemical accelerators unless otherwise accepted in mix designs.

Figure 4-4. Contractors may be required to provide temporary heat in cold locations.

The contractor also provides weather protection for the construction project as required, including protection from snow, sleet, rain, and wind. Weather elements are taken into consideration for the entire duration of the construction project. The potential needs for weather protection and heating are determined in conjunction with the project manager using the project schedule.

Cost information is available from a variety of sources. Suppliers of heating units (either for purchase or rental) can provide costs for equipment and operational expenses such as natural gas or electrical requirements. Weather protection materials include temporary walls or roof structures and plastic sheets or tarps applied around areas to be heated. Costs associated with weather protection include materials and labor for construction, maintenance, and removal of the temporary weather protection. Standard labor and material costs for wall framing may be helpful in calculating estimates for these structures. Contractors may also collect historical data related to costs for temporary weather protection in a particular geographic location.

Plumbing. Local and state laws and applicable work rules must be considered in estimating the cost of the temporary plumbing required during a construction project. At least one toilet facility may be required for a given number of tradesworkers to meet sanitation needs, shelter them from view, weather, and falling objects. The temporary facilities are commonly rented from a local supplier on a cost per month basis. The estimator must determine the length of the project and the number of workers to calculate the temporary plumbing and sanitation costs.

Communication Equipment

On large construction projects, production can be increased by using various types of communication equipment. Communication equipment allows field personnel at the job site to share current information concerning job-site conditions with each other and the office staff. Communication equipment also helps all members of the construction team stay up-to-date on current job-site conditions. Communication equipment includes voice communication and the use of various Internet-based tools.

Voice Communications. A variety of voice communication equipment is used on construction job sites. Cellular telephones (cell phones) are the most common voice communication equipment. Cell phones are most commonly provided to key personnel by the construction company, depending on the equipment and the construction company policy. Hardwired landline telephones may be provided in the job-site trailer or on-site office. The estimator may be required to include voice communication equipment and ongoing usage costs as a communication cost.

Internet-Based Tools. Use of the Internet to quickly share job-site information is critical to today's construction projects. Electronic mail (email), text messaging, and other Internet transmissions are used to quickly communicate information related to job-site materials, work hours, costs, project schedule, change orders, and a variety of other information. Webcams may be installed on larger construction projects for security and job-site observation. **See Figure 4-5.**

A project-specific web site uses Internet technology to allow owners, architects, contractors, and suppliers to access many types of job information through a web site established specifically for a construction project. Using a password, all parties involved in the project can share information concerning project schedules, change orders and plan addenda, requests for information, progress reports, and any other necessary project information. Project-specific web sites are commonly priced based on the number of users and the length of the project. The cost for a project-specific web site should be included in the communication costs.

iBeam Systems, Inc.

Figure 4-5. Webcams may be used to monitor progress on a construction project.

Support Services

In addition to temporary job-site facilities, job-site support services are provided to ensure a successful construction project. Job-site support services include office support staff; security personnel for materials and workers; job-site material handling, storage, and distribution; and provisions for worker protection during construction. Each of these is a potential overhead cost to be considered by the estimator.

Office Support Staff and Facilities. The basic costs of the contractor and subcontractor office staff and office facilities are prorated across all jobs performed by the contractor or subcontractor. These overhead costs include management, marketing, clerical services, and other costs associated with owning and operating an office, such as office supplies, utilities, communication equipment, insurance, accounting and financial operations, and custodial services. Construction company management adds a proportionate percentage onto the total bid of each job to pay for the fixed overhead costs of doing business.

Job-Site Security. On many jobs, it is necessary for a contractor to provide adequate security personnel for general day and night patrolling of the job site, including weekends and holidays. Security personnel help prevent the theft of tools and materials and also prevent trespassing on the site that can lead to injuries for which the general contractor may be held liable. Providing sufficient security may also help reduce insurance rates. Webcams and other electronic monitoring equipment may also be a part of job-site security in some situations.

Material Handling. Materials for a construction project, such as gypsum board, hollow metal doors, finish hardware, and millwork, require proper handling and storage prior to installation. Estimators may analyze the job-site layout and include additional labor costs for unloading, storing, protecting, and distributing these materials.

Material handling on large jobs can amount to a substantial expense. The general contractor and subcontractors must agree on responsibilities for the unloading of various materials and the coordination of schedules for material deliveries. This is an important clarification and should be included in the scope-of-work letter and clearly understood by estimators prior to calculating labor and material-handling costs. Significant costs can be associated with improper material delivery, scheduling, and coordination.

Worker Protection. Proper safety equipment and environmental protection for all job-site personnel is an expense to be included in the estimators overhead costs. The costs for personal protective equipment (PPE) such as hard hats, eye protection, hearing protection, respiratory protection, and fall protection devices are all included in the estimator's calculation. The number of workers on the job site and job-site conditions should be evaluated in conjunction with the project manager to allocate sufficient expenses for PPE.

In hot weather areas and conditions, provisions must be made for supplying and distributing water or other hydrating liquids to workers on a job site. Costs include containers such as coolers and ice chests, cups, ice, and labor to distribute, collect, and clean the containers. In extremely cold weather, provision of a heated space may be required to allow workers a place to warm up between work periods.

Environmental Protection

When estimating a project, an estimator must consider the environmental regulations that pertain to the project area. The specifications may require trash removal, recycling, general job-site cleaning, erosion protection, and stormwater pollution control such as sedimentation protection. Estimators add these costs to overhead for the length of the job.

Trash Removal and Recycling. Trash removal in most areas is governed by city and state codes. A careful study of codes, plans, and the construction site helps in determining the costs for trash removal. The specifications may state that each contractor shall clean and gather all debris from the construction site daily, place the debris in containers or dumpsters provided by the contractor at the entrance of the building unit or units, and have the debris removed by a trash-removal contractor.

Where sustainability requirements are in place, there may be requirements related to the recycling of various construction materials. The United States Green Building Council (USGBC) has a variety of standards that may be utilized by owners and architects related to types of construction material recycling. The estimator should be familiar with any costs associated with recycling for various material types.

On some jobs, it may be necessary to erect a trash chute to convey waste materials and trash from the upper levels of a multistory structure to a dumpster or other trash or recycling receptacle at ground level. The cost of erecting a trash chute for multistory buildings increases as the height of the building increases. Costs associated with the removal of debris from a construction site are included in the estimate to ensure these costs are included in the overall bid.

Cleaning. Cleaning includes final preparation of the building or structure for use before turning it over to the owner. This item should be verified by the estimators because there can be significant final cleanup costs in some instances. For example, during construction windows may be marked with an "X" using a grease pencil for safety reasons. When construction is complete and before the final building is turned over to the owner, the contractor may be required to clean and wash all windows. An estimator should ensure this cost is in the bid as part of the final cleanup cost.

Stormwater Pollution Control. Many localities require contractors and subcontractors to provide various types of systems to reduce or eliminate sedimentation run-off at the job site. Stormwater silt fences consisting of various geotechnical fabrics or other materials may be required around the perimeter of a job site as part of a stormwater pollution prevention plan. **See Figure 4-6.** Estimators should check the specifications and local, state, and applicable federal regulations related to stormwater pollution control requirements.

Figure 4-6. Costs for environmental protection, such as silt fences, must be included in final estimates where required.

MATERIAL QUANTITY TAKEOFF

An estimator calculates the quantities of the materials that a construction company will use to complete a construction project. Items for quantity takeoffs are located in the specifications and on the prints. The quantity for each item is calculated and entered into a ledger sheet, spreadsheet, or estimating program. In some building information modeling (BIM) applications, quantity takeoff may be provided for some building components.

Measurement Systems

Depending on the type of material, material quantities are calculated in square feet, cubic feet, linear feet, square yards, cubic yards, acres, metric quantities such as meters or hectares, or individual units. The correct unit of measure must be used for each type of material. Estimators check the specifications and prints to determine whether English or metric units of measure are used. For standard ledger sheets, spreadsheets, and estimating programs, the unit of measure for each material may be listed next to the material description to assist the estimator in using the proper unit for each takeoff item.

Estimators must also be familiar with quantity units for each material, rounding up as necessary. For example, where a calculation for cubic yards of concrete is required, there are no partial portions of a cubic yard included because the estimator rounds up to the next full cubic yard.

Area/Volume. Square or cubic units of measure, such as square feet or cubic yards, are common for many materials. Estimators must be familiar with the units of measure used for various materials, which can be determined by checking the bid sheet, estimating program, or the manufacturer or supplier literature.

Interior and exterior finish materials such as suspended ceilings, flooring, gypsum board, roofing materials, and masonry are taken off and priced based on square feet or square yards. Square feet and square yards are units of measure for area and are determined by multiplying two dimensions. The two dimensions must use the same units of measure before being multiplied together. For example, when calculating the area of a room in square feet, two dimensions with foot units of measure are multiplied together.

Calculating Area

Determine the area (in square yards) of Room 255, which measures 8′-10″ × 23′-6″.

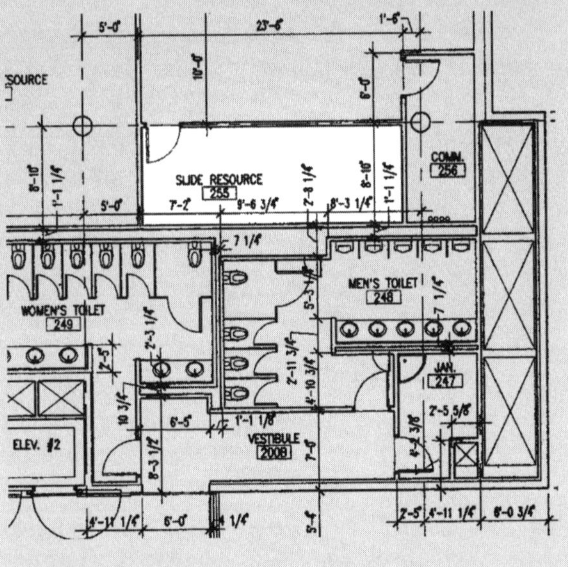

1. Convert values (if necessary).

8′-10″ = 8.83′

23′-6″ = 23.5′

2. Perform calculations.

area = width × length

$A = w \times l$

$A = 8.83' \times 23.5'$

$A = 207.5$ sq ft

3. Convert to required unit of measure and round up.

$A = 207.5 \div 9$ (9 sq ft = 1 sq yd)

$A = 23.06$ sq yd

$A = 24$ sq yd

In some situations, calculations can be performed for more than one type of material simultaneously. For example, the area of 208 sq ft can be used for a suspended ceiling system quantity takeoff while the area of 24 sq yd can be used for a carpet quantity takeoff.

Earthwork is calculated in cubic yards of material to be moved or supplied. For example, takeoff and pricing for excavation of a retention pond is based on the volume (in cubic yards) of earth to be moved. Cubic feet and cubic yards are units of measure for volume and are determined by multiplying three dimensions. The three dimensions must have the same units of measure before being multiplied together. For example, when calculating the volume of earth to be removed for the retention pond, the three dimensions, with units of measure in feet, are multiplied together.

Calculating Volume

Determine the volume of concrete (in cubic yards) required for a footing measuring 9″ × 1′-6″ × 135′-0″.

1. Convert values (if necessary).

9″ = 0.75′

1′-6″ = 1.5′

135′-0″ = 135′

2. Perform calculations.

volume = thickness × width × length

$A = t \times w \times l$

$A = 0.75' \times 1.5' \times 135'$

$A = 151.875$ cu ft

3. Convert to the required unit of measure and round up.

$A = 151.875 \div 27$ (27 cu ft = 1 cu yd)

$A = 5.625$ cu yd

$A = 6$ cu yd

Quantity. Items such as doors, windows, electrical receptacles and switches, luminaires (lighting fixtures), water closets, drinking fountains, air conditioners, boilers, and a variety of other fixtures are taken off by counting them individually. For example, eight doors are required for Corridor 203. **See Figure 4-7.** A thorough review of the specifications and accurate markup of the prints are required to properly calculate quantities of individual items and ensure all items are included in the quantity takeoff.

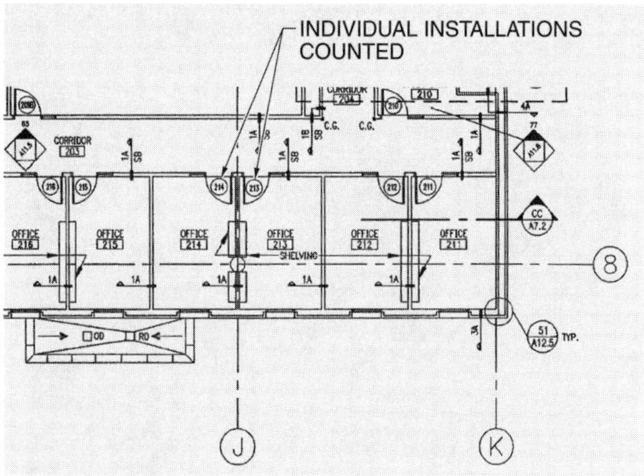

Figure 4-7. Items, such as doors and windows, are taken off by counting them individually.

LABOR PRICING

Labor pricing is determined by multiplying the number of labor hours required (labor quantity) by the cost of labor per hour (labor rate). The labor quantity required for various operations can be related to the takeoff quantities. Labor rates are available from a variety of sources. Estimators should include all costs associated with labor quantity including wages, fringe benefits, taxes, and all associated overhead.

Labor Quantities

Labor quantities are the number of labor hours required for various construction operations. One common method of calculating labor quantities is to equate a certain number of labor hours with specific material quantities. For example, it may take 0.75 hr for a tradesworker to set a metal doorjamb in place and install the metal door. The estimator calculates the number of metal doors to be installed for a particular construction project and uses this information to determine the labor quantity. If 12 metal doorjambs and doors are to be set and installed, 9.0 labor hours are included as the labor quantity in the bid for this item (12 × 0.75 hr = 9.0 hr).

Contractors often maintain their own labor quantity information based on historical data captured and recorded from past jobs performed. Historical data is the most reliable source of labor quantity information available to a company.

Estimators work with project managers during construction to track data regarding actual labor quantities required for various construction operations on each job. As contractors complete more jobs and build a broader labor quantity history, the historical data becomes more reliable. Construction companies can then develop various crew management systems and methods, resulting in labor quantity information for various construction operations. The historical labor quantity data then provides a basis for subsequent overall labor costs for each material quantity.

For estimating programs, labor quantities may be entered into a quantity takeoff sheet directly linked to a database. **See Figure 4-8.** As the estimator enters quantities of materials or number of units, the labor quantity associated with the particular material is retrieved from the database and this amount is entered into the spreadsheet.

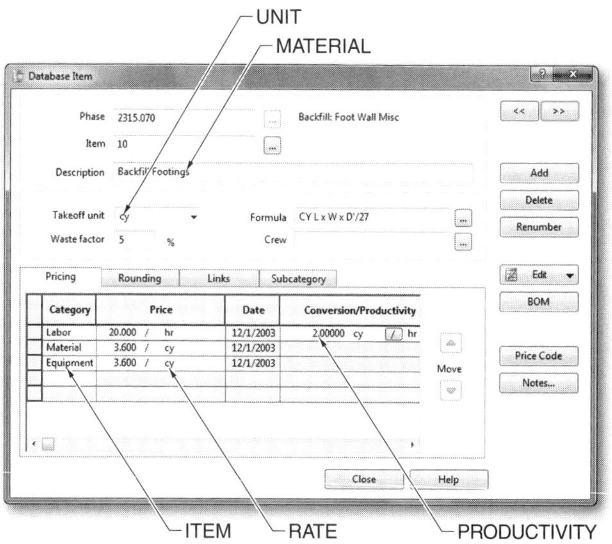

Figure 4-8. Estimating programs allow labor quantities to be entered into a quantity takeoff sheet that is linked to a database.

Costs for various types of labor are stored in another portion of the database. Labor quantities and labor rates are linked to the quantity takeoff in many estimating programs.

Standard Labor Quantity Information. Standardized labor-unit (time) tables for various materials are available for purchase from various third-party providers to help contractors determine labor quantities where no personal historical labor quantity is available. **See Figure 4-9.** Labor-unit tables, which are based on industry practices, provide information that equates various materials to the number of labor hours required for installation.

An estimator may use labor unit tables to determine labor quantities based on a variety of measures such as linear feet, square feet, cubic yards, or items of materials. Standardized labor unit tables are also available in electronic format for use with estimating programs.

Estimators should not rely solely on either standard or historical labor quantity data since there are many situations where job-site conditions impact the time required for various operations. An estimator should consult with the project manager, visit the construction site, and/or work with various members of the construction project team to determine areas where the labor quantity required for various operations may be either greater or less than either standard or historical data.

For example, the actual labor time required at the job site for a particular type of construction operation may be greater than either standard or historical data if access to the job site is difficult. Adjustments to the estimated labor quantity should be made accordingly. Where tradesworkers are required to enter a secure area that requires security screening, additional labor time should be included in the estimate.

Labor Rates

Various labor rates apply to all job-site personnel costs, including wages, taxes, insurance, and fringe benefits. Labor rates vary greatly depending on geographic area, job-site requirements, market conditions, and the type of work being performed. Labor rates are typically one of the higher risk items in the calculation of a construction bid.

WIRE INSTALLATION LABOR UNITS*

Size	Copper					Aluminum	
	TW	THW	THHN	XHHW	USE	THW	XHHW
14	5.5	6	6	6	6	—	—
12	6.5	7	7	7	7	—	—
10	7.5	8	8	8	8	—	—
8	9.5	9	9	9	9	—	—
6	11.5	10	10	10	10	9	9
4	13.5	12	12	12	12	10	10
2	15.5	13	13	13	13	11	11
1	—	14	14	14	14	12	12
1/0	—	15	15	15	15	13	13
2/0	—	16	16	16	16	14	14
3/0	—	17	17	17	17	15	15
4/0	—	18	18	18	18	16	16
250	—	20	20	20	20	18	18
300	—	23	23	23	23	19	19
350	—	24	24	24	24	20	20
400	—	25	25	25	25	21	21
500	—	26	26	26	26	22	22
600	—	30	—	30	30	23	23
750	—	32	—	32	32	25	25
1000	—	46	—	36	36	30	30

* hours per thousand feet for one conductor, solid or stranded, 600 V

Figure 4-9. Estimators may use standard industry information concerning labor units for installing various materials.

Labor rates may be higher or lower in various geographic locations depending on labor availability, types of prevalent fringe benefits, production levels, quality, and skills. Labor availability is impacted by the number of workers available in a particular area and the volume of construction work being performed. Information about other projects in the market must be taken into consideration when determining job-site labor costs. Production levels for various types of work are commonly based on experience in the market. Sources for standardized labor-rate information include local trade associations and various private and government sources.

Information Sources. Local construction trade associations or chambers of commerce may provide information concerning labor rates, benefits, and tax rates in their geographic areas. **See Figure 4-10.** Estimators may also rely on labor wage-rate tables provided by government agencies such as the Department of Commerce or Department of Labor.

Where prevailing wage laws apply, the estimator must utilize labor rates determined by local labor market surveys. Government agencies regularly collect information from employers in local areas and publish prevailing wage rates for individual counties. Estimators must be familiar with any prevailing wage requirements for a particular project when determining wage rates.

As with labor quantities, many construction companies may also have their own historical data concerning labor rates from past construction jobs their company has completed. Historical data is the most reliable source of labor rate information, as it is based on past experience of various trades in a given geographic area.

Multiplication of labor quantities and labor rates produces a labor cost for each portion of the construction project. Labor costs are entered into the ledger sheet, spreadsheet, or estimating program along with material costs to determine a total cost per item. The duration of a construction project may affect labor rates since allowances must be made for any applicable wage, tax, or benefits raises during the construction process. On lengthy construction projects with many tradesworkers, estimators should include wage-rate increases in the overall labor costs. Equipment and overhead may also be included in these calculations.

QUALITY ACA FIRST
Building America Better

WAGE RATE SURVEY

Craft	Rate
Asbestos Workers	$29.64
Supervisor	$30.64
General Supervisor	$33.64
Fringes and Increases	P $7.86, WF $4.45, V $3.90 included in hourly rate, 9¢ short term disability incl. hrly. rate, I&P 17¢
Boilermakers	$28.19
Supervisor	$30.69
General Supervisor	$32.69
Fringes and Increases	P $6.50, WF $6.62, V $1.05 plus 5% of gross wages included in hourly rate, annuity $3.50 A 37¢
Bricklayers	$27.75
Supervisor	$30.73
General Supervisor	$31.23
Fringes and Increases	P $3.10, WF $5.12, VS $2.00, A 40¢, MPF 28¢, MCAT 28¢, 56¢ union dues and building fund included in hourly rate, BMSP $2.30

Figure 4-10. Local construction trade associations provide information concerning labor rates, benefits, and tax rates in their areas.

EQUIPMENT

Equipment used on a construction project includes power equipment such as earthmoving, hauling, and lifting equipment; temporary job-site equipment; and communication equipment. **See Figure 4-11.** The quantities and costs of each piece of equipment are included in the overall estimate. In some instances, these costs may also include mobilization expenses. *Mobilization* is the process of moving the required construction equipment and support systems to the current construction site.

Depending on the type of project and company capacity, the company owner and estimator may determine whether a particular piece of equipment is to be purchased or rented/leased. If the size and scope of the project and future company plans will make additional use of a piece of equipment, purchasing it may be a more advantageous option than renting or leasing it. If the company is using a very specialized piece of equipment for a single project, renting or leasing may be the most economical choice. The various options should be discussed with equipment dealers, and distributors can help clarify these decisions prior to completing the final bid.

JCB Inc.

Figure 4-11. Estimates include costs for necessary equipment.

Power Equipment

Power equipment includes all machinery used in the construction process. An estimator determines the costs of owning, operating, and maintaining power equipment to determine direct costs and overhead for a project. A variety of factors are considered when calculating the hourly cost of owning, operating, and maintaining equipment. These costs can be allocated to a standard material quantity such as the cost per cubic yard for earthmoving equipment or cost per ton for trucks.

The total costs for hourly equipment operation include owning costs, operating costs, maintenance, and wages. Owning costs are fixed costs that are ongoing costs regardless of whether or not the equipment is in operation. Owning costs include depreciation, interest, insurance, and taxes. Operating and maintenance costs are variable costs based on the amount of equipment use and include fuel and lubricants, repairs (including parts and mechanic costs), and tires.

Wages for equipment operators depend upon local labor conditions and the type of work to be performed. If prevailing wage laws apply to the project, the equipment-operator wages are available from the wage and hour division of the local labor department. Where prevailing wage laws do apply, local union agreements should be reviewed where applicable for wage and benefit information. For jobs without prevailing wage requirements or labor agreements, individual contractors establish wage rates based on local labor demand. Construction associations or labor providers may be helpful in determining local wage conditions.

Equipment overhead is not included in owning and operating costs. Equipment overhead includes indirect costs such as supervision, mobilization, storage, and insurance. Estimators may also choose to include items such as fuel and lubricants, parts, and repair technicians in overhead rather than operating costs. Equipment overhead may be tied into operator labor costs and rated by cost per hour of operation.

John Deere Construction & Forestry Company

The cost of owning, operating, and maintaining power equipment must be calculated when determining direct costs and overhead for a project.

Depreciation. *Depreciation* is the accounting practice of reducing the value of equipment by a fixed amount each year. Each major piece of equipment owned by a construction company experiences depreciation. The depreciation cost varies according to the original purchase price of the equipment and the method of estimating depreciation.

The most common depreciation methods are the straight-line and double-declining balance methods. With the straight-line depreciation method, the total cost of a piece of equipment is divided equally over the number of years in its expected life, and the appropriate reduction in value is accounted for as an expense each year. For example, a $50,000 piece of equipment depreciated over 10 years using the straight-line depreciation method is reduced in value by $5000 each year.

With the double-declining balance method, the depreciation calculation begins by determining the straight-line depreciation percentage and multiplying the percentage by 2. This result is then used to calculate the remaining depreciation each year. For example, the straight-line depreciation of a $50,000 piece of equipment depreciated over 10 years is $5000, which is 10% of the total cost of the equipment. The percentage is multiplied by 2 to determine the percentage of depreciation for the double-declining balance method (10% × 2 = 20%). The percentage of depreciation is deducted from the total equipment purchase cost ($50,000 – 20% = $40,000). The same percentage (20%) is depreciated from the balance the following and subsequent years (year 2 = $8000 depreciation, year 3 = $6400 depreciation, year 4 = $5120, etc.).

Major equipment is selected for the specific tasks it can perform. The equipment is normally rented or leased for the time it is needed on a project.

Management and the company auditors determine the depreciation method used. Regardless of the depreciation method used, a depreciation cost is included in equipment cost. The company auditor provides an annual depreciation cost for each piece of equipment. The depreciation cost can be prorated based on the amount of time each piece of equipment is used on a particular job.

Rental/Leasing. *Equipment rental* is the use of equipment for a fixed cost per a given unit of time. *Equipment leasing* is the use of equipment based on a contract that stipulates a set amount of time and a set cost for the equipment use. Renting or leasing equipment may be more economical for a construction project than purchasing the equipment. Renting or leasing allows for more flexibility in the equipment costs, allows a company to obtain specialized equipment for a specific job, and minimizes equipment overhead between projects when the equipment is idle. The costs of owning and maintaining equipment, including depreciation, are included in the rental or lease price.

When equipment is rented or leased, the estimator works closely with the project manager and field personnel to determine the amount of time each piece of equipment is needed on the job site and coordinates this information with the costs of renting or leasing. The time unit used may be daily or monthly.

The rental or lease agreement should be checked to determine all of the specific items and services that are included in the contract agreement and those that are excluded. For example, the estimator should determine if the equipment rental or lease agreement includes an operator in the rental or lease rate, as an operator labor cost may be included in the hourly rate. Also, truck rental rates often include the cost of the driver in the daily rental rate. Truck rental or usage rates may be priced on unit price per ton.

Fuel. Equipment fuel includes gasoline, diesel fuel, or other energy sources used to power a piece of equipment. Each piece of equipment should have information available from manufacturer literature or contractor historical data describing the amount and type of fuel consumed per hour of use. In a manner similar to labor costs, the quantity of fuel is multiplied by the current market cost of the fuel to determine overall fuel costs per piece of equipment.

In recent years, fuel costs have been a high-risk item in estimates due to the escalating price of various types of fuel.

Cost Allocation. *Cost allocation* is the designation of overall company equipment costs to particular pieces of equipment. Cost allocation of equipment as overhead to a project may be based on an internal rental rate. This internal rental rate may be based on company historical data or standard industry resources. Internal rental cost allocation rates may be based on hourly, daily, or weekly rates. The internal rental cost allocation rate may include factors such as the cost of purchasing the equipment minus any expected salvage or resale value, the overall length of use of the equipment, the estimated idle and productive use of the equipment during its overall use time, mobilization, storage, maintenance, and insurance costs.

OVERHEAD

Overhead is any business expense that is not chargeable or readily allocated to a particular part of a construction project. In addition to material quantities and labor costs, contractors incur a wide range of overhead costs. As a bid is generated, a certain percentage of the overhead costs are allocated to each construction project to ensure the company can continue operations. Overhead costs include items such as temporary utilities, insurance and safety, bonds, permits, licenses, inspection fees, taxes, and benefits.

Temporary Utilities

Electric, natural gas, liquefied petroleum (LP) gas, and other utility services are necessary on a job site. Electrical power is provided for the field office, power-tool operation, lighting, and security. Natural or LP gas may be necessary to heat a structure during construction. Costs for temporary utilities are included in the bid as overhead.

Light and Power. A contractor provides temporary light and electrical power for the various trades and subcontractors under the dictates of the contract. The temporary light and power service must meet with approval of the county, township, or city in which the construction is taking place. This cost is determined by the price in relation to the size and duration of the project. The electrical contractor calculates costs for installation based on the amount of temporary power needed and the distance from the closest power source. Costs include the installation of meters, breaker panels, service boxes, and proper-height electrical power poles.

Insurance and Safety

A contractor should not begin work until all required insurance has been obtained and the insurance policies have been approved by the architect and/or owner. The policies should be of an amount that is sufficient to cover any potential loss occurrences during construction. A contractor may be required to show receipts for complete payment of premiums to the architect and/or owner to indicate that the insurance policies are up to date.

Many types, forms, and limits of insurance coverage are available. The estimator must take into account the costs to secure insurance policies that are required by the project contract and/or local, state, or federal law. The insurance coverage for various policies affects the final cost of construction. **See Figure 4-12.** The cost of these policies varies with the size, location, type, and hazards of the construction job. The unit cost applied to the estimate varies in percentage on each project.

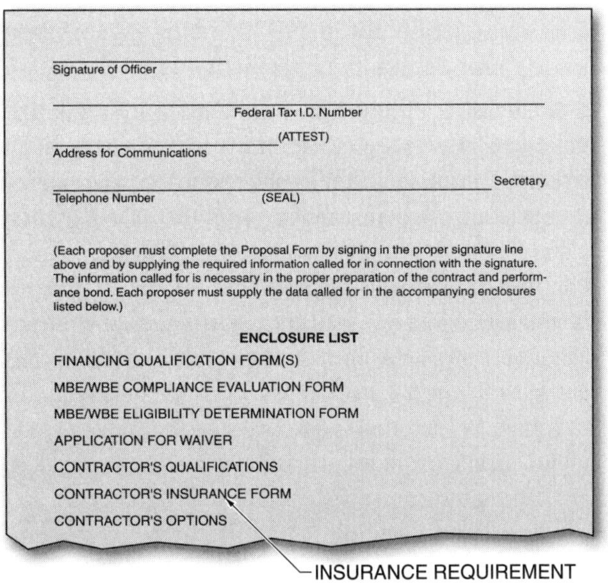
Signature of Officer

Federal Tax I.D. Number

(ATTEST)

Address for Communications

Secretary

Telephone Number (SEAL)

(Each proposer must complete the Proposal Form by signing in the proper signature line above and by supplying the required information called for in connection with the signature. The information called for is necessary in the proper preparation of the contract and performance bond. Each proposer must supply the data called for in the accompanying enclosures listed below.)

ENCLOSURE LIST

FINANCING QUALIFICATION FORM(S)

MBE/WBE COMPLIANCE EVALUATION FORM

MBE/WBE ELIGIBILITY DETERMINATION FORM

APPLICATION FOR WAIVER

CONTRACTOR'S QUALIFICATIONS

CONTRACTOR'S INSURANCE FORM

CONTRACTOR'S OPTIONS

Figure 4-12. Insurance coverage for various policies affects the final construction cost.

Contractor public liability insurance, fire and extended coverage insurance, and coinsurance are three types of insurance to consider during preparation of a bid. Other types of insurance included may be flood insurance, railroad protective insurance, or other specialized insurance. A reputable insurance agent who is knowledgeable about the specifics of construction insurance for each project should be consulted in order to determine insurance needs.

Contractor Public Liability Insurance. The contractor and subcontractor must carry public liability insurance covering potential bodily injury or death that could be suffered as a result of an accident occurring from or by reason of, or in the course of operation of, the construction project. This includes accidents occurring by reason of omission or acts of the contractor or any subcontractor or by people employed by the contractor or subcontractor. The amount of required coverage is usually determined by potential liability resulting from the size and type of the project and/or by the architect or the owner.

Fire and Extended Coverage Insurance. The contractor should carry insurance covering possible loss by fire, wind, tornado, or other natural disasters. This insurance covers a wide range of items and equipment and any damages done while work on the project is in progress. Special clauses are usually written into the policy by an insurance agent with a background in construction liability coverage. Fire protection insurance rates vary according to the rating organization having jurisdiction in a particular area.

Coinsurance. Contractors may choose to divide the insurance coverage among different companies in an attempt to obtain lower insurance rates. The use of owner-controlled insurance programs (OCIP) may also change the insurance structure. With an owner-controlled insurance program, the owner establishes insurance coverage for all contractors and subcontractors on the project. Contractors and subcontractors who are bidding on the work must check to make sure all coverage is included, add coverage that may not be included, and exclude coverage that may be duplicated.

Contractor-controlled insurance programs (CCIP) use the same principle as OCIP, with the general contractor providing a single insurance plan for themselves and the subcontractors. The estimator includes the actual coinsurance rates in the overhead.

Safety Programs. Contractors may provide a safety program for tradesworkers that covers job-site operations. Safety programs are an overhead cost that may be effective in reducing other overhead costs, such as workers' compensation insurance rates. Contractors may employ their own safety-training personnel or contract this service to a private firm or contractor association.

Many contractors conduct weekly safety meetings on the job site, referred to as toolbox meetings or toolbox talks, to meet requirements of the Occupational Safety and Health Administration (OSHA) or the Mine Safety and Health Administration (MSHA). Overhead costs for a safety program may include employment costs for personnel, such as wages and fringe benefits for both the trainer and the tradesworkers during training, and training materials, or may include contract costs based on the safety-training services provided. Any OSHA or MSHA certification costs required are also included as a safety expense.

Personal protective equipment (PPE) is used to protect against safety hazards on a job site. PPE includes protective clothing, head protection, eye and face protection, hearing protection, hand and foot protection, back protection, knee protection, fall protection, and respiratory protection.

Bonds

A *bond* is an insurance agreement for financial loss caused by the act or default of an individual or by some contingency over which the individual has no control. Bonding provides an assurance to the owner that the construction project will be completed as described in the specifications and prints. Bonding companies issue bonds for various amounts based on the contractor assets, qualifications, and experience.

Construction firms with good performance records may obtain bonds at a lower rate than inexperienced firms. Some large construction firms may be self-bonded, meaning that they do not purchase bonds from a third party, but guarantee the completion of the project with their own assets. Bond types include performance, completion, and street bonds.

Performance Bonds. A *performance bond* is a short-term insurance agreement guaranteeing that a contractor will execute a construction project in the manner described in the contract documents. The owner and the architect of a project must be assured that the contractor who is awarded the work will fully complete the work. To ensure that the contractor accepts the contract as bid, the architect and/or owner may require that the contractor or subcontractor post a performance bond. A performance bond can be for the entire cost of the job or a percentage of the job. **See Figure 4-13.** The amount of bonding required is noted in the specifications. Costs for performance bonds are obtained from a bonding agency.

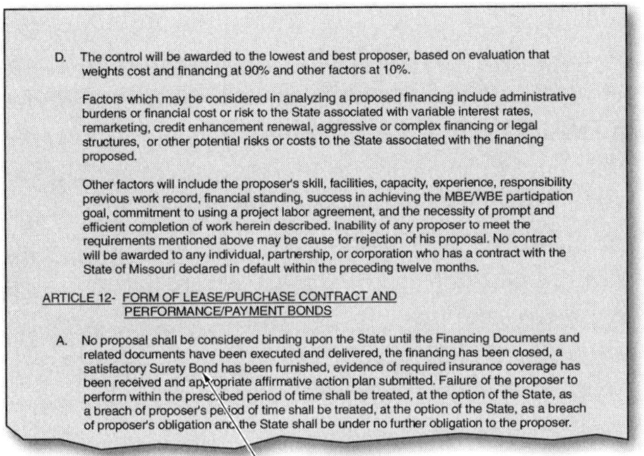

D. The control will be awarded to the lowest and best proposer, based on evaluation that weights cost and financing at 90% and other factors at 10%.

Factors which may be considered in analyzing a proposed financing include administrative burdens or financial cost or risk to the State associated with variable interest rates, remarketing, credit enhancement renewal, aggressive or complex financing or legal structures, or other potential risks or costs to the State associated with the financing proposed.

Other factors will include the proposer's skill, facilities, capacity, experience, responsibility previous work record, financial standing, success in achieving the MBE/WBE participation goal, commitment to using a project labor agreement, and the necessity of prompt and efficient completion of work herein described. Inability of any proposer to meet the requirements mentioned above may be cause for rejection of his proposal. No contract will be awarded to any individual, partnership, or corporation who has a contract with the State of Missouri declared in default within the preceding twelve months.

ARTICLE 12- FORM OF LEASE/PURCHASE CONTRACT AND PERFORMANCE/PAYMENT BONDS

A. No proposal shall be considered binding upon the State until the Financing Documents and related documents have been executed and delivered, the financing has been closed, a satisfactory Surety Bond has been furnished, evidence of required insurance coverage has been received and appropriate affirmative action plan submitted. Failure of the proposer to perform within the prescribed period of time shall be treated, at the option of the State, as a breach of proposer's period of time shall be treated, at the option of the State, as a breach of proposer's obligation and the State shall be under no further obligation to the proposer.

Figure 4-13. Bonding protects a project owner if a contractor or subcontractor is unable to complete a construction project.

Completion Bonds. A *completion bond* is a short-term insurance policy ensuring the owner that a construction project will be fully completed and free from encumbrances and liens when turned over to the owner. Many contingencies such as inclement weather, labor problems, and variances of material availability may motivate the owner of a project to obtain additional protection, ensuring that the contractor, once building has started, guarantees completion of the job. The additional protection can be provided by posting a completion bond.

A completion bond gives the owner or bonding company the funds to hire another firm to complete the project if the contractor fails to complete construction as stated in the contract. The total amount of a completion bond is usually equal to the cost of the total estimate of a particular job. The amount of bonding required is noted in the specifications. Costs for completion bonds are obtained from a bonding agency.

Street Bonds. A *street bond* is a bond that guarantees the repair of streets adjacent to a construction project if they are damaged during construction. A street bond is purchased by the general contractor. The amount of bonding required is noted in the specifications. Costs for street bonds are obtained from a bonding agency.

Permits, Licenses, and Inspection Fees

A variety of government permits, licenses, and inspection fees are required for construction projects. The permits, licenses, and inspection fees required vary depending on the type of project and the location. An estimator should check the specifications and be knowledgeable about local, state, and federal government codes and laws applicable to the project to determine the applicable permits, licenses, and inspections and the cost for each.

Permits. A *permit* is written permission granted by the governmental authority having jurisdiction over the construction process to perform the work described in the prints and specifications according to local regulations. Every construction project requires that permits be obtained from the government agency having jurisdiction. Permits are obtained from the local building department after plan review and approval. For example, one necessary permit is the fire permit for the job site required by the local fire district.

After the permits are obtained, inspectors track construction to ensure that the work is performed according to local building codes and the approved plans. Inspectors also issue final approval at the completion of the project prior to use or issuance of an occupancy permit.

Licenses. A *license* is a privilege granted by a state or other government agency to an individual or company to perform work or provide professional services. Contractors and subcontractors are often required to pay fees for a business or trade license. For example, electrical licenses and plumbing licenses are required by various localities. **See Figure 4-14.** Different localities have different requirements for licensing. The estimator should ensure that subcontractor bids are solicited only from firms meeting any applicable licensing requirements.

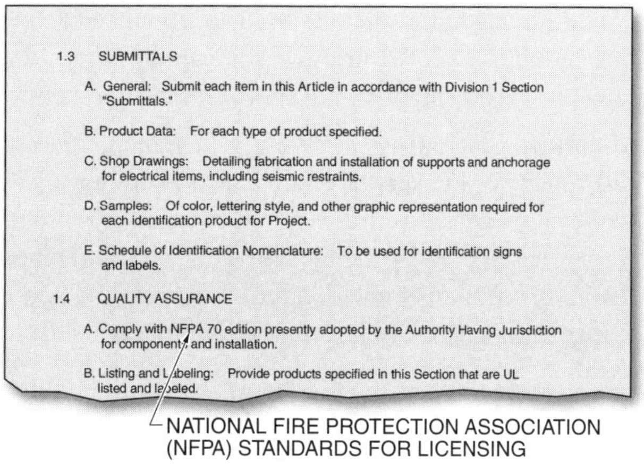
1.3 SUBMITTALS

A. General: Submit each item in this Article in accordance with Division 1 Section "Submittals."

B. Product Data: For each type of product specified.

C. Shop Drawings: Detailing fabrication and installation of supports and anchorage for electrical items, including seismic restraints.

D. Samples: Of color, lettering style, and other graphic representation required for each identification product for Project.

E. Schedule of Identification Nomenclature: To be used for identification signs and labels.

1.4 QUALITY ASSURANCE

A. Comply with NFPA 70 edition presently adopted by the Authority Having Jurisdiction for components and installation.

B. Listing and Labeling: Provide products specified in this Section that are UL listed and labeled.

Figure 4-14. Various trades and contractors require licensing to perform work.

Inspection and Testing. A contractor may be expected to furnish facilities and assistance for inspection, examination, and tests that inspectors may require. The contractor must also secure access for inspectors to factories and plants in which materials are being manufactured or prepared and to all parts of the construction project area. Contractors may also be required to give inspectors advance notice of preparation, manufacture, or shipment of any materials. Some overhead costs may be associated with these testing facilities and with acquiring access. Concrete strength, soil compaction and composition, and torque on connecting bolts are examples of several items that may require inspection or testing.

In some instances, the contract documents may require the owner or contractor to contract a third-party provider of special inspections to ensure construction is performed in accordance with all plans and specifications. These third-party special inspectors are not affiliated with any government agency or requirement but are part of the construction team assisting to ensure construction is completed properly by all contractors and subcontractors.

Taxes and Benefits

In addition to labor costs, a variety of payroll taxes and benefits are included in the overhead costs for tradesworkers and office staff. These overhead costs include Social Security; Medicare; workers' compensation; and health, welfare, and pension benefits. Additional state and local taxes are taken into consideration by the estimator when calculating labor costs and overhead. Sales tax rates on material purchases are also included in overhead cost calculations.

Social Security. Contractors pay the assessments for Social Security insurance and unemployment benefits for their employees. The amount paid is based on the wages received by the employees of the contractor or subcontractor. The general contractor usually releases the owner from any responsibility for assessments that may be defaulted by the contractor or the subcontractors.

Workers' Compensation. The United States and the Canadian provinces require workers' compensation insurance protection. Federal laws cover federal employees and private employees in the District of Columbia. Workers' compensation provides employees with certain benefits in the event of injury or death and in the case of some diseases. A number of states have severe penalties for contractors and subcontractors who fail to carry workers' compensation insurance. Costs for workers' compensation insurance vary depending on current market conditions and the safety record of the construction company applying for the insurance. The estimator should consult an insurance agent to obtain costs for workers' compensation insurance coverage.

Baldor Electric Co.

Equipment must be inspected and tested on a regular basis.

Health, Welfare, and Pension. In addition to wages, other contributions paid by contractors to tradesworkers include health benefit coverage and pension plans. Costs for these coverages for tradesworkers are based on local labor agreements or company-specific benefit plans. Health and pension costs are commonly calculated on a cost per hour basis and are added to the labor cost and unit pricing. **See Figure 4-15.**

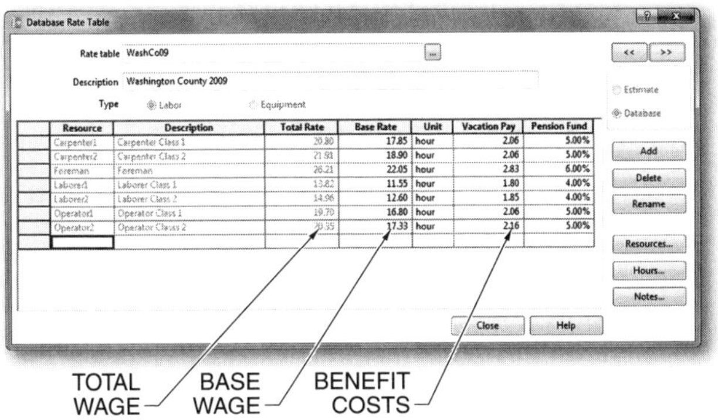

Figure 4-15. Benefit costs are added to the labor cost and unit pricing.

Sales Tax. States and municipalities use varying percentage rates for sales tax. Estimators should include sales tax as overhead. Construction materials for nonprofit organizations may be exempt from sales tax. These exemptions vary according to state regulations. Estimators should consult the state taxing agencies for information concerning sales tax applicability on the construction project.

PROFIT

Profit is the amount of money remaining at the completion of a project in excess of all project and company expenses. As with any business, the purpose of undertaking a construction project is to make a profit for the company after consideration of all project and overhead costs. Factors taken into consideration when estimating the amount of profit to be added to a bid include the overall workload within the company, the local construction market workload and competitive outlook, the number of bidders on a particular project, and the risks inherent in a particular construction project.

Companies may have a tendency to add a higher profit when the company has a large workload and will incur additional costs of doing business to perform any additional work. A lower number of bidders on a project may allow a company to take a risk on a higher profit margin. High-risk projects also require higher profit margins because there is a higher chance of losing money. Three methods for calculating profit include percentage basis, cost-plus basis, and time and material basis. Companies may use other methods depending on company history and industry trends. For example, integrated project delivery (IPD) may be used because it has a shared profit/loss structure.

Percentage Basis

The *percentage basis* is a method of determining profit in which a percentage is added to the final estimated cost. The profit percentage is added after all materials, labor, overhead, and equipment costs are calculated. This results in a fixed dollar amount being shown on the bid documents. For example, an estimator may add 3% onto the total project costs for profit. This number may be changed by the management of the company based on a number of factors. The owner and architect have generated preliminary estimates of the cost of the project and the risk for profit or loss is with the contractor. Estimators may also add varying percentages of profit to individual bid items.

Risk Percentage. Contractors may measure the historical accuracy of their bids to help them compute risks for various types of work and determine proper profit percentages. A direct link exists between the level of profit in a construction project and the level of risk. A percentage of variance can be calculated by tracking a company history of estimated bids and comparing these figures to the actual job costs. Accuracy in actual job costs is based on proper reporting of costs for materials, labor, and equipment.

A contractor using this system may break out the historical estimate accuracy on various items and assess risk and profit accordingly. For example, by reviewing historical information, a contractor may determine that estimates for equipment have varied from actual costs by an average of 3.5%. This tells the contractor that a minimum of 3.6% profit should be added to equipment pricing to balance profit and risk.

Profit is determined at the completion of the project after all of the construction and company expenses have been paid.

Cost-Plus Basis

The *cost-plus basis* is a method of determining profit in which costs are submitted for payment for labor, material, equipment, and overhead, and then an amount or percentage is added automatically for profit. Owners and architects may propose projects for contractors on a cost-plus basis. This ensures a profit for the contractor, but does not ensure a fixed price for the owner.

Time and Material Basis

The *time and material basis* is a method of determining profit in which the owner pays a fixed rate for time spent on a project and pays for all materials. Small jobs may be performed on a time and material basis. Profit is built into the labor cost for time. This ensures a profit for the contractor, but does not ensure a fixed price for the owner.

Integrated Project Delivery

Profit sharing is a common model in integrated project delivery (IPD). Members of the construction team work collaboratively to determine a profit margin and then work together to meet or exceed various performance criteria throughout the project. As savings on the project are realized, they are shared among all members of the IPD construction team. Conversely, if there are losses, they are also shared.

SAGE TIMBERLINE ESTIMATING . . .

QUICK TAKEOFF

Quick Takeoff is a function of Sage Timberline Estimating software that enables an estimator to enter quantities directly into the software spreadsheet. Quick Takeoff is the easiest way to place items into the estimate. The results are shown immediately since work is performed directly on the spreadsheet. When using Quick Takeoff, the calculations are performed using the cost information and formulas stored in the database.

1. Open the **Quick Takeoff** dialog box by picking the **Quick Takeoff** button in the toolbar.

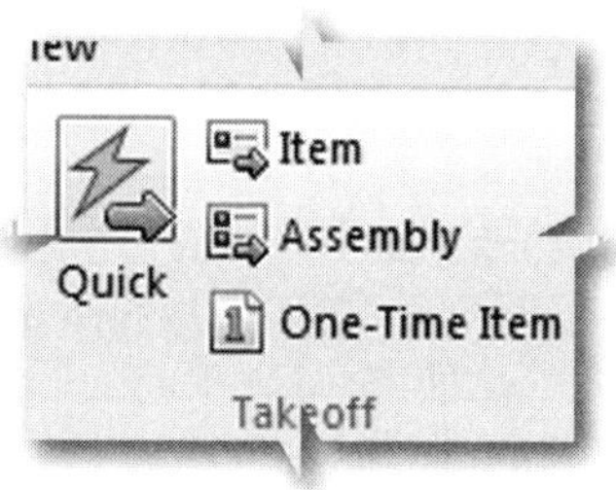

. . . SAGE TIMBERLINE ESTIMATING

2. Select the items to be added to the spreadsheet by double-clicking through each level of the database (group, phase, and item). Double-click Group **1000.00 General Requirements**, Phase **1300.010 Personnel: Supervision**, and Item **10 Superintendent**.

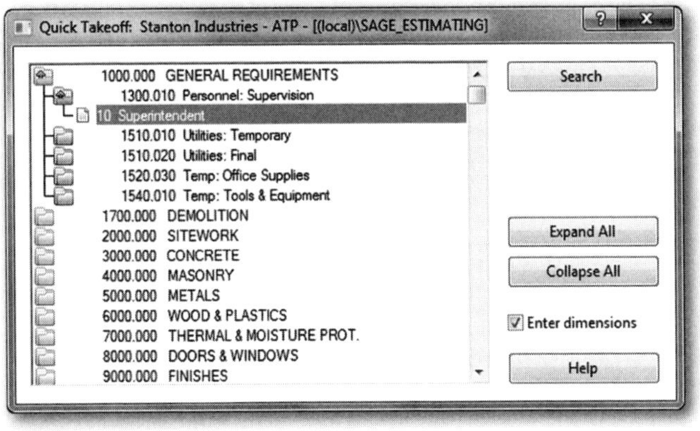

3. Close the **Quick Takeoff** dialog box by picking the Close button in the upper-right corner of the dialog box.
4. Enter **48** in the **Takeoff Quantity** cell for the superintendent item and press **Enter**.

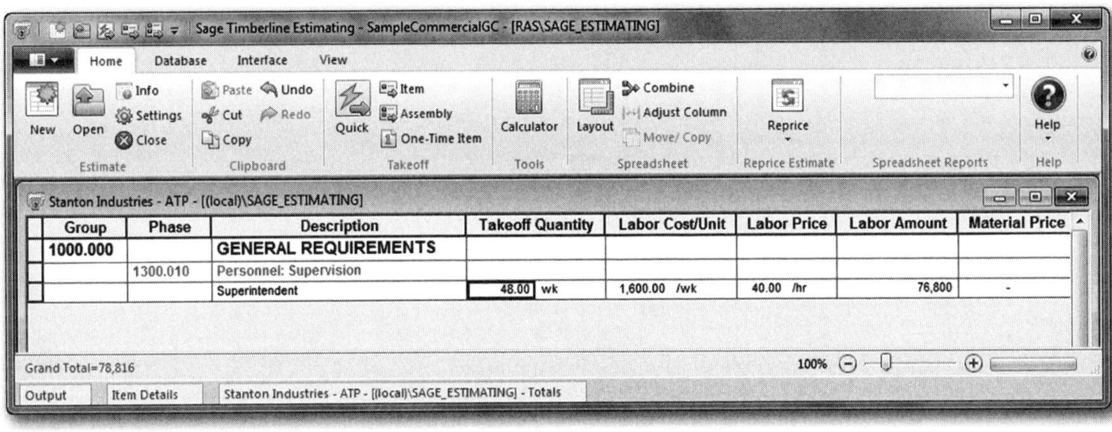

Quick Quiz®

Quick Quiz®

Refer to the CD-ROM for the Quick Quiz® questions related to chapter content.

Key Terms

Illustrated Glossary

- *access road*
- *bond*
- *commissioning*
- *completion bond*
- *cost allocation*
- *cost-plus basis*
- *depreciation*
- *equipment leasing*
- *equipment rental*
- *license*
- *mobilization*
- *percentage basis*
- *performance bond*
- *permit*
- *profit*
- *street bond*
- *time and material basis*

Web Links

Web Links

Minority Business Development Agency
www.mbda.gov

U.S. Department of Commerce
www.commerce.gov

U.S. Department of Labor
www.dol.gov

U.S. Social Security Administration
www.ssa.gov

Contracting Requirements and General Requirements 4

Review Questions

Estimating

______________ **1.** Information related to the structure of the contracting requirements on a construction project is included in the ___ section of the construction documents.

______________ **2.** ___ is the accounting practice of reducing the value of equipment by a standard amount each year.

______________ **3.** Equipment ___ is the use of equipment for a fixed cost per a given unit of time.

A. rental
B. leasing
C. allocation
D. none of the above

______________ **4.** Equipment ___ is the use of equipment based on a contract that stipulates a set amount of time and set cost for the equipment use.

A. rental
B. leasing
C. allocation
D. none of the above

______________ **5.** Cost ___ is the designation of overall company equipment costs to particular pieces of equipment.

______________ **6.** A(n) ___ road is a temporary road installed by a general contractor to facilitate transportation of material, equipment, and tradesworkers to and from a job site.

______________ **7.** ___ costs are expenses that are not attributable to any one specific operation.

______________ **8.** A(n) ___ is an insurance agreement for financial loss caused by the act or default of an individual or by some contingency over which the individual has no control.

______________ **9.** A(n) ___ is a privilege granted by a state or other government agency to an individual or company to perform work or provide professional services.

______________ **10.** A(n) ___ is written permission granted by the governmental authority having jurisdiction over the construction process.

______________ **11.** ___ is the amount of money cleared in excess of project and company expenses.

T F **12.** Interior finish materials such as suspended ceilings, flooring, and gypsum board are taken off and priced based on square feet or square yards.

T F **13.** Earthwork is calculated in cubic yards of material to be moved or supplied.

______________ **14.** Items such as doors, windows, and fixtures are taken off by counting ___ installations.

______________ **15.** Labor pricing is determined based on the labor ___.
- A. quantity
- B. rate
- C. allocation
- D. none of the above

______________ **16.** A(n) ___ bond is a short-term insurance policy ensuring the owner that a construction project will be fully completed and free from encumbrances and liens when turned over to the owner.

______________ **17.** The ___ is a method of determining profit in which a percentage is added to the final estimated cost.

T F **18.** Labor rates include costs such as wages, taxes, and fringe benefits.

T F **19.** The project owner provides light and electrical power for the various trades and subcontractors under the jurisdiction of the contract.

Short Answer

1. Identify items in the General Requirements section of the specifications that may affect the final estimate for a project.

____

____

____

____

____

____

2. Describe the technology that can be used by field staff to communicate with staff in the main office.

____

____

____

____

3. List items included in labor rates.

____

____

____

Estimating Labor Units

Refer to the wire installation labor unit chart to answer the following questions.

T F **1.** Aluminum conductors can be installed in less time than comparable copper conductors.

______________ **2.** ___ hours of labor are required to install 500′ of No. 4 Cu USE conductors.

T F **3.** No. 14 through No. 8 Cu TW conductors require the same number of labor units as No. 14 through No. 8 Cu THHN conductors.

T F **4.** The labor units chart may be used for determining labor units for installing solid or stranded conductors.

T F **5.** Wire installation labor units are expressed per 100′ in the chart.

WIRE INSTALLATION LABOR UNITS*							
Size	Copper					Aluminum	
	TW	THW	THHN	XHHW	USE	THW	XHHW
14	5.5	6	6	6	6	—	—
12	6.5	7	7	7	7	—	—
10	7.5	8	8	8	8	—	—
8	9.5	9	9	9	9	—	—
6	11.5	10	10	10	10	9	9
4	13.5	12	12	12	12	10	10
2	15.5	13	13	13	13	11	11
1	—	14	14	14	14	12	12
1/0	—	15	15	15	15	13	13
2/0	—	16	16	16	16	14	14
3/0	—	17	17	17	17	15	15
4/0	—	18	18	18	18	16	16
250	—	20	20	20	20	18	18
300	—	23	23	23	23	19	19
350	—	24	24	24	24	20	20
400	—	25	25	25	25	21	21
500	—	26	26	26	26	22	22
600	—	30	—	30	30	23	23
750	—	32	—	32	32	25	25
1000	—	46	—	36	36	30	30

* hours per thousand feet for one conductor, solid or stranded, 600 V

Contracting Requirements and General Requirements 4

Activities

Activity 4-1—Ledger Sheet Activity

Refer to Print 4-1 and Quantity Sheet No. 4-1. Take off the linear feet of welded-frame tubular scaffold required for the north and east walls of the Commons Building.

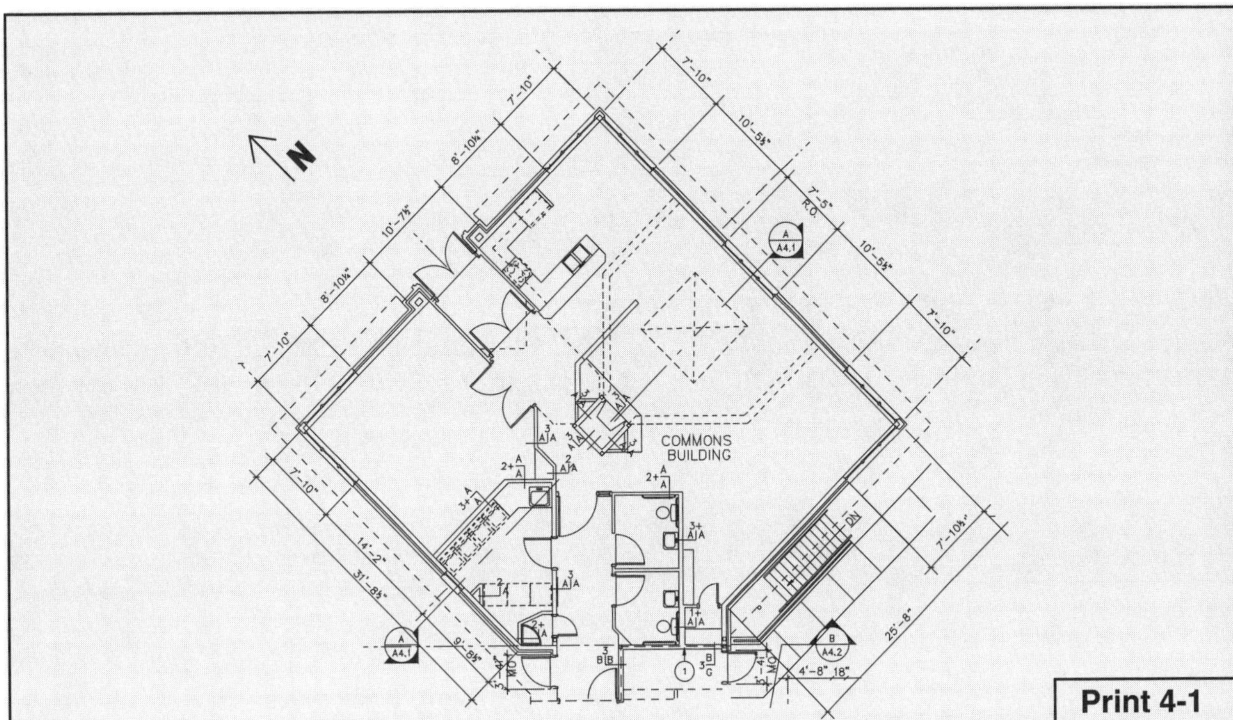

Print 4-1

QUANTITY SHEET

Project: ______________ Sheet No. 4-1
Estimator: ______________ Date: ______________
Checked: ______________

No.	Description	Dimensions L	W			Unit		Unit		Unit		Unit
	Scaffold											
	North wall											
	East wall											
	Total											

Activity 4-2—Spreadsheet Activity

Refer to the cost data shown here and on Estimate Summary Sheet No. 4-2 on the CD-ROM. A new construction site requires a chain-link fence, an access road, and an office trailer. A 6′ high chain-link fence will enclose the 180′ × 100′ property. The access road will be 18′ wide by 150′ long and filled with 4″ of gravel. An 8′ × 20′ office trailer is needed for the project for a total of 2 yr. Based on the following cost data, determine the total material, labor, and equipment cost for the chain-link fence, access road, and office trailer.

COST DATA				
Material	**Unit**	**Material Unit Cost***	**Labor Unit Cost***	**Equipment Unit Cost***
Chain-link fence	lf	3.71	5.04	—
Access road, gravel	sq yd	1.21	1.68	0.34
Office trailer rental per month	ea	147	—	—

* in $

Activity 4-3—Sage Timberline Estimating Activity

Create a new estimate and name it Activity 4-3. Perform a quick takeoff of items **1300.010 10 Superintendent**, **1510.010 10 Temp Electricity**, **1540.010 10 Tools & Equipment**, and **1740.010 10 Current Cleanup**. These items are required for 4 months. Print a standard report.

5 Earthwork

Key Concepts

- Site preparation variables that affect the final bid and estimating procedures for a construction project include the types of soil, rock, and water at and below the surface of the property, the natural or existing contours of the property, existing structures, and the possible existence of hazardous materials.
- Earthwork estimating calculations include the amount of excavation, cut and fill, final grading, and trenching required on a job site.
- Excavation costs are calculated based on the cubic yard.
- Quantities calculated for demolition estimation include the sizes and types of structures to be removed, access to the demolition site, protection of surrounding structures, and structural analysis of the structure to be demolished to determine the demolition sequence and possible need for temporary shoring.
- Items to be considered for a shoring estimate include the linear feet, type, and required strength of the shoring and the depth of the excavation.
- Estimate considerations for site utilities include the linear feet of any required excavation, depth of the excavation, existing job-site conditions, utility connection requirements, and inspection.
- Quantity takeoff for foundation materials depends on the materials used and the site conditions.
- The primary information needed for pavement takeoff is the exterior dimensions of the paved area and the dimensions of any voids such as islands.
- Landscaping costs include finish grading, sodding, seeding, excavation for trees and shrubs, landscaping materials, and landscaping protection.

Introduction

Preparation of a job site includes analyzing the existing site at grade level and below grade level, comparing the present site with the final site requirements, and determining the materials and work needed to change the property from the existing to the planned condition. Soil engineers determine the load-bearing capacities and quality of the soil below grade. A structure that cannot be supported by surface soil requires footings and a foundation. After the majority of the structure is constructed, the surrounding areas are finished to provide the appropriate access and landscape design.

Portions of five divisions of the MasterFormat™ are related to earthwork: Division 02—Existing Conditions; Division 31—Earthwork; Division 32—Exterior Improvements; Division 33—Utilities; and Division 35—Waterway and Marine Construction. While portions of other divisions may pertain to general exterior sitework functions, Divisions 02, 31, 32, 33, and 35 are the primary areas pertaining to preparation and completion of the exterior areas of a construction project.

SITE PREPARATION MATERIALS

Site preparation variables that affect estimating procedures and the final bid for a construction project include the types of soil, rock, and water at and below the surface of the property, the natural or existing contours of the property, existing structures, environmental regulations and requirements, and the possible existence of hazardous materials. Factors involved in the preparation of a job site include analysis of the existing site at grade and below grade, compatibility of the present site with the final site requirements, and the materials and work processes necessary to change from the existing to planned conditions.

An estimator calculates the materials and amount of work necessary to make the planned changes. The surface and subsurface materials taken into consideration in this portion of the estimate include soil, rock, water, concrete, chemicals, foundation materials, plant materials, and paving systems. Various existing and planned subsurface and surface conditions impact job costs.

Soil and Aggregate

Soil and aggregate include materials found at grade (surface) and below grade (subsurface) of a job site. **See Figure 5-1.** Materials found at a job site include rocks, boulders, gravel, sand, clay, silt, and soil. Solid rock typically provides a stable bearing foundation. Decayed rock may be compact and hard or fully decayed and soft. Loose rock has been detached from the rock layer in which it was originally formed.

SOIL								
Division	**Symbols**			**Description**	**Value as Foundation Material**	**Frost Action**	**Drainage**
	Letter	**Graphic**	**Color**				
Gravel and gravelly soils	GW		Red	Well-graded gravel or gravel-sand mixture; little or no fines	Excellent	None	Excellent
	GP		Red	Poorly graded gravel or gravel-sand mixture; little or no fines	Good	None	Excellent
	GM		Yellow	Silty gravels, gravel-sand-silt mixture	Good	Slight	Poor
	GC		Yellow	Clayey gravels, gravel-clay-sand mixtures	Good	Slight	Poor
Sand and sandy soils	SW		Red	Well-graded sands or gravelly sands; little or no fines	Good	None	Excellent
	SP		Red	Poorly graded sands or gravelly sands; little or no fines	Fair	None	Excellent
	SM		Yellow	Silty sands, sand-silt mixtures	Fair	Slight	Fair
	SC		Yellow	Clayey sands, sand-clay mixtures	Fair	Medium	Poor
Silts and clays	ML		Green	Inorganic silts, rock flour, silty or clayey fine sands, or clayey silts with high plasticity	Fair	Very high	Poor
	CL		Green	Inorganic clays of low to medium plasticity, gravelly clays, silty clays, lean clays	Fair	Medium	Impervious
	OL		Green	Organic silt-clays of low plasticity	Poor	High	Impervious
Highly organic soils	Pt		Orange	Peat and other highly organic soils	Not suitable	Slight	Poor

Figure 5-1. Various subsurface materials have different properties.

Rock is classified into three general categories including soft rock such as shale, medium-hard rock such as slate, and hard, sound rock such as schist or other metamorphic bedrock. Subsurface exploration should be performed and the resulting information should be studied to determine if rock is present under a site because it may result in additional excavation costs if removal is necessary. For estimating purposes, rock is classified as soft rock, medium-hard rock, and hard, sound rock.

A *boulder* is a rock that has been transported by geological action from its formation site to its current location. *Gravel* is pieces of rock that are smaller than boulders and larger than sand. *Sand* is loose granular material consisting of particles that are smaller than gravel but coarser than silt. Sand is classified as fine, medium, or coarse based on grain size. *Silt* is fine granular material that has been produced from the disintegration of rock. *Clay* is natural mineral material that is compact and brittle when dry but plastic when wet.

Soil is measured for its depth and compaction at the job site and indicated on various geotechnical reports included in the construction documents. Soil is classified into four primary types: light soil, medium soil, heavy soil, or hardpan. *Light soil* is soil that is easily shoveled by hand without the aid of machines. Light soil includes gravel and sand.

Medium soil is soil that can be loosened by picks, shovels, and scrapers. Medium soil includes cohesive soil such as clay and adobe. When clay is encountered, an excavation is usually made several inches deeper than for sand, sandy soil, or gravel to allow for fill materials, such as crushed gravel, to provide a stable base.

Heavy soil is soil that can be loosened by picks but is hard to loosen with shovels. Equipment such as a backhoe is used to move heavy soil. Heavy soil includes compacted gravel, small stones, and boulders.

Hardpan is a hard, compacted layer of soil, clay, or gravel. Hardpan includes boulders, clay, cemented mixtures of sand and gravel, and other substances difficult to loosen with a pick. Light blasting is required to loosen hardpan.

Angle of Repose. The *angle of repose* is the slope that a material maintains without sliding or caving in. Different soil types vary greatly in their ability to remain undisturbed and in place at the edge of an excavation. An *excavation* is any construction cut, cavity, trench, or depression in the earth's surface formed by earth removal. Clay soil, unless subjected to the action of running water, can stand almost vertically at the edge of an excavation with a high angle of repose. Soil with a high sand content requires a low angle of repose, resulting in long sloping bank along the excavation. **See Figure 5-2.**

ANGLE OF REPOSE			
Soil	**Dry***	**Moist***	**Wet***
Sand	20 to 35	30 to 45	20 to 40
Earth	20 to 45	25 to 45	25 to 30
Gravel	30 to 48	—	—
Gravel, sand, and clay	20 to 37	—	—

* degrees from horizontal

Figure 5-2. The angle of repose for various soil types affects excavation estimates.

Estimators must take the angle of repose into account when calculating the amount of material to be excavated and when determining the need for temporary shoring during excavation and construction activities. The depth of the excavation, soil type, angle of repose, adjoining structures, and safety requirements are used to determine shoring needs. Various government regulations, such as OSHA or MSHA regulations, may apply to shoring requirements for various depths of excavation.

Soil in its natural state is generally closely compacted. When soil is disturbed by excavation and loaded or piled, its volume increases due to swell. *Soil swell* is the volume growth in soil after it is excavated. Soil swell can increase the volume of soil by up to 50%. **See Figure 5-3.**

SOIL SWELL		
Soil	**Before Excavation***	**After Excavation***
Earth clay	100	125
Sand and gravel	100	150
Broken stones	100	150
Free stone	100	125
Rock	100	150

* in percent of initial volume

Figure 5-3. Soil swell adds to the cubic yard calculations for excavation. Soil swell must be considered when determining the number of truckloads of soil excavated from a job site.

OSHA 1926 Subpart P—*Excavations* summarizes the safety requirements for excavations, such as sloping, shoring, or shielding requirements.

Geotextiles. A *geotextile material* is any synthetic material in the form of a fabric sheet that stabilizes and retains soil or earth in position on slopes or in other unstable conditions. Geotextile material is applied after rough grading is complete. Geotextile material may be covered with a thin layer of soil, rock, or other material to hold it in place. Geotextile material can be used to divert groundwater and provide subsurface drainage and erosion control.

Hazardous Materials. A *hazardous material* is a material capable of posing a risk to health, safety, or property. Hazardous material handling includes the safe removal and proper disposal of any identifiable hazardous materials on a job site. Soil analysis may indicate the existence of hazardous materials on the construction site.

Hazardous materials include solid waste, toxic chemicals, and radiation. The Environmental Protection Agency (EPA) has identified specific hazardous waste materials and classified them according to corrosivity, ignitability, reactivity, and toxicity. Special remediation steps must be taken to properly contain or safely remove hazardous material when found. *Remediation* is the act or process of correcting situations where hazardous materials exist and returning a site to environmentally safe conditions. The locations of contaminated soil areas and locations for installation of geotextiles and groundwater monitoring wells are indicated on the site plan and in the specifications. **See Figure 5-4.**

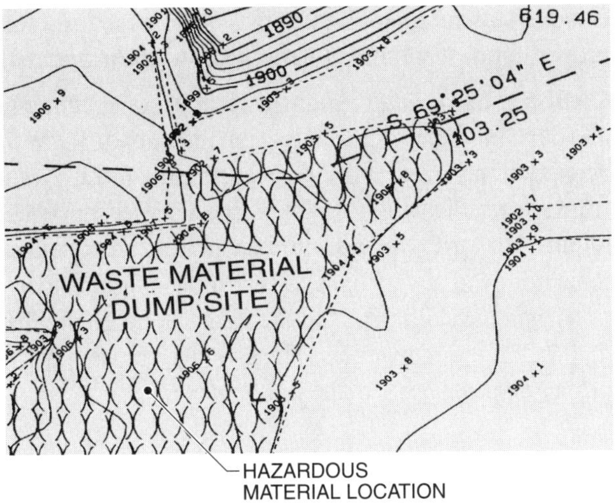

Figure 5-4. Site plans show locations of hazardous materials to be monitored or removed.

Soil Amendments. In some cases, existing soil characteristics do not allow for achievement of specified soil stabilization requirements. A *soil amendment* is a material added to existing job site soil to increase its compaction and ability to maintain the desired angle of repose or slope at the completion of construction. Soil amendments such as clay, slurry, and other locally available materials may be mixed into existing soil to achieve the required results. *Slurry* is a mixture of water and fine particles such as cement that improves soil stability.

SITE PREPARATION METHODS

Soil engineers determine the bearing capacities and quality of the soil below grade by drilling a series of exploratory holes (test borings) and taking the appropriate measurements and readings related to the drilling. Rock removal may require heavy blasting. Any removal of trees and shrubs is planned, as well as protection of trees and shrubs that are to remain. Operating engineers grade, remove, add, and compact the soil on the site in preparation for new construction. Below-grade and at-grade drainage systems are then laid out and installed.

> The International Union of Operating Engineers (IUOE) represents heavy equipment operators, mechanics, and surveyors in the construction industry.

Subsurface Investigation

Subsurface investigation is the analysis of all materials below the surface of a job site to a predetermined depth. Soil engineers analyze samples taken from test borings to determine the soil's composition. A report from the soil engineer to the owner and architect describes the various subsurface materials at specific points on the construction site. Determination of the composition of the subsurface materials provides valuable information for structural engineers when selecting and designing foundation systems.

Structural engineers calculate the live and dead loads that the structure will place on the bearing soil at the job site. A *live load* is the total of all the dynamic loads a structure is capable of supporting including wind loads, human traffic, and any other imposed loads. A *dead load* is a permanent, stationary load composed of all construction materials, fixtures, and equipment permanently attached to a structure. Loading information is compared to the soil samples to determine the steps required to ensure that the footings and foundation provide adequate support for all live and dead loads.

Costs for obtaining subsurface information and performing a soil analysis are included in overall cost estimates. Foundation support information is used by estimators when estimating any necessary drilling or excavation work. Information relevant to the estimating process includes the number, depth, and diameter of holes to be drilled and rock and soil types.

ELE International, Inc.

The California Bearing Ratio (CBR) test is a type of subsurface test that was developed to measure the load-bearing capacity of soils used for road building.

Core Drilling. *Core drilling* is the process of making holes in the ground to retrieve soil samples for analysis. Soil engineers determine a layout for drilling at the job site. A core drill is used to drill holes into the earth at predetermined points. The layers of soil, rock, and other subsurface materials are measured for depth and analyzed. The holes are often made to determine the depth at which a solid bearing is reached.

Groundwater. *Groundwater* is water present in subsurface material. A *well point* is a pipe with a perforated point that is driven into the ground to allow for the removal and pumping of groundwater. Well point pumps are used to dewater an excavation by pumping the water up vertical (riser) pipes. Horizontal pipes are connected to the vertical pipes to discharge the water at a location away from the work area. This maintains a water level (water table) that is low enough to perform any necessary excavation. The piping may be permanent or temporary depending on the type of construction.

Groundwater can create construction or structural problems unless the water flow is properly controlled. Improper handling of groundwater can cause sand excavations to be much more costly than rock excavations. Removing groundwater is a cost included in the estimate. Dewatering contractors bid on dewatering systems by calculating items such as cost for driving well points, installing and maintaining vertical and horizontal pipes, maintaining and fueling pumps, and labor involved in operating the system. Removal of a temporary dewatering system may also be included in the dewatering cost.

Demolition

A job site may have existing structures that need to be demolished and removed. *Demolition* is the organized destruction of a structure. The cost of any demolition is included in the bid along with the cost of removing and disposing of the debris. Demolition costs include heavy equipment operation, equipment ownership or leasing, labor, and overhead. Removal costs include loading of trucks, truck rental or leasing, labor, dumping fees, and overhead.

Demolition contractors may need to analyze the structure being removed to determine if hazardous materials exist in the structure, in which case additional costs will be incurred to comply with the appropriate regulations for proper hazardous material removal. For example, if asbestos insulation or roofing materials are present in an existing structure, additional costs will be incurred for proper removal and disposal of the asbestos materials.

Clearing. *Clear and grub* is the removal of vegetation and other obstacles from a job site in preparation for new construction. Trees and large vegetation are removed from areas indicated on the site plan. A site visit may be required to verify the types and quantities of vegetation to be removed and actual site conditions. Various locations may have differing requirements pertaining to the disposal of these materials. The estimator must include proper disposal costs in the bid.

Shoring

Shoring is a system of wood or metal members used as temporary support for sides of excavations or construction components such as concrete formwork. **See Figure 5-5.** Unstable soil, deep excavations, nearby structures, or other factors may create the need for temporary or permanent shoring at a job site. Shoring may be accomplished with removable boxes or jacks, various types of piles, grouting and other soil stabilization techniques, or soil and rock anchors.

Figure 5-5. Shoring is wood or metal members used to temporarily support formwork or other structural components.

Estimators must ensure that the shoring system meets industry and government agency requirements for the application selected. Shoring must be sufficient to provide complete protection for adjoining structures, workers, and materials based on the depth of the excavation and the materials being supported.

Sheet Pile. A *pile* is a structural member installed in the ground to provide vertical and/or horizontal support. A *sheet pile* is an interlocking vertical support driven into the ground with a pile-driving rig to form part of a continuous structure. Sheet piles are designed to retain the horizontal thrust of the earth that could cause excavation walls to collapse. Vertical retaining walls of wood or steel may be used when excavating adjacent to an existing structure or where protection from a possible cave-in is necessary.

The most common forms of sheet piles include wood planking (lagging) placed horizontally between vertical I-beams driven into the ground or interlocking steel sheets. Precast concrete sheet piles may be driven when sheet piling is to become a permanent part of a structure.

Where the earth is soft enough to permit driving, a pile-driving rig is used to drive piles to the point of refusal. **See Figure 5-6.** This prevents the sheet piles from being moved by horizontal pressure or lateral thrusts. Sheet piles may also be installed in a previously excavated hole. Sheet piles installed in a previously excavated hole are braced, shored, anchored to the bank, or backfilled.

Figure 5-6. Sheet piles provide support around the perimeter of an excavation.

Temporary Soil Stabilization. *Temporary soil stabilization* is the process of holding existing surface and subsurface materials in place during the construction process. Various methods are used to hold the soil adjoining an excavation in place. Cementitious grout or slurry may be shot or pumped into adjoining areas to stabilize the soil. Grout may be shot onto the surface of banks following excavation. Holes may be drilled prior to excavation and pumped with grout or slurry. After the grout or slurry sets, excavation may take place with the bank stabilized. Where water and soil conditions permit, refrigeration systems may be used to freeze the soil to retain a stable bank.

Permanent Soil Stabilization. *Permanent soil stabilization* is the process of providing long-term stabilization of the ground at a job site through mechanical compaction or the use of infill materials. *Infill material* is material, including sand, gravel, and crushed rock, used to increase the stability of the soil.

Mechanical methods of permanent soil stabilization include dynamic compaction and vibratory consolidation. Infill materials are introduced into the soil or used to replace existing soil where existing soil conditions are such that the required compaction level cannot be reached. The gradation, quality, and final compaction of the infill material affect the final stability. Common infill materials include sand, gravel, and crushed rock with a maximum particle size of 2.5″, and concrete and soil-cement mixes.

> The production rate for placing soil stabilization materials, such as riprap or gravel, varies by material. Riprap (5 cu yd to 7 cu yd pieces) can be placed at an average of 9.4 cu yd/hr, while pea gravel can be placed at 12 cu yd/hr.

Cofferdams. A *cofferdam* is a watertight enclosure used to allow construction or repairs to be performed below the surface of the water. Cofferdams are composed of a series of sheet piles that are driven to form a cell or interlocking cells. Cofferdams provide protection for projects built below grade or below the water surface. The material inside the cell, such as soil, rock, or water, is removed to allow workers access during construction. Cofferdams are commonly used in marine construction to allow access to river and lake bottoms in the construction of bridges and dams. **See Figure 5-7.**

Figure 5-7. Cofferdams are watertight enclosures that allow foundation construction in rivers, lakes, and other marine situations.

Site Utilities

Site utilities for a project include storm sewers, sanitary sewers, water service, fire protection service, natural gas service, electrical service, and telephone service. Each site utility is identified on the site plans. **See Figure 5-8.** Site preparation for utility installations must take place prior to placement of foundations and streets to ensure full accessibility.

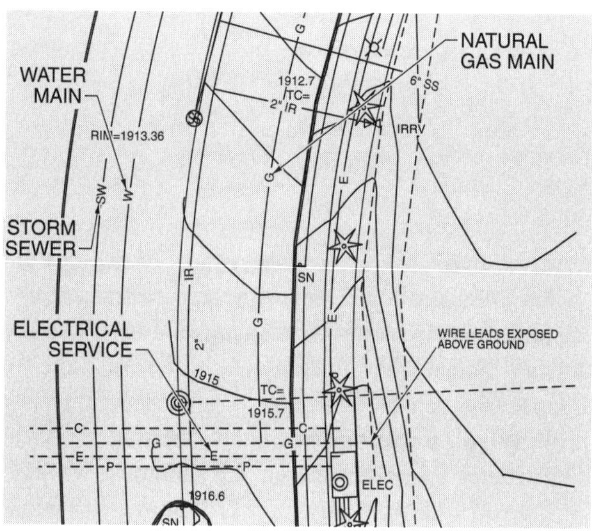

Figure 5-8. Site plans indicate the location of utilities, such as electrical service, storm sewers, natural gas mains, and water mains.

Prior to excavation, all existing site utilities must be accurately identified and located on the job site. Additional costs and construction delays can be encountered in instances where excavation activities sever utility services, such as underground telephone lines or water mains. In many areas, the utility company or a third-party contractor is required, when contacted, to mark the job site to ensure that safe excavation may proceed.

Storm Sewers. A *storm sewer* is a piping system designed to collect stormwater and channel it to a retention pond or other means of removal. Storm sewers are commonly made of precast concrete pipe. The sections of precast concrete pipe are set in place according to the prints and specifications. Excavations are made for placement of the pipe. Gravel or rock may be placed in the trench and shoring may be required. Cranes and other excavation equipment lift pipe sections and set them in place. The pipe is then covered with additional gravel and/or soil.

Retention Ponds. A retention pond is an earthen area that includes berms built to collect and retain stormwater so that it can be released at a controlled rate while improving the quality of the water. To meet stormwater release requirements, various local regulations may require the construction of retention ponds on a building site. The estimator should include costs for excavation of the retention pond, any required pond lining or soil treatment, vegetation requirements in or around the edges of the pond, and water release or overflow systems.

Sanitary Sewers. A *sanitary sewer* is a piping system designed to collect liquid and solid waste from a building and channel it to the appropriate removal or treatment system. Sanitary sewer systems may be formed of precast concrete or plastic pipe. All sanitary sewer connections must be secure at the building, throughout the piping system, and at the point of connection to the treatment system or removal piping.

Water, Fire, and Gas Mains. Piping and connection information for water supply piping, fire hydrants, and natural gas piping is shown on site plans. Water supply piping from the main service line to the structure may be copper, cast iron, or plastic. Fire hydrant piping is most commonly made of cast iron. Natural gas piping may be copper, cast iron, or plastic (polyethylene). Site plans and specifications provide information concerning water, fire hydrant, and natural gas piping. Local building and fire codes should be consulted to ensure that the specified piping is allowed. Various types of backflow control devices and check valves may be required to ensure safe installations of these utility services.

Electric and Telephone Service. Electric and telephone service may be underground or overhead. Connection points and possible transformer requirements are shown on the site plan and possibly the electrical plans and specifications. For underground installations, trenches are excavated prior to the completion of other surface construction such as pavement and sidewalks. Trenchless technology may be used to install electric and telephone cables by means of drilling through the soil without the need for further excavation. **See Figure 5-9.**

The Charles Machine Works, Inc., Manufacturer of Ditch Witch® Products

Figure 5-9. Trenchless technology may be used to install electric and telephone cables with only minor excavation at the surface.

EARTHWORK QUANTITY TAKEOFF

After the existing conditions, materials, and locations are determined, the estimator takes off the new construction information from the specifications and site plans. New construction information includes placement of new structures, utilities, surface grading, paving, curbs, walks, and landscaping. In addition, areas requiring environmental remediation are indicated on site plans and detailed in the specifications.

Site plans can be referred to as a site survey, site map, site drawings, or civil drawings. Site plans include topographic, paving, landscape, and detail drawings for drainage and piping. Site plans are often noted with the prefix C (for civil drawings) in a large set of prints. For example, drawing sheet C1.1 is the first page of the site plans.

Earthwork Estimation

Earthwork estimating calculations include the amount of excavation, cut and fill, final grading, and trenching required on a job site. Estimators should review the specifications, including the geotechnical report, and the site plans prior to beginning excavation quantity takeoff. Studying the site, topographic maps, plans, and specifications may reveal opportunities for increased efficiency in earthwork operations.

The quantity takeoff for excavation is most commonly expressed in cubic yards. The quantity takeoff may be calculated in cubic feet when the excavation requires the removal of a few inches of soil or work performed by hand. The volume of material to be removed or added must be calculated accurately for proper pricing of all costs involved.

Additional earthwork items that may be taken into account include the costs involved in backfilling, utility line excavation, and cradling. *Cradling* is the temporary support of existing utility lines in or around an excavated area to properly protect the lines during construction. Cradling is accomplished by placing timbers or shores under utility lines and jacking up the timbers or shores as the excavation goes deeper to keep the lines at their original level. Utility companies may also be contacted to install protective devices on overhead electrical lines to prevent arcing. **See Figure 5-10.**

Figure 5-10. Utility protection costs, such as electrical arc protection, are included in earthwork estimates.

Gridding. *Gridding* is the division of a topographic site map of large areas to be excavated or graded into smaller squares or grids. Grids may be drawn on an overlay sheet of tracing paper, directly on the site plan drawing, or charted on a computer screen when CAD drawings are used. The size of the grid is determined by the nature of the terrain. Grids may represent squares of 100′ on each side if the terrain is gradually sloped. Grids may represent squares of 25′ on each side if the terrain is irregular. The smaller grid squares give a more accurate calculation of the cut and fill needed for a particular site. **See Figure 5-11.**

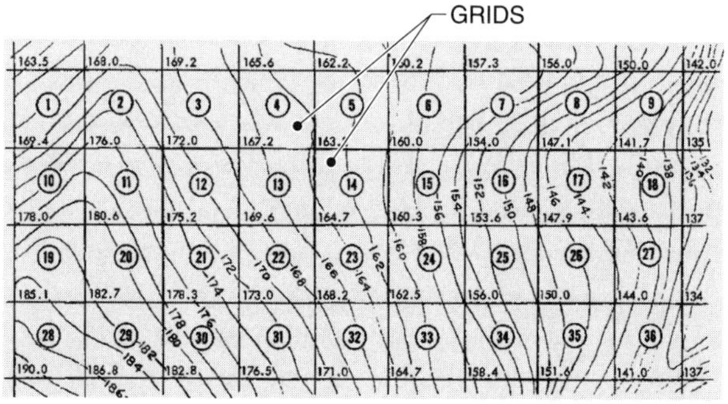

Figure 5-11. Gridding is the division of a topographical site map into small squares or grids to assist in cut and fill calculations.

The approximate elevation at each corner of each grid is established by using the nearest contour line. A *contour line* is a dashed (existing grade) or solid (finished grade) line on a site plan used to show elevations of the surface. Interpolation or averaging between adjoining contour lines may be required when the corner of a grid does not directly align with a contour line.

Cut and fill averages are calculated for each grid. The difference between the totals for the cut and fill indicate the amount of fill or cut that must be done on the site. Cut and fill numbers are entered in a ledger sheet or into a spreadsheet or estimating program. **See Figure 5-12.** The total cut or fill value is calculated and multiplied by the excavation cost. Cutting is typically represented with a negative number or a number enclosed in parentheses. Soil and rock types may also be taken into consideration when determining a final cost for excavation, with heavy soil or rock possibly requiring additional excavation time and equipment.

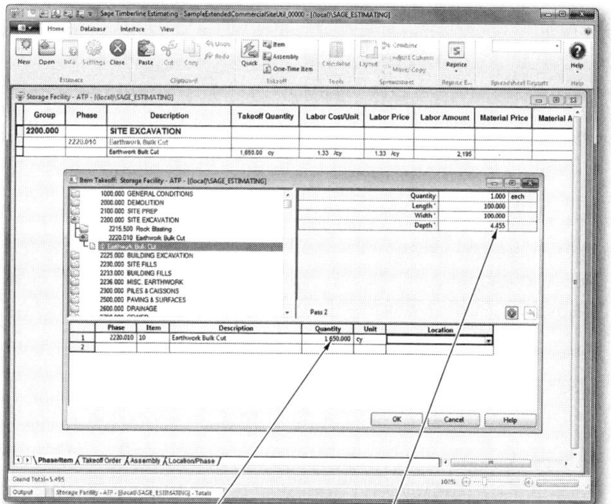

Figure 5-12. Gridding calculations determine the amount of cubic yards for cut and fill operations.

Excavation. Excavation is any construction cut, cavity, trench, or depression in the earth's surface formed by earth removal. Excavation is required for structures such as footings, foundations, basements, tunnels, ramps, exterior stairways, areaways, pipe trenches, retention ponds, and pits. General excavation includes all excavation, other than rock and water removal, that can be done by mechanical equipment or any general piece of equipment such as a bulldozer, clamshell crane, backhoe, scraper, power shovel, or loader. The use of trucks to remove the excavated material is also included in this portion of the estimate calculations.

Calculating the volume of excavation material may be accomplished using the cross-section method or the average end-area method. The cross-section method is used when the shape of the excavated area is roughly square or rectangular. The average end-area method is used when the sides of the excavation are irregularly shaped and not parallel. The estimator determines the method used based on the shape of the excavation.

When using the cross-section method to calculate the volume of excavation material, the volume excavated (in cubic yards) is calculated by determining the average depth of excavation from various known points and multiplying it by the total surface area. The excavation volume for a grid square is calculated by determining the difference between the existing elevation and the proposed elevation at each corner of the grid.

Excavation Takeoff Using the Cross-Section Method

Determine the volume of soil to be excavated (in cubic yards) from a 75′ × 75′ square grid with corners at elevations of 90.85′, 95.02′, 87.28′, and 85.13′ that require excavation to 80.5′.

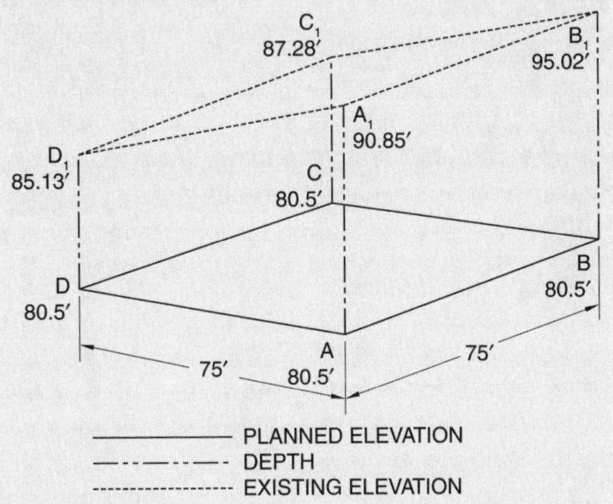

1. Compute the difference between the existing elevation and planned elevation at each corner.

difference at point A = existing elevation − planned elevation
point A = 90.85′ − 80.5′
point A = 10.35′

difference at point B = existing elevation − planned elevation
point B = 95.02′ − 80.5′
point B = 14.52′

difference at point C = existing elevation − planned elevation
point C = 87.28′ − 80.5′
point C = 6.78′

difference at point D = existing elevation − planned elevation
point D = 85.13′ − 80.5′
point D = 4.63′

2. Calculate the average excavation depth.

average excavation depth =

$$\frac{point\ A + point\ B + point\ C + point\ D}{4}$$

$$D_{ave} = \frac{10.35' + 14.52' + 6.78' + 4.63'}{4}$$

$$D_{ave} = \frac{36.28}{4}$$

$$D_{ave} = 9.07'$$

3. Determine the volume of the excavation (in cubic feet).

volume = area of square grid × average excavation depth

$V = (75' \times 75') \times 9.07'$

$V = 5625 \times 9.07$

V = 51,018 cu ft, or round to 51,000 cu ft

4. Determine the volume of the excavation (in cubic yards).

$$volume\ (cubic\ yards) = \frac{volume\ (cubic\ feet)}{27}$$

$$V = \frac{51{,}018.75}{27}$$

$$\mathbf{V = 1889.58\ cu\ yd}$$

When using the average end-area method to calculate the volume of excavation material, the area of each end is first calculated. The areas are then added and divided by 2 to obtain the average area of the ends. The average end-area is then multiplied by the excavation length to obtain the volume of the excavation in cubic feet. The cubic foot value can then be converted to its equivalent cubic yard value.

Excavation Takeoff Using the Average End-Area Method

Determine the volume of soil to be excavated (in cubic yards) from a 50′ long excavation area that has a 12′ width at one end and 6′ width at the other end. Both edges of the excavation area require 10′ of cut.

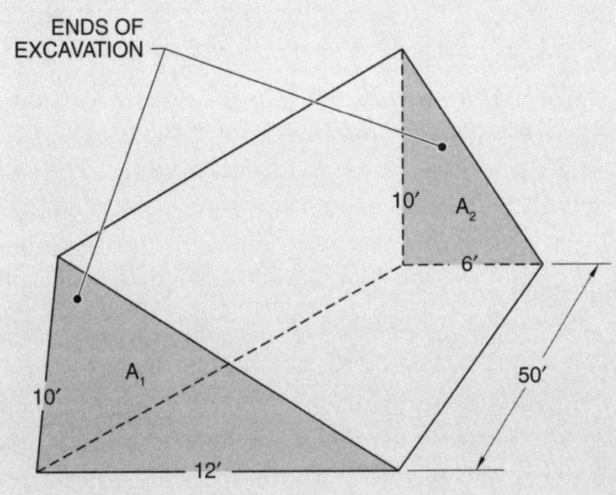

1. Determine the area of the triangular ends of the excavation.

area of triangle 1 = ½ × *base* × *height*

$A_1 = ½bh$

$A_1 = ½ \times 12' \times 10'$

$A_1 = 60$ sq ft

area of triangle 2 = ½ × *base* × *height*

$A_2 = ½bh$

$A_2 = ½ \times 6' \times 10'$

$A_2 = 30$ sq ft

2. Determine the average end-area.

$$\textit{average end-area} = \frac{\textit{area of triangle 1} + \textit{area of triangle 2}}{2}$$

$$\textit{average end-area} = \frac{60 \text{ sq ft} + 30 \text{ sq ft}}{2}$$

$$\textit{average end-area} = 45 \text{ sq ft}$$

3. Determine the volume of the excavation (in cubic feet).

volume = *excavation length* × *average end-area*

$V = 50' \times 45$ sq ft

$V = 2250$ cu ft

4. Determine the volume of the excavation (in cubic yards).

$$\textit{volume (cubic yards)} = \frac{\textit{volume (cubic feet)}}{27}$$

$$V = \frac{2250}{27}$$

$\mathbf{V = 83.33}$ **cu yd**

When a subsurface structure is to be built, the excavated area is most often excavated larger than the actual structure to provide space for tradesworkers to get to the work area for activities such as setting and stripping concrete forms. *Overdig* is the amount of excavation required beyond the dimensions of a structure to provide an area for construction activities. This additional work area must be backfilled with earth at the completion of construction. *Backfilling* is the replacing of soil around the subsurface structure after construction has been completed. An estimator must ensure that the material specified for backfill does not contain large boulders or debris that could damage the new construction. Backfilled areas must also be compacted during and at the conclusion of the backfill project. The required compaction levels are stated in the specifications.

Excavation done by hand, with specialized equipment, by blasting, or by using special methods is considered special excavation. An example of special excavation is a pipe trench required for a water line. A pipe trench is typically excavated with a trencher. A trencher is a machine that uses a toothed conveyor system to remove the soil as it digs a long, narrow groove into the earth. The groove is usually 8″ to 12″ wide and several feet deep, depending on soil conditions and the depth of trench required. Another example of special excavation is the installation of telephone, electrical, and utility poles. Pole excavation may require a vertical boring rig.

A *payline* is a measurement used by excavation contractors to compute the amount of required excavation as stated in the specifications. The payline establishes the exact amount of excavation for which the contractor is responsible. For example, the amount of excavation may be described in the specifications as follows: "The payline shall be 2′ from the top edge of the footing out, and then up on a slope as required by OSHA excavation standards for the proper soil type."

A *slope cut* is an inclined or sloped wall excavation. A slope cut is used for deep excavations in soft, shifting soil. The angle of a slope cut is determined by the angle of repose for the soil in the area according to standard safety guidelines.

Excavated earth is stockpiled on the site and represents a cost. The cost of moving earth around a job site is added to the other excavation costs, including loading and hauling. The unit cost and total cost depend on the type, size, and efficiency of the equipment used and site accessibility. Additional costs may be associated with the installation and maintenance of sedimentation collection systems around the perimeter of stockpiled soil.

When excavations are required in rock, a series of regularly spaced holes is drilled into the rock layers. Explosive charges are carefully placed in the holes. **See Figure 5-13.** The spacing, number of holes, and amount of explosives in each blasting charge hole is determined by an analysis of the type and strength of the rock to be removed. After blasting, the loose rock is removed by the appropriate excavating equipment. Blasting costs include drilling equipment, blasting caps and explosive charges, and removal of the blast material.

Figure 5-13. Blasting with blasting caps and explosive charges loosens rock for excavating.

Excavation costs are calculated based on the cubic yard. Unit cost per cubic yard of excavation, trenching, or other site preparation activity is based on expenses that include the cost of equipment ownership or rental, fuel, labor to operate machines, possible cost of transporting the equipment during the course of construction, and other equipment overhead costs such as depreciation.

The total expense (equipment, labor, transportation, and overhead) is divided by the number of cubic yards of excavation to determine the unit cost of excavation per cubic yard. The unit cost of excavation per foot may also be determined from industry resource materials or company historical records. Excavation estimate values are entered in a ledger sheet, spreadsheet, or estimating program. Maintenance of current information related to these costs in company historical records can help provide for additional accuracy in future excavation estimates for various soil types and job conditions.

Grading. *Grading* is the process of lowering high spots and filling in low spots of earth at a job site. The estimator reviews the site plans and uses a grid to determine earthwork calculations for cut and fill using the cross-section method of volume calculation. Negative numbers are used to indicate fill in areas where the planned elevation is above the existing elevation. A *zero line* is a line connecting points on a topographic map where existing and planned elevations are equal. Depending on the grid layout, some grids may require all cut or all fill, and some grids may require cut and fill. The estimator may calculate these as separate portions of the takeoff to simplify overall cut and fill calculations.

The contractor may be required to buy (borrow) fill material to bring the level to the specified grade if the amount of fill needed is greater than the amount of cut. Additional costs are incurred for purchasing the fill material and transporting it to the site. Hauling and removal of additional cut materials is also a cost if required. These costs are based on cost per cubic yard calculations derived from industry resource materials or company historical data. *Final grading* is the process of smoothing and sloping a site for paving and landscaping. Final grading quantities are taken off in square feet or square yards of surface to be graded.

Depending on the soil and surface condition, steps may be required to control the creation of dust and to minimize soil erosion during the grading process. Costs associated with this include water spray on dusty areas and installation and maintenance of temporary stormwater-erosion protection of plastic fencing or straw bales during the course of construction.

Tunneling. Tunneling costs include tunnel boring, shoring, tunnel wall lining, and removal of earthen debris. Tunnel boring costs vary depending on the material through which the tunnel is bored (rock, soil, or other material). Tunnel walls may require a steel or concrete lining, which will be shown on the prints or referred to in the specifications. Other tunnel construction costs include any required blasting, dewatering installation, maintenance, and possible removal, and worker safety provisions including air supply and personal protective equipment (PPE).

Recent developments in tunnel construction utilize systems in which a large conduit is put in place and is followed by a carrier pipe. Other trenchless technology systems allow for placement of underground piping without the need for tunneling. Costs for trenchless underground installation are commonly calculated based on the length of the installation in linear feet.

Demolition Estimation

Quantities calculated for demolition estimation include the sizes and types of structures to be removed, access to the demolition site, protection of surrounding structures, and structural analysis of the structure to be demolished to determine the demolition sequence and possible need for temporary shoring.

In addition, analysis may be required for the possible presence of hazardous materials, removal and proper disposal of these materials, calculation of the amount (in cubic yards) of demolition material to be removed, hauling distance from the site to the nearest disposal site, dumping fees, and overhead costs. Each of these items is calculated based on a study of the structure to be demolished through personal observation and the use of any existing prints. Pavement removal is taken off as the number of square yards of pavement to be removed.

In some instances, asphalt pavement or concrete materials can be ground at the worksite and recycled into new aggregate for new pavement materials or concrete structures. Cost variables for removal or recycling of pavement and aggregate materials should be considered by the estimator.

Selective demolition of individual components of a structure is taken off using a variety of methods. Quantities may be taken off in linear, square, or cubic measure, or components may be taken off as individual units. Selective demolition takeoff is a specialized area that requires experience and expertise.

Implosion is used to demolish large structures when economically feasible. Structural engineers analyze the existing structure and determine locations for the placement of explosive charges. Certain structural elements may be weakened or strengthened prior to activating the explosive charges to control the direction of debris movement during implosion and to ensure a safe collapse of the structure. Implosion contractors specialize in performing this analysis and in estimating costs for this type of work.

Costs include the structural analysis, explosive materials, labor for structural modifications prior to explosive placement and blasting, debris removal or recycling, and overhead. Due to the specialized nature of this work, implosion contractors use estimating records and information that is custom-designed for this type of work.

Clearing and Grubbing. Clearing and grubbing costs are determined by the acre (hectare for metric calculations). Local environmental regulations concerning burning, burying, or hauling of the removed vegetation are key elements in determining costs. Costs include removal equipment, labor, and disposal costs. Provisions may also need to be made for a temporary erosion control fence or straw bales to prevent loose soil from being washed into the groundwater supply.

Hazardous Material Handling. Additional costs are incurred when hazardous materials must be removed or remediated. Removal and remediation must be performed by a company that meets all applicable local, state, and federal regulatory requirements for licensing, worker training, and recordkeeping.

Costs included in hazardous-material handling include meeting licensing and training requirements, owning and operating specialized equipment for removal and cleanup, providing specialized PPE for workers, such as respirators and hazardous material suits; additional fees associated with encapsulating and safely transporting hazardous materials; and fees required for safe disposal of removed material. All of these costs vary depending on the severity of contamination and the location of the job site. Specifications and geotechnical reports are reviewed by the estimator to determine where these costs will be allocated.

Shoring Estimation

Items to be considered for a shoring estimate include the linear feet, type, and required strength of the shoring and the depth of the excavation. Small excavations may be shored by the installation of trench boxes. **See Figure 5-14.** A *trench box* is a reinforced wood or metal assembly used to shore the sides of a trench. Trench boxes are set into an excavation by a crane to prevent worker injury and equipment damage in case of a cave-in. Costs for trench boxes include the ownership or rental costs and installation costs. Trench boxes are typically set in place and removed with a crane. Costs vary for pile installations and for soil stabilization.

Piles. Items considered during takeoff of piles used for shoring include the subsurface analysis to determine the required pile depth, the linear feet of piles to be installed, the type of piles (steel I-beam and wood lagging, steel sheet, or precast concrete), and site access. Vibratory systems may also be used in place of piles. The linear feet of piles and the pile depth determine the pile quantities.

Figure 5-14. Small excavations may be shored temporarily with trench boxes.

For steel S-shapes (American Standard I-beam) and wood lagging, the S-shapes (beams) are driven into the ground with a pile-driving rig, and wood lagging is placed between the flanges of the S-shapes. *Lagging* is horizontal planks stacked on edge to support the earth on the side of an excavation. For steel S-shapes and lagging, the number, length, and type of beams as well as the length and square feet of lagging materials are determined. For steel sheet piles, the number of pieces, lengths, and designs are determined. A variety of steel sheet pile designs are available, depending on the strength requirements. **See Figure 5-15.**

For precast concrete piles, the design and length of the pile is calculated based on specifications and linear feet of pile required. In addition to material costs, equipment and labor costs include a crane outfitted with the proper type of pile-driving hammer and lead system, and operating engineers and labor necessary for transporting, lifting, setting, and driving the pile to the proper depth. Company historical data is commonly used for these types of estimating calculations.

For vibratory systems, holes are drilled and filled with compacted stone materials that are vibrated into place in such a manner as to match the bearing capacity of more traditional driven piling. Contractors that specialize in vibratory systems provide bids based on the soil analysis, number of piles to be provided, and the depth and diameter of the required compacted piles.

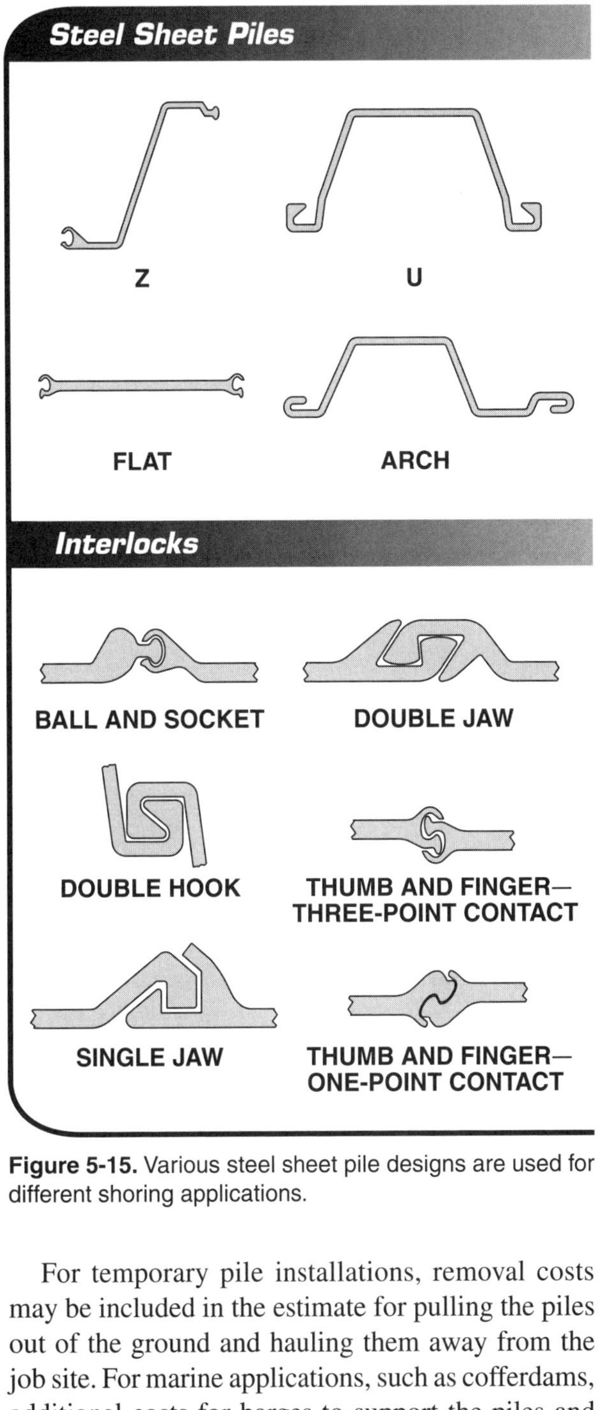

Figure 5-15. Various steel sheet pile designs are used for different shoring applications.

For temporary pile installations, removal costs may be included in the estimate for pulling the piles out of the ground and hauling them away from the job site. For marine applications, such as cofferdams, additional costs for barges to support the piles and pile-driving equipment, support boats to move labor and materials between the shore and the marine installation location, and other access and anchoring equipment is required and included in the estimate.

Soil Stabilization. Temporary soil stabilization may be required to hold an excavated bank in place depending on the angle of repose. Estimators calculate the linear feet of area to be stabilized and the depth of any necessary excavation. When chemicals or slurry are used for temporary or permanent soil stabilization, costs include chemical material and application, equipment and labor costs for drilling of slurry holes, and material costs for concrete slurry.

When excavated banks are shot with a surface covering of cementitious materials, costs include the ownership or rental of the shotcrete equipment, labor, and concrete material. Cementitious materials remain in place after construction is completed. Costs of mechanical stabilization include equipment costs for rental, leasing, or owning and maintaining, and labor costs for operation of the equipment.

Refrigerant systems for soil stabilization are complex and specialized. Estimating calculations for refrigerant systems include installation and removal of piping, cost of the refrigerant medium, and maintenance and operation of pumping equipment.

Site Utility Estimation

Estimate considerations for site utilities include the linear feet of any required excavation, depth of excavation, existing job site conditions, utility connection requirements, and inspection. For piped utilities, such as stormwater and sanitary sewers, potable and utility water, and natural gas, the pipe, pipe diameter, and required fittings and connections are taken off for the estimate. These calculations are commonly made using the linear feet of each pipe and the individual fitting requirements. Some allowances may be included in the estimate for excavations in areas where existing utilities and other subsurface conditions already exist and could create additional excavation expenses.

Utility Excavation. Excavation takeoff for site utilities is based on the linear feet of trenching required to install the utility. The existing condition of the job site at the time of the utility installation can affect the amount of trenching required and installation accessibility. It is optimum for site utilities to be installed immediately after preliminary clearing and grubbing and after site grading and compaction to avoid conflicts with other activities scheduled later in the construction process. Pipe placed in areas that have not reached proper compaction levels may shift and result in failure of the excavation and final utility installation. A site visit and coordination with other excavation contractors is essential to properly plan and determine estimates for site utility excavations.

Connection. Estimators base utility pipe and connection information on the linear feet of pipe shown on the prints, connection fittings and related connection labor expenses, and local building code requirements. Inspectors check connections prior to final approval of the completed work. In cases such as connections to water service, telephone service, and electrical service, crews from the utility company may make the final connection. Connection fees for these services must be included in the estimate.

> Site utility excavation typically includes trenching operations around a job site. Two factors in determining trench excavation costs are soil type and trench width. Based on an 18″ wide trench, 55 lf of light soil can be excavated per hour using a 55 hp wheel loader with a backhoe attachment, while only 30 lf of heavy soil can be excavated per hour. As the trench width increases, the linear feet of trench that can be excavated in one hour decreases.

FOUNDATION MATERIALS AND METHODS

A structure that cannot be fully supported by surface soil requires a footing and foundation. A *footing* is the section of a foundation that supports and distributes structural loads directly to the bearing surface of the soil, piles, or rock below the structure. A *foundation* is the primary support for a structure through which all imposed loads are transmitted to the footings or the earth. Footings and foundations hold the structure in place and provide stability. Foundation systems and materials included in the foundation portions of the CSI MasterFormat™ include deep foundation systems consisting of piles in Titles 31 60 00, 31 62 00, and 31 63 00, and cast-in-place concrete footings and walls in Titles 03 30 00 and 03 31 00.

Foundation materials and methods are described in the specifications and shown on site plans, foundation plans, and detail drawings. Specifications indicate the engineering requirements of the foundation support system and the materials to be installed. **See Figure 5-16.** Site plans show foundation dimensions, locations, and quantities. Detail drawings show items such as the diameter of reinforced concrete piles and steel reinforcement, and footing and foundation wall widths.

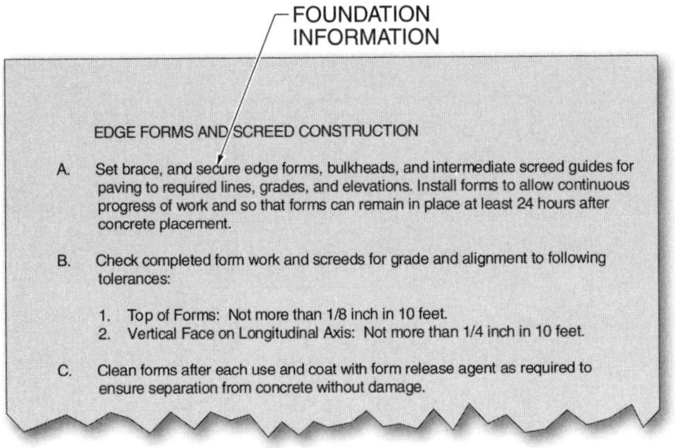

EDGE FORMS AND SCREED CONSTRUCTION

A. Set brace, and secure edge forms, bulkheads, and intermediate screed guides for paving to required lines, grades, and elevations. Install forms to allow continuous progress of work and so that forms can remain in place at least 24 hours after concrete placement.

B. Check completed form work and screeds for grade and alignment to following tolerances:

1. Top of Forms: Not more than 1/8 inch in 10 feet.
2. Vertical Face on Longitudinal Axis: Not more than 1/4 inch in 10 feet.

C. Clean forms after each use and coat with form release agent as required to ensure separation from concrete without damage.

Figure 5-16. Specifications indicate the engineering requirements of the foundation support system and the materials to be installed.

Piles

A *caisson* is a large-diameter cast-in-place concrete pile created by boring a hole and filling it with concrete. Piles or caissons may be required by the structural engineer and architect for deep foundations where soil bearing is poor. Piles include driven piles and bored piles.

Driven Piles. A *driven pile* is a steel or wood member that is driven into the ground with a pile-driving rig. Driven piles include pipe piles, HP-piles, timber piles, and precast concrete piles. A *pipe pile* is pile made of steel pipe that is driven into the ground and is supported by friction against the earth. A conical point is placed on the tip of the pipe to push the earth aside while the pile is being driven and prevent the pipe from filling with soil during driving. The pipe is filled with concrete after the proper depth is reached.

An *HP-pile* is a steel HP-shape (beam) driven into the ground and is supported by a subsurface layer of earth. A *timber pile* is a pile made of tapered, treated timber poles driven into the ground to a predetermined point of refusal. A *precast concrete pile* is a pile cast with reinforcing steel and is designed to withstand the impact of a pile-driving hammer during installation. Precast concrete piles are commonly square in cross section. All driven piles are designed to be cut off at a given elevation after the desired amount of foundation support is obtained.

A pile is considered defective if it is driven out of position or bent along its length.

Bored Piles. A *bored pile* is a foundation support formed by drilling into the ground to a predetermined depth and diameter and filling the hole with concrete. Bored piles are usually stepped to various diameters that are specified for a certain depth. Bored piles are also referred to as auger cast piles.

Cast-in-place concrete pile sides are formed by the surrounding soil, by a casing that is removed as concrete is placed, or by steel pipe. For concrete placed into the surrounding soil, a hole is drilled to a specified depth, reinforcing steel is set in place, and the hole is filled with concrete. The bottom of the pile may be belled out to create an enlarged bearing surface where additional soil bearing is needed. Metal sleeves are set into the hole as drilling proceeds in areas where the sides of the drilled hole are not stable enough to remain in place until concrete is placed. Reinforcing steel is hoisted into the drilled hole and secured in place during concrete placement. The metal sleeve is removed immediately after the concrete is placed in the hole.

Pile Caps. A *pile cap* is a large unit of concrete placed on top of a pile or group of piles. A pile cap distributes the load of the structure to a series of individual piles. Dimensions for pile caps are shown on structural drawings and include the width, length, depth, and reinforcing requirements. **See Figure 5-17.** Pile caps may be very large and require many cubic yards of concrete.

Portland Cement Association

Formwork for a pile cap is installed around the driven pipe piles.

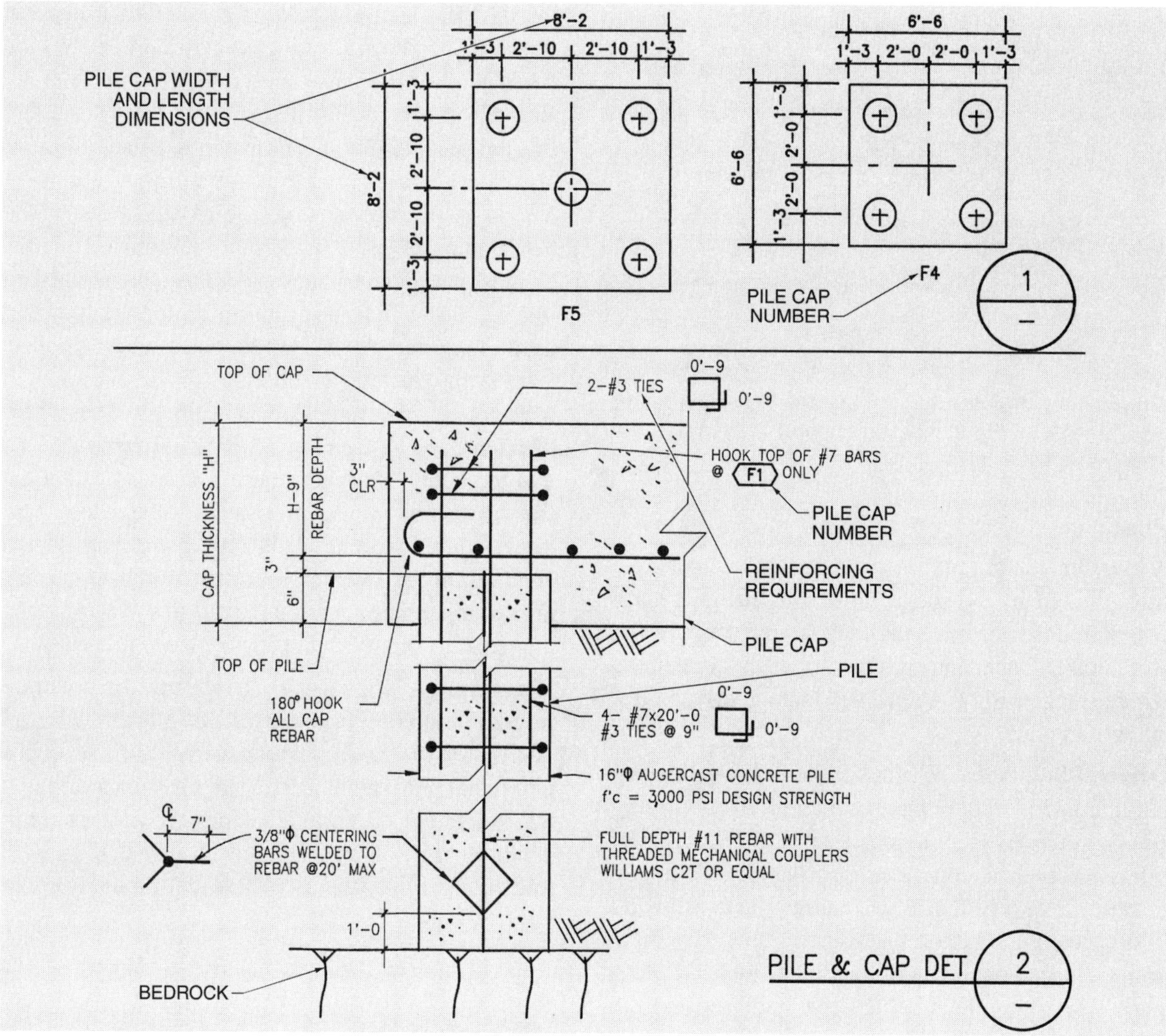

Figure 5-17. Pile caps are cross-referenced to foundation plan and detail drawings for information concerning concrete and reinforcing requirements.

PILE QUANTITY TAKEOFF

Quantity takeoff for foundation materials depends on the materials used and the site conditions. Various types of deep foundation support systems require calculations for the depth needed to reach solid bearing and the type of structure to be supported.

Pile Estimation

A grid of letters and numbers is provided at regular intervals on a foundation plan in a set of prints. The intersections of the grid lines provide references for pile placement. Piles are indicated on foundation plans with information including the elevation to the top of the pile cap, the elevation of the top of the pile, and the depth of the lower tip of the pile. **See Figure 5-18.**

Bored Piles. Cylindrical volume calculations are required to determine the amount of concrete required for cast-in-place concrete piles. The area of the pile cross section is multiplied by the hole depth to determine the volume. The area of the pile cross section is determined by multiplying 0.7854 (a constant) by the diameter squared ($A = d^2 \times 0.7854$). The area is then multiplied by the depth of the boring to determine the volume of concrete required for the hole. The volume (in cubic feet) is divided by 27 to determine cubic yards and rounded up to the next full cubic yard total.

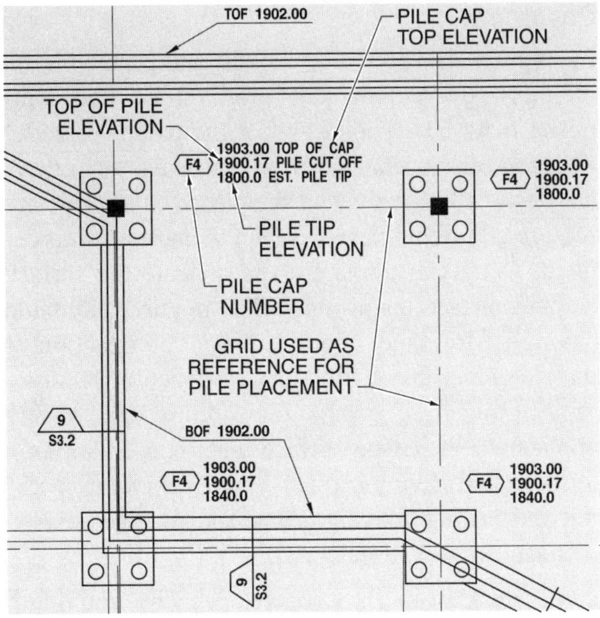

Figure 5-18. Foundation plans provide information concerning pile depths and elevations.

Calculating Cylindrical Pile Volume

Determine the volume of concrete (in cubic yards) required for a 36″ diameter pile that extends 56′-0″ deep.

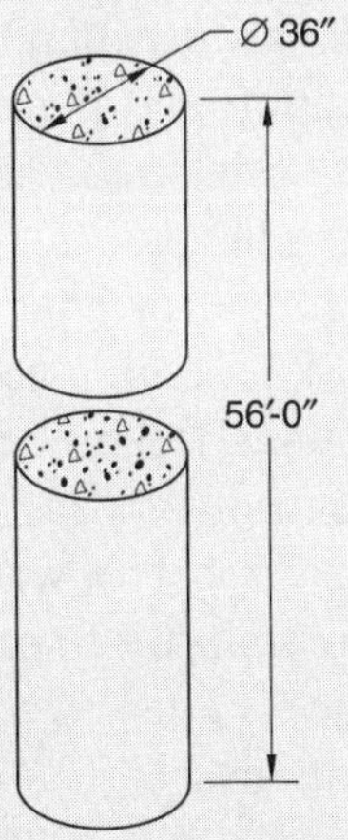

1. Determine the cross-section area of the pile.

area = 0.7854 × *diameter squared*

$A = 0.7854 \times d^2$

$A = 0.7854 \times (3')^2$

$A = 7.07$ sq ft

2. Determine the volume of the pipe pile (in cubic feet).

volume = *area* × *depth*

$V = A \times d$

$V = 7.07 \times 56'$

$V = 395.92$ cu ft, or round to 396 cu ft

3. Determine the volume of the pipe pile (in cubic yards).

$$volume\ (cubic\ yards) = \frac{volume\ (cubic\ feet)}{27}$$

$$V = \frac{395.92}{27}$$

$\mathbf{V = 14.66}$ **cu yd**

Precast Concrete Piles. Specifications for precast concrete piles provide information about concrete strength, reinforcing steel, and pile placement. Takeoff quantities are calculated by counting the number of precast concrete piles on site plans or foundation plans. Depths may be noted on the site plans or provided in a schedule. Similar to cast-in-place concrete piles, the volume of concrete required for precast concrete piles is based on pile diameter and the depth of each pile. Labor cost tables indicate labor costs based on pile type and depth. Equipment costs for cranes, leads, and the proper pile hammers are included in the cost estimate. Any additional costs should also take into account factors such as site accessibility for pile placement.

When using a spreadsheet, the number and type of piles are entered into the proper cell. Formulas may be integrated in the spreadsheet to calculate concrete quantities based on the depth and diameter. When using an estimating program, concrete piles can be entered as an item into an item listing in a manner that can allow the estimator to enter the diameter and depth, use the database to calculate concrete and reinforcing material quantities and labor costs, and enter them into the proper cell.

Steel Piles. Pipe piles and HP-piles are taken off by counting the number of each pile required. Depths are determined from site plans or are provided in a schedule. Estimators note the number of each length of steel pile required. As with concrete piles, equipment and labor costs are based on driving method, pile type, depth, and job-site accessibility.

Timber Piles. Timber piles are sized by diameter and length. The estimator determines the number of piles of each diameter and length to develop takeoff quantities and costs. Equipment and labor costs are based on the number of piles, driving method, and the depth of pile penetration required. Metal shoes may be used to protect the points of timber piles when the piles must be driven into hard strata. Metal shoes should be accounted for in the pile estimate.

Pile Caps. The volume of concrete required for pile caps is based on the pile cap length, width, and height. The volume of concrete (in cu ft) is divided by 27 to determine the volume in cubic yards. For example, the volume of concrete (in cu yd) required for a pile cap measuring 20′-0″ long by 3′-0″ wide by 2′-6″ high is calculated by multiplying the pile dimensions (20′ × 3′ × 2.5′ = 150 cu ft). The cubic foot value is divided by 27 to obtain the volume in cubic yards (150 cu ft ÷ 27 = 5.56 cu yd). The 5.56 cu yd value is rounded up to 6 cu yd.

PAVEMENT AND LANDSCAPING MATERIALS AND METHODS

After the majority of a building is constructed, the surrounding areas are finished to provide the appropriate access and landscape design. Many local building codes require a certain amount of landscaping and green space to be provided on new construction projects. The paving portion of the site plan provides the dimensions and locations for planters and open areas.

The subgrade must be properly prepared prior to placement of concrete or asphalt pavement.

Pavement

Paving material for walks, roads, and parking areas includes concrete, asphalt, brick, stone, and composite materials. The design of the paving material depends on the planned use. Heavy-use driveways are designed to withstand great loads and typically require additional pavement depth and/or reinforcement. Details in the site plans indicate the surface and subsurface paving materials required. Suitable paving performance requires proper compaction of the subgrade, installation and compaction of structural fill material, installation of reinforcement, and application of paving surface materials. In addition to the actual paving surface, curbs, pavement marking and sealing, fencing, gates, and signage are also included in the estimate.

Concrete. Cast-in-place concrete, which is composed of cement, water, and fine and coarse aggregate, is used for all types of paving. Aggregate is granular material such as gravel, sand, vermiculite, or perlite that is added to a cement and water mixture to form concrete, mortar, or plaster. Different aggregate may be used and various surface finishes applied to create variety. Coloring and/or surface stamping may be added to give the concrete a brick or stone appearance.

Pervious concrete pavement is a concrete pavement that is designed to allow surface water to penetrate through the pavement surface and minimize stormwater run-off. Additional costs may be associated with subsurface preparation and placement of pervious concrete pavement, which estimators should take into account in preparing the bid.

Shallow excavations for concrete pavement are made and formwork is set in place according to site plans and specifications. Reinforcement is placed into position within the formwork and concrete is placed. Steel reinforcing bars (rebar) or welded wire fabric are commonly set in place prior to concrete placement. The concrete surface is finished according to the site plans and specifications.

Asphalt. *Asphalt* is dark-colored pitch primarily composed of crushed stone and bituminous materials. A layer of asphalt is commonly placed on top of a subsurface of compacted limestone or other material that provides a solid base. After initial placement, asphalt is compacted to achieve a finished surface. After the bituminous materials that bind the crushed stone in the asphalt have set, striping and other markings are painted or placed onto the paving surface as shown on the site plans. Pavement markings may be raised from the surface, such as those used for tactile warning surfaces for the visually impaired. Similar to pervious concrete pavement, pervious asphalt products are available and the estimator should take considerations related to subsurface treatment requirements and costs.

Landscaping

Landscaping is the process of excavating, treating and preparing soil, and installing plants, including shrubs and trees, in specified locations. Landscaping plans provide information about plants, planting methods, and final surface treatments. Locations for exterior signage and irrigation systems may also be shown on detail drawings or on the topographic portion of the site plans.

Lawns. Unpaved and unlandscaped areas are sodded, seeded with grass seed, or left in their natural state. When sod is laid, the surface is prepared by grading and rough raking. Additional surface treatment of the soil may be required where seed is used. Surface protection such as straw, hay, or biodegradable synthetic material may be required to allow germination time for the seed. Specifications may require a certain type of grass planting. Treatment of these areas is also indicated on landscape plans.

Trees and Shrubs. Various trees, shrubs, and ground covers are shown on landscape plans, site plans, or in a schedule in the specifications. Specialized water-garden-type plants may be specified around or near retention ponds. Existing plants at a job site are noted as to whether they are to remain or be removed. The plants to remain may require protection to avoid damage during construction.

Irrigation Systems. Site plans indicate sprinkler piping, size, and head location. **See Figure 5-19.** Site plans and detail drawings show piping connections, valve boxes, and sprinkler head details. Irrigation systems are commonly made of high-grade plastic pipe that is placed slightly below the surface of the ground. Installation of irrigation systems occurs after all heavy construction operations are completed at the job site to prevent damage to the systems and the sprinkler heads.

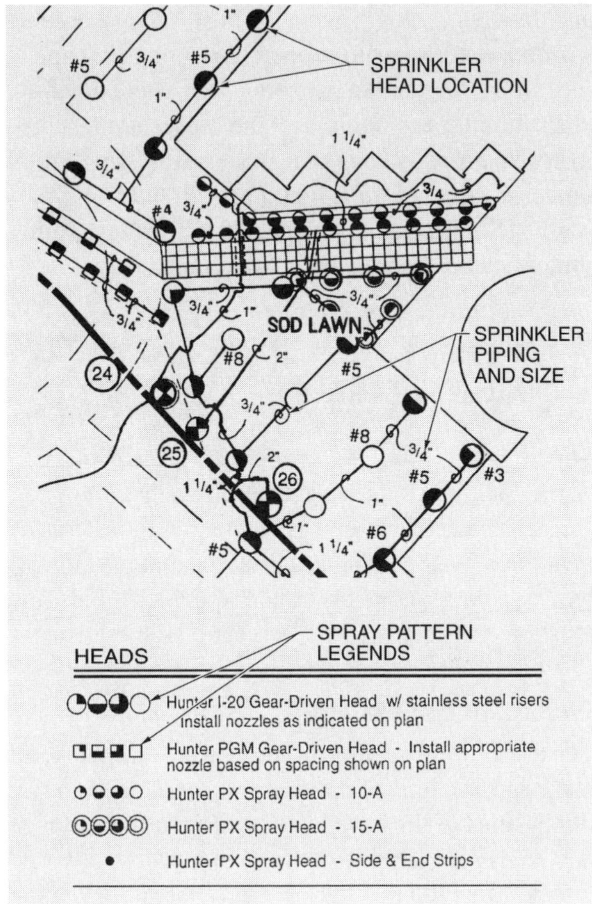

Figure 5-19. Sprinkler piping, sizing, and head location are determined using multiplication.

Site Amenities

Other finish items on a site include fountains, fences, gates, retaining walls, concrete median barriers, guide rails, seating, tables, shelters, athletic and recreational screening, and various traffic control devices and signs. Information on these items is included on site and foundation drawings.

Fencing. Fencing includes wood, chain-link, ornamental metal, and plastic fence. Each type of fence is indicated on site plans by a unique line symbol. Section 32 31 00 of the MasterFormat™ refers to permanent fencing designed to remain after completion of construction and does not include temporary security fencing placed during construction.

Fences placed on unpaved ground are fastened to posts spaced at regular intervals. Posts are commonly set in concrete to a depth below the frost line in locations where there is a possibility of freezing and thawing cycles moving the posts. Posts are set at a distance sufficient to support horizontal rails and at locations necessary to support gates. After stretching chain-link fences into place, the fences are fastened to horizontal rails and vertical posts with wire. Wood fence is fastened to horizontal rails with nails or screws. Some ornamental metal fencing may require prefabricated sections or welding in place.

PAVEMENT AND LANDSCAPING QUANTITY TAKEOFF

Streets, parking lots, walks, and curb takeoff information is included in the specifications, detail drawings, and site plans. The area is calculated to determine quantities for the various subsurface and paving materials as well as labor. Landscaping materials and locations are noted in the specifications and site plans.

Pavement Estimation

The primary information needed for pavement takeoff is the specific type of pavement material, the exterior dimensions of the paved area, and the dimensions of any voids such as islands. These dimensions are used to determine the number of square feet of pavement to be put in place, subgrade preparation, pavement thickness and possible reinforcement, surface finish, and possible surface protection.

Subgrade preparation includes spreading a granular subbase of crushed gravel or sand. In some cases, stabilization fabric may also be required. The area of the paved area (in square feet) is calculated by multiplying its length and width. Deductions are made for any sizable internal void areas such as planters or signage islands. Some estimating programs allow for length and width dimension entry and then perform automatic item calculation of materials, reinforcing, and labor costs. With the proper hardware and software, a digitizer may be used when taking off large and irregularly shaped paving areas containing curves that may require additional geometric calculations.

The volume of crushed gravel or stone subbase required is calculated in cubic yards and converted by the quarry into the appropriate number of tons of material. Costs include the subbase material, hauling, and the labor for spreading and compacting the subbase in preparation for asphalt application.

Concrete Pavement. Concrete pavement is estimated based on the volume of pavement required in cubic yards. The cubic yards of concrete required determines the material cost. The number of square feet of surface area determines the labor cost. The primary items considered for concrete pavement include the type of concrete mix, thickness of the pavement, the surface finish, and protection to be provided after placement of the concrete. Specifications provide information concerning concrete mix, compressive strength, and required reinforcement.

Calculating Concrete Pavement Volume

Determine the volume of concrete (in cubic yards) required for a 6″ thick driveway measuring 140.9′ long by 20′-0″ wide.

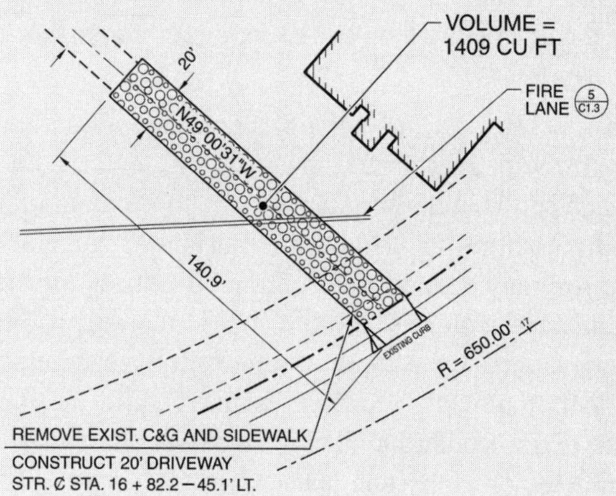

1. Determine the volume of concrete required (in cubic feet).

$volume = thickness \times width \times length$

$V = t \times w \times l$

$V = 0.5' \times 20' \times 140.9'$

$V = 1409$ cu ft, or round to 1400 cu ft

2. Determine the volume of concrete required (in cubic yards).

$$volume\ (cubic\ yards) = \frac{volume\ (cubic\ feet)}{27}$$

$$V = \frac{1409\text{ cu ft}}{27}$$

$\mathbf{V = 52.18}$ **cu yd**

When steel reinforcement is required, the number of each type of rebar is determined by calculating the spacing in each direction and the number of bars required in a given area. For example, when rebar is spaced 16″ (1.33′) OC each way, a 16′ square section of concrete pavement requires 24 rebar ([16 ÷ 1.33] × 2 = 24 rebar). Additions are made for the amount of bar overlap required in the specifications or detail drawings. Welded wire fabric is calculated by the number of square feet of coverage required.

Asphalt. Many different types and designs of asphalt mixtures are available. The design of the asphalt pavement is based on the traffic loads imposed and climatic conditions. Selection of the proper asphalt mixture for traffic loads is made by the architect and engineer and is included in the specifications or noted on the site plans. The area covered by the asphalt and the depth of the asphalt determine the amount of asphalt required. The amount of asphalt required is calculated by multiplying the coverage area (in square feet) by the depth (in feet). Asphalt material is priced by the ton of material delivered to the job site. Conversion amounts vary based on the type of mixture specified.

Items taken into consideration when estimating costs and quantities for hot-mix asphalt (HMA) include mixture design and type, surface preparation required, application, screeding methods, joint construction, compaction, and spreading and compacting equipment used. The number of square feet determines the labor cost. Labor costs are based on industry standard information or company historical data.

Curbs and Gutters. Various curb designs are used depending on the need to match existing curbs or to protect against damage in heavy-use conditions. Estimators calculate the number of linear feet of each type of curb from the site plans. Standard costs per linear foot of curb include materials such as concrete (including possible reinforcement) and asphalt as well as labor. The amount of formwork required for curbs and gutters depends on radiuses, curves, aprons, and tolerances permitted horizontally and vertically.

> Curbs and gutters can be formed manually using wood or metal forms that are secured in place with stakes. Concrete is placed in the forms and the curb and gutter contour is created by cement masons. Curbs and gutters can also be formed using mechanical curb-forming equipment.

Landscaping Estimation

Landscaping costs include finish grading, sodding, seeding, excavation for trees and shrubs, landscaping materials, and landscaping protection. Specifications, schedules, and detail drawings show required planting methods and materials, including the depth of planting, soil amendments, and mulch. In some instances, specifications may include requirements for landscaping materials to grow and remain alive for a given period of time after the completion of construction, which may involve additional landscape maintenance costs.

Lawns. Specifications indicate if the lawn areas are to be sodded or seeded. Costs are calculated based on the area of the lawn in square feet or acres. Possible costs include finish grading and final surface preparation, laying of sod or spreading of seed, fertilizer, watering, and straw to protect the surface during germination. Hydroseeded areas are estimated based on the number of square feet of coverage.

Trees and Shrubs. The scientific name, common name, and size of each plant are indicated on a plant schedule along with the quantity required. **See Figure 5-20.** Plant nurseries commonly provide a cost per plant that includes plant material, planting, and staking. Estimators should check the specifications for protective barriers placed around trees and shrubs during and following construction. Any ongoing maintenance obligations are also noted in the specifications and cost allowances may be added.

Portland Cement Association

Mechanical curb-forming equipment can be used to form curbs and gutters.

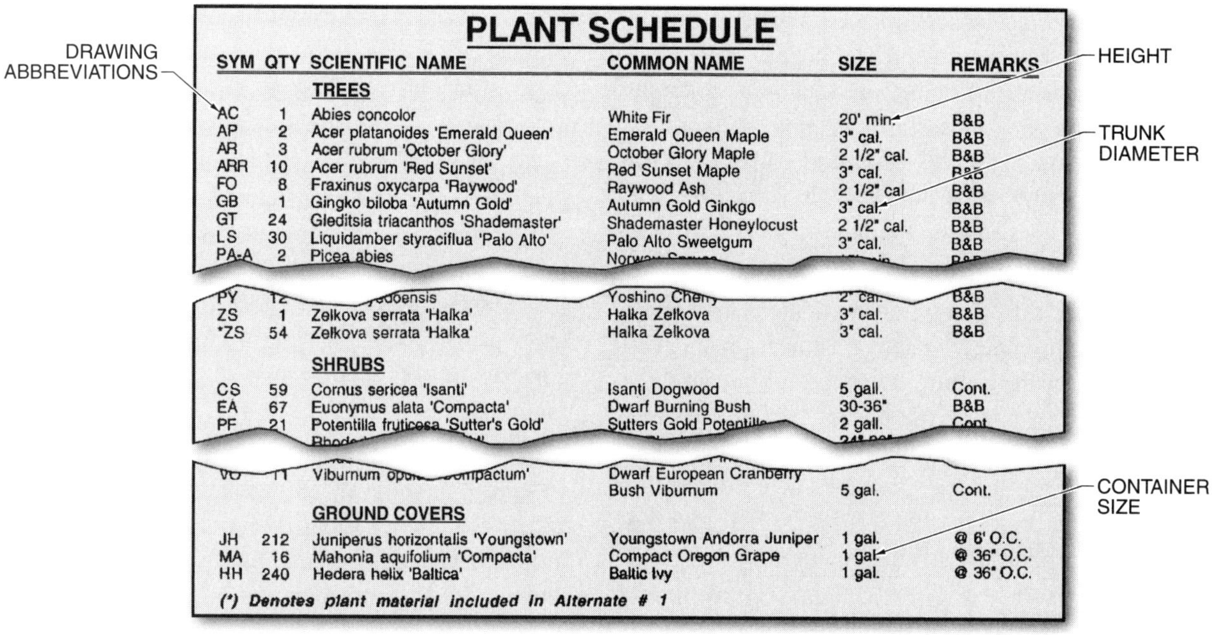

PLANT SCHEDULE

SYM	QTY	SCIENTIFIC NAME	COMMON NAME	SIZE	REMARKS
		TREES			
AC	1	Abies concolor	White Fir	20' min.	B&B
AP	2	Acer platanoides 'Emerald Queen'	Emerald Queen Maple	3" cal.	B&B
AR	3	Acer rubrum 'October Glory'	October Glory Maple	2 1/2" cal.	B&B
ARR	10	Acer rubrum 'Red Sunset'	Red Sunset Maple	3" cal.	B&B
FO	8	Fraxinus oxycarpa 'Raywood'	Raywood Ash	2 1/2" cal	B&B
GB	1	Gingko biloba 'Autumn Gold'	Autumn Gold Ginkgo	3" cal.	B&B
GT	24	Gleditsia triacanthos 'Shademaster'	Shademaster Honeylocust	2 1/2" cal.	B&B
LS	30	Liquidamber styraciflua 'Palo Alto'	Palo Alto Sweetgum	3" cal.	B&B
PA-A	2	Picea abies			
PY		...oensis	Yoshino Cherry		B&B
ZS	1	Zelkova serrata 'Halka'	Halka Zelkova	3" cal.	B&B
*ZS	54	Zelkova serrata 'Halka'	Halka Zelkova	3" cal.	B&B
		SHRUBS			
CS	59	Cornus sericea 'Isanti'	Isanti Dogwood	5 gall.	Cont.
EA	67	Euonymus alata 'Compacta'	Dwarf Burning Bush	30-36"	B&B
PF	21	Potentilla fruticosa 'Sutter's Gold'	Sutters Gold Potentilla	2 gall.	
		Viburnum ...ompactum'	Dwarf European Cranberry Bush Viburnum	5 gal.	Cont.
		GROUND COVERS			
JH	212	Juniperus horizontalis 'Youngstown'	Youngstown Andorra Juniper	1 gal.	@ 6' O.C.
MA	16	Mahonia aquifolium 'Compacta'	Compact Oregon Grape	1 gal.	@ 36" O.C.
HH	240	Hedera helix 'Baltica'	Baltic Ivy	1 gal.	@ 36" O.C.

(*) Denotes plant material included in Alternate # 1

Figure 5-20. Estimates for landscaping material can be calculated from plant schedules.

Irrigation Systems. Takeoff for irrigation systems includes calculation of the linear feet of each size of pipe and the number of each valve and spray head in the system. Control valves, timers, drains, and other specialty items are also included in the quantity takeoff. Labor costs are based primarily on company historical data for installation times and labor rates.

Site Amenity Estimation

Costs for site amenity items such as shelters, seating, or athletic screening are obtained from suppliers on an as-needed item basis. Costs for site amenities such as traffic control devices and signage are calculated on an item basis. Estimators determine the number and types of each sign, traffic device, or other item required.

Fencing. The linear feet of fence is calculated from information included on the site plans. Estimators determine the linear feet of each type of fence required. Variables for the estimator include the fence material (wood or metal), height, finish, and items such as swinging or sliding gates. Material costs per linear foot include fence posts, railings, and fencing. Material and labor rates for fencing and gates are commonly based on company historical data or industry standards.

SAGE TIMBERLINE ESTIMATING . . .

QUICK TAKEOFF ENTER DIMENSIONS FEATURE

When using the Quick Takeoff function of the Sage Timberline Estimating program, the Enter Dimensions feature controls whether or not the **Enter Dimensions** dialog box opens when selected items are added to the spreadsheet. The Enter Dimensions feature is turned on and off by selecting the **Enter dimensions** option in the **Quick Takeoff** dialog box or by right-clicking in the **Quick Takeoff** dialog box and selecting **Enter Dimensions**. Leaving the feature unselected loads the selected items from the **Quick Takeoff** dialog box directly into the spreadsheet. Quantities can then be entered in the **Takeoff Quantity** column of the spreadsheet.

. . . SAGE TIMBERLINE ESTIMATING

1. Ensure the **Sample Ext Commercial GC** database is open. Open the **Quick Takeoff** dialog box by picking the **Quick Takeoff** button.
2. Select the **Enter dimensions** option in the dialog box or right-click in the **Quick Takeoff** pane and select **Enter Dimensions**. A checkmark is displayed next to the option, indicating that the **Enter Dimensions** feature is enabled.

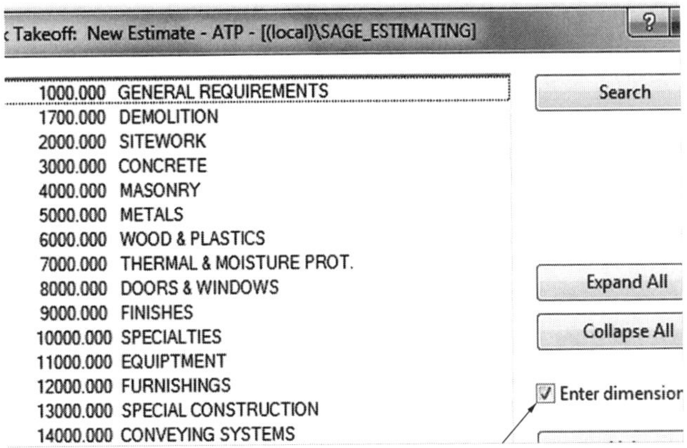

3. Double-click on **2000.00 Sitework** and **2315.021 Earthwk: Excav Foot/Misc**. Select item **20 Excavate Footing by Machine** by double-clicking on the item or dragging and dropping the item into the spreadsheet.
4. Item 20 has an associated formula. The Enter Dimensions feature prompts the user with the appropriate variables based on the associated formula, which is displayed in the title bar of the dialog box. In the dialog box, enter **1** for **Quantity**, **50** for **Length**, **4** for **Width**, and **4** for **Depth**. Pick **OK** to add the item to the spreadsheet.

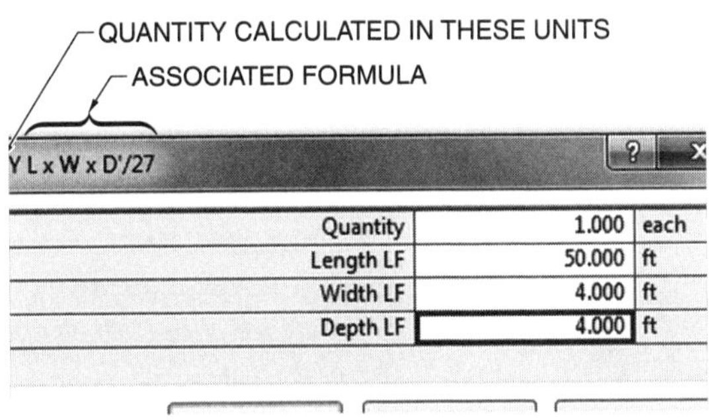

5. Next, disable the Enter Dimensions feature by deselecting the **Enter dimensions** option in the **Quick Takeoff** dialog box or by right-clicking in the **Quick Takeoff** pane and deselecting **Enter Dimensions** in the shortcut menu. The checkmark is no longer displayed.

Quick Quiz®

Quick Quiz®

Refer to the CD-ROM for the Quick Quiz® questions related to chapter content.

Key Terms

Illustrated Glossary

- angle of repose
- asphalt
- backfilling
- bolder
- bored pile
- caisson
- clay
- clear and grub
- cofferdam
- contour line
- core drilling
- cradling
- dead load
- demolition
- driven pile
- excavation
- final grading
- footing
- foundation
- geotextile
- grading
- gravel
- gridding
- groundwater
- hard pan
- hazardous material
- heavy soil
- HP pile
- infill material
- lagging
- landscaping
- light soil
- live load
- overdig
- payline
- permanent soil stabilization
- pervious concrete pavement
- pile
- pile cap
- pipe pile
- precast concrete pile
- remediation
- sand
- sanitary sewer
- sheet pile
- shoring
- silt
- slope cut
- slurry
- soil amendment
- soil swell
- storm sewer
- subsurface investigation
- temporary soil stabilization
- timber pile
- trench box
- well point
- zero line

Web Links

Web Links

American Fence Association
www.americanfenceassociation.com

American Gas Association
www.aga.org

Asphalt Institute
www.asphaltinstitute.org

Association of Soil & Foundation Engineers
www.asfe.org

Deep Foundations Institute
www.dfi.org

Engineering & Utility Contractors Association
www.euca.com

Institute of Makers of Explosives
www.ime.org

International Society of Trenchless Technology
www.istt.com

International Union of Operating Engineers
www.iuoe.org

National Asphalt Pavement Association
www.hotmix.org

National Demolition Association
www.demolitionassociation.com

National Stone, Sand and Gravel Association
www.nssga.org

National Utility Contractors Association
www.nuca.com

North American Society for Trenchless Technology
www.nastt.org

Pile Driving Contractors Association
www.piledrivers.org

Pipe Line Contractors Association
www.plca.org

Professional Landcare Network
www.landcarenetwork.org

Earthwork 5

Review Questions

Estimating

__________ **1.** A(n) ___ is a rock that has been transported by geological action from its formation site to its current location.

__________ **2.** ___ is loose granular material consisting of particles that are smaller than gravel but coarser that silt.

__________ **3.** ___ is natural mineral material that is compact and brittle when dry but plastic when wet.

T F **4.** Hardpan is a hard, compacted layer of soil, clay, or gravel.

T F **5.** Geotextile material is applied after rough grading is complete.

__________ **6.** A soil ___ is a material added to existing job-site soil to increase its compaction and ability to maintain the desired angle of repose or slope at the completion of construction.

__________ **7.** ___ is water present in subsurface material.

__________ **8.** ___ is the organized destruction of a structure.

__________ **9.** A(n) ___ load is the total of all the dynamic loads a structure is capable of supporting including wind loads, human traffic, and any other imposed loads.

__________ **10.** A(n) ___ load is a permanent, stationary load composed of all construction materials, fixtures, and equipment permanently attached to a structure.

__________ **11.** A(n) ___ is a structural member installed in the ground to provide vertical and/or horizontal support.

T F **12.** Shoring is a system of wood or metal members used to permanently support formwork or other structural components.

T F **13.** Clearing and grubbing is the removal of vegetation and other obstacles from a construction site in preparation for new construction.

__________ **14.** A(n) ___ is a watertight enclosure used to allow construction or repairs to be performed below the surface of the water.

__________ **15.** A ___ sewer is a piping system designed to collect stormwater and channel it to a retention pond or other means of removal.

A. sanitary
B. storm
C. drainage
D. none of the above

_______________ **16.** ___ is the temporary support of existing utility lines in or around an excavated area to properly protect the lines during construction.

_______________ **17.** A(n) ___ line is a dashed or solid line on a site plan used to show elevations of the surface.

_______________ **18.** ___ is the process of lowering high spots and filling in low spots of earth at a job site.

_______________ **19.** ___ horizontal planks stacked on edge to support the earth on the side of an excavation.

_______________ **20.** A(n) ___ is a large-diameter cast-in-place concrete pile created by boring a hole and filling it with concrete.

_______________ **21.** ___ is granular material, such as gravel, sand, vermiculite, or perlite, that is added to a cement and water mixture to form concrete, mortar, or plaster.

T F **22.** Landscaping is the process of excavating, treating and preparing soil, and installing plants, including shrubs and trees, in specified locations.

T F **23.** Angle of reprise is the slope that a material maintains without sliding or caving in.

_______________ **24.** Soil ___ is the volume growth in soil after it is excavated.

_______________ **25.** A(n) ___ material is a material capable of posing a risk to health, safety, or property.

_______________ **26.** ___ is a mixture of water and fine particles such as cement that improves soil stability.

_______________ **27.** ___ material is material, including sand, gravel, and crushed rock, installed to increase the stability of soil.

_______________ **28.** ___ is the division of a topographical site map for large areas to be excavated or graded into smaller squares or grids.

_______________ **29.** A(n) ___ pile is a steel or wood member that is driven into the ground with a pile-driving rig.

_______________ **30.** A(n) ___ pile is a foundation support formed by drilling into the ground to a predetermined depth and diameter and filling the hole with concrete.

_______________ **31.** ___ is the act or process of correcting situations where hazardous materials exist and returning a site to environmentally safe conditions.

_______________ **32.** ___ is a dark-colored pitch primarily composed of crushed stone and bituminous material.

_______________ **33.** Temporary soil ___ is the process of holding existing surface and subsurface materials in place during construction.

Cross-Section Volume Calculation

1. Using the cross-section method, determine the volume of soil (in cubic yards) to be excavated from the given area.

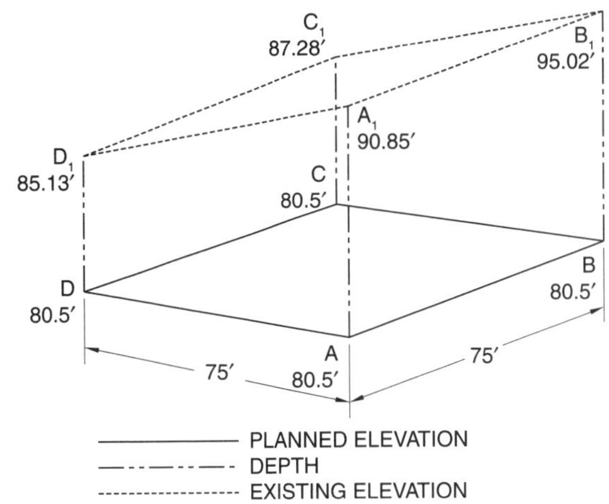

Excavation

______________ 1. Overdig

______________ 2. Backfilling

______________ 3. Trencher

______________ 4. Payline

______________ 5. Slope cut

A. An inclined or sloped wall excavation

B. A measurement used by excavation contractors to compute the amount of necessary excavation as stated in the specifications

C. A machine with a conveyor to remove the soil as it digs a narrow, long trench into the earth

D. The amount of excavation required beyond the dimensions of a building to provide an area for construction activities

E. The replacing of soil around the outside foundation walls after the walls have been completed

Short Answer

1. Explain why the angle of repose must be taken into consideration when calculating the amount of material to be excavated.

2. Describe applications for sheet piles and why they may be preferred over other types of piles.

Earthwork 5

Activities

Activity 5-1—Ledger Sheet Activity

Refer to the cost data, Print 5-1, and Estimate Summary Sheet No. 5-1. Take off the labor and equipment costs for lawn seeding of the site and the asphalt required for the driveway. Include 25% of the subtotal for overhead and mobilization. If the subtotal for the lawn seeding and asphalt pavement is less than $2000, include a minimum overhead and mobilization cost of $500 for the lawn seeding and asphalt pavement.

COST DATA			
Material	**Unit**	**Labor Unit Cost***	**Equipment Unit Cost***
Lawn seeding 175 lb/acre	acre	649.00	413.00
Asphalt pavement	sq yd	8.32	7.43

* in $

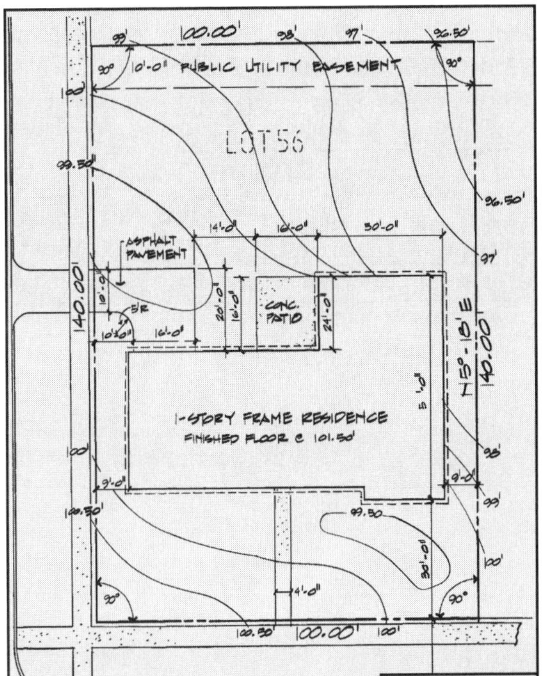

Print 5-1

ESTIMATE SUMMARY SHEET

Project: ______
Estimator: ______

Sheet No. 5-1
Date: ______
Checked: ______

No.	Description	Dimensions			Quantity		Labor		Equipment		Subtotal	
						Unit	Unit Cost	Total	Unit Cost	Total	Unit Cost	Total
	Lawn seeding											
	Asphalt pavement											
	Overhead—lawn seeding											
	Overhead—asphalt pavement											
	Total											

Activity 5-2—Spreadsheet Activity

Refer to the cost data (below) and Estimate Summary Sheet No. 5-2 on the CD-ROM. A job requires the demolition of a 1′ × 6′ × 20′ concrete-masonry-unit wall with brick veneer and the excavating and backfilling of a 3′ × 3′ × 60′ trench. Take off the cubic feet of masonry wall demolition, cubic yards of excavating, and cubic yards of backfilling. Determine the total labor and equipment cost for the demolition, excavating, and backfilling.

COST DATA			
Material	**Unit**	**Labor Unit Cost***	**Equipment Unit Cost***
Masonry wall demolition	cu ft	0.81	0.64
Excavating	cu yd	2.93	1.61
Backfilling	cu yd	0.38	1.02

* in $

Activity 5-3—Calculating Cutting and Filling

Refer to the cost data and Print 5-3 (below), and Estimate Summary Sheet No. 5-3 on the CD-ROM. Determine the labor and equipment costs for cutting and filling the job site to an elevation of 98.0′. A 25′ grid is superimposed over the site plan for ease in calculating. The bulldozer used to perform the operation can move 40 cu yd of material per hour.

COST DATA			
Material	**Unit**	**Labor Unit Cost***	**Equipment Unit Cost***
Cut and fill	cu yd	2.85	90/hr

* in $

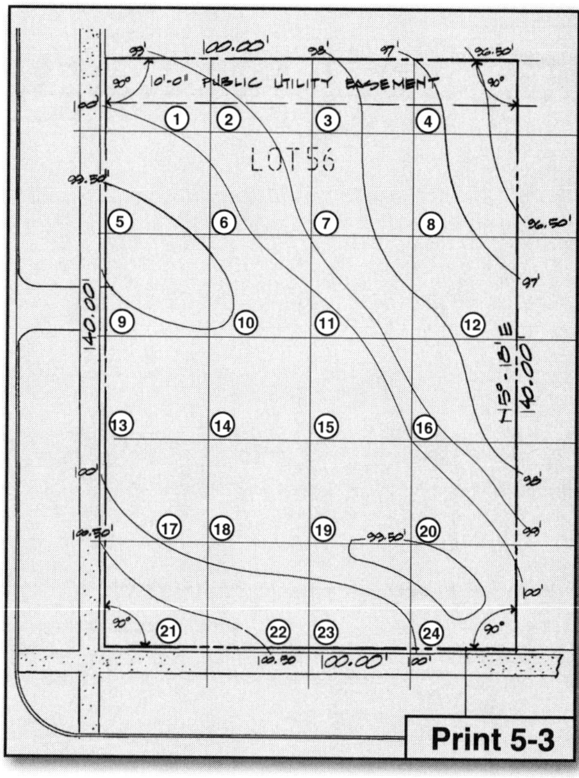

Print 5-3

Activity 5-4—Sage Timberline Estimating Activity

Create a new estimate and name it Activity 5-4. Perform a quick takeoff of item **2315.070 10 Backfill Footings** for a 275′ long by 3′ wide by 4′ deep footing excavation. Take off item **2740.030 20 H Asphalt Material** for a 100′ long by 20′ wide by 6″ deep (thick) road. Print a standard estimate report. *Note:* The Enter Dimensions feature must be turned on.

Concrete 6

Key Concepts

- For large construction projects, information concerning the type and design of formwork is typically provided on a set of formwork drawings supplied by the concrete form supplier.
- An estimator must calculate the amount of formwork materials required when formwork drawings are not provided.
- Costs for formwork materials and installation labor for concrete footings and foundation walls are a major portion of the estimate calculations for concrete footings and foundation walls.
- Foundation plans, elevations, and detail drawings provide the quantity takeoff information for footings and foundation walls.
- Concrete volume is taken off and expressed in cubic yards.
- Estimators consult with the casting plant to obtain costs for producing plant-precast members.

Introduction

Concrete consists of portland cement, aggregate, and water that solidifies into a hard, strong mass. Architectural drawings and specifications contain notes and information concerning specific concrete performance requirements. Concrete components and members may be cast-in-place or precast. Concrete quantity takeoffs take into account the type and volume of concrete required, reinforcing, transportation, surface finish, and possible weather protection. Formwork is the system of support for fresh concrete. Formwork may be job-built or patented. Concrete formwork design takes into account all forces to be placed on the forms.

FORMWORK MATERIALS AND METHODS

Concrete is a mixture of portland cement, fine and coarse aggregate, and water that solidifies through a chemical reaction called hydration into a hard, strong mass. *Reinforced concrete* is concrete that has increased tensile strength due to tensile members, such as steel rods, bars, or mesh, embedded in the concrete. Reinforced concrete members may be cast-in-place, precast on the job site, or precast at a casting plant.

Concrete is commonly used in conjunction with other construction methods such as wood or metal framing, masonry, and structural steel. Specifications and architectural and structural drawings contain information for reinforced concrete construction. The majority of reinforced concrete information is shown on foundation, floor, and structural plans, and in elevations and detail drawings.

Formwork is the entire system of support for fresh concrete including the forms, hardware, and bracing. Many methods are used to construct formwork for cast-in-place concrete. The formwork used depends on the overall size of the concrete structure.

A *form* is a temporary structure or mold used to retain and support concrete while it sets and hardens. Forms may be set on or in the ground for piles, footings, and slabs. Forms may be set on concrete slabs to support walls, columns, and above-grade slab systems. Elevated forms may be set for above-grade slabs. Each cast-in-place concrete member requires specific print information such as dimensions of the finished concrete, reinforcing requirements, inserts, and concrete properties.

> Formwork must be constructed using safe construction procedures such as those established by the American Concrete Institute (ACI).

Footings and Foundation Walls

Materials used for structural footings and foundation walls include wood, masonry, and concrete. Concrete is the most commonly used material for commercial and residential footings and foundation walls. Treated wood and concrete masonry units (CMUs) are more commonly used for residential or light commercial applications. A *footing* is the section of a foundation that supports and distributes structural loads directly to the bearing surface of the soil, piles, or rock below the structure.

Wood or steel forms are set to retain cast-in-place concrete while the concrete is in a plastic state. The forms are set according to elevations and dimensions noted on the foundation plans. Reinforcing steel is set between the footing and foundation wall forms. Concrete for footings is placed prior to concrete walls. After the concrete has reached an initial set, the forms are removed to allow construction to proceed.

Footing and foundation wall formwork is typically reused many times. Fasteners and pins used to secure the formwork are removed, the faces of the forms are cleaned, and a form-release agent is applied to the form face to allow the form to properly release from the footing or wall surface. Insulating concrete forms (ICFs) are created from a series of interlocking foam block forms and are used in light-commercial and residential construction. ICFs are permanent forms that stay in place and provide additional insulation to the concrete foundation wall.

Wall Forms

Cast-in-place concrete walls are constructed on footings or slabs to provide proper support. Wall forms are made of wood or metal faces that are reinforced with wood or steel frames. **See Figure 6-1.** Wall forms are aligned with walers, strongbacks, and braces.

Gates & Sons, Inc.

Figure 6-1. Wall forms may be made of wood or metal faces reinforced with wood or steel frames.

A *waler* is a horizontal member used to align and brace concrete forms or piles. A *strongback* is a vertical support attached to concrete formwork behind the horizontal walers to provide additional strength against form deflection during concrete placement. A *brace* is a structural piece, either permanent or temporary, designed to resist the weight or pressure of a load. Reinforcing steel and blockouts for doors and windows are set before the opposing side of a wall form is completed. A *blockout* is a frame set between concrete form walls to create a void in the finished concrete structure.

A *form tie* is a metal bar, strap, or wire used to hold concrete forms together at a predetermined distance and resist the outward pressure of fresh concrete. Different types of form ties are used to hold the two wall forms at the proper distance and distribute concrete loads during placement. The type of form tie used depends on the width and design of the wall. Form ties include snap ties, she bolts, and coil ties.

A *snap tie* is a concrete form tie with ends that are designed to be snapped off after the forms are removed and the concrete is set. A *she bolt* is a steel rod that threads into the ends of a tie rod. A *tie rod* is a steel rod that extends between form walls to hold them together after concrete is placed. The she bolt is removed when the concrete has set and forms are removed. The tie rod remains embedded in the concrete. A *coil tie* is a type of concrete form hardware used in heavy construction in which a bolt screws into the tie and is removed during form removal. For large commercial projects with complex forming requirements, concrete form manufacturers provide detailed drawings of forms, ties, and bracing placement to ensure proper form construction.

For applications such as abutments, slurry walls, or retaining walls, one side of the wall is formed and the opposite side is created by soil or rock. An *abutment* is a supporting structure at the end of a bridge, arch, or vault. A *slurry wall* is a wall built around an excavation to retain groundwater. A *retaining wall* is a wall constructed to hold back earth. Reinforcement is set in place and the forms are set and braced. Additional bracing of one-sided wall forms is necessary to support the hydrostatic concrete pressure (weight) that is not offset by the other side of the wall as with a conventional form-tie system for a two-sided formed concrete wall.

Blockouts are placed in wall forms to create openings for later placement of various building items such as doors and windows. Blockout placement is determined from elevations, foundation plans, and floor plans. Large or numerous blockouts in concrete forms affect labor and material costs and concrete volume. Concrete wall forms include job-built and patented wall forms.

Job-Built Wall Forms. Job-built wall forms are constructed with 2 × 4 framing members and faced with ½″ or ¾″ plywood. **See Figure 6-2.** Job-built wall form panels typically measure 4′ × 8′ in size to make maximum use of plywood materials. Studs are spaced 12″ or 16″ OC (on center). Holes are drilled at a regular spacing to allow for placement of snap ties.

Figure 6-2. Job-built wall forms are constructed with 2 × 4 framing members that are faced with plywood. Snap ties maintain the proper spacing between form walls and reinforcing steel reinforces the concrete.

Job-built wall forms may be built for reuse or built for a single application. Individual wall forms may be assembled into a large unit and fastened together to form a ganged panel form. A *ganged panel form* is a large form constructed by joining a series of small panels into a larger unit. Ganged panel forms may be constructed of either job-built or patented wall forms. Ganged panel forms are set in place by cranes and reused in several applications. Ganged panel forms help reduce the labor cost of moving and assembling many smaller forms repeatedly.

Patented Wall Forms. Many manufacturers provide patented wall-form systems that incorporate a variety of panel designs, tie systems, and form-fastening systems. Panels are manufactured with metal frames faced with plywood or metal faces. Panels are available in standard sizes to allow these systems to form walls of various lengths, heights, and widths. **See Figure 6-3.** Patented-form tie systems vary according to the manufacturer and the wall being formed.

Symons Corporation

Figure 6-3. Patented wall forms are manufactured with metal frames faced with plywood or metal faces.

Job-built and patented forms are typically braced using 2 × 4s or 2 × 6s and a variety of anchoring and support methods. Bracing systems range from the simple system of a stake driven into the ground and a 2 × 4 nailed to the back of the form and stake to a complex system with metal channels bolted to the back of the forms, with additional bracing of screw jacks fastened to metal or wood braces. Design of wall-form bracing systems can be a complex portion of estimating concrete form material and labor costs.

Slabs

Concrete slabs may be placed on the surface of the ground to be used as a foundation system in light load-bearing applications. Slabs on grade are also used for paving and basement and ground-level floors. Paving, such as highways and roads, is covered in Titles 32 13 13 and 32 12 16 of the CSI MasterFormat™. Estimates for highways and roads are typically performed by contractors and subcontractors who specialize in road construction. Information used in creating an estimate is obtained from the prints and specifications, including treatment of the area below the slab, items placed in the slab, blockouts, elevations, and finishes.

Slabs on Grade. A *slab on grade* is a concrete slab that is placed directly on the ground. A *grade beam* is a reinforced short, vertical concrete wall placed below the slab at ground level and supported by piles or piers. A grade beam provides support and protection against movement during freezing and thawing cycles. Spread footings or piles are necessary where heavy load support is required. A *spread footing* is a foundation support with a wide base designed to spread a load over a wider area. A *pile* is a structural member installed in the ground to provide vertical and/or horizontal support.

Forms may be built for perimeter footings and slabs where they project above the surface of the soil. **See Figure 6-4.** Some sections of a slab may be thickened to support other loads. Locations of the thickened sections are commonly indicated with dashed lines on plan views and are shown with additional sections.

Forms are not required where a slab on grade is placed between existing walls or other structural members, such as a basement slab placed between foundation walls. Expansion joints, including isolation joints and control joints, may be installed at various locations around a concrete slab. An *isolation joint* is a separation between adjoining sections of concrete used to allow movement of the sections without affecting the adjacent section. A piece of ½″ or ¾″ asphalt-impregnated material or various designs of vinyl strips are placed in the joint. The locations and types of isolation joints are indicated on detail drawings. **See Figure 6-5.**

A *control joint*, also known as a contraction joint, is a groove made in a vertical or horizontal concrete surface to create a weakened plane and control the location of cracking due to imposed loads. Control joints are typically shown on foundation and floor plans and detail drawings. Control joint spacing is critical to properly controlling random slab cracking. Control joint spacing varies depending on the dimensions of the concrete slab. These are most commonly sawn into the surface of the slab after the concrete has reached an initial set, but may be troweled in during concrete slab placement.

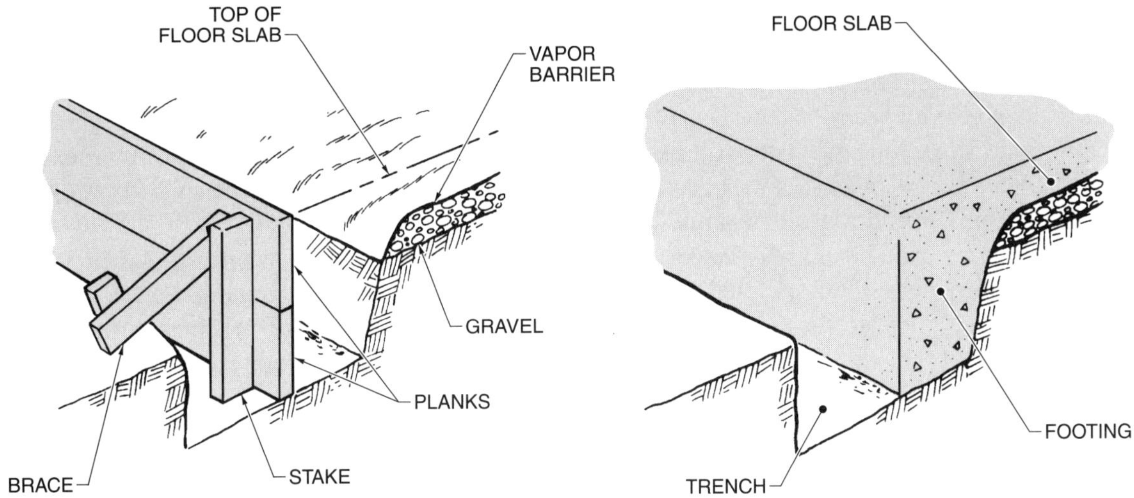

Figure 6-4. Footings provide additional foundation support at the perimeter of a slab on grade.

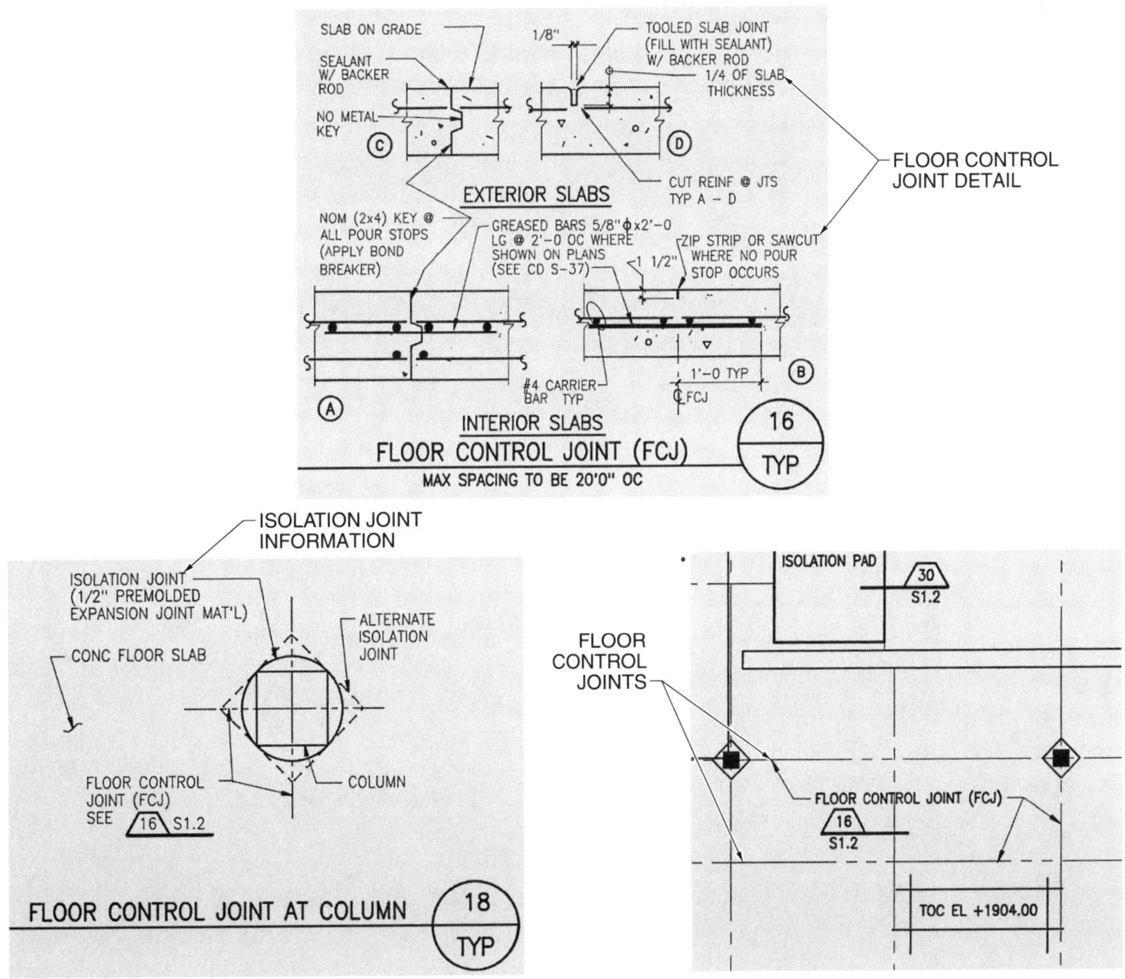

Figure 6-5. Isolation joints in a slab on grade require additional labor and material costs.

Elevated Slabs. Concrete slabs placed above grade may be constructed using several different methods. Three types of elevated slabs include a monolithic design, with the concrete for beams and the slab placed at the same time; supporting the slab with precast concrete or structural steel beams; or supporting the slab with steel beams and corrugated decking.

Steel or precast concrete beams may be designed to be placed independently of an elevated concrete slab. Detail drawings indicate the thickness of the slab and placement of the reinforcement. Forms are supported by beams after columns, walls, and beams are in place. Hangers are placed on the beams to allow temporary joists and bracing brackets to be supported and set to the proper slab height. Threaded rods are lubricated and used to support brackets for temporary form joists. Deck forms are placed on the joists.

The slab is tied to the beams by projecting reinforcing steel from precast concrete beams or welded studs on top of structural steel beams. Edge forms are built around the perimeter of the deck. Electrical piping, mechanical blockouts, and reinforcement are set in place prior to the concrete being placed and finished. The temporary joists, bracing brackets, and deck forms are removed from below after the concrete has reached its design strength. Threaded lubricated rods are removed and the remainder of the hanging hardware for the concrete deck remains in place.

Perimeter forms for a concrete slab on grade typically require minimal bracing. Forms for a slab on grade are usually only several inches high, and stakes and wood bracing provide adequate support.

Elevated slabs use a system of shoring and decking to form a flat surface at the proper elevation. *Shoring* is a system of wood or metal members used as temporary support for sides of excavations or construction components such as concrete formwork. Shoring consists of wood posts, steel pipes, or panelized shoring systems. Beams and joists are placed on the shoring to support deck forms. A *deck* is a floor made of light-gauge metal panels (decking). Dome forms are used to form voids in the concrete slab that create the beams in either one or two directions. **See Figure 6-6.**

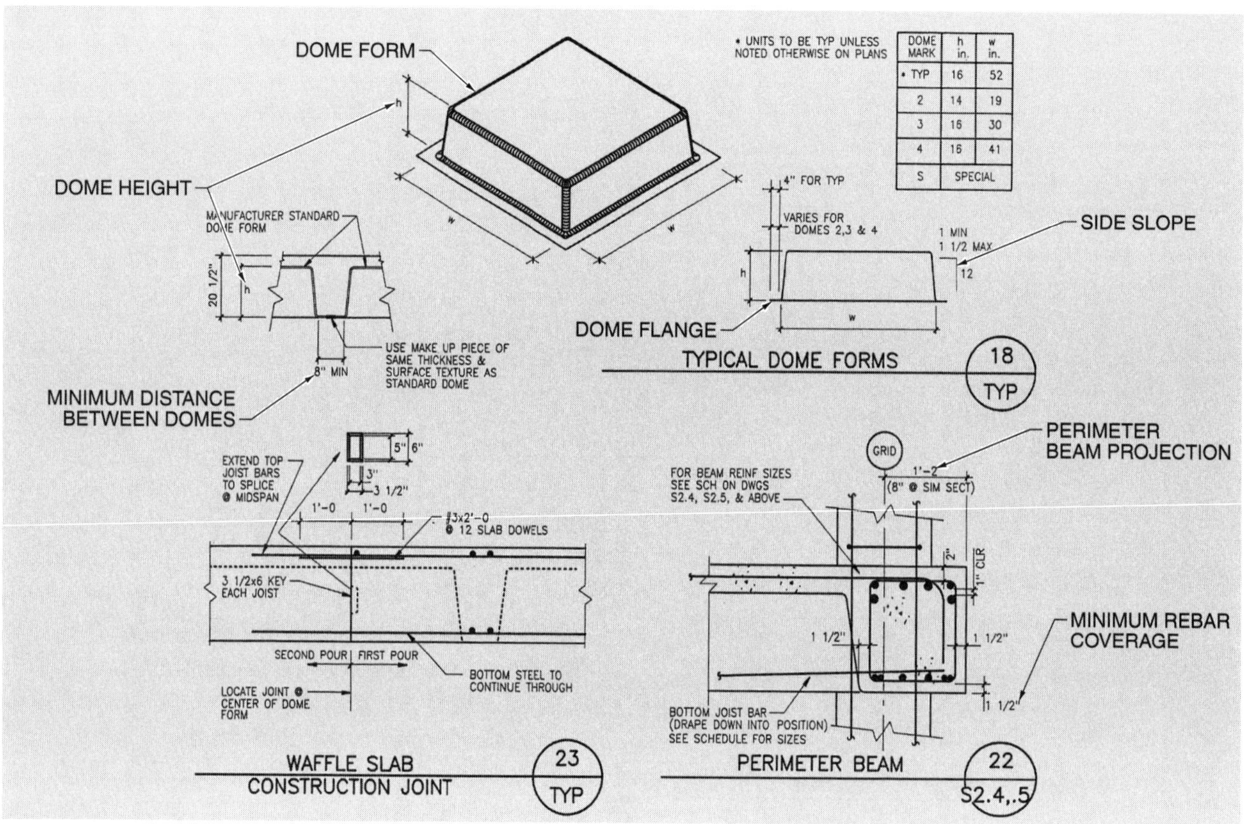

Figure 6-6. Steel or fiberglass dome forms are used to form voids in the concrete slab that create the beams in either one or two directions.

A *spandrel beam* is a beam in the perimeter of a structure that spans several columns and typically supports a floor or roof. Spandrel beams are formed along the outside edge of a deck. Perimeter edge forms hold the concrete in place and keep the top of the slab at the proper elevation.

Formwork for elevated slabs is designed by specialists to ensure that the proper amount of shoring is in place and the proper amount of camber has been designed into the forming system. *Camber* is the slight upward curve in a structural member designed to compensate for deflection of the member under load. Shoring must support the weight of the form itself; the fresh concrete; all reinforcing, electrical, and mechanical materials; and the tradesworkers who are placing and finishing the concrete. **See Figure 6-7.** Shoring is set on the slab below and erected to the proper elevation. Beams, joists, decking, edge forms, and other blockouts are set in place on the shoring.

Screw jacks or other adjustment mechanisms are generally installed in the shoring system to allow for final height adjustment prior to concrete placement. Large construction projects may use a system in which the shoring and possibly part of the formwork is designed as large units that can be moved as a single structure. This allows shoring, decking, and bracing to be set up and removed as a single unit, reducing labor costs. Shoring and bracing costs are significant for elevated slab construction.

Portland Cement Association

Figure 6-7. Shoring must support the weight of the form itself; the fresh concrete; all reinforcing, electrical, and mechanical materials; and the tradesworkers who are placing and finishing the concrete.

FORMWORK QUANTITY TAKEOFF

For large construction projects, information concerning the type and design of formwork is typically provided on a set of formwork drawings supplied by the concrete form supplier. **See Figure 6-8.** Formwork drawings are developed by formwork design specialists. Formwork design is based on dimensions provided on the prints. Formwork drawings indicate manufacturer form identification numbers and type, placement, form-fastening systems, form ties, and shoring and bracing information. Concrete formwork design takes into account all forces to be placed on the form.

The information pertaining to structural concrete foundations is contained in the site plans, foundation plans, and related detail drawings, and in Title 03 31 00 of the specifications. Estimators should consult these sources and other appropriate sections of the specifications and the complete construction documents to obtain information about foundation types, sizes, and materials to determine the amount of equipment, labor, and materials required for the foundation system.

An estimator must calculate the amount of formwork materials required when formwork drawings are not provided. Estimates for formwork should include calculations for footing and foundation wall forms, slabs on grade and elevated slabs, columns, beams, and the related shoring.

Formwork materials and installation labor for concrete footings and foundation walls are a major portion of the estimate calculations for concrete footings and foundation walls. Formwork materials include consumable materials such as snap ties and anchoring hardware, and reusable materials such as plywood, studs and bracing for job-built systems, and prefabricated forms for patented systems.

Form ties and anchoring hardware are taken off on a per piece basis. Calculations for formwork materials are based on the number of square feet of contact area (SFCA) for footings and foundation wall forms, and foundation wall thickness. For job-built forms, lumber quantities are determined in a manner similar to wood-framed walls. Any allowances for reuse of job-built form materials are calculated in conjunction with project field personnel and company historical records.

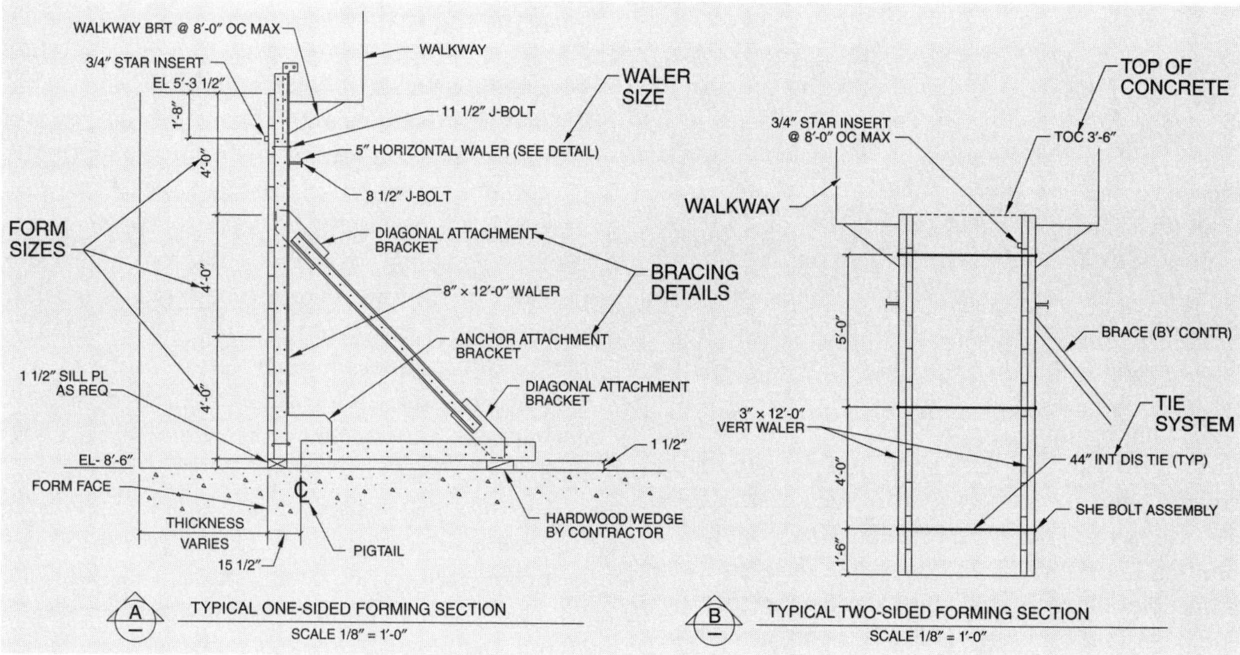

Figure 6-8. Formwork drawings provided by the form supplier aid the formwork estimating process.

Patented form systems may be either owned or rented. When owned, costs include depreciation and overhead. When rented, rental costs of the forms are the primary expense. In either case, labor expenses for transporting, setting, bracing, stripping, and cleaning are included in concrete footing and foundation expenses. Standard industry labor data, company historical data, or manufacturer recommendations are utilized to determine the labor time required for various concrete form applications.

The sequence of the concrete placement and schedule of the job will have an impact on the amount of forming materials required. The estimator should obtain information from the project superintendent or project manager to gain an understanding of the sizes and schedules for concrete placement. This will impact the amount of forming materials that will be required at any one time and allow for the appropriate reuse of forming materials according to this job site schedule.

Footings and Walls

Footing form information used by estimators includes the thickness, width, and length of the footings. Cast-in-place concrete wall information used in estimating formwork materials includes elevations to the top and bottom of the wall, wall thickness, placement of wall surface features, all embedded items, and reinforcing steel. Length and width dimensions for footings and foundation walls are noted on foundation plans, elevations, and sections. Concrete formwork is commonly taken off in linear feet or square feet of contact area. Estimators should check foundation plans, floor plans, and elevations for blockouts that could require additional form materials and labor.

Snap ties are calculated as a specified number of ties per square foot of wall area. Snap tie spacing is determined from the specifications, elevations, patented form manufacturer recommendations, or the snap tie loading information available from the manufacturer. For example, the specifications may state that snap ties are required 24″ OC vertically and horizontally for an 8″ thick wall. Each 4′ × 8′ section of wall (32 sq ft) requires 8 ties with an allowance for adjoining forms, resulting in one snap tie for each 4 sq ft of wall area (32 sq ft ÷ 4 sq ft/snap tie = 8 snap ties). **See Figure 6-9.** The total square feet of foundation wall contact area is divided by 4 to determine the number of snap ties needed for the foundation walls.

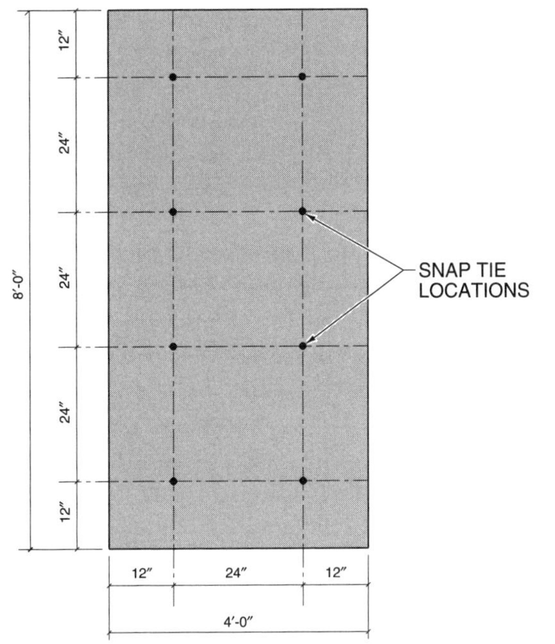

Figure 6-9. Snap ties are calculated as a specified number per square feet of wall area.

Labor costs for wall form construction are usually based on the SFCA. These costs are available from industry resource materials or company historical data. Variables for labor costs include the type of formwork used (such as job-built, patented, or ganged panel forms), lifting equipment for large formwork (such as cranes), the height of the wall that may require scaffold or fall protection equipment, job-site accessibility for erecting and stripping formwork, scheduling of the work in relation to other construction activities, special architectural details, and weather conditions.

Footings. Estimators utilize detail drawings, sections, and elevations to determine the footing depth, thereby determining the required heights of the footing forms and the required forming materials. The total linear feet of footings is multiplied by 2 to calculate the linear feet of footing form materials required for both sides of the footing. Additional materials may be required for stakes or bracing as required.

Job-Built Walls. The amount of formwork required for a concrete foundation wall is based on the length and contact area of the wall forms. The contact area is determined by multiplying the wall length (in ft) by wall height (in ft) and then multiplying this total by 2 to account for both sides of the wall. The number of 4′ × 8′ panels used for the form walls is determined by dividing the contact area by the area of a panel (32 sq ft).

> Deflection of wall forms occurs when panels of inadequate thickness are used or when studs or walers are spaced too far apart. Wall forms must be constructed to hold the concrete to a straight and true surface and contain the outward pressure of the concrete.

Determining Quantities of Wall Form Panels

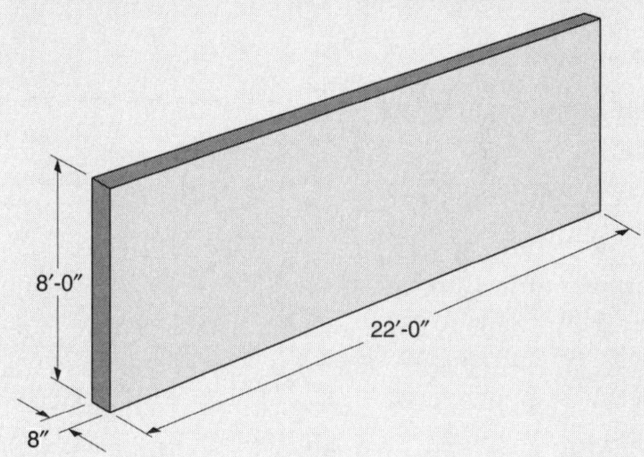

Determine the number of 4′ × 8′ wall form panels required for a wall measuring 8″ wide by 8′-0″ high by 22′-0″ long.

1. **Determine the contact area on one side of the wall forms.**

   contact area = height × length

   $A = h \times l$

   $A = 8' \times 22'$

   $A = 176$ sq ft

2. **Determine the number of panels required on one side of the wall forms.**

   no. of panels = contact area ÷ panel area

   no. of panels = 176 sq ft ÷ (4′ × 8′)

   no. of panels = 176 ÷ 32

   no. of panels = 5.5 panels, use 6 panels

3. **Determine the number of panels required for both sides of the wall forms.**

   no. of panels = no. of panels on one side × 2

   no. of panels = 6 × 2

   ***no. of panels* = 12**

The number of snap ties required for a wall form is based on the snap tie spacing and contact area of one side of the wall form.

Determining Quantities of Snap Ties

Project specifications state that snap ties for the 8′-0″ by 22′-0″ wall form should be spaced 12″ OC horizontally and 24″ OC vertically. Each 4′ × 8′ section of wall panel requires 16 snap ties (1 tie per 2 sq ft). Determine the number of snap ties required for the wall form.

1. Determine the number of snap ties.

no. of snap ties = *contact area of one side* ÷ *no. of square foot per tie*

no. of snap ties = 176 sq ft ÷ 2 sq ft

***no. of snap ties* = 88 ties**

Framing lumber is calculated separately and is based on the spacing of the wall form studs. The number of wall form studs is determined by dividing the wall length by the on-center distance between the studs. The number of wall form studs is doubled when there are two studs at each position. The number of wall form studs is also doubled to determine the total number of form studs required for both sides of the wall. Additional studs are included in the estimate for the ends of the wall form.

Determining Quantities of Wall Form Studs

The wall form studs for the 8′-0″ high by 22′-0″ wide wall are spaced 16″ OC. Two studs are specified at each location on each side of the wall. An extra set of studs is added for the ends of the wall forms on each side. Determine the number of wall studs required for the wall form.

1. Determine the number of wall studs required on one side of the wall form.

no. of wall form studs on one side = [(*length of wall form* ÷ *stud spacing (in feet)*) × *no. of studs at each location*] + *extra studs*

no. of wall form studs on one side = [(22′ ÷ 1.33′) × 2] + 2

no. of wall form studs on one side = 35 studs

2. Determine the number of wall studs for both sides of the wall form.

no. of wall form studs on both sides = *no. of wall form studs on one side* × 2

no. of wall form studs on both sides = 35 × 2

***no. of wall form studs on both sides* = 70 studs**

Top and bottom plates may also be part of the form panel design, similar to a wood-framed wall. The linear feet of plates is determined by multiplying the wall form length by 2 for a single top and bottom plate. This value is then multiplied by 2 to account for plates on both sides of the wall form. The wall form length is multiplied by 4 when double top and bottom plates are required.

Determining Quantities of Plates

Based on the specifications and prints, single top and bottom plates are required for the 22′-0″ wall form. Determine the linear feet of plates required.

1. Determine the linear feet of plates required for one side of the wall form.

top and bottom plates (in linear feet) on one side = *wall form length* × 2

top and bottom plates (in linear feet) on one side = 22′ × 2

top and bottom plates (in linear feet) on one side = 44′

2. Determine the linear feet of plates required for both sides of the wall form.

top and bottom plates (in linear feet) on both sides = *top and bottom plates (in linear feet)—one side* × 2

top and bottom plates (in linear feet) on both sides = 44′ × 2

***top and bottom plates (in linear feet) on both sides* = 88′**

Form walers are spaced horizontally at the specified on-center distance, which is determined by the concrete wall width, rate of concrete placement, and other structural considerations in formwork design. Lower rows of walers may be doubled because of the greater hydrostatic pressure at the bottom of the form. As the height of the form increases, a greater number of double rows of walers is required due to increased hydrostatic pressure.

The number of walers needed is calculated by dividing the height of the wall form by the on-center spacing to determine the number of rows of walers. The number of rows is multiplied by the number of walers in each row. This value is multiplied by the wall length to determine the linear feet of walers required for one side. If walers are required on both sides of the wall form, the linear feet is multiplied by 2.

Determining Quantities of Walers

Based on the formwork drawings, the 8′-0″ high by 22′-0″ long concrete wall has double walers spaced 2′-0″ OC. Determine the linear feet of walers required for the wall.

1. **Determine the number of rows of walers.**

   no. of rows = *wall height* ÷ *waler spacing (in feet)*

   no. of rows = 8′ ÷ 2′

   no. of rows = 4 rows

2. **Determine the number of walers in each row.**

   no. of walers in each row = *no. of rows* × 2 (for double walers)

   no. of walers in each row = 4 × 2

   no. of walers in each row = 8 walers

3. **Determine the linear feet of walers for one side of the wall form.**

   walers (in linear feet) = *no. of walers in each row* × *wall length*

   walers (in linear feet) = 8 × 22

   walers (in linear feet) = 176′

4. **Determine the linear feet of walers for both sides of the wall form.**

   walers total (in linear feet) = *walers (in linear feet)* × 2

   walers total (in linear feet) = 176′ × 2

   ***walers total (in linear feet)* = 352′** (an allowance may be made for overlap at corners)

Patented Wall Forms. Calculating the number of patented wall forms typically requires a wall form elevation and a schedule of the forms needed for each wall form to be developed. The formwork schedule indicates the size and type of the forms and the quantity of each required for a specific job. An estimator must also calculate the amount of fastening hardware required based on the fastening method that is specified. Snap ties and walers are estimated in the same manner as job-built forms.

Slabs

A slab on grade usually requires a minimal amount of formwork. Formwork may be more complex in special applications, such as ice rinks, stepped seating, or freezer slabs. The estimator must take into consideration the proper subsurface preparation and reinforcement when developing an estimate. Elevated slab takeoff requires more planning and information because of the equipment and materials required for the formwork.

Formwork for slabs depends on the type of slab and whether the sides of the slabs are to be open or framed. The slab formwork is an extension of the exterior wall formwork and calculated as such if the slab is to be a continuous member and placed as part of the walls.

Slabs on Grade. Formwork for a slab on grade consists of forms placed along the slab perimeter and bracing of wood or metal stakes. Calculating the linear feet around the perimeter of the slab results in the linear feet of forms needed. **See Figure 6-10.** The slab thickness determines the width of lumber or steel forms needed. Wood or metal stakes are used to maintain alignment of the forms. Additional formwork may be required for open areas in the slab.

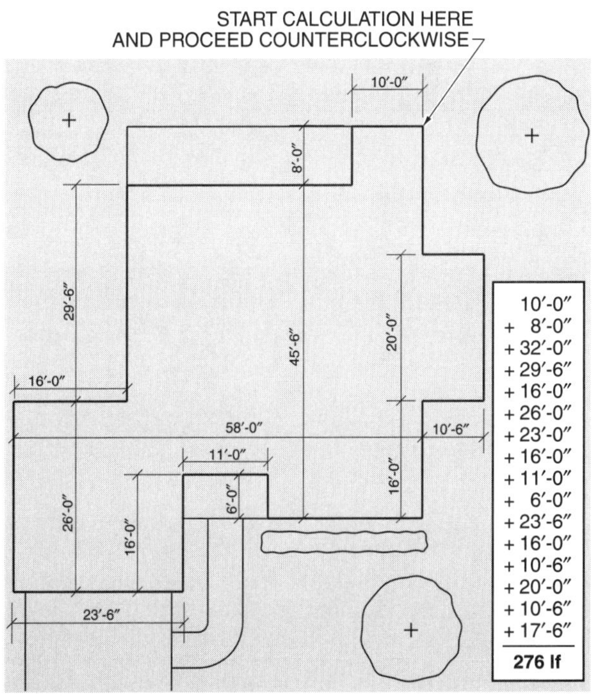

Figure 6-10. Plan views provide perimeter information for takeoff of slab formwork.

Elevated Slabs. Formwork components for elevated slabs include beams, columns, decking, dome forms, and shoring. Formwork for elevated slabs may be calculated as the SFCA of the underside of the slab.

Formwork for beams is calculated as the SFCA of exposed surface, beam sides, and beam bottom. An outside beam side includes a slab edge form. An intermediate beam has formwork on three sides (beam sides and bottom). A spandrel beam has formwork on three sides. To calculate the amount of formwork needed for a spandrel beam, for example, the widths of the three exposed sides are added, and the total width is multiplied by the length of the beams that size.

When estimating column formwork, an estimator must note that column dimensions vary. The first step in determining formwork requirements for columns is to develop a schedule of column dimensions and heights and determine the total number of each type of column form to be filled at each placement. **See Figure 6-11.** Steel column formwork is commonly removed and reused.

For columns, formwork is calculated in a similar manner to wall formwork. The contact area is calculated in square feet or the number of patented forms required for each. For square or rectangular columns, the column perimeter is multiplied by the column height to determine the square feet of formwork needed. For job-built wood column forms, additional vertical lumber bracing may be required based on the form design. Circular columns are commonly formed using fiberboard or metal forms. The height and diameter of each column should be calculated and entered into the estimate. For decking, the square feet of slab area is determined by multiplying the slab length by slab width.

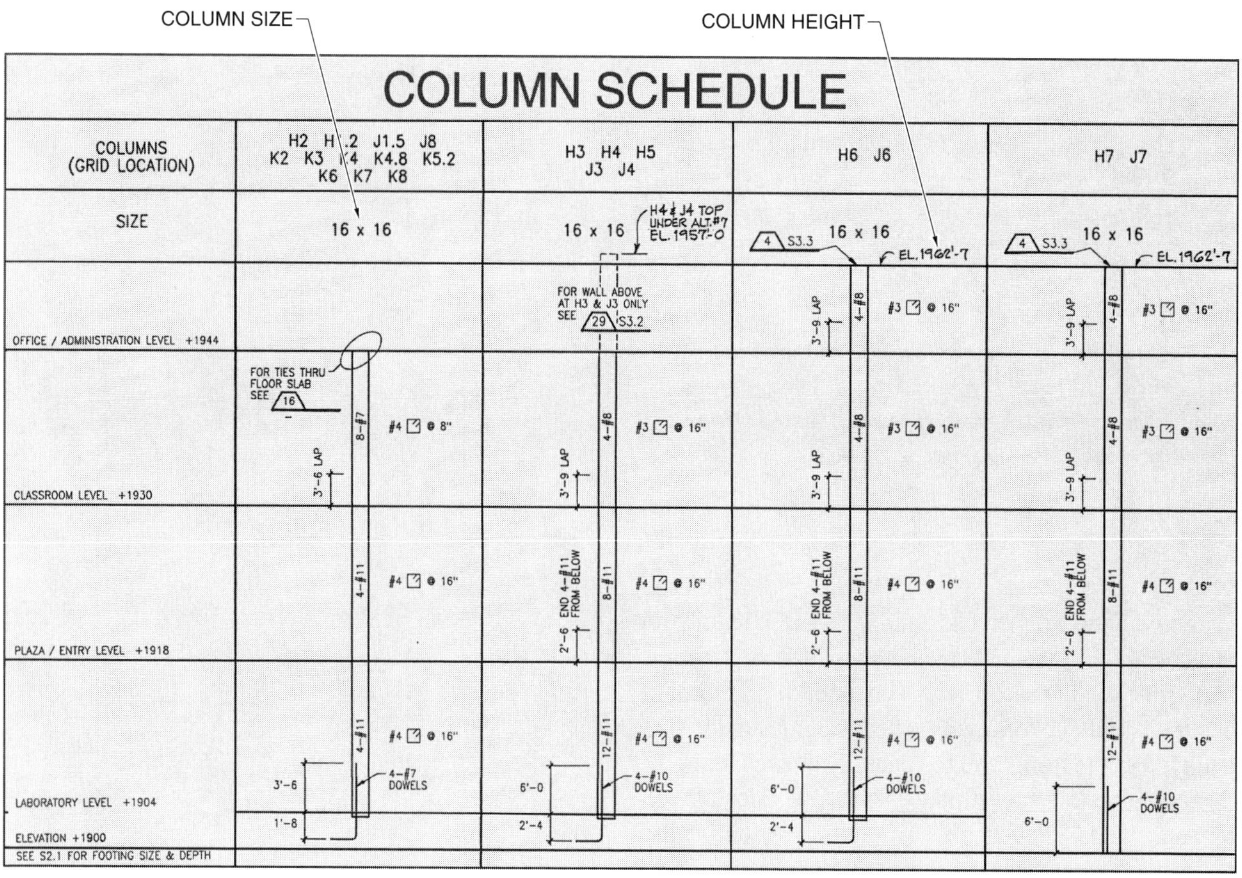

Figure 6-11. A column schedule specifies column profile and height information for calculating formwork.

Where dome forms are used for a one-way or two-way beam system, the estimator should develop a schedule in a manner similar to a column schedule. Using the prints, an estimator determines the total number of each type of dome form required and enters the totals in a ledger sheet, spreadsheet, or estimating program cell.

Shoring requirements are specified by an engineer based on the live and dead loads placed on formwork. Estimators should consult shoring manufacturers and formwork designers for detailed information and costs concerning loading capabilities of various shoring systems. Costs per square foot for shoring depend on the height between floors for an elevated slab, the thickness of the slab, and the concrete placement method. Shoring placement, removal, and reuse will be dependent of the concrete placement schedule and should be coordinate with the project manager. These costs are based on manufacturer and supplier information or company historical data.

CAST-IN-PLACE CONCRETE MATERIALS

The prints and specifications contain notes and other information concerning specific concrete performance requirements, including 28-day strength, size of aggregate, slump, water-cement ratio, air entrainment, and admixtures. **See Figure 6-12.** Overall concrete strength is stated as pounds per square inch (psi) of pressure that the concrete can withstand after 28 days.

Concrete

Concrete is composed of portland cement, fine and coarse aggregate such as sand and gravel, water, and sometimes admixtures. By varying the quantity and type of each component, concrete properties such as workability, strength, and setting time are affected. Admixtures may be introduced into the mixture to alter the properties of the concrete. An *admixture* is a substance other than water, aggregate, and portland cement that is added to concrete to modify its properties.

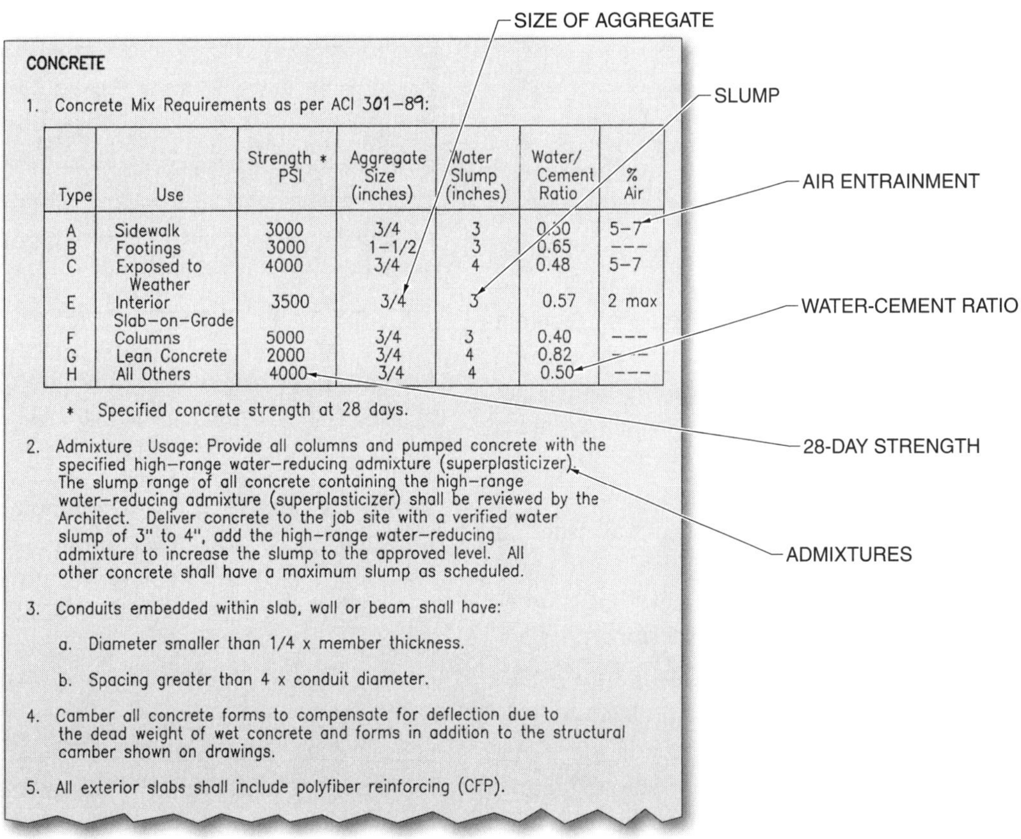

CONCRETE

1. Concrete Mix Requirements as per ACI 301–89:

Type	Use	Strength * PSI	Aggregate Size (inches)	Water Slump (inches)	Water/ Cement Ratio	% Air
A	Sidewalk	3000	3/4	3	0.50	5–7
B	Footings	3000	1–1/2	3	0.65	---
C	Exposed to Weather	4000	3/4	4	0.48	5–7
E	Interior Slab–on–Grade	3500	3/4	3	0.57	2 max
F	Columns	5000	3/4	3	0.40	---
G	Lean Concrete	2000	3/4	4	0.82	---
H	All Others	4000	3/4	4	0.50	---

* Specified concrete strength at 28 days.

2. Admixture Usage: Provide all columns and pumped concrete with the specified high–range water–reducing admixture (superplasticizer). The slump range of all concrete containing the high–range water–reducing admixture (superplasticizer) shall be reviewed by the Architect. Deliver concrete to the job site with a verified water slump of 3" to 4", add the high–range water–reducing admixture to increase the slump to the approved level. All other concrete shall have a maximum slump as scheduled.
3. Conduits embedded within slab, wall or beam shall have:
   a. Diameter smaller than 1/4 x member thickness.
   b. Spacing greater than 4 x conduit diameter.
4. Camber all concrete forms to compensate for deflection due to the dead weight of wet concrete and forms in addition to the structural camber shown on drawings.
5. All exterior slabs shall include polyfiber reinforcing (CFP).

Figure 6-12. Concrete design specifications affect concrete costs and placement requirements.

Portland Cement Association
A slump test is conducted on the job site to measure the consistency of concrete.

The amount of concrete slump may be noted in the specifications. A slump test is used to determine concrete slump. A *slump test* is a test that measures the consistency, or slump, of concrete. In general, concrete slump can be used as an indicator of water content in fresh concrete. Water content in a concrete mixture is important in determining final strength. The proper amount of water is required to fully hydrate the cement, while excess water weakens the mixture. The *water-cement ratio* is the weight of the water divided by the weight of the cement contained in one cubic yard of concrete.

Portland Cement. Portland cement is classified by the ASTM International as Type I, Type II, Type III, Type IV, and Type V. Type I cement is general-purpose cement used where no special design properties are required and there is no exposure to sulfates. Type II cement produces slightly lower temperatures during hydration than Type I cement and provides moderate protection in areas where concrete may be exposed to sulfates.

Type III cement is used where forming needs to be removed rapidly, cold temperatures are prevalent during placement, or high strength is needed at an early stage. Type IV cement generates a low amount of heat during hydration and is suitable for large placements of concrete. Type V cement has a high resistance to sulfates.

Aggregate. *Aggregate* is granular material such as gravel, sand, vermiculite, or perlite that is added to a cement and water mixture to form concrete, mortar, or plaster. Aggregate is categorized as fine or coarse, with particles less than ⅜″ graded as fine, and those equal to or greater than ⅜″ graded as coarse. Types of material used for aggregate include sand, crushed stone or gravel, shale, slate, clay, or slag. Aggregate, including sand and gravel, represents approximately 70% of the overall volume of concrete.

Information in the specifications regarding aggregate affects final concrete strength and finishing properties. Specifications may describe the type and maximum size of aggregate used, which affects the overall costs.

Admixtures. Admixtures are added to concrete to affect air content, setting time, freeze resistance, and workability. The addition of admixtures for air entrainment creates minute air pockets in concrete that increase workability and resistance to freezing and thawing, and decrease separation of the concrete ingredients. **See Figure 6-13.** Information concerning use and types of admixtures is included in the specifications.

RECOMMENDED AIR CONTENT*		
Nominal Maximum Size of Coarse Aggregate†	Exposure	
	Mild	Extreme
⅜	4.5	7.5
½	4.0	6.0
¾	3.5	6.0
1	3.0	6.0
1½	2.5	5.5
2	2.0	5.0
3	1.5	4.5

* in %
† in in.

Figure 6-13. Aggregate and exposure variables affect the air content in concrete.

Reinforcement. The most common method of reinforcing piles, roadways, footings, columns, walls, and beams is with steel reinforcing bars, which have deformed surfaces. These reinforcing bars, or rebar, are tied together in various configurations to reinforce large concrete structures. Deformations are manufactured into the surface of rebar to improve its adhesion to the surrounding concrete. Bars sized #3 through #8 are produced in increments of ⅛″ diameter. For example, a #4 bar is 4/8″ or ½″ in diameter. **See Figure 6-14.**

Deformed Steel Reinforcing Bars

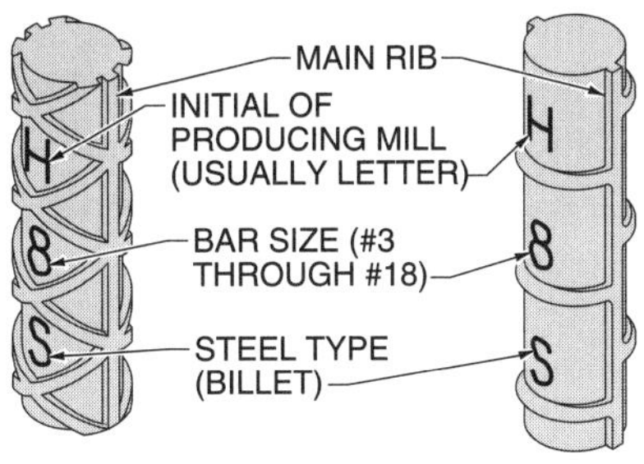

STEEL GRADE	STEEL TYPE
S	BILLET
I	RAIL
A	AXLE
W	LOW ALLOY

STANDARD REBAR SIZES			
Bar Size Designation	Weight Per Foot*	Diameter†	Cross-Sectional Area Squared†
#3	0.376	0.375	0.11
#4	0.668	0.500	0.20
#5	1.043	0.625	0.31
#6	1.502	0.750	0.44
#7	2.044	0.875	0.60
#8	2.670	1.000	0.79
#9	3.400	1.128	1.00
#10	4.303	1.270	1.27
#11	5.313	1.410	1.56
#14	7.650	1.693	2.25
#18	13.600	2.257	4.00

* in lb
† in in.

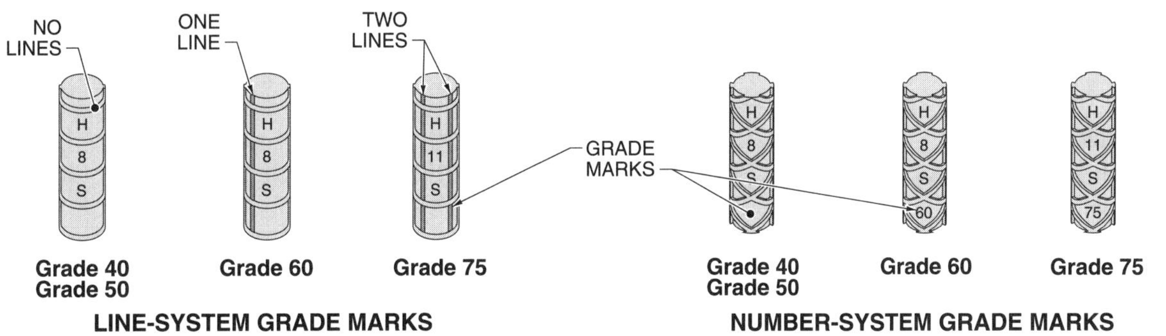

REINFORCING STEEL STRENGTH AND GRADE		
ASTM Specification	Minimum Yield Strength*	Ultimate Strength*
Billet Steel ASTM A615 Grade 40 Grade 50 Grade 75	 40,000 50,000 75,000	 70,000 90,000 100,000
Rail Steel ASTM A616 Grade 50 Grade 60	 50,000 60,000	 80,000 90,000
Axle Steel ASTM A617 Grade 40 Grade 60	 40,000 50,000	 70,000 90,000
Deformed Wire ASTM A496 Welded Fabric	 70,000	 80,000
Cold-Drawn Wire ASTM A82 Weded Fabric < W1.2 Size ≥ W1.2	 56,000 65,000	 70,000 75,000

* in psi

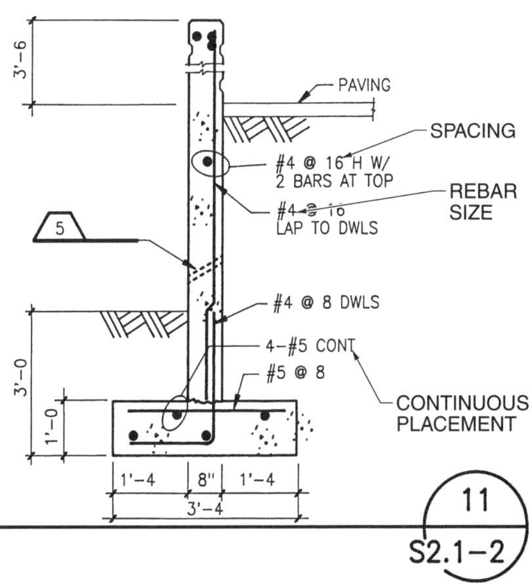

Figure 6-14. Rebar is identified with a standardized system denoting size and strength characteristics.

Rebar is manufactured from steel graded 40, 50, 60, and 75. Grade 40 has minimum yield strength of 40,000 psi; Grade 50, a minimum yield strength of 50,000 psi; Grade 60, a minimum yield strength of 60,000 psi; and Grade 75, a minimum yield strength of 75,000 psi. Epoxy-coated rebar is commonly used for exterior applications, such as road and bridge construction, to minimize the effects of corrosion on the steel.

Welded wire reinforcement is also used to reinforce concrete slabs. Rolls or mats of welded wire reinforcement are stretched out and set in place to reinforce slabs such as sidewalks and driveways. Welded wire reinforcement size is designated by W number (spacing and cross-sectional area) or wire gauge. **See Figure 6-15.** Wires may be smooth or deformed. Welded wire reinforcement has yield strength between 60,000 psi and 70,000 psi.

Architectural, structural, and detail drawings for concrete structures provide reinforcing steel information. Cages of vertical pieces of rebar with intermediate stirrup ties are formed for columns. The size of the rebar and stirrup ties and the spacing of the bars in both directions are indicated on the column schedule. **See Figure 6-16.** A variety of rebar sizes and shapes are required for beams. A schedule for sizes, shapes, and lengths is provided on the beam schedule. Schedules for the placement of rebar in cast-in-place concrete joists and slabs are also included in the structural drawings.

CAST-IN-PLACE CONCRETE METHODS

Proper transportation, handling, and placement of concrete are required for the production of the final concrete structure. Estimators must ensure that equipment and provisions are included in the bid for accessibility of the concrete source to the final placement site, equipment required for moving the concrete from the supply source to the forms, proper placement in the forms, consolidation after placement, and possible weather protection if additional heating, cooling, or hydration water is required.

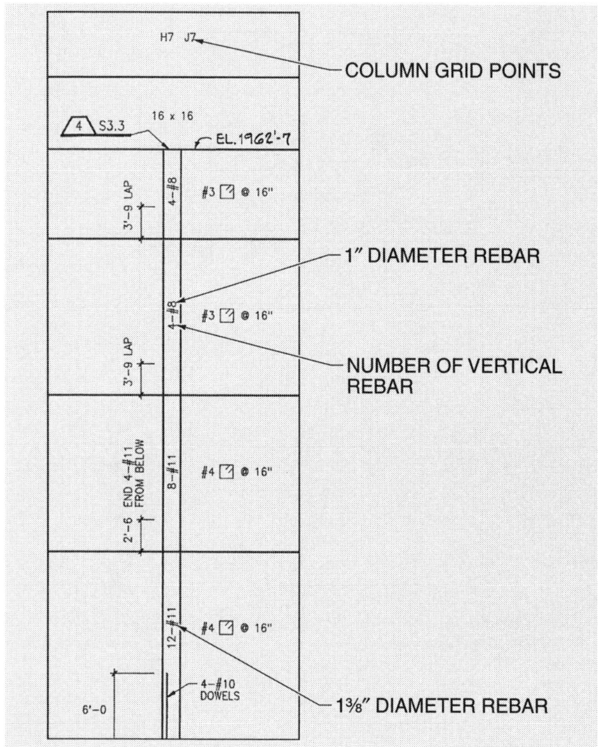

Figure 6-16. A column schedule provides information concerning rebar used in concrete columns.

COMMON STOCK STYLES OF WELDED WIRE REINFORCEMENT				
New Designation (W Number)	Old Designation (Wire Gauge)	Steel Area*		Weight†
		Longitudinal	Transverse	
6 x 6 – W1.4 x W1.4	6 x 6 – 10 x 10	0.028	0.028	21
6 x 6 – W2.0 x W2.0	6 x 6 – 8 x 8	0.040	0.040	29
6 x 6 – W2.9 x W2.9	6 x 6 – 6 x 6	0.058	0.058	42
6 x 6 – W4.0 x W4.0	6 x 6 – 4 x 4	0.080	0.080	58
4 x 4 – W1.4 x W1.4	4 x 4 – 10 x 10	0.042	0.042	31
4 x 4 – W2.0 x W2.0	4 x 4 – 8 x 8	0.060	0.060	43
4 x 4 – W2.9 x W2.9	4 x 4 – 6 x 6	0.087	0.087	62
4 x 4 – W4.0 x W4.0	4 x 4 – 4 x 4	0.120	0.120	85

* in sq in./ft
† in lb per 100 sq ft

Figure 6-15. Floor slabs, walkways, and driveways are often reinforced using welded wire fabric.

Placement

A transit-mix or ready-mix truck is typically used to deliver concrete to a job site for use in walls and slabs. Concrete is mixed, or batched, and transported to the construction site. It may be more economical to rent or construct a batch plant at the construction site for large projects due to the size of the job or accessibility from existing batch plants. Estimators should determine the most economical method for delivery and placement of concrete into forms for walls and slabs.

Additional labor calculations are required for concrete placement. Where transit-mix trucks have ready access to the forms, concrete placement into forms can be performed with minimal labor. In other situations, a crane and bucket or concrete pumping equipment may be required to reach remote locations. These additional costs include equipment ownership or rental costs related to concrete transportation and placement, labor needed to operate the equipment, and additional labor for final concrete placement and equipment maintenance. These labor rates and costs are available from standard industry labor data related to the number of cubic yards of concrete to be placed or company historical data.

Walls. For most foundation walls, concrete can be discharged from the transit-mix truck directly into the wall forms using a chute. Where walls are far below or above grade or cannot be accessed by a transit-mix truck, a concrete pump is used to place concrete and prevent long drops that could result in separation of the cement paste and aggregate. **See Figure 6-17.** For slabs or walls above grade, a concrete pump or bucket attached to a crane may also be used to place concrete.

Slabs. Concrete for a slab on grade may also be discharged into motorized buggies or wheelbarrows for placement. For slabs on grade and elevated slabs, concrete pumps or buckets attached to cranes are also used where height or accessibility due to reinforcement or other obstacles does not permit access by other means.

Finishes

Estimators should consult the elevations and specifications to determine the finishes for exposed concrete surfaces of walls, slabs, columns, beams, and decks. **See Figure 6-18.** The surface finish required can affect the formwork required, form removal times, and cost per square foot for surface finish.

Portland Cement Association

Figure 6-17. Concrete pumps deliver concrete to areas that are inaccessible to concrete transit-mix trucks.

Portland Cement Association

Figure 6-18. Concrete slab finish requirements affect labor costs and are described in the specifications.

Walls. Exposed concrete wall surfaces may have a variety of finishes including smooth, textured, and various architectural designs. Smooth concrete is produced when the surface of the concrete is rubbed and stoned smooth after form removal. Any voids are patched and filled to produce a smooth surface. A rough-textured concrete is commonly produced by inclusion of decorative aggregate and/or sandblasting of the concrete surface.

Architectural designs can be produced by placing wood, metal, or plastic form liners inside of the concrete forms. Designs such as vertical or horizontal grooves or a simulated brick pattern can be produced. Specialized patented forms may also be utilized to create unique surface finishes. **See Figure 6-19.** Architectural designs and specialized finishes increase the cost per square foot for formwork and finishing of wall surfaces.

Figure 6-19. A variety of finishes may be used for exposed concrete wall surfaces.

Slabs. Slab surfaces may have smooth, exposed-aggregate, patterned, or superflat finishes. Broom-finish slabs are used for common applications including sidewalks and roadways. Concrete is placed, screeded to grade, and finished with trowels or brooms as noted in the specifications to produce the required smooth finish. Exposed aggregate surfaces are produced by including decorative aggregate in the concrete and washing and brushing the cement paste off the slab surface prior to reaching a final set.

Patterned concrete slabs are created by using rubber or metal patterns to create brick, stone, or other decorative finishes. Colorants may be added to the concrete to create special effects. Superflat floors are used in industrial applications where automated equipment or machinery requires a very flat surface to ensure proper operation after installation. Laser screeds are used to create an extremely flat surface during finishing of the concrete. The cost for the surface finish of slabs must be included in the estimate.

Elevated-slab components also include columns, beams, and the undersides of slabs. In a manner similar to the process for concrete walls, the specifications are checked to ensure provisions are included in the estimate for surface finishing of these items.

CAST-IN-PLACE CONCRETE QUANTITY TAKEOFF

Items to consider for concrete quantity takeoff include the type and volume of concrete required, type of size of reinforcement, transportation of concrete and other materials, surface finish, and possible weather protection. Each of these items is typically unit-priced based on the standard concrete volume measurement of a cubic yard. Costs for waiting time during concrete delivery and discharge into forms, as well as items in the concrete mixture such as special sand or cement, must be included in the final concrete estimate.

Footing and Foundation Wall Takeoff

Foundation plans, elevations, and detail drawings provide the quantity takeoff information for footings and foundation walls. The length, width, and height of footings and foundation walls are found in these sources.

Volume Calculations

Concrete volume is taken off and expressed in cubic yards. Knowledge of basic geometric volume formulas is essential when calculating concrete volume. (Volume formulas for a variety of shapes are included in the Appendix.) In some estimating programs, basic dimensional information is entered into the proper cell and the concrete volume is automatically calculated. Concrete estimates must include divisions for various concrete mixtures used and variables in the mixtures where changes in aggregate, cement, admixtures, or reinforcement may affect costs.

Footings and Walls. Concrete foundation walls and sizes are often grouped together by the estimator. The volume of concrete for walls and footings is calculated by multiplying the wall or footing thickness, width, and length dimensions (in feet) and dividing by 27 (1 cu yd = 27 cu ft). Using some estimating programs, an estimator can enter the concrete width, length, and thickness and have the program calculate the volume of concrete. **See Figure 6-20.**

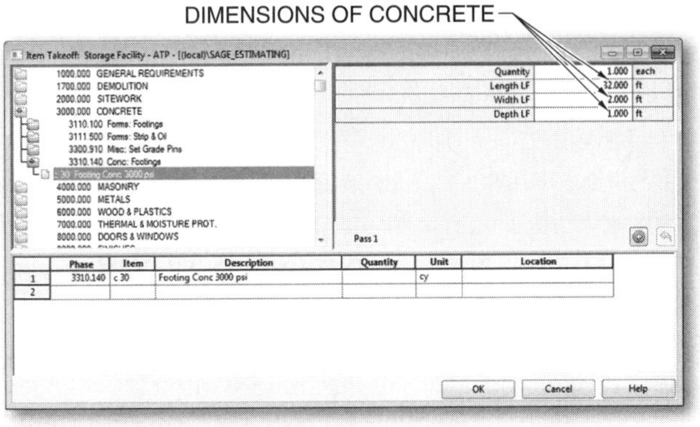

Figure 6-20. Estimating software allows the estimator to calculate the volume of concrete by entering the dimension information.

Determining Concrete Volume for Walls

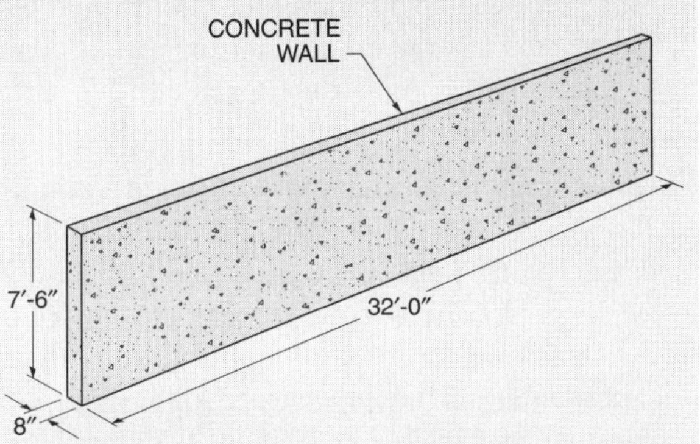

Determine the volume of concrete (in cubic yards) required for the 8″ × 7′-6″ × 32′-0″ wall.

$$V = \frac{t \times w \times l}{27}$$

$$V = \frac{0.67' \times 7.5' \times 32'}{27}$$

$$\mathbf{V = 5.96\ cu\ yd}$$

Each type of concrete should be priced separately according to the cost per cubic yard from the supplier. Concrete volumes may be reduced for applications containing large or numerous blockouts. A 2% to 5% waste allowance should be added to each concrete calculation to allow for spillage during placement. Where concrete is being placed in a footing form in contact with the earth, a higher waste percentage up to 10% should be added to compensate for potential unevenness of the excavation. It is best to add additional amounts onto the amount of concrete for a project to ensure that sufficient quantities are ordered at the time of placement to keep crews working.

Reinforcing steel placement, spacing, and sizes are determined from detail drawings. The linear feet of each type of rebar is determined by multiplying the linear feet of wall or footing by the reinforcement required per linear foot.

Slabs. The volume of concrete for slabs is determined by multiplying the surface area (in square feet) by the slab thickness. An estimator can multiply the slab width, length, and thickness to calculate the volume of concrete.

Determining Concrete Volume for Slabs

Determine the volume of concrete (in cubic yards) required for a 6″ thick slab measuring 10′-0″ × 20′-0″.

$$\text{volume (in cubic yards)} = \frac{(\text{width} \times \text{length}) \times \text{thickness (in feet)}}{27}$$

$$V = \frac{(w \times l) \times t}{27}$$

$$V = \frac{(10' \times 20') \times 0.5'}{27}$$

$$\mathbf{V = 3.7\ cu\ yd}$$

The amount of each type of rebar needed is determined by calculating the spacing in each direction and calculating the amount of rebar required in a given number of square feet. Additions are made for the rebar overlap noted in the specifications or detail drawings. Rebar is commonly priced by the cost per ton of steel, with costs varying with the diameter of the rebar and type of steel required. Welded wire fabric is calculated based on the amount of coverage area required (in square feet).

Placement of expansion joints, electrical conduit, plumbing pipes, and any other items below the slab must be completed prior to concrete placement. Proper placement and protection of these items during concrete placement can impact the difficulty or ease of concrete placement. Costs for expansion joint material are typically included in this portion of the bid. These quantities are determined by the linear foot.

Columns. The volume of concrete required for columns is based on the cross-sectional area of the column and the column height.

Determining Concrete Volume for Columns

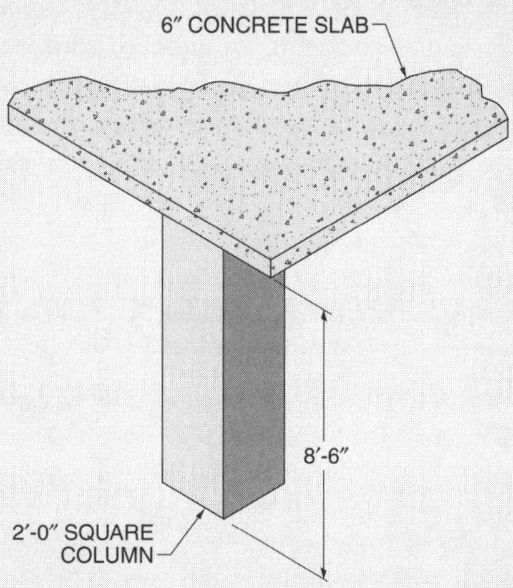

Determine the volume of concrete (in cubic yards) required for the 2′-0″ square by 8′-6″ high column.

$$\textit{volume (in cubic yards)} = \frac{\textit{cross-sectional area} \times \textit{height}}{27}$$

$$V = \frac{A \times h}{27}$$

$$V = \frac{(2' \times 2') \times 8.5'}{27}$$

$$\mathbf{V = 1.26\ cu\ yd}$$

When estimating columns using a ledger sheet or spreadsheet, estimators enter the total for each type of column and calculate the total number of each type of column. With some estimating programs, entry of the column dimensions and the number of columns results in a final volume calculation in cubic yards. Estimators must verify the column heights and cross-sectional dimensions for each column since differences may occur.

Schedules and notes on the prints provide information related to rebar for columns. The quantities and sizes of rebar for each column are determined and entered in a ledger sheet, spreadsheet, or estimating program. The total number of each type of rebar is calculated based on the individual totals.

Stairways. Concrete volume for stairways is calculated by multiplying the number of treads by the width of the stairway. The detail drawings should be consulted for dimensions pertaining to concrete-stairway volume calculations. A rough estimate of concrete volume can be made by allowing one cubic foot of concrete per linear foot of tread. For example, to calculate the volume of concrete required for a concrete stairway with ten 3′-6″ wide treads, multiply the number of treads by the tread width (10 × 3.5′ = 35 cu ft; 35 cu ft ÷ 27 = 1.3 cu yd).

Labor

Various labor costs are incurred with concrete forming and placement and finishing. Estimators can obtain labor information from standard industry tables or company historical data. Variables related to concrete forming and placement should be coordinated with the project manager to ensure that any unique conditions related to the job site are taken into account in the final estimate.

Forming. Labor costs related to concrete formwork for footings, foundation walls, slabs, and columns include transporting and distributing concrete formwork to the proper location on the job site, erection of the forms, blockout installation, application of form-release agents, and bracing and securing the forms in preparation for concrete placement. Additional labor costs are incurred for placement of reinforcing steel. After the concrete reaches the required set, labor costs include form removal, form cleaning, and transportation of formwork from the job site to storage areas or the next job site.

The amount of labor required for each of these operations is based on the forming system utilized and job site conditions. Form suppliers, standard industry sources, and company historical data are used to determine labor costs associated with forming.

Concrete Placement and Finishing. The primary labor costs related to concrete placement include transportation of the concrete to the forms, spreading and screeding the concrete to the proper height and thickness, consolidation of the concrete through vibration, and initial surface finishing. Finishing costs include any additional surface treatments to the concrete following form removal, such as surface rubbing, sandblasting, or concrete sawing for control joints.

Where weather protection is necessary during concrete curing, additional labor costs may be incurred related to installation, maintenance, and removal of the weather barriers. Labor costs for concrete placement are typically based on the cubic yard. Labor costs for surface finishing are based on the surface area (in square feet). Standard industry labor rates and company historical data are utilized to accurately determine these costs.

PRECAST CONCRETE MATERIALS

Developments in concrete design and lifting equipment have led to the widespread use of precast concrete members. Precast concrete members are used for beams, walls, slabs, piles, and a variety of other structural applications. Precast concrete members may be precast at the job site and lifted into place or formed at a casting yard, transported to the job site by truck or barge, and lifted into position. Precast members require special reinforcement to ensure they meet the stresses of transportation and installation in addition to being able to support the final structural loads. **See Figure 6-21.** Concrete members that are precast and transported to the job site include beams, piles, exterior wall finish panels, paving materials, and various piping and masonry units.

Portland Cement Association

Figure 6-21. Precast concrete members minimize the need for extensive job-site formwork for walls or other concrete members.

Portland Cement Association

Precast concrete members, such as this double-T roof panel, are commonly formed at a casting yard, transported to the job site by truck, and lifted in position.

Beams

Precast concrete beams are commonly used for building and bridge construction. Precast concrete beams include rectangular, L-shaped, and inverted-T beams. **See Figure 6-22.** Structural engineers design the placement of reinforcing steel based on loads to be imposed on the beam. Beams and other precast members are commonly reinforced by pretensioning or post-tensioning.

Pretensioning is a method of prestressing reinforced concrete in which the steel reinforcing cables (tendons) in the structural member are tensioned before the concrete has hardened. In a pretensioned beam, the tendons are put in tension before the concrete is placed. After the concrete is set, the tension on the cables is transferred to the concrete, putting the beam in tension. *Post-tensioning* is a method of prestressing concrete in which the reinforcing cables (tendons) are tensioned after the concrete has hardened.

SAFE IMPOSED LOADS FOR PRECAST CONCRETE BEAMS*													
Beam	**Designation**	**Strand No.**	**H†**	**H1/ H2†**	**Span‡**								
					18	**22**	**26**	**30**	**34**	**38**	**42**	**46**	**50**
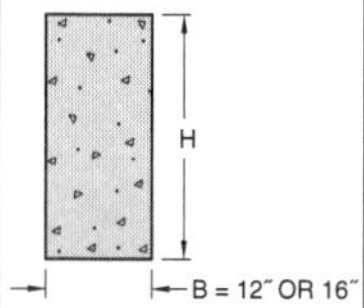 RECTANGULAR	12RB24	10	24	—	6726	4413	3083	2248	1684	1288	1000	—	—
	12RB32	13	32	—	—	7858	5524	4059	3080	2394	1894	1519	1230
	16RB24	13	24	—	8847	5803	4052	2954	2220	1705	1330	—	—
	16RB32	18	32	—	—	—	7434	5464	4147	3224	2549	2036	1642
	16RB40	22	40	—	—	—	—	8647	6599	5163	4117	3332	2728
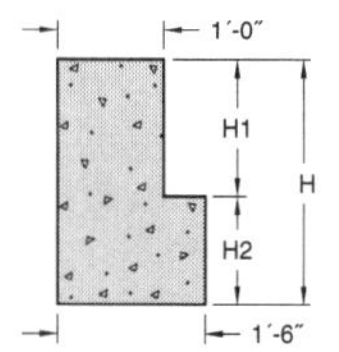 L-SHAPED	18LB20	9	20	12/8	5068	3303	2288	1650	1218	—	—	—	—
	18LB28	12	28	16/12	—	6578	4600	3360	2531	1949	1524	1200	—
	18LB36	16	36	24/12	—	—	7903	5807	4405	3422	2706	2168	1755
	18LB44	19	44	28/16	—	—	—	8729	6666	5219	4166	3370	2754
	18LB52	23	52	36/16	—	—	—	—	9538	7486	5992	4871	4007
	18LB60	27	60	44/16	—	—	—	—	—	—	8116	6630	5481
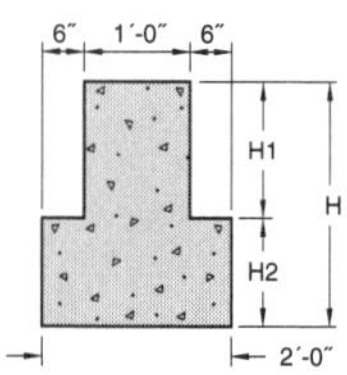 INVERTED-T	24IT20	9	20	12/8	5376	3494	2412	1726	1266	—	—	—	—
	24IT28	13	28	16/12	—	6951	4848	3529	2648	2030	—	—	—
	24IT36	16	36	24/12	—	—	8337	6127	4644	3598	2836	2265	1825
	24IT44	20	44	28/16	—	—	—	9300	7075	5514	4378	3525	2868
	24IT52	24	52	36/16	—	—	—	—	—	7916	6326	5132	4213
	24IT60	28	60	44/16	—	—	—	—	—	—	8616	7025	5800

* safe loads shown indicate 50% dead load and 50% live load and 800 psi top tension requiring additional top reinforcement
† in in.
‡ in ft

Figure 6-22. Different precast beam designs are used depending on building load requirements.

Rectangular beams are the most commonly used beam design as they are easy to form and cast. L-shaped beams are incorporated as perimeter beams for spandrels and around openings. Inverted-T beams are used for spanning between column caps, beams, or girders in bridge construction. Inverted-T beams may also be formed as a double-T with two vertical sections. Inverted-T beams may be joined side-by-side to create an integrated unit for bridge beams and road surfaces.

Precast slabs are available in a variety of thicknesses and designs. Precast slabs are used for above-grade floor slab systems and for roof panels. Precast slabs may have a hollow or solid core. **See Figure 6-23.**

The types and sizes of precast slabs are noted on floor plans and elevations, and may be indicated by the architect as a separate panel schedule with individual drawings of each panel included. The thickness of a precast slab member is expressed in inches preceded by "HC" for hollow core or "FS" for flat slab. For example, a slab designated as FS6 is a solid flat slab 6″ thick. Identification may also be provided in the form of a manufacturer code number or name. Methods for fastening the precast slabs to each other and to supporting members are shown on detail drawings and sections. Fastening may be accomplished by bolting or welding steel inserts.

Walls

The three primary types of concrete precast walls are structural, finish, and curtain wall panels. Structural precast wall panels are most commonly cast on the job site and tilted up into place. Finish precast wall panels are typically cast off-site with various concrete colors, aggregate sizes and types, and surface finishes. A *curtain wall* is a non-load-bearing prefabricated panel suspended on or fastened to structural members.

Structural precast wall panels are cast with door and window openings in place. Lifting and bracing devices are embedded in the wall. Many surface textures for precast concrete panels are used for structural, finish, and curtain walls. Steel angles or plates are embedded in the panels to allow them to be welded to other members such as structural steel columns and beams or reinforced concrete members with adjoining steel inserts. Precast wall panels are lifted into place with a crane, braced, and secured in place.

PRECAST CONCRETE CONSTRUCTION METHODS

Precast concrete may be precast at a plant or precast on-site. Concrete members precast at an off-site plant are designed according to industry standards or project specifications. Plant-precast concrete members are transported to the job site and set in place using a crane. Concrete members precast on-site are typically too large to be transported. Site-precast concrete members are cast at a location on the job site that minimizes formwork and concrete movement; many precast wall panels are cast directly on the concrete slab that will be used as the floor of the building. The panels are lifted or jacked into place after obtaining the required strength specifications.

Precasting at Plant

Concrete pipe, certain specialized walls, curtain walls, beams, and pavers may be precast at a plant and transported to the job site. Estimators determine from the prints and specifications whether the architect has used standard-size components or if special casting arrangements are required.

Precasting On-Site

Components precast on-site include walls and floor slabs. Walls are precast and tilted up into place with cranes. Floor slabs are set with perimeter forms and jacked into place. Site-precast construction methods include tilt-up and lift-slab construction.

Tilt-Up Construction. *Tilt-up construction* is a method of concrete construction in which concrete members are cast horizontally at a location close to their final position and tilted in place after the forms are stripped. A *tilt-up panel* is a concrete slab that is precast and lifted into place at the job site. **See Figure 6-24.** Tilt-up construction is commonly used for construction of low-rise concrete structures of one to three stories. Print information used by estimators for tilt-up panels includes wall and opening dimensions, placement and types of lift anchors and reinforcement, and interior and exterior wall finish.

After the first floor slab has set sufficiently, low forms are set on the slab to create the perimeter of the tilt-up panel. The height of the forms is equal to the thickness of the wall to be precast. The outside perimeter of the forms is set to the dimensions of the wall. Provisions are made around the perimeter of the form for reinforcing steel or any other attachment hardware which may project outside the forms. A bond breaker is applied to the floor slab to ensure concrete placed for the wall panel does not adhere to the floor slab. Blockouts are set in place for door and window openings and beam pockets.

PRECAST SLABS			
Hollow Core		**Solid Core**	
Thickness*	**Designation**	**Thickness***	**Designation**
6	4HC6	4	FS4
8	4HC8	6	FS6
10	4HC10	8	FS8
12	4HC12		

* in in.

Figure 6-23. Precast slabs are available in a variety of thicknesses and may have a hollow core or solid core.

Figure 6-24. Precast tilt-up wall panels allow large-scale sections of concrete wall to be precast and lifted into place at the job site.

Any surface treatment to be provided on the exterior face of the wall panel is set in place. Other items, such as welding plates for connecting adjoining panels or form liners for creating decorative finishes, are also placed into the proper position.

Detail drawings for reinforcing steel for precast tilt-up wall panels are similar to those used for cast-in-place concrete. Lift anchors are set in the concrete wall for attachment of lifting hardware and crane cables. Brace inserts are also set in the concrete wall to allow for attachment of braces. Special symbols on detail drawings indicate the type and placement of lift anchors and brace inserts.

A crane with a spreader bar to facilitate lifting is used to set the finished tilt-up wall panels in place after the concrete has set properly. Cables are fastened to the lift anchors embedded in the wall. The wall is lifted and pivoted into final position. Braces are then bolted to the brace inserts embedded in the wall. Each wall section is braced for final positioning. Braces for tilt-up wall panels remain in place until panels are fastened together and floors and roofs are in place. Floors and roofs tie the tilt-up wall panels together and provide structural support and stability to the entire structure.

Tilt-up wall panels are fastened together using flush-cast pilasters, cast-in-place columns, precast columns, steel columns, or flush steel plates. **See Figure 6-25.** Detail drawings indicate the panel-joining method used.

Lift-Slab Construction. *Lift-slab construction* is a method of concrete construction in which concrete slabs are cast on top of each other, lifted into position using jacks, and secured to columns at the desired elevation. Forms and support used for forming precast concrete members are set on the lower floor slab. A bond breaker is applied to the surface of the lower slab to allow for the upper slab to be lifted without adhering to the lower slab. A series of slabs are cast on top of each other, with the appropriate blockouts for columns and mechanical equipment runs.

After the slabs are cast and the concrete has set properly, the slabs are jacked to the proper elevation and fastened to supporting columns and beams. Lift-slab construction minimizes the shoring costs associated with conventional elevated-slab construction. Details for reinforcing steel and lifting anchors are included in the detail drawings.

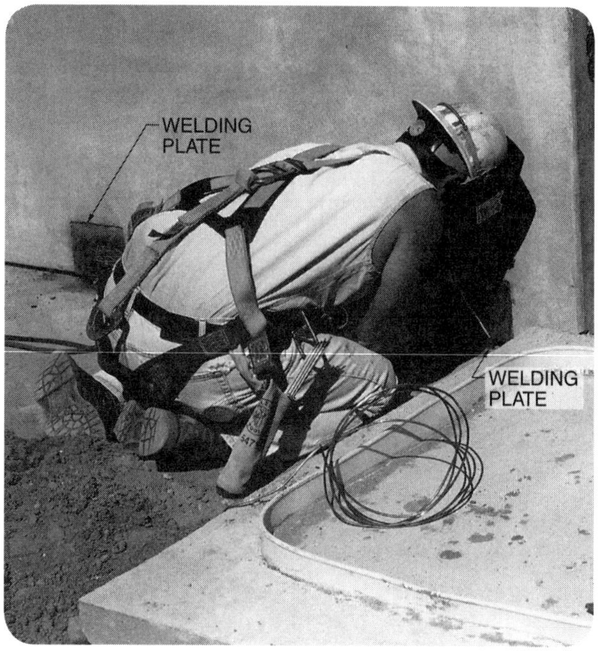

Porland Cement Association

Welding plates, used to connect adjoining members, are embedded in precast members.

Tilt-Up Panel Joining Methods

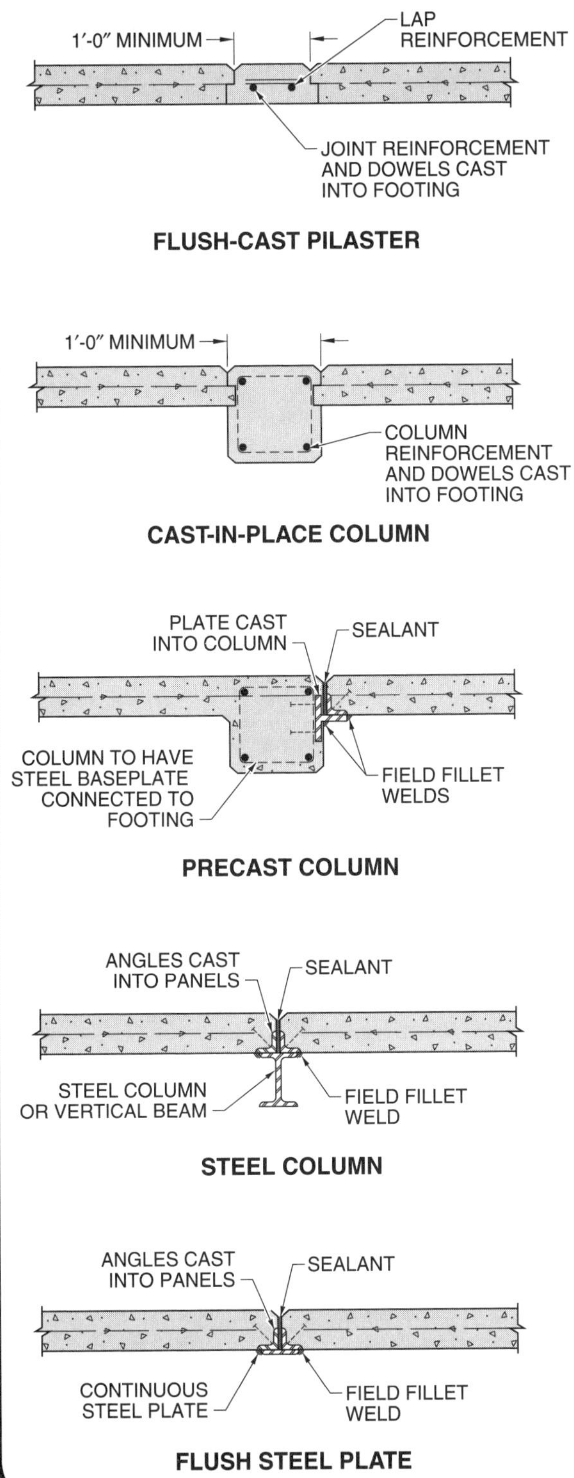

Figure 6-25. Flush-cast pilasters, cast-in-place columns, precast columns, steel columns, and flush steel plates may be used to join tilt-up wall panels after placement.

PRECAST CONCRETE QUANTITY TAKEOFF

The estimator begins precast concrete quantity takeoff by consulting a schedule for the various precast members. Each precast member is taken off separately with entries for cost information, including materials and labor.

Plant-Precast Members

Estimators consult with the casting plant to obtain costs for producing plant-precast members. Costs for the casting plan include concrete, forming materials, and labor. These are stated for the project estimator as a cost per unit and may include transportation of the units to the jobsite. Placement costs related to labor at the job site such as lifting, setting, and joining precast units are not included in plant-precast costs and must be added to the cost of the precast members.

Site-Precast Members

Material quantities for site-precast members are determined in a manner similar to those for cast-in-place concrete members. Concrete volume, reinforcement, and surface finish are calculated in the same way as for cast-in-place slabs and walls. Differences in takeoff for site-precast members include forming costs and lifting, bracing, and joining costs.

Tilt-Up Panels. Concrete for tilt-up panels is placed in a similar manner as for cast-in-place concrete slabs, so finishing labor costs are similar to finishing labor costs for slabs. Additional expenses related to the use of tilt-up panels include costs for lifting and temporary bracing of the wall until support columns are set or roof members are placed between the tilt-up panels. Forming costs are generally lower for tilt-up panels, as high walls can be cast with minimal forming expenses when compared to cast-in-place walls. Lifting costs include crane ownership or rental, operator expenses, and bracing and attachment of the wall panels or slabs.

For large construction projects, multiple tilt-up wall panels may be of the same design. Estimators may create a schedule of tilt-up panels to minimize duplication of efforts in takeoff for identical wall panels.

Lift Slabs. Concrete volume, reinforcement, and finishing for lift slabs are calculated in the same way as for a slab on grade. Additional costs are incurred for the additional reinforcement required for lifting the slab, jacking- and lifting-equipment operations, and shoring during placing and setting support columns.

SAGE TIMBERLINE ESTIMATING

MULTIPLE-ITEM QUICK TAKEOFF

The Sage Timberline Estimating Quick Takeoff function can be used to take off multiple items and enter dimensions for all of the items simultaneously. Multiple-item quick takeoff is accomplished by applying the following standard procedure:

1. Ensure the **Sample Ext Commercial GC** database is open. Open the **Quick Takeoff** dialog box.
2. Select items **3110.100 10 Footing Forms, 3110.100 50 Keyway In Footing**, and **3310.140 c 30 Footing Conc 3000 psi** by holding down the **Ctrl** key and picking the items individually. Drag and drop the items into the spreadsheet.

SELECTED ITEMS

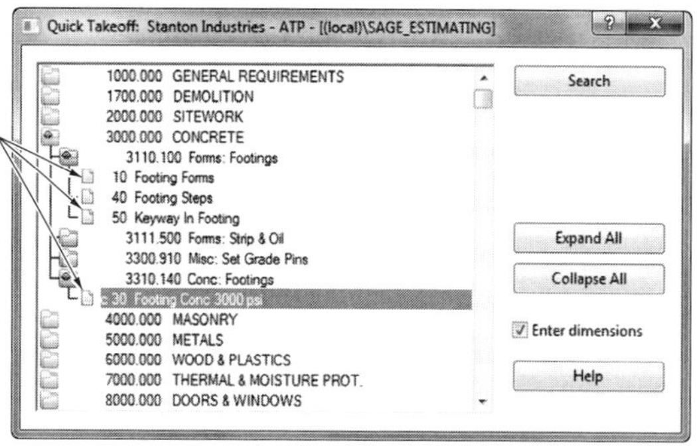

SELECTED ITEMS ADDED TO SPREADSHEET

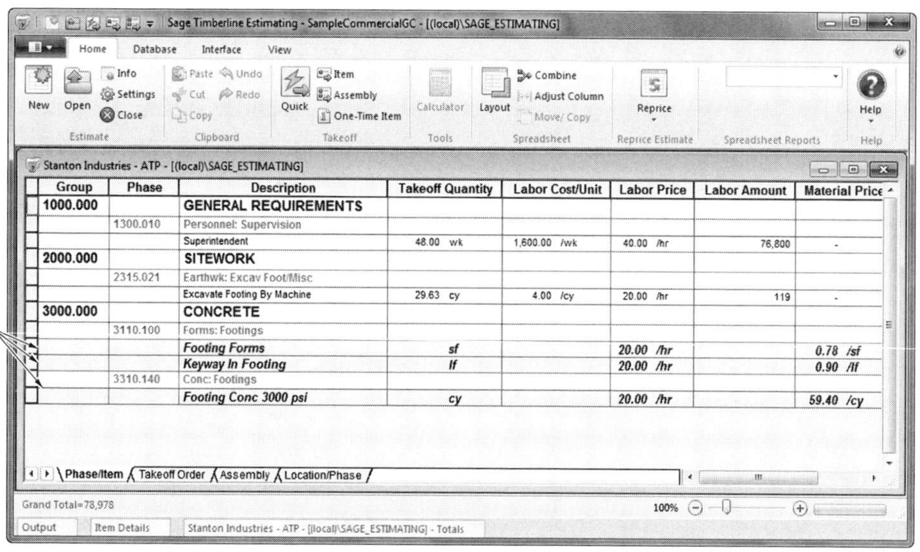

SAGE TIMBERLINE ESTIMATING

3. With all items added to the spreadsheet, the dimensions can be entered for all items simultaneously. Position the cursor in the **Takeoff Quantity** cell for the **Footing Forms** and while keeping the mouse button pressed, drag the cursor down to the last item taken off (**Footing Conc 3000 psi**). Once these cells have been highlighted, right-click anywhere in the block of cells and select **Enter Dimensions...** from the shortcut menu. Enter **50** for **Length**, **3** for **Width**, and **4** for **Depth of Concrete** in the **Enter Dimensions** dialog box. Each selected item has an associated formula, which relies on values input in the **Enter Dimensions** dialog box.

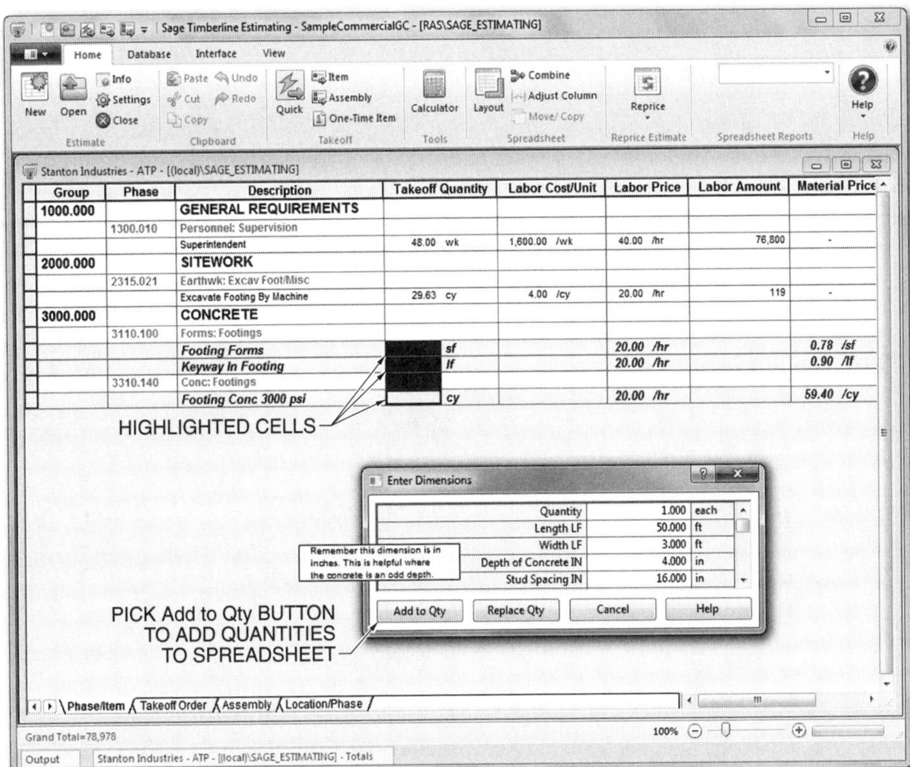

4. Pick the **Add to Qty** button to add the quantities to the spreadsheet.

Quick Quiz®

Quick Quiz®

Refer to the CD-ROM for the Quick Quiz® questions related to chapter content.

Key Terms

Illustrated Glossary

- *abutment*
- *admixture*
- *aggregate*
- *blockout*
- *brace*
- *camber*
- *coil tie*
- *concrete*
- *control joint*
- *curtain wall*
- *deck*
- *footing*
- *form*
- *form tie*
- *formwork*
- *ganged panel form*
- *grade beam*
- *isolation joint*
- *lift-slab construction*
- *pile*
- *post-tensioning*
- *pretensioning*
- *reinforced concrete*
- *retaining wall*
- *she bolt*
- *shoring*
- *slab on grade*
- *slump test*
- *slurry wall*
- *snap tie*
- *spandrel beam*
- *spread footing*
- *strongback*
- *tie rod*
- *tilt-up construction*
- *tilt-up panel*
- *waler*
- *water*
- *water-cement ratio*

Web Links

Web Links

American Concrete Institute
www.concrete.org

American Concrete Pumping Association
www.concretepumpers.com

Architectural Precast Association
www.archprecast.org

ASTM International
www.astm.org

Canadian Precast/Prestressed Concrete Institute
www.cpci.ca

Concrete Reinforcing Steel Institute
www.crsi.org

Insulating Concrete Form Association
www.forms.org

National Precast Concrete Association
www.precast.org

National Ready Mixed Concrete Association
www.nrmca.org

Operative Plasterers' and Cement Masons' International Association
www.opcmia.org

Portland Cement Association
www.cement.org

Post-Tensioning Institute
www.post-tensioning.org

Precast/Prestressed Concrete Institute
www.pci.org

Tilt-Up Concrete Association
www.tilt-up.org

Wire Reinforcement Institute
www.wirereinforcementinstitute.org

Concrete 6

Review Questions

Estimating

______________ 1. ___ is a mixture of portland cement, fine and coarse aggregate, and water that solidifies through a chemical reaction called hydration.

______________ 2. A(n) ___ is the section of a foundation that supports and distributes structural loads directly to the soil or piles.

______________ 3. A(n) ___ is a temporary structure or mold used to retain and support concrete while it sets and hardens.

______________ 4. A(n) ___ bolt threads into the ends of a tie rod.

______________ 5. A(n) ___ tie is a concrete form tie that is snapped off after the forms are removed and the concrete is set.

______________ 6. A(n) ___ tie is a concrete form tie used in heavy construction in which a bolt screws into the tie and is removed during form removal.

T F 7. A blockout is a structural piece, either permanent or temporary, designed to resist weights or pressures of loads.

T F 8. A brace is a frame set in a concrete form to create a void in the finished concrete structure.

______________ 9. A(n) ___ test measures the consistency, or slump, of concrete.

______________ 10. ___ is granular material such as gravel, sand, vermiculite, or perlite that is added to a cement and water mixture to form concrete, mortar, or plaster.

______________ 11. ___ is a method of prestressing reinforced concrete in which the steel reinforcing cables (tendons) in the structural member are tensioned before the concrete has hardened.

______________ 12. ___ is a method of prestressing concrete in which the tendons are tensioned after the concrete has hardened.

T F 13. A curtain wall is a non-load-bearing precast panel suspended on or fastened to structural members.

T F 14. Formwork is the entire system of support for fresh concrete including the forms, hardware, and bracing.

______________ 15. A(n) ___ is a concrete slab that is placed directly on the ground.

______________ 16. A(n) ___ is a horizontal member used to align and brace concrete forms or piles.

______________ **17.** A form ___ is a metal bar, strap, or wire used to hold concrete forms together at a predetermined distance and resist the outward pressure of fresh concrete.

______________ **18.** A(n) ___ wall is a wall constructed to hold back earth.

______________ **19.** A(n) ___ wall is a wall built around an excavation to retain groundwater.

______________ **20.** A(n) ___ is a supporting structure at the end of a bridge, arch, or vault.

______________ **21.** A(n) ___ form is a large form constructed by joining a series of small panels into a larger unit.

______________ **22.** ___ is the slight upward curve in a structural member designed to compensate for deflection of the member under load.

______________ **23.** A(n) ___ beam is a beam at the perimeter of a structure that spans columns and commonly supports a floor or a roof.

______________ **24.** A(n) ___ is a floor made of light-gauge metal panels (decking).

______________ **25.** ___ construction is a method of concrete construction in which concrete members are cast horizontally at a location close to their final position and tilted in place after the forms are stripped.

Short Answer

1. Identify five items that may be indicated on formwork drawings.

____

____

____

2. Explain why lower rows of walers may be doubled.

____

____

____

3. Identify five items that must be considered when calculating concrete quantities.

____

____

____

4. List sources of labor costs related to formwork for footings, foundation walls, slabs, and columns.

____

____

____

Wall Form Calculations

____________ 1. A total of ___ 4′ × 8′ panels is required.

____________ 2. At eight snap ties per panel, ___ snap ties are required.

____________ 3. At two studs per position with studs spaced 16″ OC, ___ studs are required for both sides of the wall. Allow for two extra studs in the calculation.

____________ 4. ___′ of top and bottom plates are required.

____________ 5. With double walers spaced 2′-0″ OC, ___ lf of walers are required for one side of the wall form.

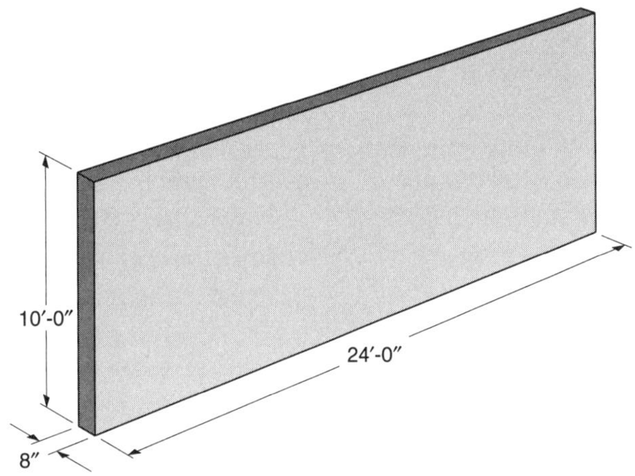

Column Concrete Volume

____________ 1. A total of ___ cu yd of concrete is required for Column A.

____________ 2. A total of ___ cu yd of concrete is required for Column B.

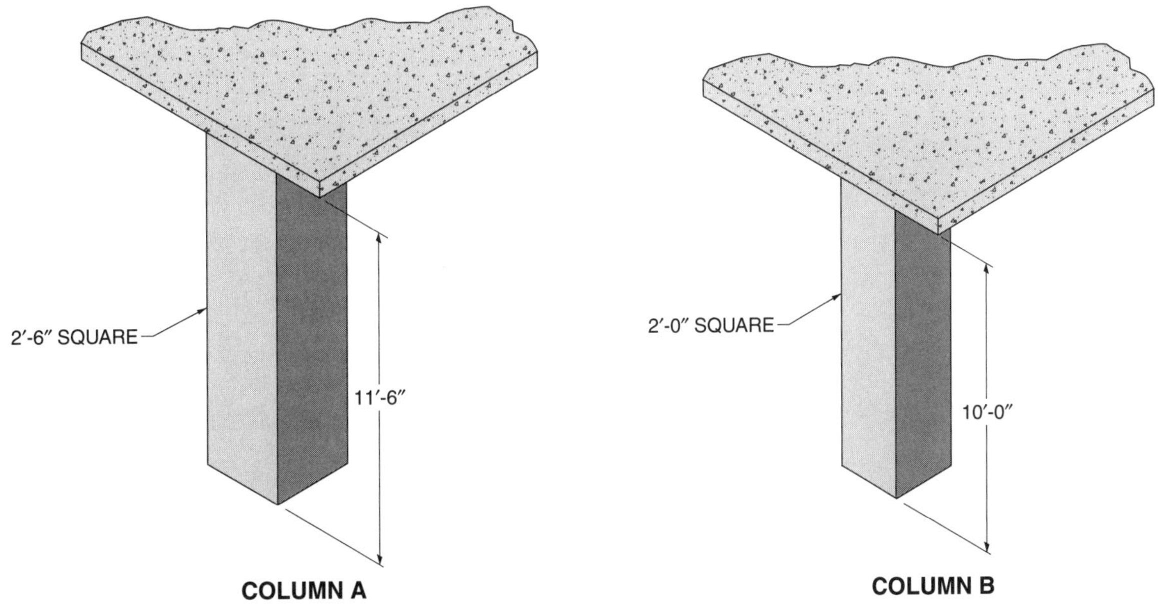

Concrete 6 Activities

Activity 6-1—Ledger Sheet Activity

Refer to Print 6-1 and Quantity Sheet No. 6-1. Take off the volume of concrete and linear feet of reinforcement required for the eight footings. The reinforcement consists of four #5 rebar 3′-0″ long equally spaced along the bottom of each footing.

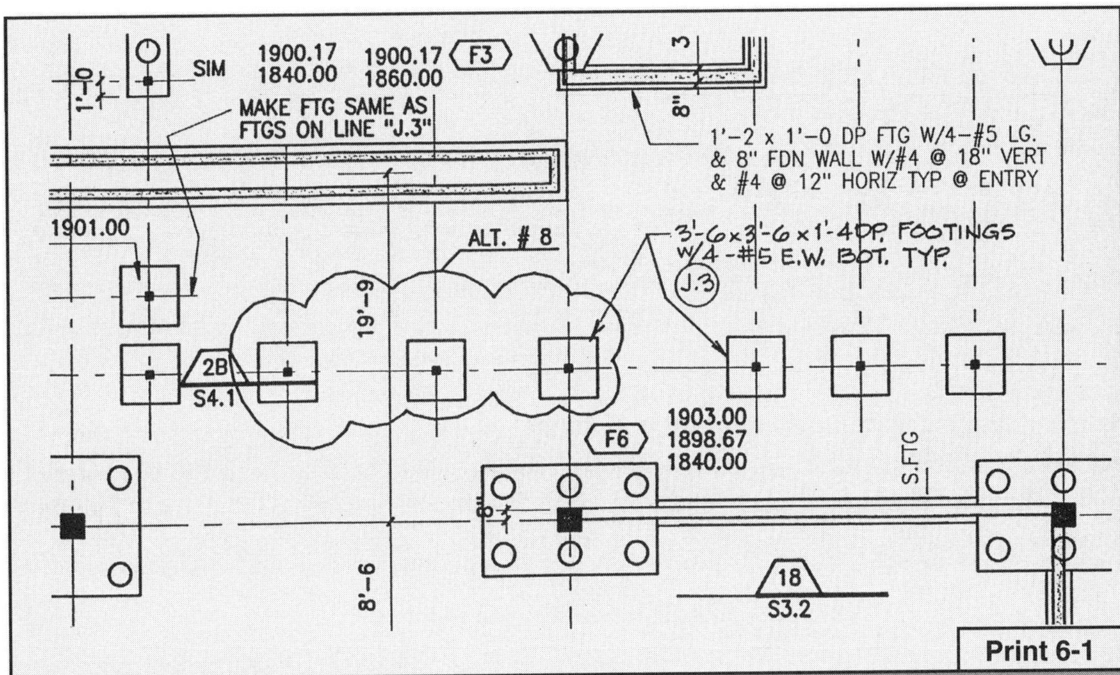

Print 6-1

QUANTITY SHEET

Project: ______________________ Sheet No. 6-1
Estimator: ______________________ Date: ______________________
Checked: ______________________

No.	Description	Dimensions				Unit		Unit		Unit		Unit
		L	W	D								
	Footings (8)											
	Concrete											
	Reinforcement (4 × 8)											

Activity 6-2—Spreadsheet Activity

Refer to the cost data, Print 6-2, and Estimate Summary Sheet No. 6-2 on the CD-ROM. Take off the square feet of wall forms and cubic yards of ready mix concrete for the walls of the elevator pit. Determine the total material and labor cost for the wall forms and ready mix concrete.

COST DATA			
Material	**Unit**	**Material Unit Cost***	**Labor Unit Cost***
Wall forms, job built	sq ft contact area	1.28	4.25
Concrete, ready mix, 3500 psi	cu yd	67	—

* in $

Activity 6-3—Footing Forms

Refer to Print 6-3 and Quantity Spreadsheet No. 6-3 on the CD-ROM. Calculate the linear feet of metal forms required for the footings. In addition, calculate the volume (in cu yd) of concrete and amount of reinforcing steel (in lf) required for the footings. Also calculate the material costs for the concrete with the unit cost of $70/cu yd. Include a 2% waste factor for the concrete.

Activity 6-4—Sage Timberline Estimating Activity

Create a new estimate and name it Activity 6-4. Perform a quick takeoff of items **3110.100 10 Footing Forms**, **3110.100 50 Keyway In Footing**, **3111.500 10 Strip/Oil Forms-Footing**, and **3310.140 c30 Footings Conc 3000 psi** for a 190′-6″ long by 30″ wide by 8″ deep footing. Print a standard estimate report.

7 Masonry

Key Concepts

- When estimating masonry materials, an estimator commonly performs the takeoff from the bottom of the structure up, and from the outside of the structure toward the inside.
- An estimator should also include the costs of cleaning materials and labor related to cleaning masonry after it is laid.
- Labor costs for masonry construction are typically based on square foot of surface area.
- Structural stone members such as lintels and sills are taken off individually, with length, width, and depth noted. Stone for walls is ordered by the ton.

Introduction

Masonry units include brick, structural clay tile, and concrete masonry units (CMUs). Masonry units are commonly used with other construction methods such as wood or metal framing, structural steel, or structural concrete. Mortar is a bonding mixture of fine aggregate, cement, and water that is used to fill voids between masonry units and reinforce a structure.

Stone is used for veneer, structural, and decorative applications. Stone may be placed dry and fitted without mortar or set with mortar similar to brick and CMUs.

MASONRY UNIT MATERIALS

Brick and concrete masonry units are available in many shapes and sizes and are made from a variety of materials such as clay, concrete, and glass. Masonry units are joined with mortar and may be reinforced with steel members. Masonry units can be used for decorative or structural purposes. **See Figure 7-1.** Masonry unit materials are fire-resistant and are commonly used as firebreaks between adjoining areas of a structure.

Portland Cement Association

Figure 7-1. A variety of brick and concrete masonry is used in masonry construction.

Brick

A *brick* is a masonry unit made of clay or shale that is formed into a rectangular shape while soft and then fired in a kiln. Brick can be used for building or paving purposes. Brick includes face and building brick, ceramic-glazed clay masonry units, clay tile, and terra cotta. Brick is manufactured in a variety of shapes and sizes and is designed for load-bearing and non-load-bearing applications. Most brick is packaged and delivered on pallets, with each pallet containing 250 bricks. Specifications and exterior elevations provide information concerning brick designs and types.

Building brick, also known as common brick, is made from local clay or shale and includes standard modular, Norman, SCR, engineer, and economy brick. **See Figure 7-2.** Each type of brick has different dimensions and architectural applications. The nominal dimensions of modular bricks are based on a standard 4″ unit, which includes the mortar joints. Norman bricks are longer than most other bricks. SCR bricks are thicker than other bricks and allow walls to be laid in a single wythe. A *wythe* is a single, continuous vertical masonry wall that is one unit thick. Engineer bricks have a greater height than other designs but retain standard thickness and length dimensions. Economy bricks are available in full 4″ modules for rough-work applications.

Face brick is brick made of select clay to impart distinctive, uniform colors and is commonly used for an exposed, finished surface. Face brick is manufactured in a variety of colors and shapes. Manufacturing of face brick creates uniform hardness, size, and strength for a quality finished surface.

Firebrick is brick manufactured from refractory ceramic materials that resist disintegration at high temperatures. Firebrick is used in areas subjected to high temperatures, such as fireplace hearths or industrial furnace linings.

Glass block is hollow translucent or transparent block made of glass. Glass block is available in many sizes and colors. Glass block cannot be used in load-bearing applications.

Structural Clay Tile

A *structural clay tile* is a hollow masonry unit composed of burned clay, shale, or a combination of clay and shale. Glazed and unglazed structural clay tiles are used for structural and decorative applications. A wide range of tile designs and shapes are available for stretchers, sills, caps, miters, corners, jambs, and cove bases. **See Figure 7-3.** Structural clay tile is used as a structural back-up material to face brick in older buildings and for interior construction applications in high-wear areas such as restaurant kitchens and public restrooms.

> Structural clay tile is made from clay, ceramic, and refractory minerals, mixed with chemicals, and baked in kilns.

BUILDING BRICK							
Designation	Nominal Dimensions*			Joint Thickness*	Actual Dimensions*		
	t	h	l		t	h	l
 STANDARD MODULAR	4	2⅔	8	⅜	3⅝	2¼	7⅝
				½	3½	2 3/16	7½
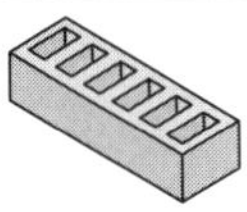 NORMAN	4	2⅔	12	⅜	3⅝	2¼	11⅝
				½	3½	2 3/16	11½
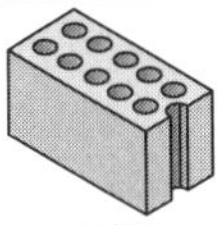 SCR	6	2⅔	12	⅜	5⅝	2¼	11⅝
				½	5½	2¼	11½
 ENGINEER	4	3⅕	8	⅜	3⅝	2 13/16	7⅝
				½	3½	2 11/16	7½
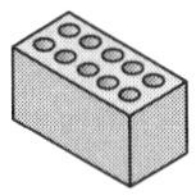 ECONOMY	4	4	8	⅜	3⅝	3⅝	7⅝
				½	3½	3½	7½

* in in.

Figure 7-2. Building brick types include standard modular, Norman, SCR, engineer, and economy, with each type having different thickness, height, and length dimensions.

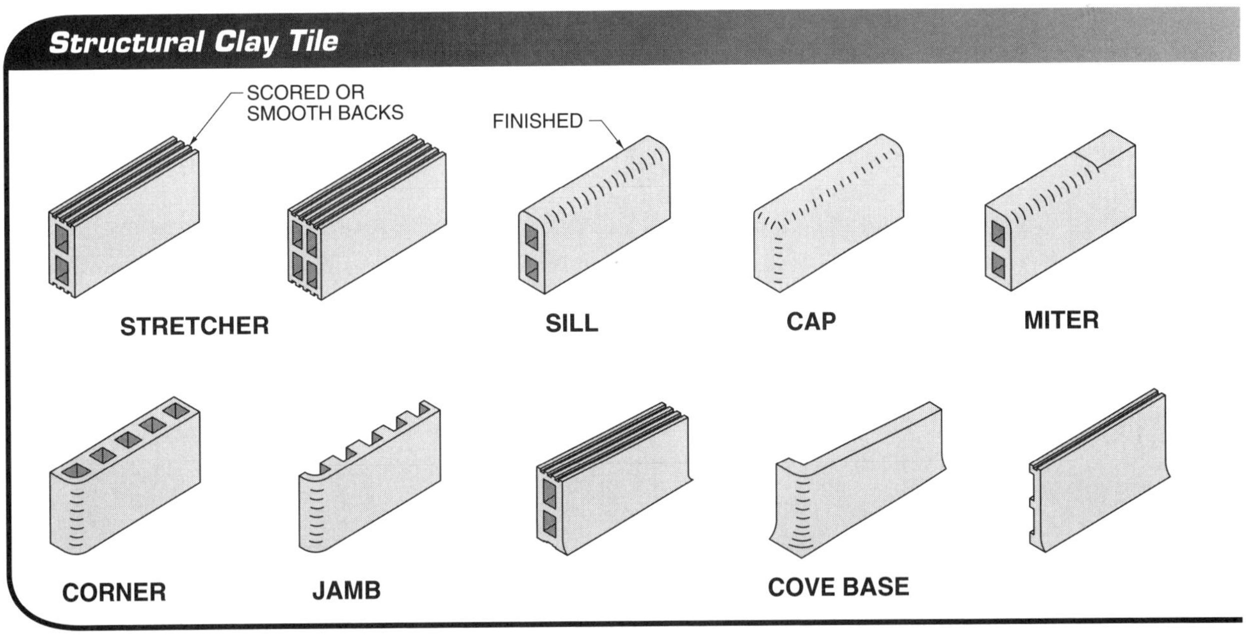

Figure 7-3. Structural clay tile is available in a variety of designs and is used in high-wear areas such as restaurant kitchens and restrooms.

CMUs

A *concrete masonry unit (CMU)* is a precast hollow or solid block made of portland cement and fine aggregate, with or without admixtures or pigments. CMUs are used for many applications, including foundations and exterior and interior walls. Actual dimensions of CMUs are 3⁄8″ to 1⁄2″ less than nominal size.

CMUs possess excellent fire-resistance qualities and are commonly used for firebreaks. Various materials, wall designs, and cavity fill materials affect the fire-resistance ratings of walls constructed with CMUs. Fire-resistance ratings are determined by the amount of heat transmission on the side of the wall opposite the heat source. Plaster and other facing materials on CMU walls also affect the fire-resistance ratings.

CMUs are cast in a variety of styles and shapes. **See Figure 7-4.** CMUs include stretcher, corner, header, bond beam, lintel, sill, column, jamb, screen, and split-face designs. Faces of CMUs may be smooth, split with a rough finish, fluted, or ribbed.

Special Masonry Materials

In addition to brick, structural clay tile, and CMUs, special masonry materials are used in some masonry installations. Special masonry materials include refractory materials and accessories such as movement joints, reinforcement, ties, flashing, shelf angles, and sealants. Accessories are noted in specifications under MasterFormat™ title 04 05 00.

Refractory Materials. A *refractory material* is a masonry unit that can withstand high temperatures without structural failure. Refractory materials include flue liners and refractory brick. As with many other masonry unit materials, flue liners and refractory brick are manufactured in standard sizes and shapes. **See Figure 7-5.** Availability of refractory materials is dependent on local suppliers. Estimators can find information concerning installation of refractory materials in areas on drawings such as fireplaces and furnace and boiler flues. Refractory materials are noted in specifications under MasterFormat™ title 04 05 00.

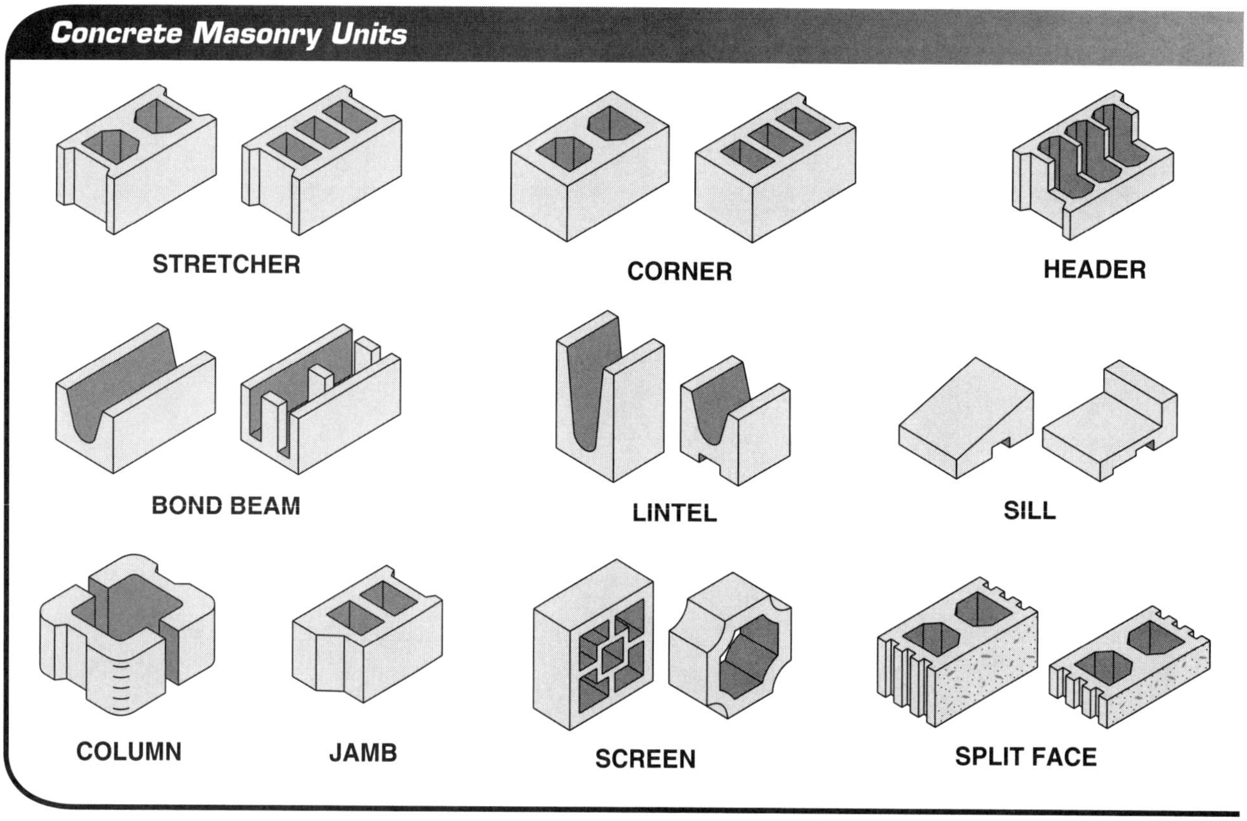

Figure 7-4. Concrete masonry units (CMUs) are widely used in building construction due to the variety of styles and shapes available.

FLUE LINERS				
Liner Type	**Area***	**Dimensions†**		
		A	**B**	**T**
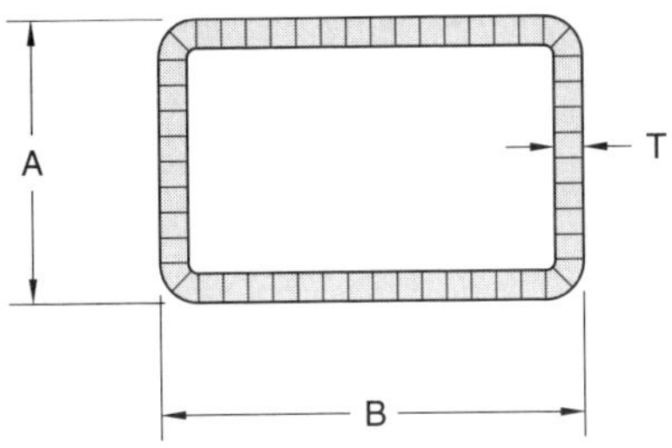**RECTANGULAR (STANDARD)**	49	8½	8½	¾
	83	8½	13	⅞
	104	8½	18	1
	127	13	13	⅞
	173	13	17¾	1
	233	17¾	17¾	1¼
	298	20	20	1⅜
	357	20	24	1½
	473	24	24	1⅝
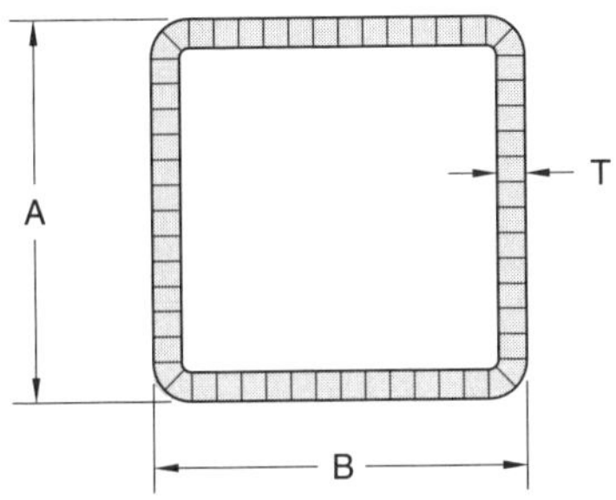**RECTANGULAR (MODULAR)**	60	7½	11½	¾
	95	11½	11½	⅞
	128	11½	15½	1
	176	15½	15½	1⅛
	221	15½	19½	1¼
	281	19½	19½	1⅜
	338	19½	23½	1½
	410	23½	23½	1⅝
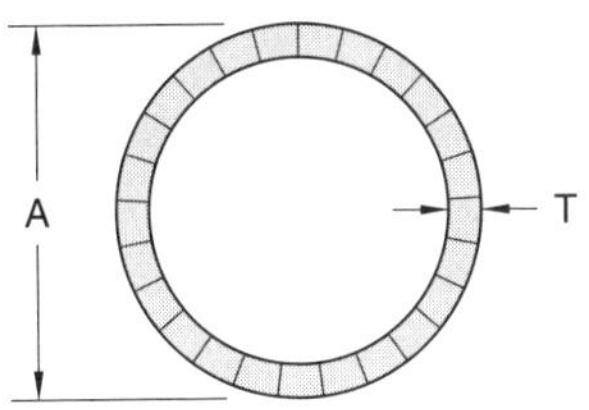**ROUND**	33	8	—	¾
	54	10	—	⅞
	79	12	—	1
	128	15	—	1⅛
	195	18	—	1¼
	234	20	—	1⅜
	338	24	—	1⅝

* in sq in.
† in in.

Figure 7-5. Rectangular and round flue liners are refractory materials available for masonry construction.

Accessories. Masonry materials expand and contract due to temperature and moisture variations. Movement joints, specialized reinforcement, ties, flashing and weep holes, shelf angles, and sealants are used in masonry construction to minimize damage due to dimensional changes in different parts of the structure.

A *movement joint* is a continuous vertical or horizontal joint in a masonry wall without mortar or other nonresilient materials that is used to accommodate expansion and contraction of the wall system. Movement joints include expansion and control joints. Backing rods, which are covered by polyurethane or silicone sealant, are commonly used to fill movement joints.

Slip channels may also be used to allow movement between masonry walls and where masonry walls abut adjoining floor and roof structures. Locations and types of movement joints are shown on elevations and detail drawings.

Reinforcement is material embedded in or attached to another material to provide additional support or strength. Reinforcement for masonry walls typically consists of steel wire shaped into various patterns and set in mortar joints to provide tensile strength to masonry construction. A *tie* is a type of reinforcement used to secure two or more members together. Ties are embedded in mortar joints to tie masonry walls to each other or to anchor masonry veneer walls to the structural backing material.

Flashing is a metal or flexible material embedded in a masonry wall to direct water out of the wall and prevent moisture penetration at exposed wall surfaces. Flashing is attached to the inner wall surface and terminates at a weep hole in the masonry wall. The weep hole may be open or may be filled with materials that allow the moisture to escape, such as wicks, weep tubes, or cellular vents.

A *shelf angle* is a steel angle attached to structural backing members to support horizontal courses of brick for high masonry walls. Shelf angles are fastened to structural members using lag bolts or other means of attachment specified by the architect.

Masonry walls must be properly braced until permanent support, such as structural roof members, are in place.

A *sealant* is a liquid or semiliquid material that forms an airtight or waterproof joint. Sealants are caulked into movement joints to allow movement of the adjacent structure while sealing the open joint. Polyurethane sealants should be used for porous materials and for joints between porous and nonporous materials. Silicone sealants should be used with nonporous materials.

Accessories and sealants are noted in the specifications, exterior elevations, or masonry detail drawings. Estimators must include these items in the overall masonry bid to ensure that all items and their proper quantities and costs are taken into account.

Mortar

Mortar is a bonding mixture of fine aggregate, cement, and water that is used to fill voids between masonry units, join them together, and reinforce a structure. Mortar selection for masonry construction is based on overall strength requirements, type of masonry units being joined, and the environmental conditions that the mortar must resist. Different mortars may be used for brick, stone, refractory, and corrosion-resistant applications. Colorants may be added to the mortar as required by the architect.

Masonry mortar is classified as Type M, S, N, O, and K, with Type N mortar the most common type of mortar specified for construction projects. Type M mortar has the highest cement content and the highest compressive strength characteristics of all the mortar types, with an allowable compressive load strength of approximately 2500 psi. Type M mortar is commonly used for below-grade applications. Type S mortar has a compressive strength of approximately 1800 psi but has a moderately high shrinkage rate. Type N mortar has a compressive strength of approximately 750 psi and has a relatively low shrinkage rate.

Type O mortar has a low resistance to freeze/thaw cycles and a relatively low strength. Type O mortar is commonly used for tuckpointing. Type K mortar, often referred to as sand/lime mortar, has low resistance to freeze/thaw cycles, low strength, and high water retention. An estimator should ensure that the proper amount, type and color of mortar is taken off for each masonry application.

MASONRY UNIT CONSTRUCTION METHODS

Construction methods for masonry walls and fireplaces depend on the design, materials used, height of the structure, environmental conditions, and exterior and interior finishes. A wall design may be hollow or solid, a veneer covering a structural member, or prelaid curtain walls. Masonry units are commonly used in conjunction with other construction materials such as wood or metal framing, structural steel, or structural concrete.

Various equipment is required in masonry construction. For example, rough-terrain forklifts and/or telehandlers (telescopic material handlers) are often necessary to move pallets of brick and CMUs at a job site. Mortar mixers are used to mix batches of mortar. Scaffolds are required to facilitate the placement of brick and block. Scaffolds include both traditional sectional metal-framed scaffolds, as well as hydraulic mast-climbing scaffolds. In cold climates, weather protection and heat may be required to allow the mortar to set properly. **See Figure 7-6.** For high, unsupported walls, temporary bracing may be required to secure masonry walls in position until roof members and other final structural wall supports are set in place.

Figure 7-6. In cold weather, protection and heat are required to allow the mortar to set properly.

Masonry unit construction begins with a solid foundation of cast-in-place concrete. The foundation may be a concrete footing, foundation wall, or thickened portion of a slab. Steel dowels are set in place in the concrete to tie the masonry and concrete together.

Brick masons lay out the work on the supporting surface to determine proper brick or CMU spacing. The corner masonry units of the structure are set in mortar and work proceeds from each corner of a structure. A mason's rule or story pole is used to lay out courses of brick or CMUs. A *course* is a continuous horizontal layer of masonry units bonded with mortar. **See Figure 7-7.** After the masonry units for the corners are laid, strings are stretched between them or story poles are used to guide the masonry work. Mortar joints are struck as work proceeds to create the joint profile. Muriatic acid and water are commonly used after the mortar has set to remove any excess mortar from the face of the masonry.

Figure 7-7. Masonry corners act as guides for masonry work on walls.

Masonry Walls

Exterior masonry walls may be constructed brick, CMUs, or a combination of brick and/or CMUs with other building materials. Masonry walls may be structural bearing walls or a veneer built around and between structural members. Information pertaining to masonry units for exterior finish is shown on detail drawings, floor plans, wall sections, and exterior elevations on the prints. **See Figure 7-8.**

Brick and CMUs are also used to construct interior firebreak walls and exposed interior walls. Finish information for interior masonry walls is similar to exterior exposed masonry information, including material type, mortar joint color and finish, and bond.

Where brick or CMUs are used for exterior walls, the exterior finish, including the bond, face finish of the CMUs, mortar color, lintels, and bonds, is noted in the specifications and on elevations. Special masonry units may be fabricated specifically for a particular job. Special brick shapes and coursing information are included on the elevations. **See Figure 7-9.**

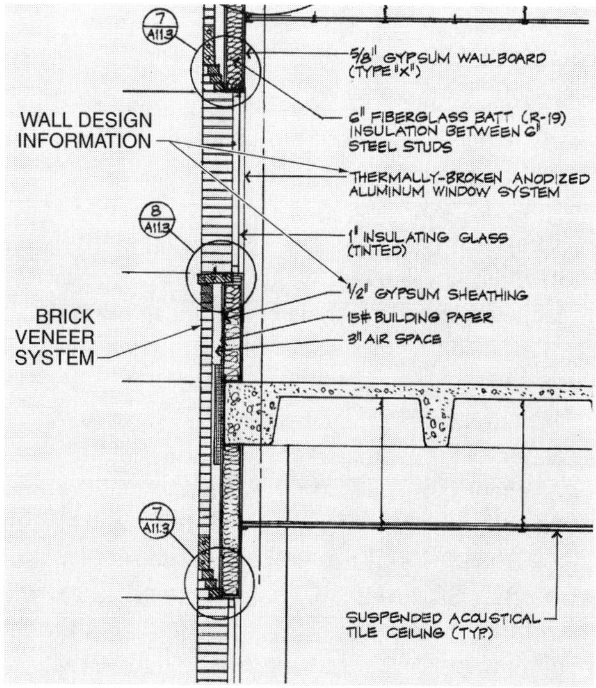

Figure 7-8. Estimators consult detail drawings, wall sections, and exterior elevations of the prints for a complete understanding of masonry installations.

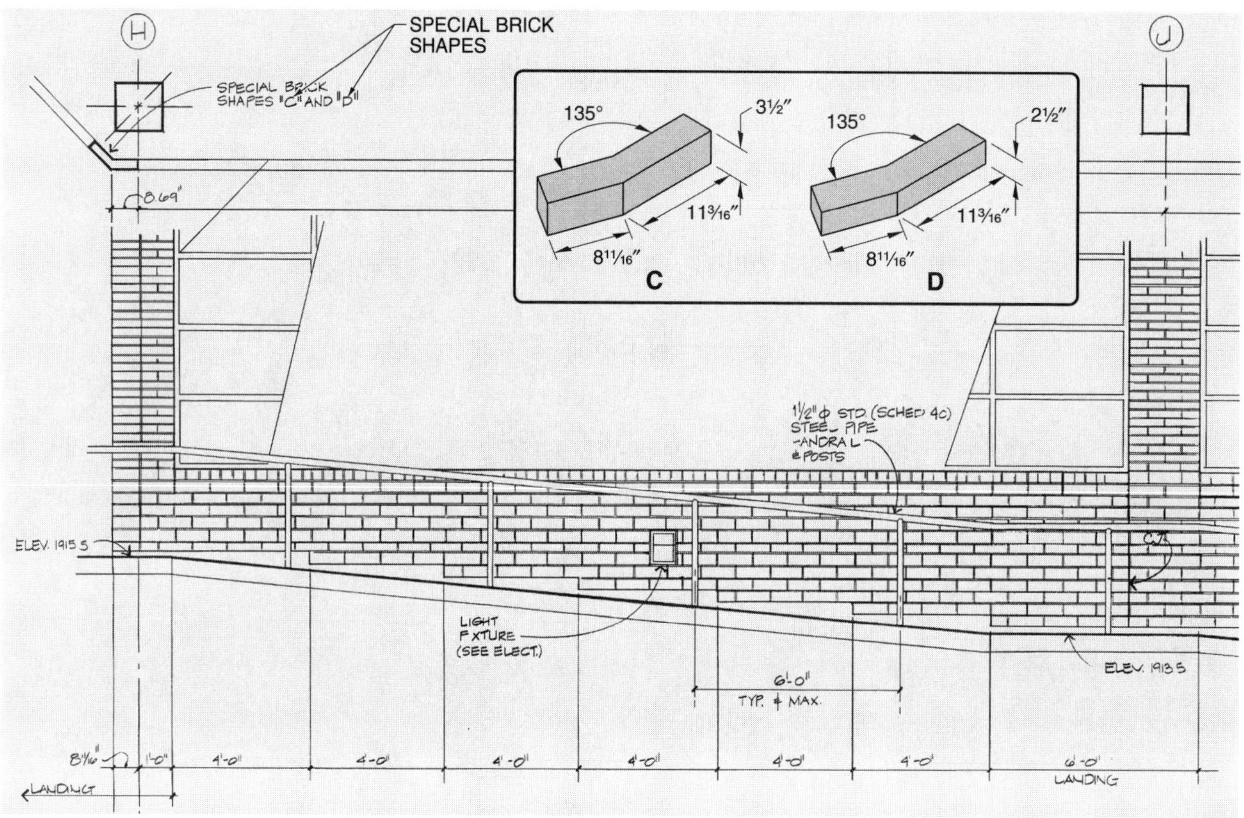

Figure 7-9. Information about special brick shapes that may be required for a structure is included on exterior elevations.

Bonds. A *bond* is the arrangement of masonry units in a wall created by overlapping the masonry units one upon another to provide a sturdy structure. Architects specify a bond and a joint finish for a masonry structure. Brick bonds include running, common, English, Flemish, stacked, and garden wall. **See Figure 7-10.** Each bond uses a combination of brick positions, such as stretchers, headers, soldiers, shiners, rowlocks, and sailors, to create a design and provide structural support.

Solid Masonry Walls. Solid masonry walls are constructed with masonry materials only and reinforced with metal ties and other reinforcement. Solid masonry walls can be formed of one or two layers of brick tied together with metal ties, brick tied to the face of CMUs, CMUs only, or structural or decorative tile set onto concrete masonry. An estimator should check architectural details and symbols to accurately determine the materials used in solid masonry wall construction in each building area.

Cavity Walls. A *cavity wall* is a vertical masonry structure in which the facing and backing wythes are completely separate except for metal ties joining the wythes. A 2″ (minimum) void is provided between the facing and backing wythes. **See Figure 7-11.** The air space is provided to improve insulation and reduce thermal transmission. Metal ties join the two separate wythes together to create a single structural unit.

Brick Bonds and Positions

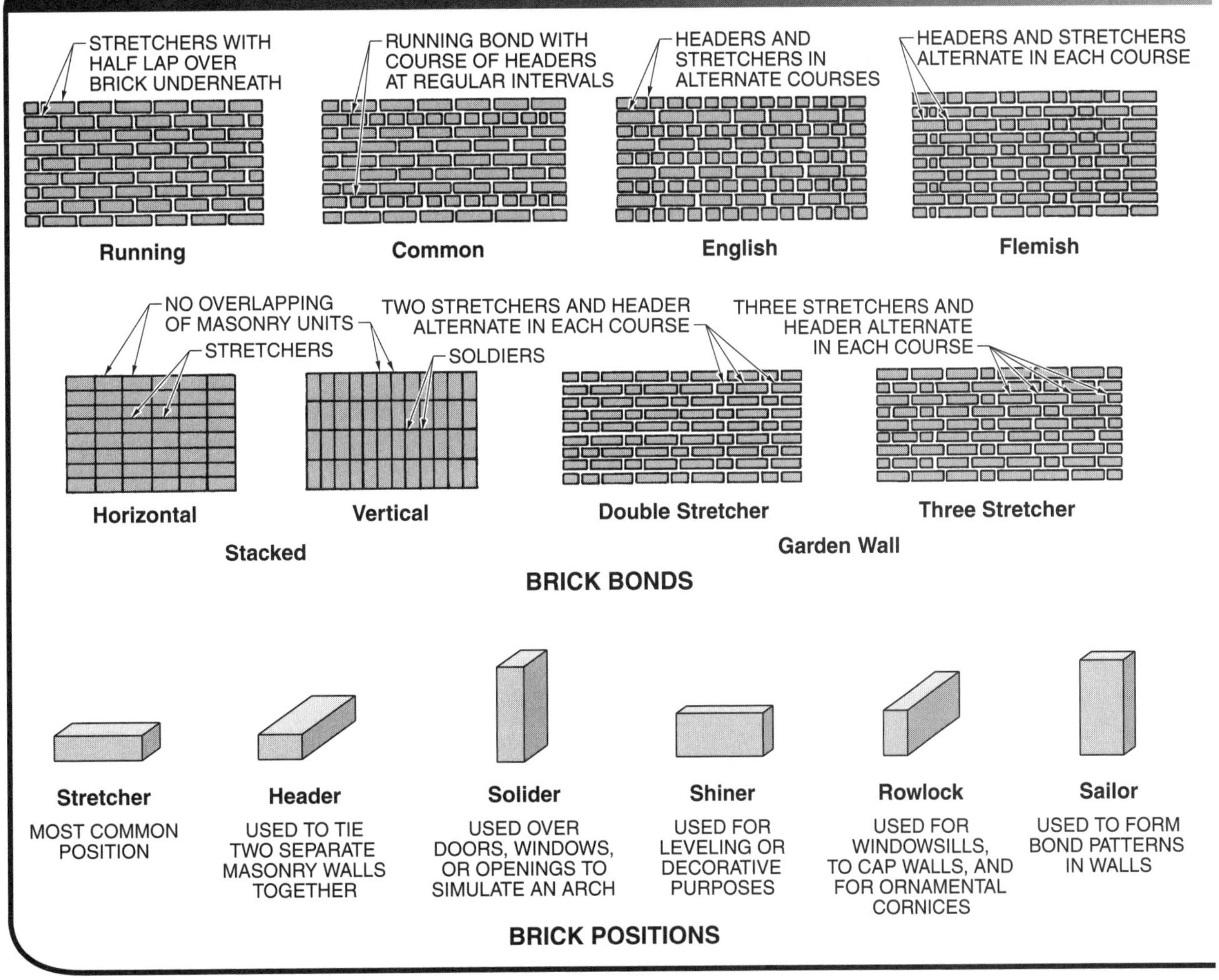

Figure 7-10. Brick bonds are created by placing bricks in various positions and overlapping them one upon another to provide a sturdy structure.

Cavity Walls

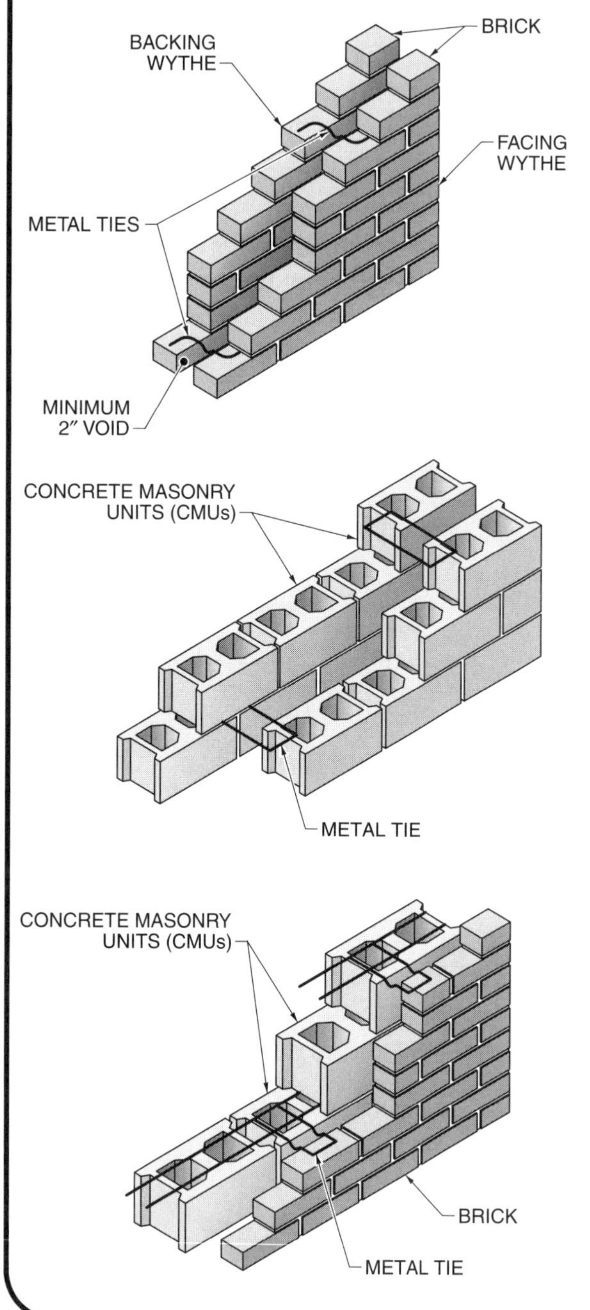

Figure 7-11. Cavity walls have a 2″ minimum void between masonry wythes.

Masonry Veneer Walls. Masonry veneer walls are used in frame, structural steel, and reinforced concrete construction. Ties are used to anchor the masonry veneer to the backing wall. Masonry veneer walls are commonly built in place. However, masonry veneer walls may also be made of prefabricated panels set in a manner similar to precast concrete panels. Prefabricated masonry veneer panels are laid with steel support frames and inserts, lifted into place, and attached to structural members. In some cases, thin face brick may be set in forms and cast into a concrete panel. These panels can be cast in larger units and set in a manner similar to other precast concrete panels.

Rain screen walls are constructed in a similar manner as masonry veneer walls, creating an air separation between the exterior finish material and the face of the exterior structural wall cladding. Moisture penetrating the exterior finish is captured by a drainage plane on the face of the exterior structural wall such as building paper or polyethylene sheets. The rain or moisture is then removed by airflow through vents at the top of the wall and weep holes at the bottom. Airflow through the separation space creates the rain screen wall.

Estimators should also check the prints for other masonry veneer applications such as columns, pilasters, and interior partitions. Construction methods for these masonry veneer applications are similar to exterior masonry veneer walls.

Fireplaces

Special factors must be considered for fire safety in the design of fireplaces. All combustible materials, including wood beams, joists, and studs, must have a minimum clearance of 2″ from the front faces and sides of a masonry fireplace and not less than a 4″ clearance from the back faces of a masonry fireplace. The clearance space should not be filled, except to provide fireblocking.

The inner hearth of the fireplace is laid first using firebrick. Refractory mortar is used to bond firebrick together. *Refractory mortar* is fire-resistant material made with high-temperature cement and carefully selected aggregate that does not expand and deteriorate when heated. Two types of refractory mortar are permitted by the International Code Council (ICC)—hydraulic-setting mortar and air-drying mortar. The sides and back of the combustion chamber are also laid with firebrick and refractory mortar. Dampers are set in place and smoke shelves are laid as designed by the architect and shown on the details. **See Figure 7-12.**

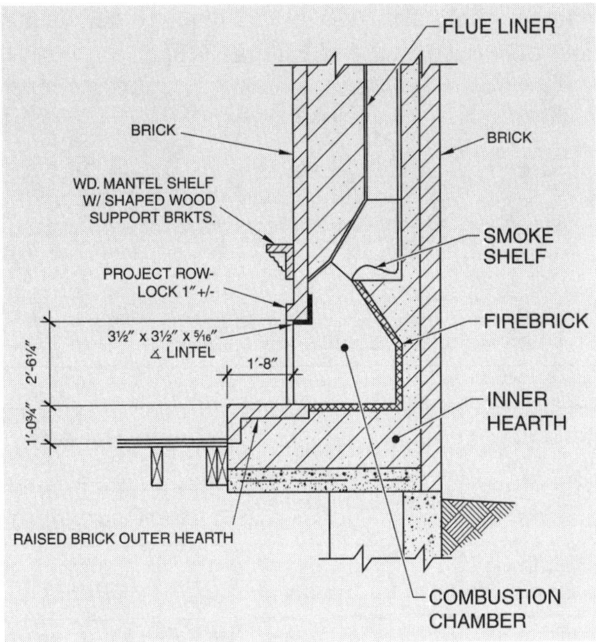

Figure 7-12. Specialized materials and construction methods are used in masonry fireplace construction.

A *smoke shelf* is a section of a chimney directly over a fireplace and below the flue liner that prevents downdrafts from blowing smoke into the living area. Refractory materials such as flue liners are set in place and masonry work proceeds up to the chimney. Face brick and other fireplace facing materials are applied after all other fireplace components are installed.

MASONRY MATERIAL QUANTITY TAKEOFF

When estimating masonry materials, an estimator commonly performs the takeoff from the bottom of the structure up and from the outside in. Estimators begin by computing the types and quantities of masonry materials used for the foundation and above-grade exterior walls. After computing exterior materials, the interior masonry materials such as fireplaces, interior masonry walls, and hearths are taken off. In general, estimators should base masonry material costs on the delivered cost of the materials.

Before masonry materials can be laid, preliminary work, such as excavation and concrete work for footings and foundations, must be performed. Some estimators may consider these items as part of the masonry cost of a structure if they are included in the contractual scope of work.

Estimators must also include equipment costs in a masonry bid. Masonry equipment costs include ownership, maintenance, and operation or rental costs for forklifts or telehandlers to move pallets of brick, CMUs, and mortar at the job site; scaffold ownership or rental, erection, and removal; required bracing for masonry walls; and possible weather protection. Each of these contingency costs must be considered by the estimator in the overall masonry cost to ensure an accurate estimate.

An estimator should also include the costs of cleaning materials and labor related to cleaning masonry after it is laid. The specifications should be checked to determine the cleaning materials and amount of cleaning required. Additional costs that may be included as a portion of the masonry work are costs for wood and metal forms or bucks for arched doorways, windows, or other special applications where temporary forms are required. Metal lintels, which are installed over doors, windows, fireplaces, and dampers, are usually computed last.

Labor costs for masonry construction are typically based on square foot of masonry surface area. Standard labor tables or company historical data can provide information about the number of square feet of surface area of each masonry application that can be set in place by a particular labor crew. The labor hours are multiplied by the cost per crew-hour to determine overall labor costs.

Masonry Wall Takeoff

Masonry material takeoff begins by determining the gross surface area (in square feet) of masonry for the construction project. The estimator first computes the square feet of surface area of the masonry structure and then deducts areas not composed of masonry, including large door and window openings, to obtain the net surface area.

Door, window, and other wall opening dimensions are determined from floor plans, elevations, and door and window schedules. The two steps yield the total area of masonry material. Common waste factors are 2% to 5% for masonry units and 10% to 20% for mortar. The waste factor is added after the net surface area calculations are completed. In estimating the quantity of face brick used as trim, the linear feet of trim is measured and multiplied by the number of brick per linear foot.

Brick and CMU Quantities. After calculating the net surface area, an estimator may refer to standardized tables to determine the number of masonry units per square foot. The total number of masonry units is computed by multiplying the area (in square feet) by the number of masonry units per square foot or by using a masonry wall material table to determine the number of standard-size bricks required per square foot of surface area. See Figure 7-13.

> Brick and CMUs are priced by the unit, which is converted into a price per square foot. Openings of less than 2 sq ft are typically ignored because savings in units are usually offset by trimming and cutting.

MASONRY WALL STANDARD SIZE FACE AND BUILDING BRICK*

Wall‡	Bonds														
	Running			Common (Header Course Every 7th Course)			English and English Cross† (Full Headers Every 6th Course)			Flemish (Full Headers Every 5th Course)			Double Headers (Alternating with Stretchers Every 5th Course)		
	Face Brick	Building Brick		Face Brick	Building Brick		Face Brick	Building Brick		Face Brick	Building Brick		Face Brick	Building Brick	
		8″ Wall	12″ Wall		8″ Wall	12″ Wall		8″ Wall	12″ Wall		8″ Wall	12″ Wall		8″ Wall	12″ Wall
1	6.16	6.16	12.32	7.04	5.28	11.44	7.19	5.13	11.29	6.57	5.75	11.91	6.78	5.54	11.70
5	31	31	62	36	27	58	36	26	57	33	29	60	34	28	59
10	62	62	124	71	53	115	72	52	113	66	58	120	68	56	117
20	124	124	248	141	106	229	144	103	226	132	115	269	136	111	234
30	185	185	370	212	159	344	216	154	339	198	173	358	204	167	351
40	247	247	494	282	212	458	288	206	452	263	230	477	272	222	468
50	308	308	616	352	264	572	360	257	565	329	288	596	339	277	585
60	370	370	740	423	317	687	432	308	675	395	345	715	407	333	702
70	432	432	864	493	370	801	504	360	791	460	403	834	475	388	819
80	493	493	986	564	423	916	576	411	904	526	460	953	453	444	936
90	555	555	1110	634	476	1030	648	462	1017	592	518	1072	611	499	1053
100	616	616	1232	704	528	1144	719	513	1129	657	575	1191	678	554	1170
200	1232	1232	2464	1408	1056	2288	1438	1026	2258	1314	1150	2382	1356	1108	2340
300	1848	1848	3696	2110	1584	3432	2157	1539	3387	1971	1725	3573	2034	1662	3510
400	2464	2464	4928	2816	2112	4576	2876	2052	4516	2628	2300	4764	2712	2216	4680
500	3080	3080	6160	3520	2640	5720	3595	2565	5645	3285	2875	5955	3390	2770	5850
600	3696	3696	7392	4224	3168	6864	4314	3078	6774	3942	3450	7146	4068	3324	7020
700	4312	4312	8624	4928	3696	8010	5033	3591	7903	4599	4025	8337	4746	3878	8190
800	4928	4928	9856	5632	4224	9152	5752	4104	9032	5256	4600	9528	5424	4432	9360
900	5544	5544	11,088	6336	4752	10,296	6471	4617	10,161	5913	5175	10,719	6102	4986	10,530
1000	6160	6160	12,320	7040	5280	11,440	7190	5130	11,290	6570	5750	11,910	6780	5540	11,700

* For other than ½″ joints, add 21% for ⅛″ joint, 14% for ¼″ joint, 7% for ⅜″ joint. Subtract 5% for ⅝″ joint, 10% for ¾″ joint, 15% for ⅞″ joint, and 20% for 1″ joint
† Quantities also apply to common bond with headers in every sixth course
‡ in sq ft

Figure 7-13. Masonry wall material tables provide information regarding the number of masonry units required for various bonds and wall thicknesses.

Calculating Brick Quantities

Calculate the number of face and building bricks required for an 8″ thick solid-brick building of common bond with exterior dimensions of 24′-6″ × 24′-6″, a wall height of 10′-0″, and two doors, each 9′-0″ wide by 8′-0″ high.

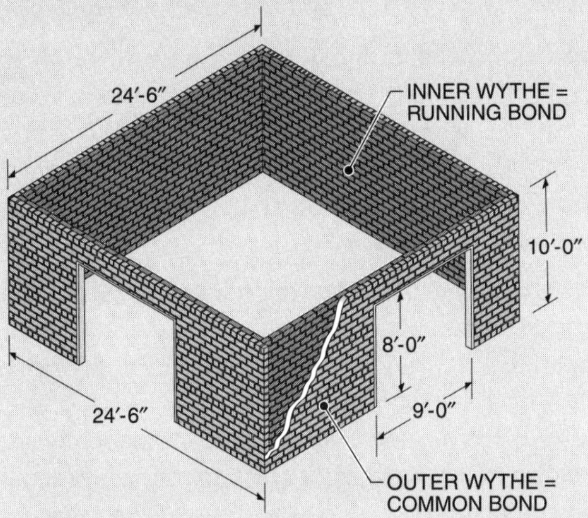

1. Compute the gross surface area of the exterior walls.

area of one wall = length × height

$A_1 = l \times h$

$A_1 = 24.5' \times 10'$

$A_1 = 245$ sq ft

area of four walls = area of one wall × 4

$A_4 = A_1 \times 4$

$A_4 = 245 \times 4$

$A_4 = 980$ sq ft

2. Compute the net area of the exterior walls.

area of openings = width × height × number of openings

$A_o = w \times l \times n$

$A_o = 9' \times 8' \times 2$

$A_o = 144$ sq ft

$A_n = A_4 - A_o$

$A_n = 980$ sq ft $- 144$ sq ft

$A_n = 836$ sq ft

3. Determine the number of bricks required.

no. of face bricks = A_n × multiplier (from masonry wall material table)

no. of face bricks = 836 × 7.04

***no. of face bricks* = 5885 bricks**

no. of building bricks = A_n × multiplier (from masonry wall material table)

no. of building bricks = 836 × 5.28

***no. of building bricks* = 4414 bricks**

Waste factors may be added to these totals and other adjustments made for mortar joint variations.

The number of CMUs required for a wall is determined by calculating the wall surface area (in square feet) and multiplying the area by a standard multiplier based on the CMU used and the size of mortar joints. Corner CMUs and other special units are calculated separately.

The number of CMUs may also be determined by dividing the wall length (in linear feet) by the length of the CMUs (in feet), and multiplying this subtotal by the number of courses in the overall wall height. A common waste factor is 5% for CMUs and 10% to 20% for mortar. The waste factor is added after the total number of CMUs is determined.

Calculating Concrete Masonry Unit Quantities

Determine the number of 16″ long by 8″ high CMUs required for a wall measuring 32′-0″ long by 7′-6″ high.

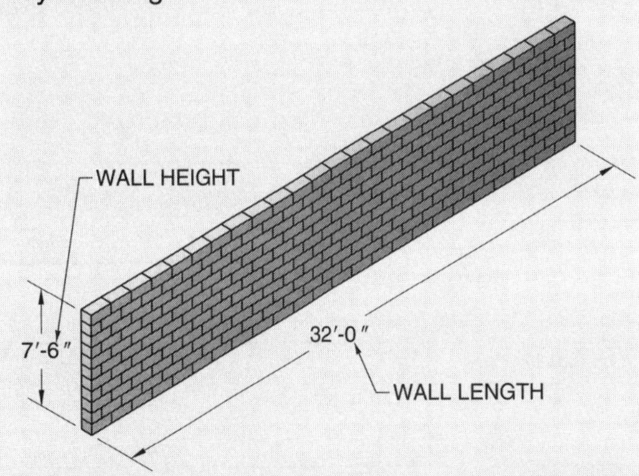

1. Determine the number of CMUs per course.

no. of CMUs per course = wall length ÷ CMU length

no. of CMUs per course = 32′ ÷ 1.33′

no. of CMUs per course = 24

2. Determine the number of CMU courses.

no. of CMU courses = wall height ÷ CMU height

no. of CMU courses = 7.5′ ÷ 0.67

no. of CMU courses = 11

3. Determine the number of CMUs required.

total number of CMUs = no. of CMUs per course × no. of CMU courses

total number of CMUs = 24 × 11

***total number of CMUs =* 264**

Mortar quantity is determined from standard mortar quantity tables based on varying wall thicknesses. **See Figure 7-14.** For example, for a cement-lime mortar, 2.40 sacks of cement, 2.40 sacks of lime, and 0.60 cu yd of sand will be used to lay approximately 1000 CMUs in a 12″ thick wall with ½″ mortar joints. (Additional tables are available for CMUs and other masonry materials.)

The linear feet of reinforcing installed in mortar joints is based on spacing between courses of reinforcing and the linear feet of wall. Below-grade masonry is commonly laid with ½″ mortar joints. Specifications are checked to determine these variables and other reinforcing information.

Elevations and floor plans are used to determine the number and types of lintels over openings, and deductions for wall openings. A *lintel* is a structural member placed over an opening or recess in a wall to support the construction above. Lintels are calculated on a per opening basis from door and window schedules. Lintels are listed as individual units on a takeoff.

When using a spreadsheet, the appropriate cell for brick or CMU information may contain a standard formula that automatically performs necessary mathematical functions or refers to a standard chart. With some estimating programs, entering wall length and height dimensions into the appropriate item screen allows the program to automatically calculate brick and CMU quantities, mortar amounts, equipment costs, reinforcing, and other items required by the estimator. **See Figure 7-15.**

Labor. Labor estimates for masonry construction include mixing and transporting mortar and masonry materials, scaffold erection and removal, laying the masonry units, striking the masonry joints, any required cleaning, and possible erection and removal of weather protection. Labor costs are related to square foot calculations of surface areas for various masonry materials. Standard labor-rate tables or company historical data are utilized for these estimates. Where additional heights or unusual construction situations make scaffold erection and material movement difficult, additional labor costs may be incurred.

MORTAR QUANTITY*

Proportion	8″ Wall			12″ Wall		
	Cement Sacks	Lime Sacks	Sand†	Cement Sacks	Lime Sacks	Sand†
Cement Mortars						
Cement 1, Sand 2	5.00	—	0.55	5.40	—	0.60
Cement 1, Sand 2½	4.20	—	0.55	4.40	—	0.60
Cement 1, Sand 3	3.70	—	0.55	4.00	—	0.60
Lime Mortars						
Lime 1, Sand 2	—	3.80	0.55	—	4.20	0.60
Lime 1, Sand 2½	—	3.30	0.55	—	3.60	0.60
Lime 1, Sand 3	—	2.70	0.55	—	3.00	0.60
Cement-Lime Mortar						
Cement 1, Lime 1, Sand 6	2.20	2.20	0.55	2.40	2.40	0.60

* Quantities given are for below-grade foundation work or other places where joints between courses of brick are filled with mortar. Per 1000 CMUs.
† in cu yd

Figure 7-14. Standard mortar quantity tables give mortar quantity and ingredient amounts based on masonry wall thickness.

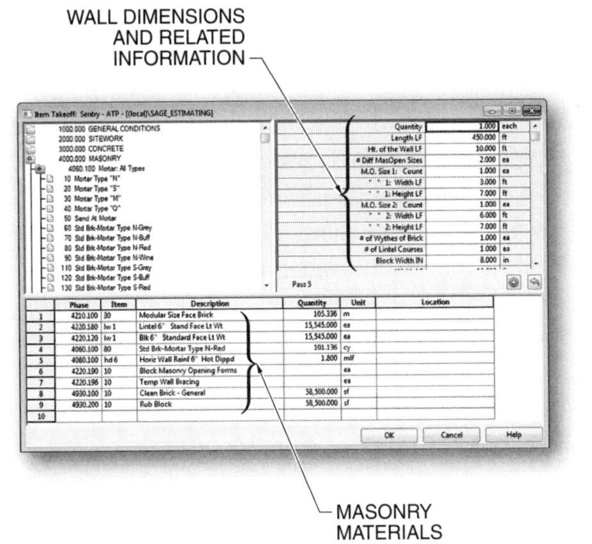

Figure 7-15. Estimating programs simplify masonry estimates by calculating quantities of several masonry materials at one time based on basic wall information.

Fireplace and Chimney Takeoff

For small chimneys, a standard table is used to determine the number of bricks per foot of chimney height. **See Figure 7-16.** The number of bricks required for a chimney is based on the size and number of flues in the chimney. When more than one flue liner is to be installed in the same chimney, masonry must be laid between the liners. The masonry must be at least 4″ thick and be bonded into the walls of the chimney.

SMALL CHIMNEY BRICKS PER FOOT OF HEIGHT		
Number of Flues	**Size of Flue***	**Number of Bricks†**
1	8 × 8	26
2	8 × 8	44
3	8 × 8	63
1	8 × 12	31
2	8 × 12	52
3	8 × 12	74
1	12 × 12	38
2	12 × 12	60

* in in.
† per foot of height

Figure 7-16. Brick quantities for small chimneys are determined based on a standard table and the height of the chimney.

For chimneys above the roof, quantity takeoff for masonry is performed by determining the linear distance around the chimney and multiplying it by the height of the chimney above the roofline. This results in the number of square feet of chimney above the roofline. Standard tables or square foot calculations, similar to those used for masonry quantity takeoffs for walls, are then used to determine the number of masonry units required.

When developing a quantity takeoff for fireplace masonry materials, an estimator calculates the surfaces of a fireplace as solid areas. Using this method, the number of bricks for the surface area is multiplied by the number of layers or tiers of brick deep. For example, face brick on the surface of the fireplace may be backed up with another interior wall, face brick, or common brick to provide additional support for flue materials.

In estimating the number of brick required for a chimney or fireplace, the easiest ledger sheet method is to calculate the cubic feet of chimney or fireplace from the prints and deduct the flue lining and hearth areas. The estimator must also add the square feet of firebrick required for the combustion chamber. The number of firebricks is determined from standard face brick tables. The required number of flue liners is based on the linear feet of flue liner and type and size required. Dampers and lintels are taken off as single unit items from elevations, detail drawings, and floor plans. Standard labor-rate tables or company historical data is utilized to determine labor costs.

STONE

Stone is used for many construction applications, including veneer, structural, and decorative applications. Three characteristics that affect the features of building stone are color, pattern, and texture. Stone color varies based on stone type and the location from which the stone was quarried. Stone color may be consistent throughout or vary within a particular stone. Patterns are variable and give special features to building stone. Texture varies from coarse to fine. Stone should be selected for use in a construction project based on moisture penetration, weatherability, price, availability, color, pattern, and texture.

Rock Classifications

The three types of stone (rock) are igneous, sedimentary, and metamorphic. *Igneous rock* is rock formed from the solidification of molten lava. Igneous rock has high compressive strength. Igneous rock includes granite and traprock. *Granite* is extremely hard natural rock consisting of quartz, feldspar, and other minerals produced under intense heat and pressure. Granite varies in color from almost white to gray and white, and retains a cut shape. Granite is used for building stone, as a stone veneer on walls, and for other decorative purposes. *Traprock* is fine-grained, dark-colored igneous rock. Traprock is commonly used as a base under or between layers of asphalt pavement.

Sedimentary rock is rock formed from sedimentary materials such as sand, silt, and rock and shell fragments. Sedimentary rock has low compressive strength but relatively high shear strength. Sedimentary rock includes limestone and sandstone. *Limestone* is sedimentary rock consisting of calcium carbonate. Limestone is white to light gray in color and is relatively soft. Limestone scratches and cuts easily and has a relatively high level of water absorption compared to other stone materials. Common applications of limestone include lintels, sills, and light decorative uses. *Sandstone* is sedimentary rock consisting of quartz held together by silica, iron oxide, and/or calcium carbonate. Sandstone has a granular texture and is primarily used as a decorative facing.

California Redwood Association
Stone can be combined with rough-sawn lumber to provide a rustic effect.

Metamorphic rock is sedimentary or igneous rock that has been changed in composition or texture by extreme heat, pressure, or chemicals. Metamorphic rock has high compressive strength but low shear strength. Metamorphic rock includes marble and slate. *Marble* is crystallized limestone. Marble is slightly harder than limestone and is commonly used as a decorative veneer. Marble is commonly available in preshaped squares or panels. Marble colors are highly varied and commonly show veins of different colors. *Slate* is a fine-grained metamorphic rock that is easily split into thin sheets. Slate is used as roofing tile and flooring material.

Many types of manufactured or precast stone materials are used for structural and masonry veneer materials. Precast stone is concrete cast in molds in a wide variety of sizes and shapes. Unlike natural stone, precast stone offers the advantages of consistency of size and structural qualities. Various face treatments on precast stone provide rough, smooth, grooved, or other surface finishes. Backs and sides of precast stone materials can be designed to provide for interlocking between individual pieces or for solid bonding with mortar. Styles and sizes of precast stone materials vary depending on local material availability and manufacturer designs.

STONE CONSTRUCTION METHODS

Many of the masonry construction methods used for brick and CMUs also apply to stone. Stone veneer must be placed on a firm footing or against a firm backing material. Equipment may be required to set stone in place where large-sized stone is specified in the construction documents. Mortar handling, scaffold requirements, and weather protection for stone construction are similar to masonry unit construction.

Rubble, squared stone, and ashlar are the most common stone classifications for construction. *Rubble* is rough, irregular stone that is used as fill or to provide a rustic appearance. Rubble is usually obtained locally as field stone or blasted from a quarry. *Squared stone* is rough stone made approximately square either at a quarry or at the job site. *Ashlar* is stone cut to precise dimensions according to the prints. When rubble or squared stone is to be installed, an architect shows the general arrangement of the stone on elevations.

As with brick and CMUs, a variety of bonds can be used for stone. Basic stone bonds include random, uncoursed roughly squared, coursed, and broken range.

See Figure 7-17. A random bond is created with various sizes of stone placed in a random arrangement without maintaining courses or ranges. A coursed bond is created with stone arranged in courses or ranges with vertical joints aligned on alternating rows. Each course of a coursed bond can consist of a different thickness of stone as long as the range remains unbroken.

Figure 7-17. Stone bonds include random, coursed, and broken range.

A *range* is a course of any thickness that extends across the face of a wall, but not all courses need to be the same thickness. A broken range bond is created with stone so the range is broken with larger stone. Horizontal mortar joints break against larger stone. When performing a quantity takeoff, an estimator should consult the specifications and elevations to determine the type of stone and bond to be used.

Placement

For structural applications, stone is set with mortar similar to brick and CMUs. Nonstaining cement mortar is used for light-colored stone. Additional metal ties are often required for structural stone applications. Care should be taken to use nonstaining metal ties such as plated steel or zinc. For some applications, such as low decorative walls and low retaining walls, stone such as limestone may be placed dry and fitted without mortar.

A variety of mastics, mortars, and clips may be used where decorative face-stone panels are fastened to structural members. The estimator should check specifications and details for fastening information pertaining to face stone to determine labor requirements.

STONE QUANTITY TAKEOFF

Structural-stone members such as lintels and sills are taken off individually, with length, width, and depth noted. Stone for walls is ordered by the ton. Estimators should consult with potential supplying quarries to determine the number of square feet of surface coverage expected per ton of stone for structural uses.

Stone Wall Takeoff

As with other masonry materials, the stone wall design affects the stone estimate. Structural stone walls incur construction costs similar to laying CMUs or brick, including mortar, transportation, scaffolding, weather protection, and temporary bracing. Square footage calculations that equate square footage with unit costs per ton of stone guide the estimator in pricing stone materials. Labor prices are also based on the price per ton calculated from standard industry guides or company historical data.

For stone-veneer walls, takeoff and cost estimates may be based on square foot or per piece costs. The decorative stone types, thicknesses, locations, and fastening methods affect material, labor, and equipment costs. Standard industry information or company historical data is referenced for labor-unit production levels for various stone-veneer applications.

Stone Takeoff

Quarrying costs, quarry location, construction-site location, and availability of stone affect stone costs. Each of the pieces for individual members, such as stone lintels, are added into the spreadsheet or database and priced individually. Estimators should contact suppliers in the geographic location of a project to check for the availability of stone required for a project. Some stone may need to be transported great distances between the quarry and the construction site, thereby adding transportation expenses to the estimate. Estimators should obtain stone costs, including delivery and transportation costs, to ensure an accurate estimate.

SAGE TIMBERLINE ESTIMATING

CALCULATOR USE

The Sage Timberline Estimating program includes a calculator into which values and operators are entered to calculate various construction material quantities. The calculator contains a single field that prefills with the value from the field in which the cursor was placed before the calculator was opened.

1. Using the Sample Ext Commercial GC database, perform a quick takeoff for item **4220.100 lw 1 Blk 12″ Standard Face Lt Wt** by picking the **Quick Takeoff** button and double-clicking through the group, phase, and item categories.
2. Close the **Quick Takeoff** dialog box by picking the **Close** button.
3. Calculate the number of concrete masonry units (CMUs) required for a 50′ × 8′ wall by picking the **Takeoff Quantity** cell for the 12″ CMU and opening the calculator. Open the calculator by picking the calculator button from the toolbar. Pick the **Clear** button to delete the contents of the calculator if required.
4. Develop the formula. Using the formula (wall length × wall width × # of blocks per sq ft), enter **50 * 8 * 1.125** in the pane. A math operator or function can also be inserted from the list by picking the **Functions** button. Note that the takeoff unit for this item is "ea" so the formula developed must calculate the total number of CMUs. The value generated by a formula or the calculator must always be in the units specified as the takeoff unit for the item. For example, if the takeoff unit is in cubic yards, the calculation must also return the value in cubic yards.

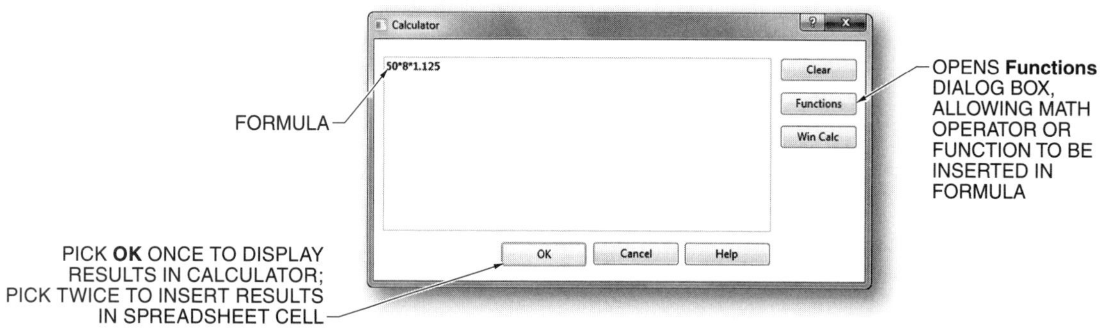

5. Press **Enter** or pick **OK** to display the results in the calculator pane.
6. Pick **OK** to insert the result into the **Takeoff Quantity** cell and close the calculator.

Quick Quiz®

Quick Quiz®

Refer to the CD-ROM for the Quick Quiz® questions related to chapter content.

Key Terms

Illustrated Glossary

- *ashlar*
- *bond*
- *brick*
- *cavity wall*
- *concrete masonry unit (CMU)*
- *course*
- *face brick*
- *firebrick*
- *flashing*
- *glass block*
- *granite*
- *igneous rock*
- *limestone*
- *lintel*
- *marble*
- *metamorphic rock*
- *mortar*
- *movement joint*
- *range*
- *refractory material*
- *refractory mortar*
- *reinforcement*
- *rubble*
- *sandstone*
- *sealant*
- *sedimentary rock*
- *shelf angle*
- *slate*
- *smoke shelf*
- *squared stone*
- *structural clay tile*
- *tie*
- *traprock*
- *wythe*

Web Links

Web Links

Adhesive and Sealant Council, Inc.
www.ascouncil.org

Brick Industry Association
www.bia.org

Building Stone Institute
www.buildingstoneinstitute.org

Cast Stone Institute
www.caststone.org

Indiana Limestone Institute of America, Inc.
www.iliai.com

International Masonry Institute
www.imiweb.org

International Union of Bricklayers and Allied Craftworkers
www. bacweb.org

Marble Institute of America
www.marble-institute.com

Mason Contractors Association of America
www.masoncontractors.org

Masonry Advisory Council
www.maconline.org

Masonry Institute of Washington
www.masonryinstitute.com

National Stone, Sand & Gravel Association
www.nssga.org

Portland Cement Association
www.cement.org

The Masonry Society
www.masonrysociety.org

Masonry 7

Review Questions

Estimating

______________ **1.** A(n) ___ is a continuous vertical or horizontal joint in a masonry wall without mortar or other nonresilient materials that is used to accommodate expansion and contraction of the wall system.

______________ **2.** A(n) ___ is a masonry unit made of clay or shale that is formed into a rectangular shape while soft and then fired in a kiln.

______________ **3.** ___ brick is brick commonly used for an exposed, finished surface.

______________ **4.** ___ is brick manufactured from refractory ceramic materials that resist disintegration at high temperatures.

______________ **5.** ___ is a hollow masonry unit composed of burned clay, shale, or a combination of clay and shale that is used for structural and decorative applications.

T F **6.** A concrete masonry unit is a precast hollow or solid block made of portland cement and fine aggregate, with or without admixtures or pigments.

T F **7.** Masonry is noted in Division 5 of the CSI MasterFormat™.

T F **8.** A refractory material is a material that can withstand high temperatures without structural failure.

______________ **9.** ___ is a bonding mixture of fine aggregate, cement, and water and is used to fill voids between masonry units and reinforce a structure.

______________ **10.** A(n) ___ is a type of reinforcement used to secure two or more members together.

______________ **11.** A(n) ___ is a liquid or semiliquid material that forms an airtight or waterproof joint.

______________ **12.** A(n) ___ is a continuous horizontal layer of masonry units bonded with mortar.

______________ **13.** A(n) ___ is the arrangement of masonry units in a wall created by overlapping them one upon another to provide a sturdy structure.

T F **14.** A cavity wall is a masonry wall with at least a ¼″ void between wythes.

______________ **15.** A(n) ___ is the fireproof floor immediately surrounding a fireplace.

Stone

________ **1.** Igneous rock

________ **2.** Granite

________ **3.** Traprock

________ **4.** Sedimentary rock

________ **5.** Limestone

________ **6.** Sandstone

________ **7.** Metamorphic rock

________ **8.** Marble

________ **9.** Slate

A. Crystallized limestone

B. Sedimentary or igneous rock that has been changed in composition or texture by extreme heat, pressure, or chemicals

C. Sedimentary rock consisting of quartz held together by silica, iron oxide, and/or calcium carbonate

D. Fine-grained metamorphic rock that is easily split into thin sheets

E. Rock formed from the solidification of molten lava

F. Extremely hard natural rock consisting of quartz, feldspar, and other minerals produced under intense heat and pressure

G. Rock formed from sedimentary material such as sand, silt, and rock and shell fragments

H. Fine-grained, dark-colored igneous rock

I. Sedimentary rock consisting of calcium carbonate

Stone Bonds

________ **1.** Uncoursed roughly squared

________ **2.** Coursed ashlar, broken bond, broken range

________ **3.** Random rubble

________ **4.** Coursed ashlar

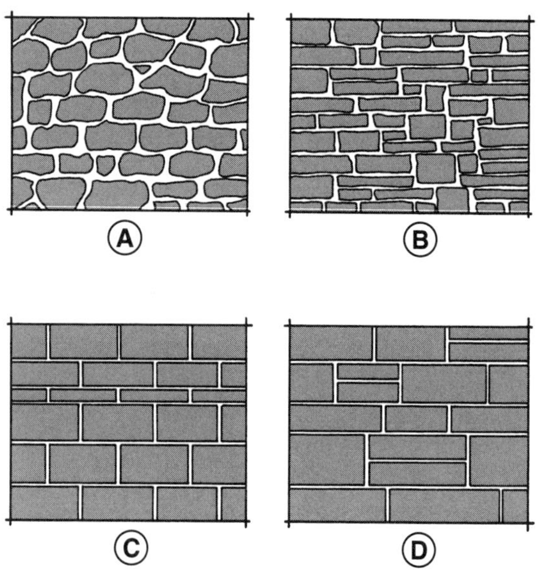

Brick Bonds

______ 1. English

______ 2. Flemish

______ 3. Stacked, horizontal

______ 4. Stacked, vertical

______ 5. Common

______ 6. Running

______ 7. Garden wall, double stretcher

______ 8. Garden wall, three stretcher

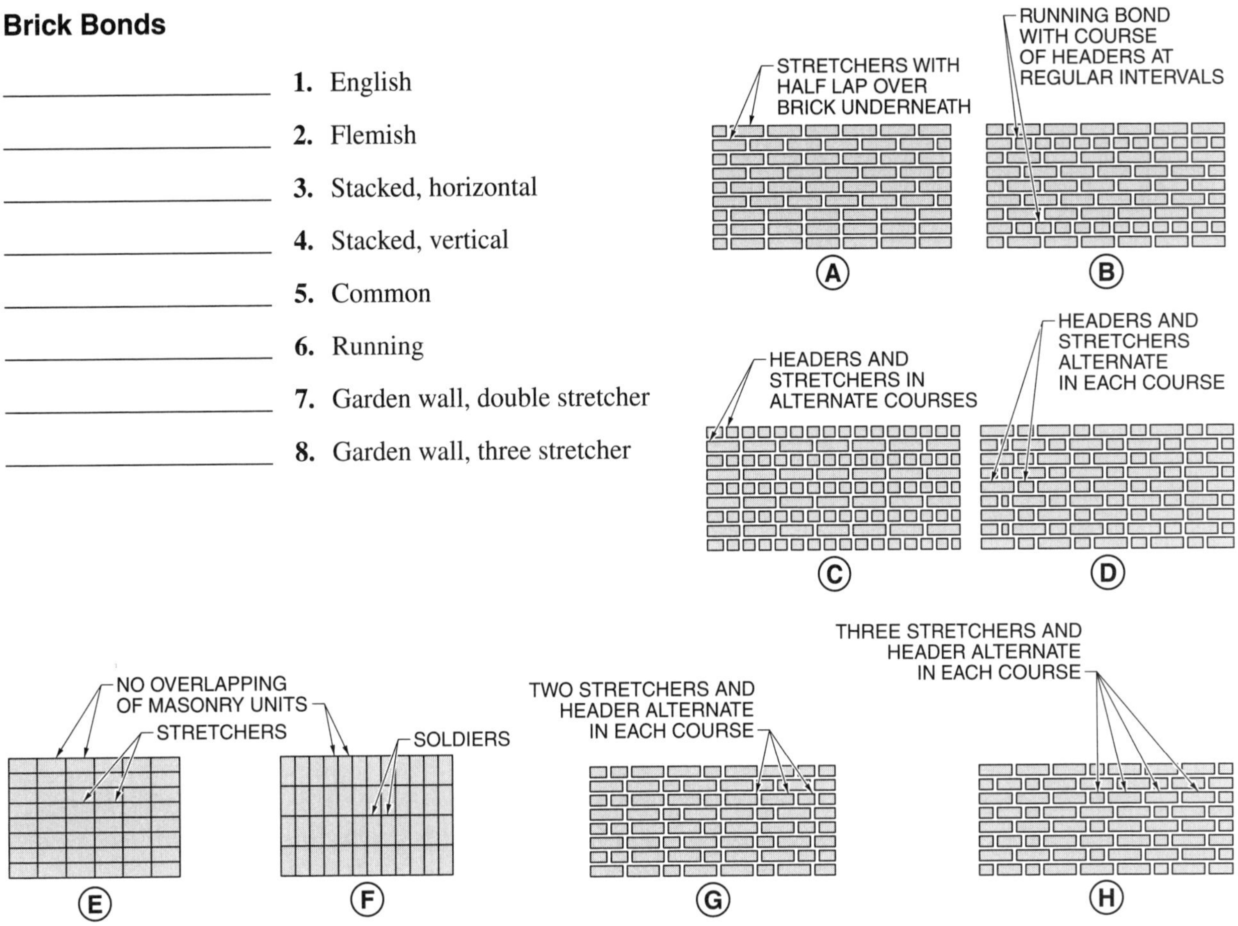

Structural Clay Tile

______ 1. Cove base

______ 2. Jamb

______ 3. Miter

______ 4. Corner

______ 5. Stretcher

______ 6. Sill

______ 7. Cap

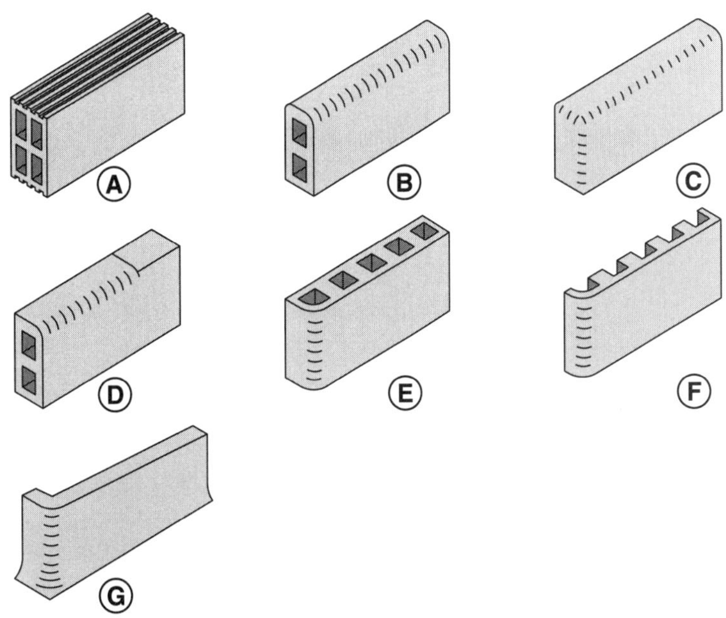

Short Answer

1. Identify the five types of masonry mortar and briefly describe each type.

2. Discuss how an estimator approaches a masonry takeoff.

3. Identify the three types of rock used for building stone and briefly list the characteristics of each type.

Masonry 7

Activities

Activity 7-1—Ledger Sheet Activity

Refer to the cost data and Estimate Summary Sheet No. 7-1. Take off the number of utility brick. Determine the total material and labor cost for the brick.

COST DATA			
Material	**Unit**	**Material Unit Cost***	**Labor Unit Cost***
Utility brick 4″ × 4″ × 12″ 3.00 brick/sq ft	1000	918.50	819.50

* in $

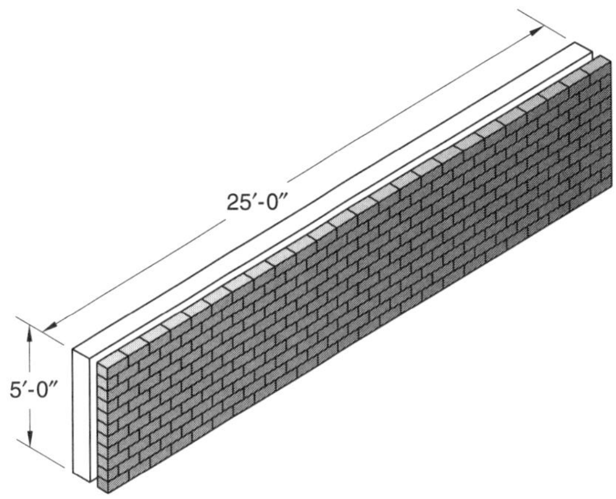

ESTIMATE SUMMARY SHEET

Sheet No. 7-1

Project: ______________________

Date: ______________________

Estimator: ______________________

Checked: ______________________

No.	Description	Dimensions			Quantity		Material		Labor		Total	
		H	L			Unit	Unit Cost	Total	Unit Cost	Total	Unit Cost	Total
	Utility brick 4″ × 4″ × 12″											
	Total											

Activity 7-2—Spreadsheet Activity

Refer to the cost data and Estimate Summary Sheet No. 7-2 on the CD-ROM. Take off the square feet of CMUs for the 6′-0″ high foundation walls of a 25′-0″ × 70′-0″ building. Determine the total material and labor cost for the CMU walls.

COST DATA			
Material	Unit	Material Unit Cost*	Labor Unit Cost*
Concrete masonry units	sq ft	1.36	2.67

* in $

Activity 7-3—Estimating Masonry Activity

Refer to Prints 7-3A and 7-3B and Quantity Spreadsheet No. 7-3 on the CD-ROM. Determine the number of 4″ × 8″ face bricks and 4″ × 8″ × 16″ split-faced blocks that are in the three sections of the exterior wall between door A (room 100) and the rear freezer at the back of room 108A, and room 112. Assume that a running bond is used for the bricks and CMUs.

Activity 7-4—Sage Timberline Estimating Activity

Create a new estimate and name it Activity 7-4. Perform a quick takeoff for a masonry wall 35′ long by 8′ high having no openings. Take off items **4210.100 30 Modular Size Face Brick**, **4060.100 10 Mortar Type "N"**, and **4930.100 10 Clean Brick-General**. The bricks are 4″ wide and the wall consists of a single wythe (vertical course) of bricks. Print a standard estimate report.

Metals 8

Key Concepts

- Due to the various sizes and shapes of many structural steel pieces, it is common for a contractor to work with a steel fabrication shop to obtain the required structural steel members.
- A schedule of the steel members is produced, with costs for material, fabrication, and transportation associated with each member.
- Estimators at the fabrication shop determine costs for each element, including material, labor, and transportation costs to the job site.
- Crew sizes for setting various structural steel members is determined based on company historical data or past job-site experience.

Introduction

Structural steel construction uses a series of horizontal steel beams and trusses and vertical columns that are joined to create large structures with open areas. The general classifications of structural steel construction are beam-and-column, long-span, moment-resisting, framed-tube, and wall-bearing. Common steel shapes required for structural steel construction include W-, S-, C-, and M-shapes and angles, tees, plates, bars, and zees. Estimators rely on bids from a steel fabricator for costs for structural steel members, and on other sources for labor information.

STRUCTURAL STEEL MEMBERS

Structural steel construction is a building process in which a series of horizontal steel beams and trusses and vertical columns are joined to create large structures with open areas. The various types and grades of steel used in structural steel construction include carbon, high-strength, high-strength low-alloy, corrosion-resistant high-strength low-alloy, and quenched and tempered alloy steel. **See Figure 8-1.**

The most common steel used for structural steel construction is designated by ASTM International as A36. A36 carbon steel has a yield strength of 36,000 lb/sq in. The uses and applications of the various types of steel depend on the engineering requirements for a particular structure.

Many steel shapes are required for structural steel construction. **See Figure 8-2.** A variety of letters and symbols are used to indicate the different steel shapes on architectural and structural drawings. Estimators must be familiar with these symbols to accurately identify and take off structural steel members. While common notations are used to represent various steel shapes, different manufacturers, steel fabricators, and architects may use slightly different notations. Estimators should check with the manufacturer, fabricator, or architect when unfamiliar symbols or notations are encountered and there is the potential for error. Structural steel members are coated with a primer, and a finish coat may be applied at the fabrication shop or job site.

A variety of steel designs are manufactured to meet all of the applications and loading requirements for steel structural members. Variables in the designs include the steel composition, shape, thickness, weight, and length. Variances in the weight of the steel shape (in pounds per square foot) create differences between nominal size classifications and actual sizes of structural steel members. To prevent failure in case of fire, structural steel members are sprayed with fireproofing materials and may be encased in concrete, masonry, or gypsum, which adds additional cost to the material and labor estimates.

A36 carbon steel is the most common steel used in the United States for structural steel members in buildings. A36 carbon steel is ductile (able to deform without fracturing) and can be welded or bolted. A44 carbon steel is the most common structural steel used in Canada.

STRUCTURAL STEEL				
Steel	**ASTM Designation**	**Minimum Yield Stress***	**Form**	**Characteristics**
Carbon	A36	36	Plates, shapes, bars, sheets and strips, rivets, bolts, and nuts	For buildings and general structures; available in high-toughness grades
	A529	42	Plates, shapes, bars	For buildings and similar construction
High-strength	A440	42 to 50	Plates, shapes, bars	Lightweight and superior corrosion resistance
High-strength low-alloy	A441	40 to 50	Plates, shapes, bars	Primarily for lightweight welded buildings and bridges
	A572	42 to 65	Several types. Some available as plates, shapes, or bars	Lightweight, high toughness for buildings, bridges, and similar structures
Corrosion-resistant high-strength low-alloy	A242	42 to 50	Plates, shapes, bars	Lightweight and added durability; weathering grades available
	A588	42 to 50	Plates, shapes, bars	Lightweight, durable in high thicknesses; weathering grades available
Quenched and tempered alloy	A514	90 to 100	Several types. Some available as shapes, others as plates	Strength varies with thickness and type

* in KSI (1000 lb)

Figure 8-1. The type of steel used for structural steel construction depends on the engineering requirements for the structure.

STRUCTURAL STEEL SHAPES				
Description	**Pictorial**	**Old Symbol**	**New Symbol**	**Use**
W-shape (wide flange)	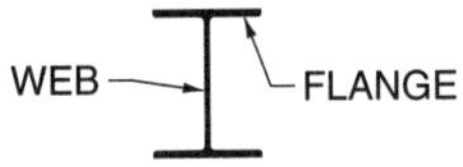	WF	W	W24×76
S-shape (I-beam)		I	S	S15×42.9
Bearing pile		BP	HP	HP14×102
C-shape (American Standard Channel)		⊔	C	C10×30
M-shape	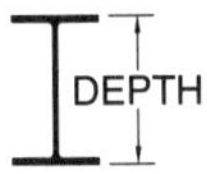	M	M	M12×10
MC-shape (channels other than C-shapes)		⊔	MC	MC12×37
Angle (equal legs)		L	L	L1×1×1/8
Angle (unequal legs)		L	L	L1 3/4×1 1/4×1/8
Structural tee (cut from W-shape)		ST	WT	WT16.5×84.5
Structural tee (cut from S-shape)		—	ST	ST10×35
Structural tee (cut from M-shape)		—	MT	MT12×36
Plate	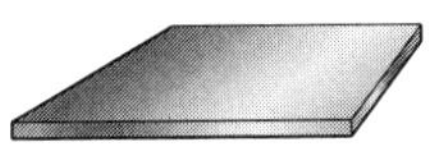	PL	PL	PL1/2×14×36
Flat bar		BAR	BAR	BAR2 7/8×1/4
Zee		Z	Z	Z3×2 11/16×2 11/16×1/2
Pipe	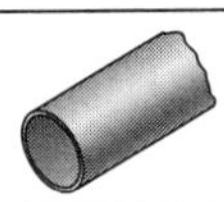	○	○	○6Ø

Figure 8-2. Standard steel shapes used for structural steel construction are represented on architectural and structural drawings using special symbols.

Beams

A *beam* is a horizontal structural member that is used to support loads over an opening. W- and S-shapes are commonly used as beams in structural steel construction. M-shapes may also be used where specified. Steel members are classified as beams and girders when they carry horizontal loads and are spaced greater than 4′-0″ OC. A *girder* is a large horizontal structural member that supports loads at isolated points along its length. Girders carry the loads of beams and joists. Girders are usually the heaviest horizontal members in a structure.

Shop drawings are often used for beam fabrication to show the beam size, type, length, cutout dimensions (to allow for intersection with other structural steel members), hole location dimensions, and any required connecting angle sizes. **See Figure 8-3.** It may be necessary to refer to plan views and elevations to acquire all information required for beam takeoff.

W-Shapes. A *W-shape,* or wide-flange beam, is a structural steel member with parallel inner and outer flange surfaces that are of constant thickness and are joined with a perpendicular web. A W-shape is designed so the web provides strength in the vertical plane and the flanges provide strength in the horizontal plane. W-shapes are noted on prints using the letter "W" followed by the web nominal depth (in inches) and nominal weight (in pounds per foot). **See Figure 8-4.** W-shapes are the most commonly used structural steel members used for beams, columns, truss members and other load-bearing applications.

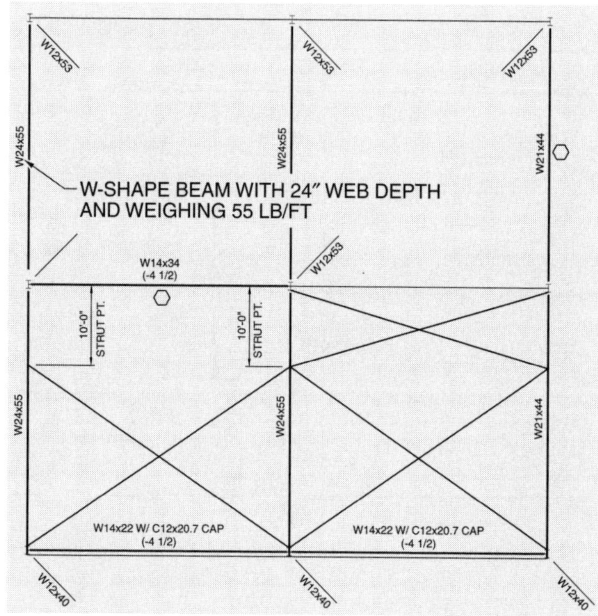

Figure 8-4. W-shapes, also known as wide-flange beams, are represented on erection plans with the letter W.

S-Shapes. An *S-shape,* or American Standard I-beam, is a structural steel member with parallel outer flange surfaces and sloping inner flange surfaces that are joined with a perpendicular web. The slope of the inner flange surfaces is approximately 17%. S-shapes are noted on prints using the letter S followed by the web nominal depth (in inches) and nominal weight (in pounds per foot). For example, a beam noted as S12×31.8 refers to an S-shape with a nominal web depth of 12″ and a nominal weight of 31.8 lb/ft. The nominal and actual depths of S-shapes are equal.

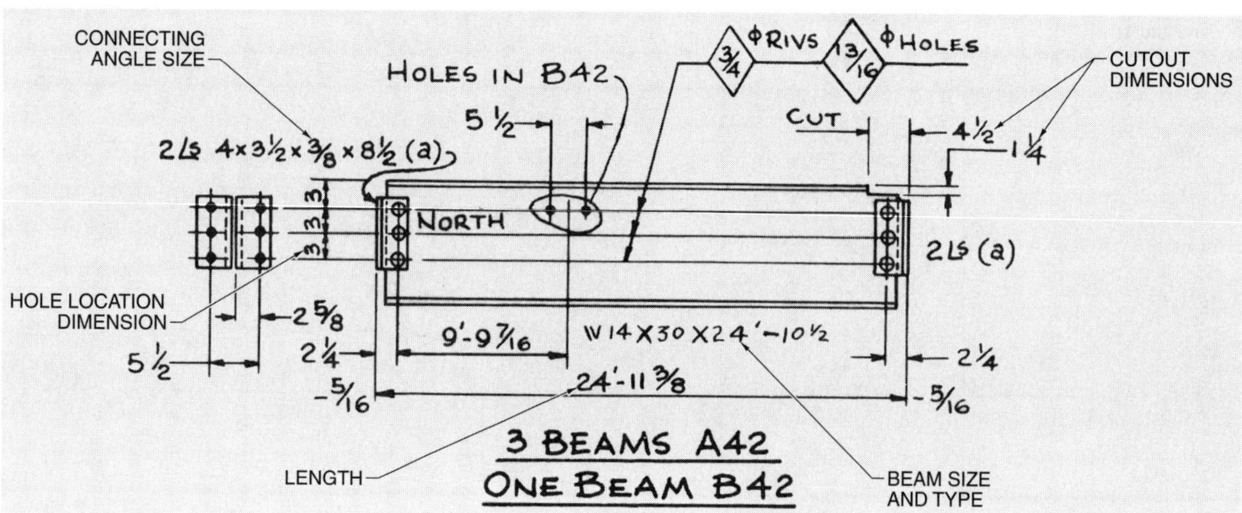

Figure 8-3. Shop drawings for structural steel members indicate fabrication information required for determining labor costs.

Columns

A *column* is a vertical structural member used to support axial compressive loads. Columns are the principal load-carrying vertical members in structural steel construction. Shop detail drawings indicate column type and dimensions, hole locations for the attachment of beams and braces, baseplate information such as plate thickness and size, and support angle location. **See Figure 8-5.**

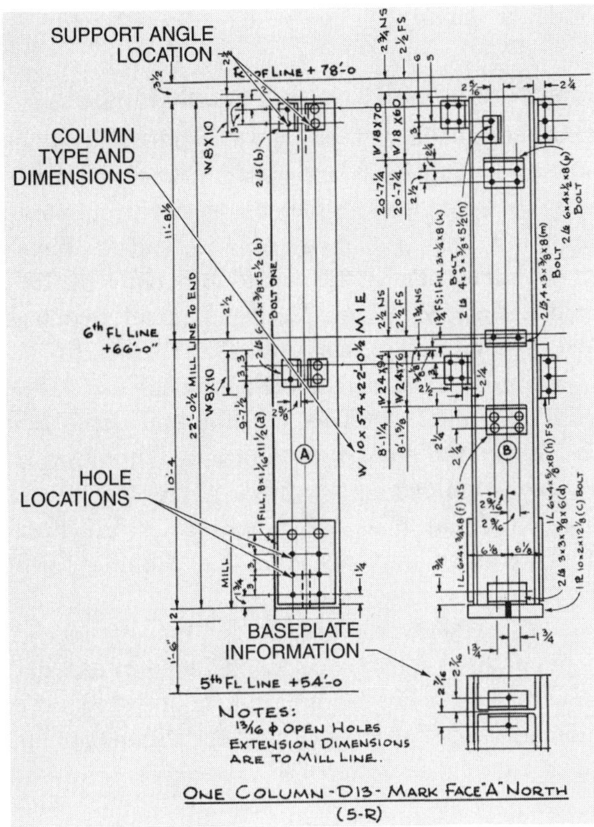

Figure 8-5. Detail drawings provide the information required for fabrication of steel columns. Cost estimates must take into account the amount of fabrication required.

Steel columns are constructed using W- and S-shapes. The size and design of the columns may be provided on a schedule and detail drawings. The load-support requirements for a column determine column size and design. Architects and engineers use standard tables to indicate sizes of steel shapes for column applications. **See Figure 8-6.** Steel columns may also be made from hollow structural shapes (HSS) such as round pipe or square tubing where loads are relatively light.

W-SHAPES				
Designation	Depth*	Flange*		Web Thickness*
		Width	Thickness	
W18×71	18½	7⅝	13/16	½
×65	18⅜	7⅝	¾	7/16
×60	18¼	7½	11/16	7/16
×55	18⅛	7½	⅝	⅜
×50	18	7½	9/16	⅜
W18×46	18	6	⅝	⅜
×40	17⅞	6	½	5/16
×35	17¾	6	7/16	5/16
W16×100	17	10⅜	1	9/16
×89	16¾	10⅜	⅞	½
×77	16½	10¼	¾	7/16
×67	16⅜	10¼	11/16	⅜
W16×57	16⅜	7⅛	11/16	7/16
×50	16¼	7⅛	⅝	⅜
×45	16⅛	7	9/16	⅜

* in in.

Figure 8-6. Notations for W-shapes include the letter W, followed by the nominal web depth (in inches) and weight (in pounds per square foot).

Joists

A *joist* is a horizontal structural member that supports the load of a floor or ceiling. Structural steel joists are lightweight beams spaced less than 4′-0″ OC. A joist may be a single structural member, such as a C-shape, or built up as a trussed or open web joist.

A *purlin* is a horizontal support member that spans across adjacent rafters or between beams, columns, or joists to carry intermediate loads, such as a roof deck or wall panels. Purlins that span between columns to carry wall panels are also referred to as girts. Purlins are formed using C-shapes or zees.

Open Web Joists. An *open web joist* is a horizontal structural steel member constructed with steel angles that are used as chords, with steel bars or angles extending between the chords at an angle. Open web joists are the most commonly used joists in structural steel construction. **See Figure 8-7.**

Standards developed by the Steel Joist Institute establish the depths, spans, and load-carrying capacities of open web joists. K series open web joists range in depth from 8″ to 24″, in 2″ increments, and are available in spans up to 60′-0″. K series open web joists are commonly used in light commercial construction. Long-span open web joists (LH series) and deep long-span open web joists (DLH series) can span up to 96′-0″ and 144′-0″ respectively, without center support. Open web joists are designated using the nominal depth (in inches), joist series, and chord diameter (in ⅛″ increments). For example, an open web joist with a designation of 24LH07 has a nominal depth of 24″, is a long-span joist, and has a chord diameter of ⅞″.

Figure 8-7. Open web joists are constructed with steel angles that are used as chords, with steel bars or angles extending between the chords at an angle.

Decks

A *deck* is a floor made of light-gauge metal panels (decking). Corrugated decking is placed horizontally and fastened to structural members to form floor or roof decks. **See Figure 8-8.** A plan view on the erection plans indicates the deck layout and opening locations. Floor plans may provide information concerning the number of floor or roof panels in each run. Detail drawings indicate methods for closing openings, applying bracing around columns, and applying edge treatments.

Roof Decks. Metal roof decks provide a high strength-to-weight ratio that reduces the amount of dead load on a structure. Variations in roof deck design include the width and height of the corrugated ribs, distance between the corrugations, and metal finish. Some steel roof decks are designed to create a finished surface with watertight joints between panels, while other roof decks are designed to be covered with additional insulation and roofing materials. Where sustainable roof structures are utilized, additional weight support may be required for the heavier imposed loads of turf, plantings, and water and moisture.

Floor Decks. A metal floor deck provides a work platform during construction as well as a form and reinforcement for concrete slabs. A common method of creating a floor in a structural steel building is to fasten corrugated steel decking to the top of the joists with self-tapping screws or by welding. Corrugated floor decking is commonly topped with several inches of concrete. The depth of the concrete varies depending on the loads to be supported. The thickness of the concrete is indicated on erection plans, floor plans, or detail drawings.

Miscellaneous Structural Steel Members

Other metal shapes used for structural applications such as bracing, joists, and truss construction include tees, bars, and zees. Structural tees are commonly made by cutting standard W-, S-, and M-shapes through the center of the web to form two tees. Tees formed from W-, S-, and M-shapes are referred to as WT, ST, and MT, respectively.

Notation systems for tees indicate the flange width first, followed by the nominal depth and the steel thickness. For example, a notation of WT6×11 indicates a tee from a W-shape, with a nominal depth of 6″ and weighing 11 lb/lf. Steel bars are noted on prints using their nominal width and thickness dimensions.

Zees are shown on drawings in the sequence of depth of the zee, first and second flange widths, and steel thickness. For example, a note of Z4×3×3×5⁄16 indicates a Z-shaped steel member with a depth of 4″, equal flange widths of 3″, and a steel thickness of 5⁄16″.

Framing Members

Metal framing members commonly used in structural steel construction include tracks, C-shapes, furring channels, and U-channels. **See Figure 8-9.** Other metal shapes are available for various applications. All of these shapes are used in different applications for structural and finish metal work.

Metal framing members, such as tracks and C-shapes, are 100% recyclable. The amount of metal framing members recycled over the past 10 years has extended the life of landfills by more than three years.

STEEL DECKS

Roof Decks

Pictorial	Span	Width*	Max. Length
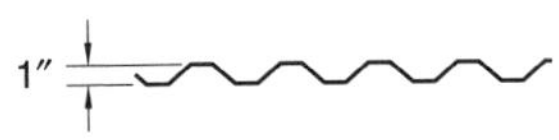	2′-6″ to 8′-0″	32 to 33	42′-0″
	4′-0″ to 11′-0″	36	42′-0″
	4′-0″ to 11′-0″	36	42′-0″
	5′-0″ to 12′-0″	36	42′-0″
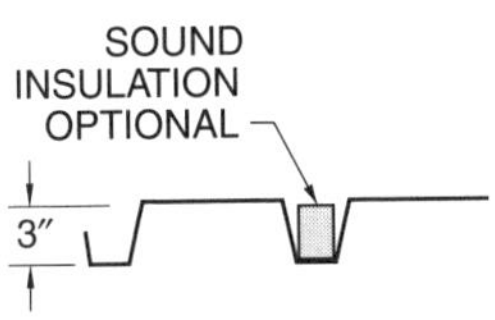	10′-0″ to 20′-0″	24	42′-0″
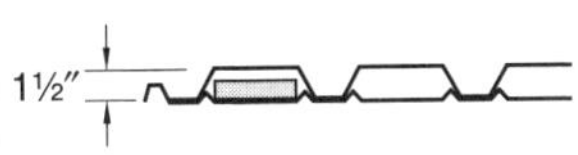	9′-0″ to 12′-0″	24	40′-0″
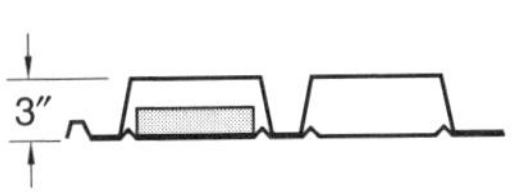	10′-0″ to 13′-0″	24	40′-0″
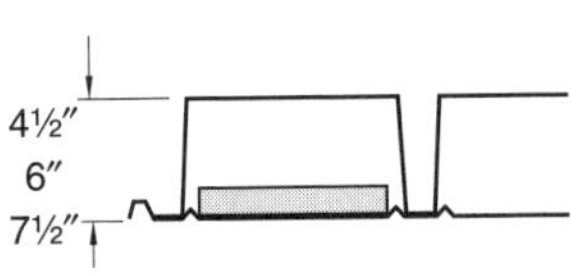	20′-0″ to 30′-0″	24	30′-0″

Composite Floor Decks

Pictorial	Span	Width*	Max. Length
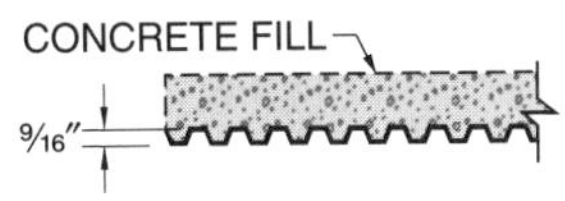	2′-0″ to 5′-6″	30 35 36	42′-0″
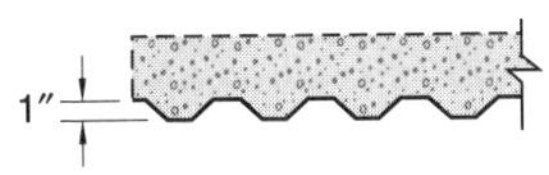	3′-0″ to 10′-0″	32 33	42′-0″
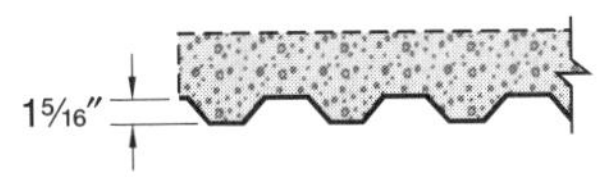	4′-0″ to 11′-0″	32	42′-0″
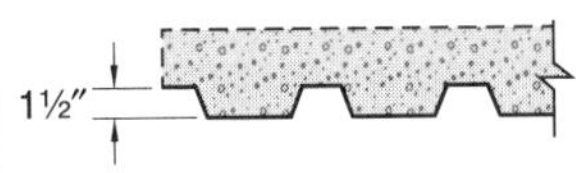	4′-0″ to 11′-0″	30 36	42′-0″
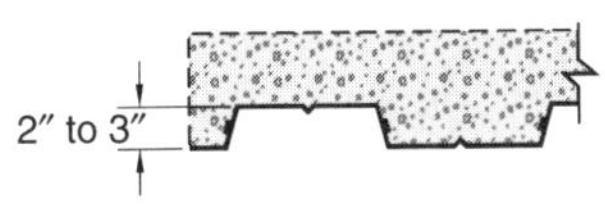	5′-6″ to 14′-0″	24 36	42′-0″

Noncomposite Floor Decks

Pictorial	Span	Width*	Max. Length
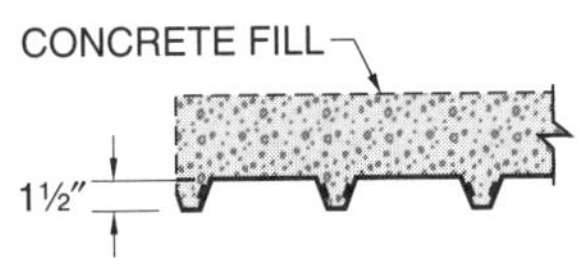	5′-0″ to 12′-0″	36	42′-0″
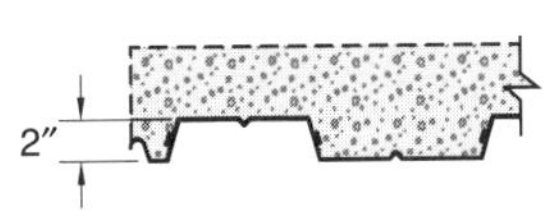	6′-0″ to 13′-0″	24 or 36	42′-0″
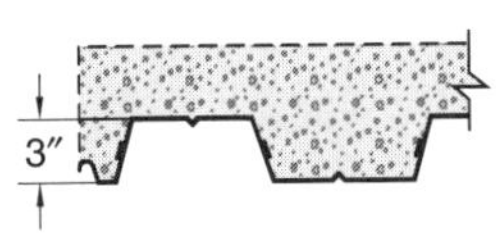	7′-0″ to 14′-0″	24 or 36	42′-0″

* in in.

Figure 8-8. Floor and roof deck panels are manufactured in various designs, widths, and lengths.

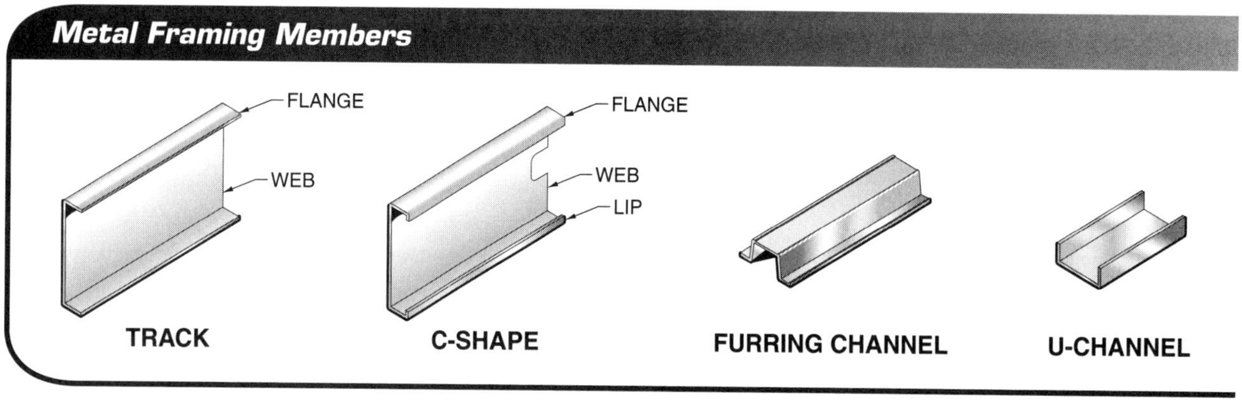

Figure 8-9. Light-gauge metal members are used to frame residential and commercial construction projects.

Studs. A *stud* is a vertical steel-framing member in a wall that extends from a bottom track to a top track. Metal studs used in light-construction applications are cold-formed and are used in the same manner as wood studs for load-bearing walls in commercial and residential construction. Metal studs and joists are most commonly made of 14 ga, 16 ga, 18 ga, or 20 ga steel. Load-bearing studs include channels, C-shapes, and nailable studs. **See Figure 8-10.**

Common channel studs are available in widths ranging from 2½″ to 6″ and in depths of 1″ and 1⅜″. Common C-shape studs are available in widths ranging from 2½″ to 8″ and in depths ranging from 1¼″ to 1$^{13}/_{16}$″. Nailable studs are available in widths of 3⅝″ and 4″ and depths of 1$^{13}/_{16}$″ and 1$^{15}/_{16}$″.

Due to the various gauges of metal used for load-bearing studs, the review and approval process of the prints is typically more rigorous than for wood-framed walls. An estimator must check stud spacing and placement on each job because stud placement may vary from job to job. In some areas, the different lighter gauges of metal studs and joists are installed by a drywall or carpentry contractor or subcontractor rather than a structural steel subcontractor. This will impact labor rate calculations.

Light-gauge steel framing members are fabricated from structural-quality sheet steel using cold-forming methods, such rolling, hammering, or stretching at low temperatures.

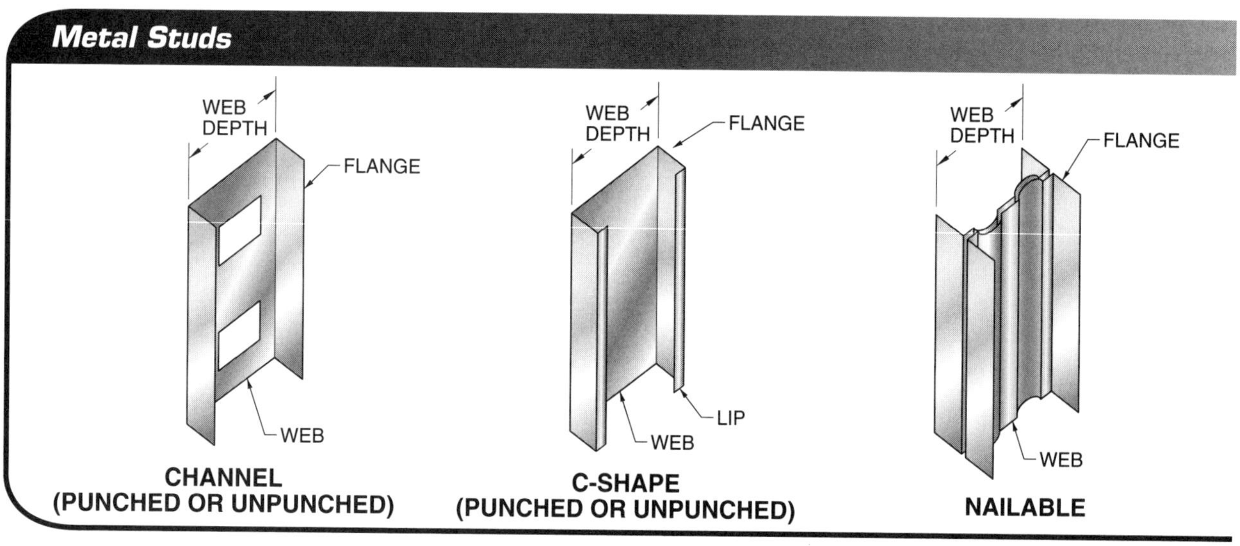

Figure 8-10. Metal studs of various gauges and designs are used for load-bearing applications.

Channels. A *channel* is light-gauge metal member used for framing and supporting gypsum board and other wall-finish materials. U- and furring channels are the most common types of light-gauge channels. Channels are shown on erection plans using the [symbol. Channel sizes are expressed as the channel depth stated first, followed by the flange width, and then the steel thickness. For example, a notation of [1⅜x3½x97 indicates a channel with a 1⅜″ depth, a 3½″ flange width, and a 97 mil steel thickness.

Light-gauge steel framing of C-channels or zees is used for purlins and girts. When used to span between columns or roof beams, C-channel or zee dimensions are noted on erection plan detail drawings.

Stairways

Metal stairway systems are installed in commercial buildings for common areas and for fire escapes, or as freestanding structures such as spiral stairs. Cast iron, steel, or aluminum may be used for metal stairways. When spiral metal stairs are installed, they are supported by a vertical pipe onto which treads are fastened. **See Figure 8-11.**

Spiral Metal Stairs

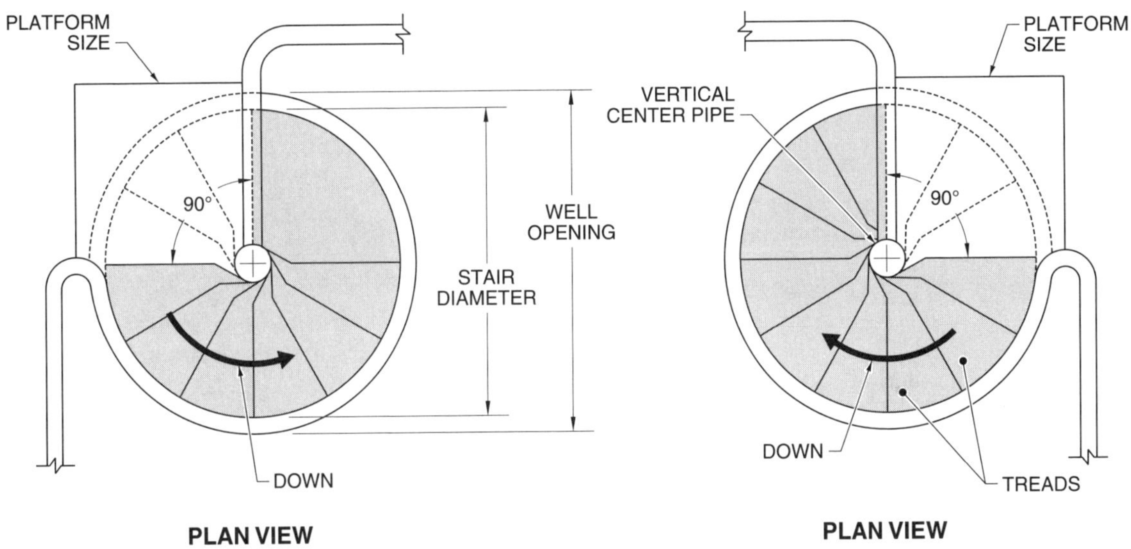

STAIR, PLATFORM, AND WELL STANDARD SIZES

Stair Diameter*	Center Pipe OD*: Cast Iron	Center Pipe OD*: Stainless Steel and Aluminum	Platform Size*	Well Opening*
42	3½	3½	22 × 22	44 × 44
48	3½	3½	25 × 25	50 × 50
54	4½	3½	28 × 28	56 × 56
60	4½	3½	31 × 31	62 × 62
66	4½	5	34 × 34	68 × 68
72	4½	5	37 × 37	74 × 74

* in in.

Figure 8-11. Spiral metal stairways are supported by a vertical center pipe onto which treads are fastened.

Treads. Metal stairway treads may be made of cast iron, checker plate steel, expanded metal grating, C-channel, angles, or bar grating depending on the tread application and location. Metal stairway treads may also be coated with an abrasive grit to provide greater traction. Metal pans or other metal designs may be combined with cast-in-place concrete to create a long-wearing, non-skid tread with a metal nosing.

Handrails. Metal handrails are made of steel or aluminum pipe and are designed to support a minimum of 200 lb at any point in any direction. Pipe is joined with mechanical connectors or by cutting and welding to form a smooth surface. Pipe handrails and posts are commonly bolted in place with special brackets designed to fit the pipe diameter, or the posts may be embedded in concrete. Metal handrails and balusters may also have wooden handrail components fastened to the top of the metal baluster/handrail frame.

The most cost-efficient steel stairs use common stock materials and standard details and are assembled at the job site using common construction techniques.

STRUCTURAL STEEL CONSTRUCTION METHODS

Structural steel construction may utilize lightweight metal members to build light commercial buildings and storage structures, or it may use large beam, column, and truss assemblies to form the skeleton of high-rise buildings and bridges. Structural steel may be used independently or combined with masonry or reinforced concrete.

In structural steel construction, a structural engineer determines all loads to be placed on a structure and then specifies the types of steel to be used and the sizes and shapes of the required steel members. Shop drawings are developed by engineers and architects in preparation for fabrication of individual structural steel components. Various steel assemblies are fabricated according to shop drawings at a fabrication shop. Takeoff of each piece of structural steel is also performed at the fabrication shop. Building information modeling (BIM) principles are common in the structural steel fabrication process. BIM allows for seamless integration from the design process into the fabrication shop. The assemblies and all other individual structural steel members are then transported to the job site, where the members are unloaded (shaken out).

Shaking out is the process of unloading steel members and assemblies in a planned manner to minimize any additional handling and moving of pieces during the erection process. Bar codes or other marking systems provided from the fabrication shop help track delivery and final placement information with the job site. Structural steel members are lifted into place with various lifting equipment. Structural steel components are erected, braced, and fastened together to create a structural frame. **See Figure 8-12.** The frame is covered with the materials for floors, walls, and roofing.

Figure 8-12. Structural steel construction uses a variety of prefabricated and job-built assemblies.

The proper and safe handling, lifting, and placement of steel members is essential in structural steel construction. Estimators should check the specifications concerning required worker fall protection, which may include a body harness and lanyard or guardrails. Titles 01 35 23 and 01 35 26 of the specifications should be consulted to determine owner and governmental safety requirements related to worker fall protection.

Varying engineering requirements and job site conditions create the need for several structural steel construction designs. General classifications of structural steel design are beam-and-column, long-span, moment-resisting, framed-tube, and wall-bearing construction. Erection plans provide information for placement of steel members for each structural steel construction method.

Beam-and-Column Construction

Beam-and-column construction is a structural steel construction method consisting of bays of framed structural steel members that are repeated to create large structures. Beam-and-column construction is the most common structural steel construction method. **See Figure 8-13.**

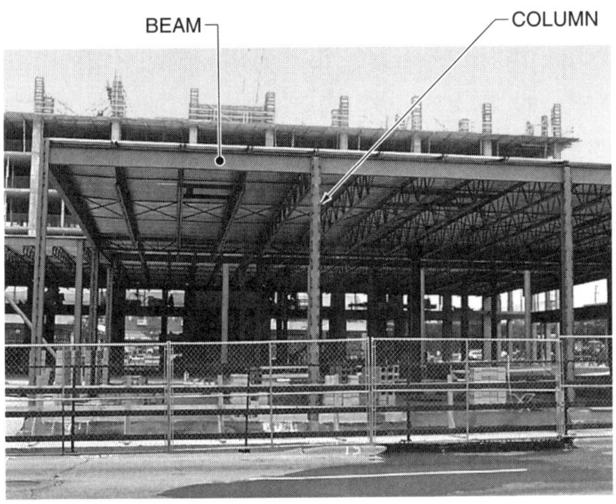

Figure 8-13. Beam-and-column construction uses structural steel beams that distribute horizontal loads onto vertical columns.

In beam-and-column construction, a series of footings, piers, or piles are constructed to support the columns. Anchor bolts and baseplates are prepared prior to setting the columns in place. When concrete foundations and footings are placed, anchor bolts used to fasten the columns to the foundation are set to specific dimensions for spacing and height according to the prints. Templates may be set on the anchor bolts and grouted in place to ensure exact elevations for steel column baseplates. Baseplates are welded onto small columns at a fabrication shop. Large column baseplates are set separately. Columns are normally the first members erected in beam-and-column construction.

Horizontal steel beams and girders are attached to the columns. Angles may be attached to columns to form a seat for the beams, or the angles may be attached to beams that are bolted to columns. Tie rods, channels, wire rope, and other temporary and permanent bracing members are used to stabilize the structural steel during erection. Joists and purlins are placed between the columns and beams to complete the structural portion of the construction.

Beams and girders are rigged and lifted into place before being bolted to columns or other girders. A *spandrel beam* is a beam in the perimeter of a structure that spans several columns and typically supports a floor or roof. Angles fastened to the columns or beams allow the spandrel beams to be fastened into place.

Bolts and nuts are commonly used for the initial fastening of columns and beams. Some structural steel components must be temporarily attached for later alignment and final attachment. Seat lugs or shear plates made of angles attached to the columns may be used to support beam-and-column connections during erection. For multimember beams, splice plates may be required to secure adjacent ends of each beam in place. As construction progresses, guy wires, turnbuckles, sag rods, sway rods, and other cross-bracing members are added to plumb, level, and secure all members at the proper elevations. Final bolting or welding of the connections of these members is required to complete the erection.

Spandrel beams span several columns and support a floor or roof.

Long-Span Construction

Long-span construction is a structural steel construction method that consists of fastening large horizontal steel members together and bracing them with angles, channels, and other steel members to create large girders and trusses. Long-span construction is used for structures such as bridges and arenas, where a series of built-up girders and trusses need to span large areas with few intermediate columns or other supports. **See Figure 8-14.**

Large girders and beams span between bridge abutments, piers, and other supports. A concrete or steel deck is supported by the large structural steel members. Detail drawings and shop drawings indicate the layout of the various angles, channels, and other members used to construct girders and trusses.

Moment-Resisting Construction

Moment-resisting construction is a structural steel construction method in which the stability of the steel frame and its wind and seismic resistance depend on a solid connection between the columns and beams. The top and bottom flanges of each beam are welded or bolted to the columns using full-length welds or high-strength bolts. **See Figure 8-15.** Moment-resisting structures are designed to act as a single unit when subjected to wind or seismic loads. Moment-resisting construction is typically used with low-rise structures.

Framed-Tube Construction

Framed-tube construction is a structural steel construction method in which the structural steel members form an exterior load-bearing wall that resembles a structural steel tube. Columns in the exterior walls are spaced close together and welded to large horizontal spandrel beams that span several columns. Framed-tube construction is the most common type of structural steel construction used with high-rise structures.

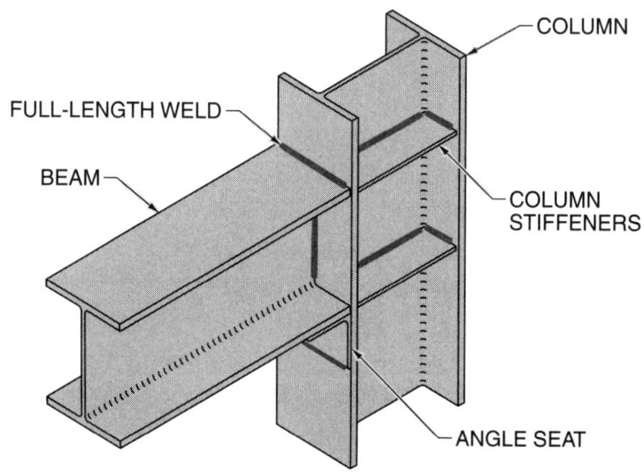

Figure 8-15. A welded or bolted connection between columns and beams is used in moment-resisting construction.

Long span construction requires careful planning for the erecting process, including the need for temporary supports.

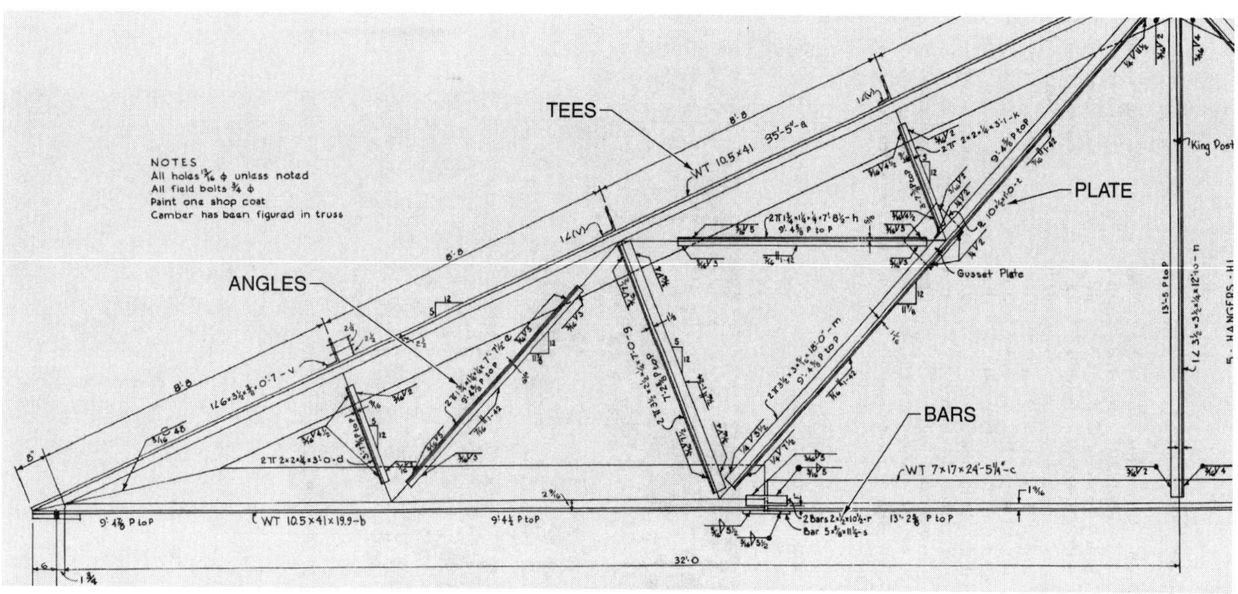

Figure 8-14. Long-span construction uses built-up steel trusses and girders to span long distances.

Wall-Bearing Construction

Wall-bearing construction is structural steel construction method in which horizontal steel beams and joists are supported and reinforced with other construction materials, such as masonry or reinforced concrete. Steel beams and joists span areas between masonry or reinforced concrete walls to create floors or roofs. The masonry or concrete walls support the vertical loads while structural steel beams and joists carry the horizontal loads. The installation of bearing baseplates on the walls provides proper load distribution at the points where the structural steel members rest on the walls.

Joists and Decks

Joists are commonly welded to their supporting members. Detail drawings indicate the manner of joist attachment to beams and girders. Print information for the installation of purlins and girts includes spacing and direction of placement. C-channels or zees are typically bolted to their supporting members. Spacing and sizes of bolts are noted in detail drawings.

After the structural frame is erected, connected, and braced, various panel and deck members are attached to the frame. For structural steel buildings, the panel and deck members create floors, walls, and roofs on the framework to complete the structure. Erection plans provide information concerning the various panel and deck members.

Metal Framing

Metal framing is a construction method in which a framework is constructed using light-gauge metal components. Studs are fastened with self-tapping screws or welded to top and bottom tracks in a design similar to wood platform framing. Door and window openings are framed using back-to-back C-shapes and cripple studs in the same manner as wood framing. As the studs and tracks are fastened together, angular strapping or bracing is also attached to provide lateral stability. Exterior finish panel and deck materials are then fastened to the metal frame.

STRUCTURAL STEEL QUANTITY TAKEOFF

Due to the various sizes and shapes of many structural steel pieces, it is common for a contractor to work with a steel fabrication shop to obtain the required structural steel members. A structural steel fabricator works with the designer, architect, and structural engineer to design and produce each piece of steel. Working drawings or shop drawings are developed that indicate baseplates, bearing members, types of welds, types of steel or other metal, locations for drilled or burned holes, and other special elements necessary for structural steel members. A schedule of the steel members is then produced, with costs of material, fabrication, and transportation associated with each member. Utilization of BIM provides a fully integrated system between the architect, structural engineer, and fabrication shop.

Installation costs must be considered during finalization of a bid. The installation items include the size and type of crane that is required for setting and holding the structural steel members in place during erection, the height of the building and any need for additional lifting or safety equipment, the job schedule and its potential impact on erection accessibility, and the season of the year, which could affect labor costs.

Structural Steel Estimation

Estimators rely on bids from a steel fabricator for costs regarding structural steel members. Estimators at the fabrication shop determine costs for each element, including material, labor, and transportation costs to the job site. **See Figure 8-16.** Because of variations in steel material costs due to the variability of material demand, estimators must work closely with steel suppliers to obtain accurate material cost data. In addition, lead times for fabrication and shipping of structural steel materials may be longer than other materials.

Figure 8-16. Structural steel members are calculated as individual units and grouped whenever possible.

Structural steel fabrication estimators rely on company historical data to determine costs for fabrication labor. Transportation prices depend on the distance from the fabrication shop to the job site, available transportation modes, and the sizes of the final pieces that are to be transported. Structural steel fabrication estimators begin with the structural drawings to determine the quantity of each type of column, beam, girder, and other structural steel members required. Designs for each include attachments for adjoining members, baseplates, and bracing.

Labor costs at the job site may be impacted by a number of situations. Job-site accessibility for steel delivery, shaking-out areas, and hoisting equipment must be coordinated with field supervisory personnel to minimize delays. Crew sizes for setting various structural steel members is determined based on company historical data or past job-site experience. Additional costs are incurred for safety equipment, including provisions for a safety net, personal fall-protection equipment, and perimeter protection that may be required for tradesworkers erecting structural steel.

Additional labor costs may be incurred when tradesworkers are required to work at heights.

Columns. Column locations are noted on erection plans based on a grid of letters and numbers. References to the letters and numbers identify each column. For example, a column located at the intersection of grid lines D and 2 is referred to as column D2. The design, size, and weight of each column may be noted at this intersection. For example, a notation of W12×53 indicates a column made of a W-shape with a 12″ web and weighing 53 lb/lf. A schedule of column information may also be included in the plans to provide information about the design, web, and weight of each column in relation to the plan grid.

Hollow steel sections used include round pipe, square tubing, and other enclosed shapes. For round pipe columns, the nominal inside diameter (in inches) and pipe schedule is indicated. For example, a round pipe column shown on prints as "8 Sch 60" indicates an 8″ diameter pipe column with schedule 60 wall thickness. For square tubing columns, the outside dimensions of the tubing are noted along with the wall thickness. For example, a notation of 3×3×¼ indicates a 3″ square tube column with a ¼″ wall thickness.

Estimators count the number of columns of each type and enter the amounts into a ledger sheet, spreadsheet, or estimating program. BIM drawings can provide this quantity information without additional calculation. Some columns may be similar in design and can be grouped. Pricing for each column must include overall height, bearing capacity, steel type, baseplate, and connections for intersecting beams or girders or additional columns for multistory buildings. Material cost variations, transportation, and labor costs must also be included in the final estimate.

Beams and Girders. Beam and girder sizes are noted on plan views and elevations. Sizes are noted along the grid lines on plan views. Lengths are obtained from the grid spacing dimensions. In a manner similar to columns, a letter and number system is commonly used to identify each beam or girder. The letters and numbers are marked on the beam or girder at the fabrication shop and correspond to letter and number notations on the erection plan drawings. A schedule of beam sizes and types may be provided in addition to the drawings.

Estimators must take material size, material type, and connection information into account when pricing structural steel beams. Estimators count the number of structural steel beams of each type and enter the number into a ledger sheet, spreadsheet, or estimating program. Material cost variations, transportation, and labor costs must also be included in the final estimate.

Joists. Unlike columns and beams, joists are not normally identified using a grid of letters and numbers. The spacing of joists and the direction for their placement is noted on the erection plans. **See Figure 8-17.** The type of joist to be installed may also be noted by a manufacturer identification code, standard classification format, or fabrication-shop code number. Open areas for stairwells or other access areas between levels of a structure are shown on the joist plan view. Open areas are indicated by dashed lines in an "X" pattern. Solid diagonal lines between joists indicate required cross-bracing.

Estimators determine the number of each type of joist of a particular length and develop a joist schedule. Each portion of a floor or roof deck may have many different types of joists depending on the spans and loads to be supported. Estimators should take off and place steel joist orders with the fabrication shop well in advance of the projected construction. Fabrication of steel joists may take time and could affect the project schedule if not ordered in a timely manner.

Information forwarded to the fabrication shop for each type of joist should include the quantity, length (span), depth, top and bottom channel design, web design, and bearing and support design at each end of the joist. Material cost variations, transportation, and labor costs must also be included in the final estimate.

Framing Members. Based on the prints and specifications, estimators identify the stud spacing, length, and gauge, and determine the linear feet of wall for which studs are required. Similar to determining the number of wood studs required, multipliers are used to determine the number of studs required. For example, a 1′-0″ OC spacing uses 1 as a multiplier, a 16″ OC spacing uses 0.75 as a multiplier, and a 24″ OC spacing uses 0.50 as a multiplier. The stud wall length is multiplied by the multiplier to determine the number of studs. Overall wall height determines stud length. The length of the wall is used to determine linear feet of top and bottom track.

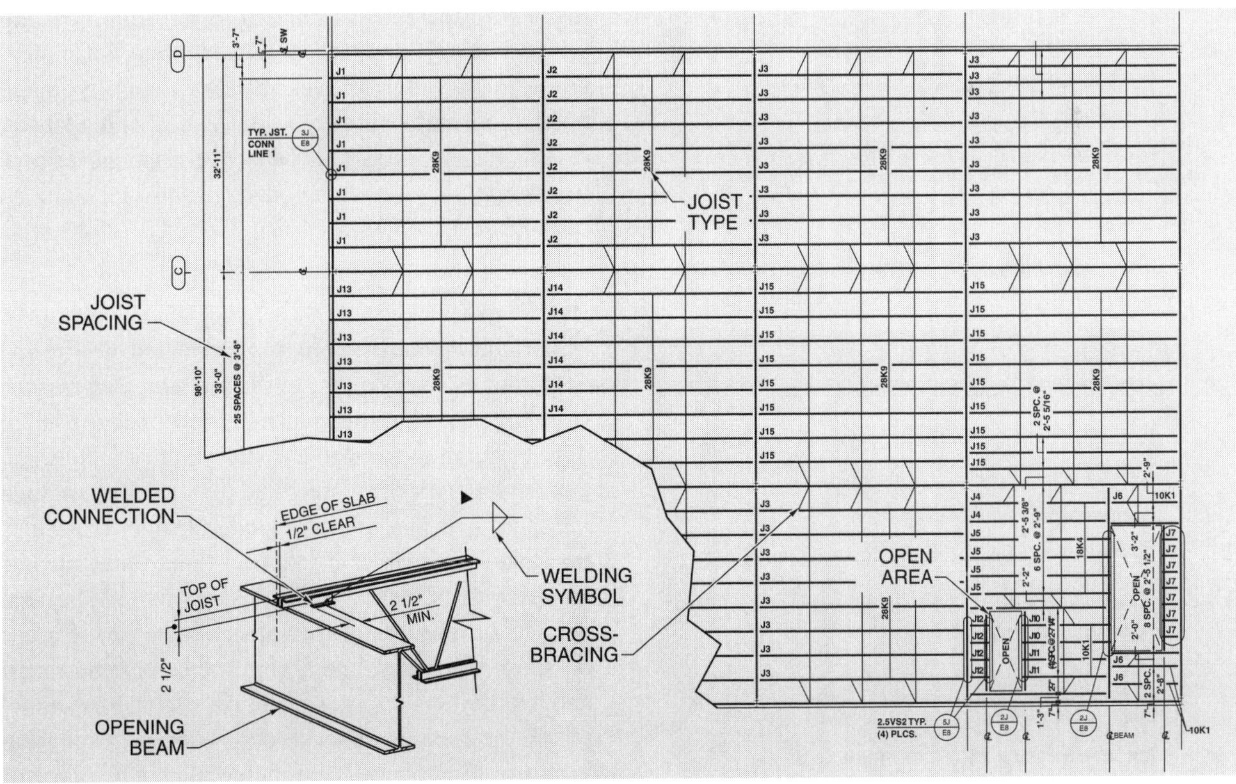

Figure 8-17. Erection plans provide estimators with information concerning joist types and spacing.

Estimators should include additional heavy-gauge studs at the sides of window and door openings and additional track material for headers and sills for window and door openings. In addition, bracing material should be added to the estimate where required. Estimators determine the number of each stud classified by stud length and metal gauge. Totals for each length and gauge classification should include approximately 3% to 5% for waste, depending on the overall size of the project. Material cost variations, transportation, and labor costs must also be included in the final estimate.

Calculating Quantities of Metal Studs and Track

Calculate the number of studs and top and bottom tracks needed for the 12′-10½″ wall section highlighted in the print. The studs are spaced 16″ OC.

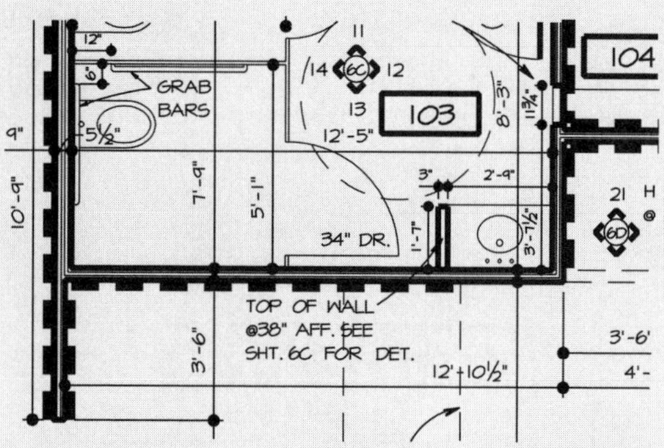

1. Convert the wall plate length to its decimal foot equivalent.

12′-10½″ = 12.875′

2. Determine the number of studs required.

number of studs = (*wall plate length* × *multiplier*) + 1

number of studs = (12.875′ × 0.75) + 1

number of studs = 9.66 + 1

number of studs = 10.66; rounded to 11 studs

3. Determine the length (in linear feet) of top and bottom track required.

length of track = *wall plate length* × 2

length of track = 12.875′ × 2

***length of track* = 25.75′; rounded up to 26′ (without waste)**

TrusSteel Div. of Alpine Engineered Products, Inc.

Metal roof trusses are fabricated offsite and lifted into position with a crane.

Roof or Floor Decks

A roof panel plan of a steel floor or roof deck is included in the erection plans. **See Figure 8-18.** A schedule of the types and sizes of floor or roof decking may be provided by the manufacturer to help take off the proper quantity of each type of panel.

Floor decking is cut to length and around unusual shapes with a cutting torch. It is then fastened to open web joists with self-tapping screws or by welding. Roof decking is attached to purlins or ceiling joists with self-sealing screws or clips. Estimators count the number of deck panels of each type and enter the number into a ledger sheet, spreadsheet, or estimating program. Material cost variations, transportation, and labor costs must also be included in the final estimate.

Stairways

Metal stairways are commonly fabricated off-site and delivered as a single unit to the job site. Dimensions from the plan view and interior elevations are used to develop estimates for metal stairways of a common pan design with concrete-filled treads. Costs include fabrication and installation labor, stringer materials, tread pans, and risers if required. There may also be handrail costs included in the stairway bid. Several designs of prefabricated treads are available for use on spiral stairways, and several types of center pipe and handrail systems are available. Estimators should check the specifications to determine if installation is part of the stairway cost included in their scope of work prior to submitting a bid.

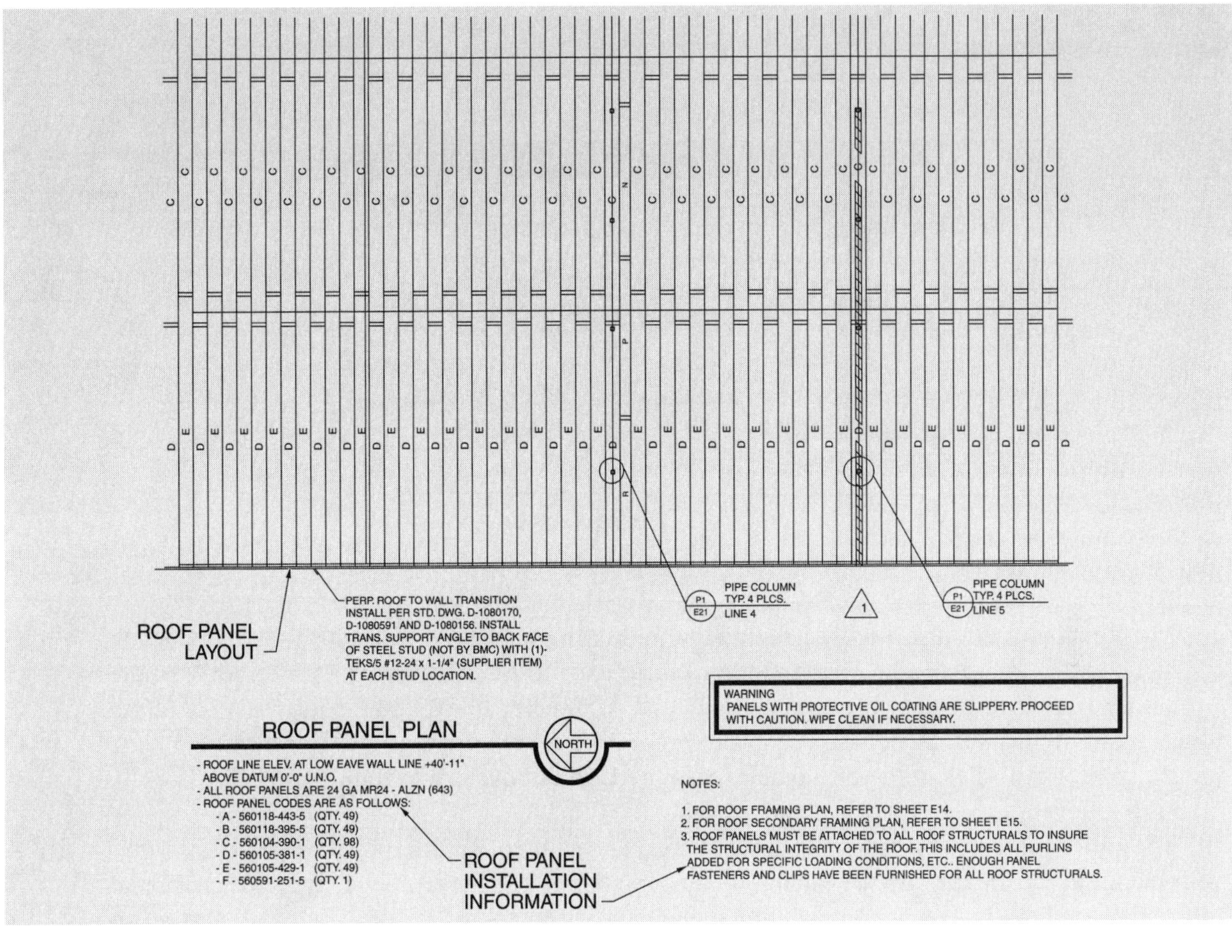

Figure 8-18. A roof panel plan provides information concerning the number of each type of roof panel as specified by a metal building manufacturer.

SAGE TIMBERLINE ESTIMATING . . .

LISTS

Lists are available throughout the Sage Timberline Estimating program to help navigate to selections and to provide ways of searching for information. Lists can be sorted by Description or by Phase/Item.

Expanding and Collapsing Lists

Lists can be expanded and collapsed by double-clicking the left mouse button.

1. Open the **Quick Takeoff** dialog box.
2. Using the mouse, double-click one of the groups to expand from the group level to the phase level. Double-click the phase to expand from the phase level to the item level. Double-click the phase to collapse back to the phase level. Double-click the group to collapse back to the group level.

. . . SAGE TIMBERLINE ESTIMATING . . .

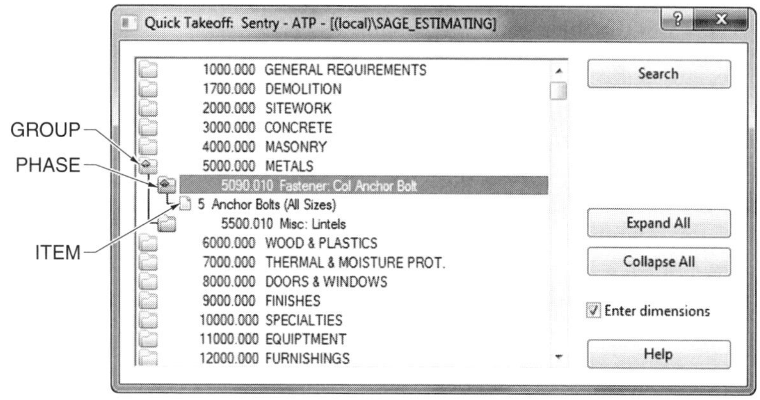

The keyboard can also be used to expand and collapse lists to navigate to another level. Select a group and press the **Right Arrow** key to expand the list from the group level to the phase level. Select the phase and press the **Right Arrow** key to expand the list from the phase level to the item level. Select the phase and press the **Left Arrow** key to collapse the list back to the phase level. Select the group and press the **Left Arrow** key to collapse the list back to the group level.

The entire contents of a database can also be expanded or collapsed using the right-click shortcut menu. In the **Quick Takeoff** dialog box, right-click in the list to display the shortcut menu. Pick **Expand All** to expand the list to view the entire contents. Right-click and pick **Collapse All** to collapse the entire list. The same results can be obtained by picking the **Expand All** or **Collapse All** button in the **Quick Takeoff** dialog box.

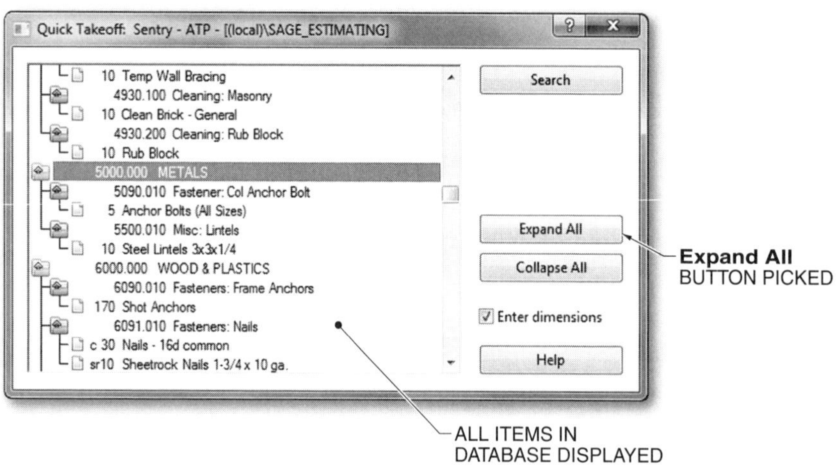

. . . SAGE TIMBERLINE ESTIMATING

Sorting Lists

The **Quick Takeoff** list can be sorted by description or phase/item. By default, the list is displayed by group, phase, and item. To select the sort order of the list, right-click in the item list to display a shortcut menu. Select **Description** or **Phase/Item** and the list is immediately re-sorted.

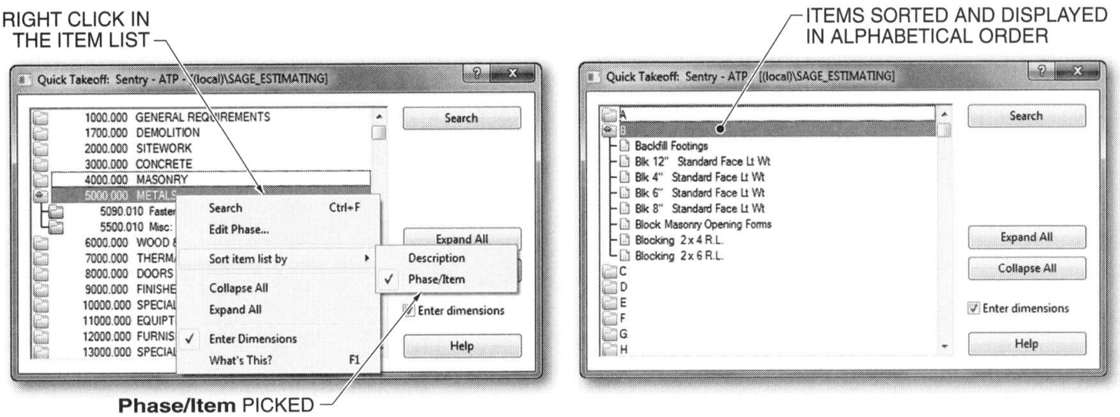

Finding Items

In any list, the quickest way to search for a specific item is by using the Search feature. Pick the **Search** button in the **Quick Takeoff** dialog box or right-click in the list and select **Search**. Enter the desired item in the input field at the top of the dialog box and pick the **Go** button. The results of the search are displayed in the lower pane. If there is no match, the cursor moves to the end of the list. The lists can be expanded and collapsed as previously discussed.

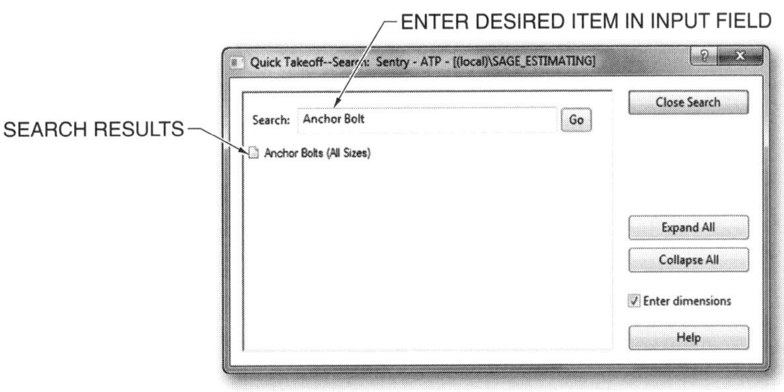

Quick Quiz®

Quick Quiz®

Refer to the CD-ROM for the Quick Quiz® questions related to chapter content.

Key Terms

Illustrated Glossary

- *beam*
- *beam-and-column construction*
- *channel*
- *column*
- *deck*
- *framed-tube construction*
- *girder*
- *joist*
- *long-span construction*
- *metal framing*
- *moment-resisting construction*
- *open web joist*
- *purlin*
- *shaking out*
- *spandrel beam*
- *S-shape*
- *structural steel construction*
- *stud*
- *wall-bearing construction*
- *W-shape*

Web Links

Web Links

American Galvanizers Association
www.galvanizeit.org

American Iron and Steel Institute
www.steel.org

American National Standards Institute
www.ansi.org

American Welding Society
www.aws.org

Association of Women in the Metal Industries
www.awmi.com

ASTM International
www.astm.org

Canadian Institute of Steel Construction
www.cisc-icca.ca

Cold-Formed Steel Engineers Institute
www.lgsea.com

Metal Construction Association
www.mca1.org

National Institute of Steel Detailing
www.nisd.org

Steel Deck Institute
www.sdi.org

Steel Framing Alliance
www.steelframingalliance.com

Steel Joist Institute
www.steeljoist.org

Steel Manufacturers Association
www.steelnet.org

Steel Recycling Institute
www.recycle-steel.org

Steel Stud Manufacturers Association
www.ssma.com

Metals 8

Review Questions

Estimating

________ **1.** In ___ construction, a series of horizontal steel beams and trusses are joined to vertical columns to create large structures with open areas.

________ **2.** A(n) ___ is a horizontal structural member that is used to support loads over an opening.

________ **3.** A(n) ___ is a large horizontal structural member that supports loads at isolated points along its length.

________ **4.** A(n) ___-shape is a structural steel member with parallel outer flange surfaces and sloping inner flange surfaces that are joined with a perpendicular web.

________ **5.** A(n) ___-shape is a structural steel member with parallel inner and outer flange surfaces that are of constant thickness and are joined with a perpendicular web.

T F **6.** A column is a vertical structural member used to support axial compressive loads.

T F **7.** A joist is a vertical structural member that supports the load of a floor or ceiling.

________ **8.** A(n) ___ is a horizontal support member that spans across adjacent rafters or between beams, columns, or joists to carry intermediate loads, such as the roof deck or wall panels.

________ **9.** A(n) ___ joist is a horizontal structural steel member constructed with steel angles that are used as chords, with steel bars or angles extending between the chords at an angle.

________ **10.** A(n) ___ is a floor made of light-gauge metal panels.

________ **11.** ___ is the process of unloading steel members and assemblies in a planned manner to minimize any additional handling and moving of pieces during the erection process.

________ **12.** ___ construction is a structural steel construction method consisting of bays of framed structural steel members that are repeated to create large structures.

________ **13.** A(n) ___ beam is a beam in the perimeter of a structure that spans several columns and typically supports a floor or roof.

________ **14.** ___ construction is a structural steel construction method that consists of fastening large horizontal steel members together and bracing them with angles, channels, and other steel members to create large girders and trusses.

________ **15.** ___ construction is a structural steel construction method in which the stability of the steel frame and its wind and seismic resistance depend on a solid connection between the columns and beams.

Short Answer

1. Discuss the relationship between the steel fabricator and the estimator.

2. Identify three items that may impact the labor costs for steel erection at a job site.

Metals 8 Activities

Activity 8-1—Ledger Sheet Activity

Refer to the cost data, Print 8-1, and Estimate Summary Sheet No. 8-1. Take off the square feet of metal decking required and determine the total material and labor cost for the deck.

COST DATA			
Material	Unit	Material Unit Cost*	Labor Unit Cost*
Metal deck, galvanized 18 gauge	sq ft	3.60	0.77

* in $

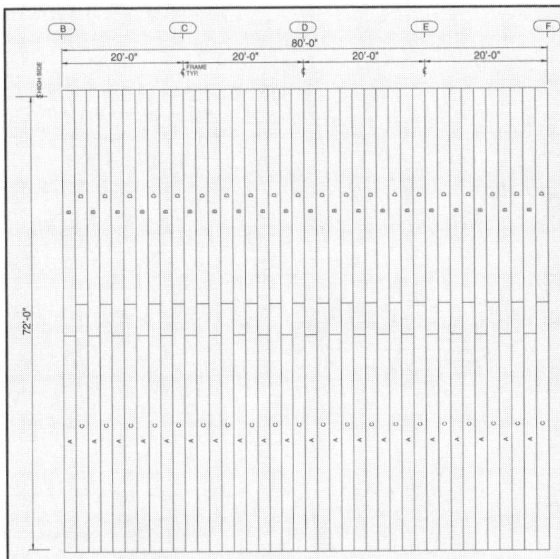

ROOF PANEL SCHEDULE					
ID	PART NUMBER	GA	LENGTH	DESCRIPTION	HOLE TO HOLE
A	560118	18	418-6	MR24 EAVE PANEL ALZN	33'-10 23/32"
B	560105	18	463-2	MR24 EAVE PANEL ALZN	37'-11 9/32"
C	560118	18	478-6	MR24 EAVE PANEL ALZN	38'-10 3/4"
D	560105	18	403-2	MR24 EAVE PANEL ALZN	32'-11 1/4"

Print 8-1

ESTIMATE SUMMARY SHEET

Project: __________ Sheet No. 8-1
Estimator: __________ Date: __________
Checked: __________

No.	Description	Dimensions			Quantity		Material		Labor		Total	
		L	W			Unit	Unit Cost	Total	Unit Cost	Total	Unit Cost	Total
	Metal deck											
	Total											

Activity 8-2—Spreadsheet Activity

Refer to the cost data (below) and Print 8-2 and Estimate Summary Spreadsheet No. 8-2 on the CD-ROM. Take off the linear feet of W16×26, W16×31, W16×36, W24×55, W24×62, W24×76, and W27×84 beams. Determine the total material, labor, and equipment cost for the steel beams.

COST DATA				
Material	**Unit**	**Material Unit Cost***	**Labor Unit Cost***	**Equipment Unit Cost***
Structural steel beams				
W16×26	lf	18.87	1.84	1.11
W16×31	lf	22.55	2.04	1.23
W16×36	lf	24.48	2.11	1.31
W24×55	lf	37.95	2.40	1.09
W24×62	lf	42.90	2.40	1.09
W24×76	lf	52.80	2.40	1.09
W27×84	lf	58.30	2.23	1.01

* in $

Activity 8-3—Estimating Structural Steel

Refer to the cost data, Print 8-3, and Estimate Summary Sheet No. 8-3 on the CD-ROM. Take off the linear feet of 28K9, 10K1, and 18K4 open web joists. Determine the total material, labor, and equipment cost for the steel joists.

COST DATA		
Material	**Unit**	**Material Unit Cost***
Open web joists		
J1	ea	375
J3	ea	400
J7	ea	30

* in $

Activity 8-4—Sage Timberline Estimating Activity

Create a new estimate and name it Activity 8-4. Perform a quick takeoff for item **5090.010 5 Anchor Bolts (All Sizes)** and item **5500.010 10 Steel Lintels 3×3×¼**. Take off 500 anchor bolts and 20 lintels. Print a standard estimate report.

9 Woods, Plastics, and Composites

Key Concepts

- Methods for specifying lumber quantities vary according to the use of the lumber. For example, structural wood framing members for platform and balloon framing are ordered according to the individual number of pieces needed and are priced based on board feet.
- Sheathing materials are based on the number of sheathing panels required to cover a specified area.
- Millwork quantities are expressed in linear feet.
- Cabinetry is noted by the dimensions, number of doors, number of drawers, and finish of each cabinet.
- Countertop calculations are based on the countertop area (in square feet).
- Wood paneling is calculated as the number of sheets.

Introduction

Information about wood and plastic materials, wood treatments, wood coatings, and wood and plastic fastenings is cited in various numbered titles of Division 6 of the specifications. Wood framing is described in Title 06 11 00. Heavy timber construction information is included in Title 06 13 00. Wood treatments are described in Title 06 05 73. Wood and plastic fastening information is contained in Title 06 05 23.

Wood is used in construction for structural support members, temporary bracing, shoring, sheathing, decking, finish moldings, paneling, and casework. Estimators calculate the number of pieces, area, number of sheets, or the linear feet of wood products depending on the material installed. Estimators rely primarily on interior elevations, floor plans, schedules, and the specifications to take off wood finish members.

MATERIALS

Wood products are available in a variety of types, sizes, shapes, and finishes. Wood is used for structural support members, temporary bracing, shoring, sheathing, decking, finish moldings, paneling, and casework. Estimators calculate the number of pieces, the area, and the number of panels or the linear feet of wood products depending on the material installed.

Wood is used in many other applications on a construction project that are included in other titles and divisions of the specifications. Wood used for concrete formwork is included in Title 03 11 00, temporary signage in 01 58 13, doors and windows in Division 8, finish wood flooring in Division 9, and casework made of hardwoods in Divisions 11 and 12. Wood included in these divisions must be included in the overall estimate, but are not part of the wood information described in Division 6.

Plastics and composites are used for many construction applications. Title 06 50 00 of the CSI MasterFormat™ pertains to structural plastic materials and title 06 70 00 pertains to structural composites used as a substitute for structural wood members, finish plastic materials such as laminates, and plastic and composite decking, railings, paneling, and trim.

Wood, plastic, and composite materials are described in many divisions and numbered titles. Division 6 of the specifications and bid documents includes wood framing (06 11 00), heavy timber construction (06 13 00), sheathing (06 16 00), shop-fabricated structural wood members (06 17 00), glued laminated (glulam) members (06 18 00), prefinished paneling (06 25 00), millwork (06 46 00), windowsills (06 46 33), and architectural woodwork including cabinetry (06 41 00), stairways (06 43 13), and finish plastic members (06 65 00).

Structural Wood Members

A structural wood member provides support for live and dead loads in a structure. A *live load* is the total of all the dynamic loads a structure is capable of supporting including wind loads, human traffic, and any other imposed loads. Live loads include moving and variable loads such as people, furniture, and equipment. A *dead load* is a permanent, stationary load composed of all construction materials, fixtures, and equipment permanently attached to a structure.

Wood is classified as hardwood and softwood. *Hardwood* is wood produced from broadleaf, deciduous trees such as ash, birch, maple, oak, and walnut. Hardwood is primarily used in construction for finish materials such as cabinetry and millwork. *Millwork* is finished wood materials or parts, such as moldings, jambs, and frames, completed in a mill or fabrication facility.

Softwood is wood produced from a conifer (evergreen) tree. Softwood is used for structural framing members and trim components and includes pine and fir. Cedar is softwood often used for exterior framing and finish applications. For exterior applications or where wood materials will be in contact with the soil, special chemical treatments may be applied to the wood at the mill. Estimators should ensure the proper type of wood, grade, and any required treatments are taken into consideration when taking off and pricing the materials.

Grading. *Grading* is the classification of various pieces of wood according to their quality and structural integrity. Wood is graded based on strength, stiffness, and appearance. *Lumber* is wood used for construction that has been sawn and sized. High grades of lumber have few knots or defects. Low grades have many loose knots. Specifications indicate the lumber grades required for various building applications. Grade marks are stamped on softwood lumber to provide information concerning grade. **See Figure 9-1.** Material costs increase for higher grades of lumber.

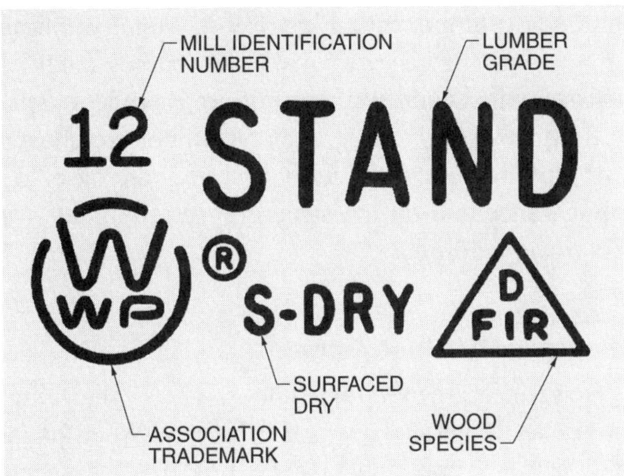

Figure 9-1. Lumber grade marks indicate lumber grade, wood species, and related information.

Sizing. The size of a structural wood framing member is indicated as nominal size. *Nominal size* is the size of rough lumber prior to planing. Nominal size varies from actual size. **See Figure 9-2.** For example, the actual dimensions of a 2 × 4 and 2 × 6 are 1½ × 3½ and 1½ × 5½ respectively, after planing. Nominal wood framing members are cut to standard lengths of 8′, 10′, 12′, 14′, and 16′.

A *common board* is a wood member that has been sawn or milled and has a nominal size from 1 × 4 to 1 × 12. *Dimension lumber* is lumber that is precut to a particular size for the construction industry, normally having nominal sizes of 2 × 4 to 4 × 6. *Timber* is heavy lumber that has a 5 × 5 or larger nominal size.

Shop-Fabricated Structural Members. Various engineered wood products are available for structural applications. Wood I-joists are composed of flanges of structural composite lumber or finger-jointed lumber with webs of oriented strand board (OSB) or plywood. Wood I-joists are commonly used as floor or roof joists.

STANDARD LUMBER SIZES

Type	Nominal Size*		Actual Size*	
	Thickness	Width	Thickness	Width
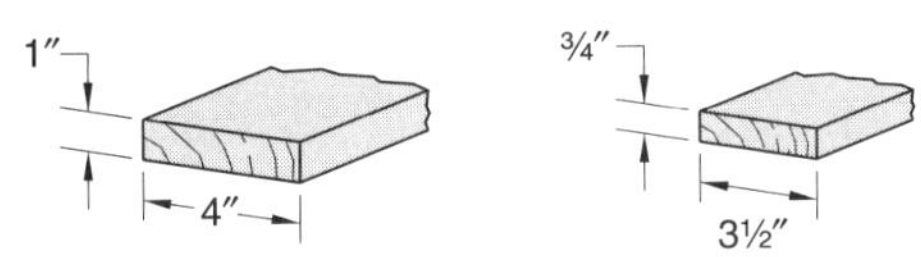 BOARD	1	2	¾	1½
	1	4	¾	3½
	1	6	¾	5½
	1	8	¾	7¼
	1	10	¾	9¼
	1	12	¾	11¼
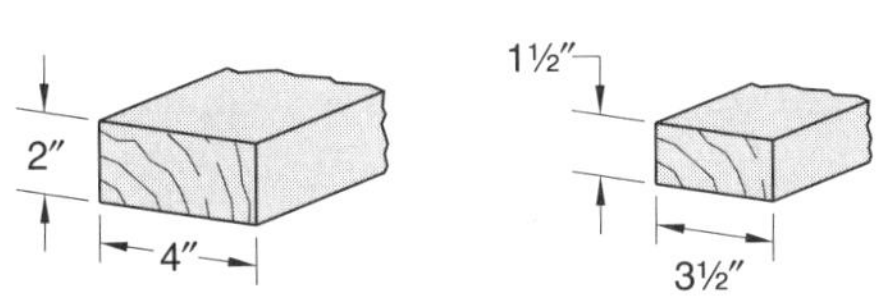 DIMENSION LUMBER	2	2	1½	1½
	2	4	1½	3½
	2	6	1½	5½
	2	8	1½	7¼
	2	10	1½	9¼
	2	12	1½	11¼
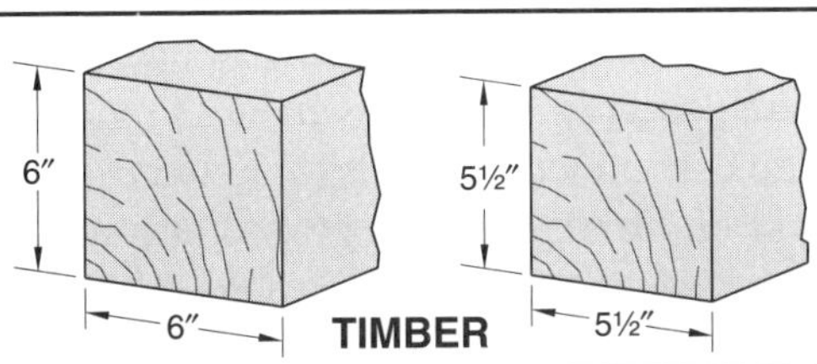 TIMBER	5	5	4½	4½
	6	6	5½	5½
	6	8	5½	7½
	6	10	5½	9½
	8	8	7½	7½
	8	10	7½	9½

* in in.

Figure 9-2. Nominal board, dimension lumber, and timber sizes noted on prints differ from actual sizes.

Parallel-strand lumber (PSL) is lumber manufactured by combining wood strands with adhesive and forming rough lumber under intense heat and pressure. The rough lumber is then trimmed to standard dimensional lumber sizes. Parallel-strand lumber members are commonly used in lieu of solid lumber for applications where long lengths are required, such as balloon-framed structures. Laminated-veneer lumber (LVL) is a layered composite of wood veneers and adhesive that can be cut into stock for headers and beams, flanges for wood I-joists, or other items. LVL is typically available in 1¾″ thickness, and long lengths are available.

Glulam Timber. *Glulam (glued laminated) timber* is an engineered lumber product composed of wood laminations (lams) that are bonded together with adhesives. Glulam members are specified on architectural drawings by the type and grade of wood in the glulam members and the exact dimensions. In many cases, various geometric shapes are specified for glulam members.

Use of glulam members also allows architects and engineers to create curved wood members that are not possible with solid timber. **See Figure 9-3.** Applications for glulam members include floor and roof beams, columns, and trusses. A *beam* is a horizontal structural member that is used to support loads over an opening. A *column* is a vertical structural member used to support axial compressive loads.

APA—The Engineered Wood Association

Figure 9-3. Glulam members are designed to span large open areas.

A *truss* is a manufactured roof or floor support member with components commonly placed in a triangular arrangement. Various configurations allow for trussed wood members to span wide areas and withstand additional levels of live and dead loading according to the design used and the sizes and grades of lumber. **See Figure 9-4.** Metal webbing may be used in these designs for flat trusses and floor trusses with wood top and bottom members. Specific types, sizes, and grades of lumber are chosen for glulam members. Lumber choices are based on appearance and on the structural requirements for span and loading.

Engineered Wood Panels

Wood panels are used for subflooring, wall sheathing, roof sheathing, shelves, and cabinetry. Wood panel products are commonly manufactured in 4′ × 8′ sheets (32 sq ft). The panels are also available in 4′ × 10′ and 4′ × 12′ sheets. Engineered wood panels include oriented strand board, plywood, particleboard, and medium-density fiberboard.

An oriented strand board or plywood subfloor is used to cover floor joists. Oriented strand board is also commonly used for wall and roof sheathing. Particleboard or plywood panels covered with plastic laminate or hardwood veneer are used in laminated casework. Medium-density fiberboard (MDF) is a nonstructural panel product used for siding, molding, and cabinets.

Plywood. *Plywood* is an engineered wood panel product made of wood layers glued and pressed together under intense heat and pressure. Plywood panels consist of an odd number of layers such as three, five, or seven. Each layer consists of one or more plies. A ply is a single veneer sheet.

The layers of a plywood panel are cross-laminated. *Cross-lamination* is a process in which each layer is placed with its grain at a 90° angle to the adjacent layer. Cross-lamination provides greater strength and stiffness in both directions while minimizing shrinkage and swelling. Plywood is graded according to appearance, size and number of repairs in the plies, and outer veneer grade. **See Figure 9-5.** Plywood is commonly available in thicknesses ranging from ¼″ to 1⅛″.

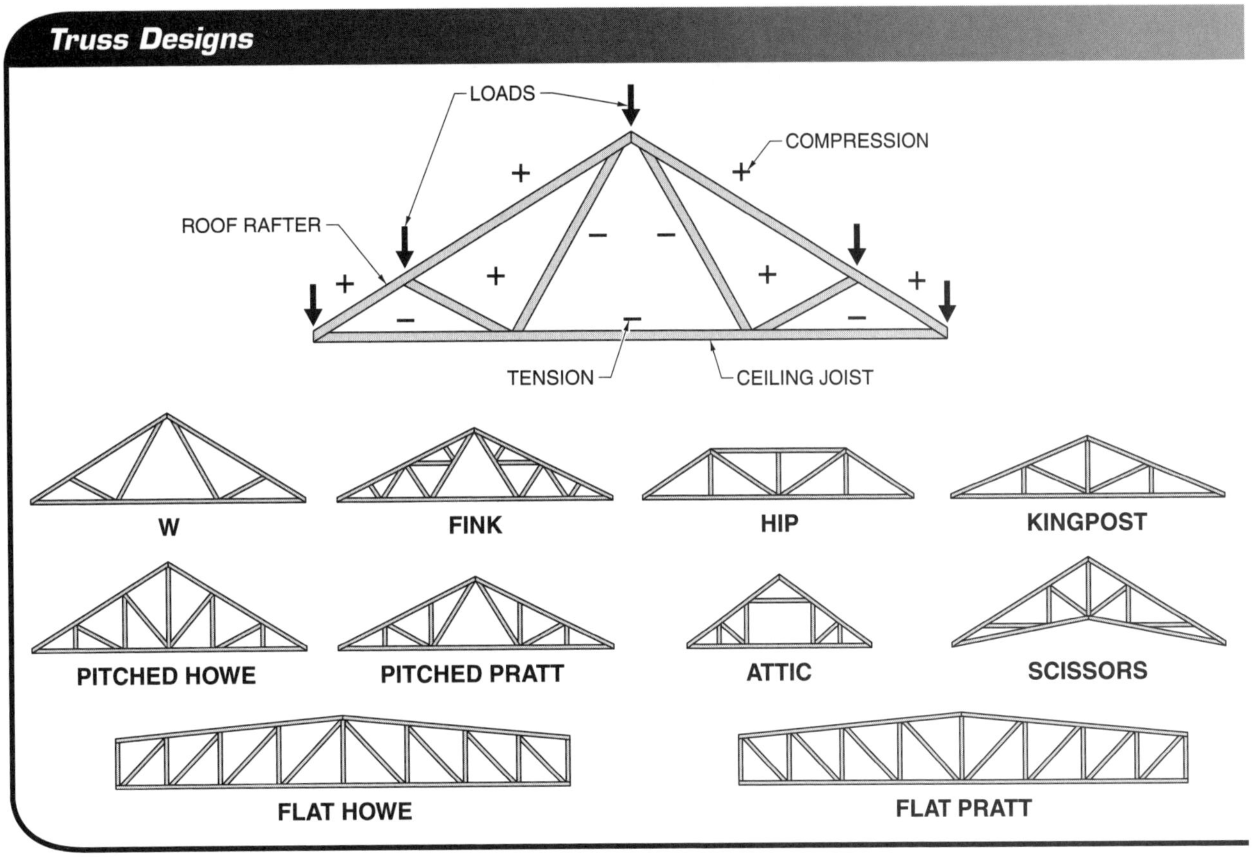

Figure 9-4. Trusses are created using common framing lumber and are available in a variety of designs.

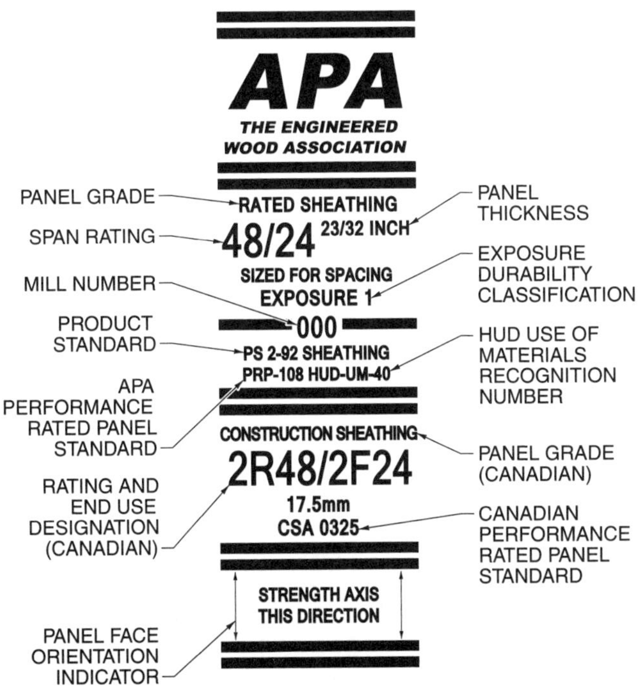

PLYWOOD-VENEER GRADES	
Grade	**Characteristics**
A	Smooth, paintable; not more than 18 neatly made repairs, boat, sled, or router type, and parallel to grain, permitted; may be used for natural finish in less demanding applications
B	Solid surface; shims, circular repair plugs and tight knots to 1″ across grain permitted; some minor splits permitted
C	Tight knots to 1½″; knotholes to 1″ across grain and some to 1½″ if total width of knots and knotholes is within specified limits; synthetic or wood repairs; discoloration and sanding defects that do not impair strength permitted; limited splits allowed; stitching permitted
C PLUGGED	Improved C veneer with splits limited to ⅛″ width and knotholes and borer holes limited to ¼″ × ½″; admits some broken grain; synthetic repairs permitted
D	Knots and knotholes to 2½″ width across grain and ½″ larger within specified limits; limited splits are permitted; stitching permitted; limited to interior (exposure 1 or 2) panels

Figure 9-5. Plywood panel grades, span rating, veneer grades, and other information is noted on the panel face.

American Hardwood Export Council
Finish wood members include millwork and visible stairway components.

Oriented Strand Board. *Oriented strand board (OSB)* is an engineered wood panel product in which wood strands are mechanically oriented and bonded with phenolic resin under heat and pressure. OSB panels are composed of strands of wood layered in a manner similar to plywood veneers, with alternating grain placement and cross-lamination. Similar to plywood, an odd number of plies are used. OSB panels are commonly available in thicknesses varying from ¼″ to ¾″.

Particleboard. *Particleboard* is an engineered wood panel product constructed of wood particles and flakes that are bonded together with a synthetic resin. Particleboard has no grain and has a smooth finish that is suitable for underlayment and for various cabinet components. Three grades of particleboard are low-density, medium-density, and high-density.

Low-density particleboard panels weigh less than 37 lb/cu ft and are of limited architectural use. Medium-density particleboard panels weigh from 37 lb/cu ft to 50 lb/cu ft and are the most commonly used type of particleboard. High-density particleboard panels weigh over 50 lb/cu ft and are used where high strength and hardness are required. Particleboard is available in thicknesses varying from ½″ to 1¾″.

Medium-Density Fiberboard. *Medium-density fiberboard (MDF)* is an engineered wood panel product manufactured from fine wood fibers mixed with binders and formed with heat and pressure. The MDF panels may be faced with wood veneer or other surfacing materials. The edge of MDF has a better appearance than particleboard. MDF is available in thicknesses varying from ⅛″ to 1½″.

Finish Wood Members

A *finish wood member* is a decorative and nonstructural wood component in a structure. Hardwood and/or softwood may be used as finish wood members. Most paneling, cabinetry, and visible stairway components are commonly made of hardwood, with millwork made from softwood and hardwood. Millwork includes trim moldings, window and door frames, and stairways. Wood ornaments specified on architectural drawings also include mantels, pediment heads, and decorative and structural columns.

Hardwood lumber used for finish applications is graded according to appearance. Hardwood lumber is graded as firsts and seconds (FAS), Select, No. 1 Common, No. 2 Common, and No. 3 Common. FAS is the highest grade of hardwood lumber in terms of appearance. FAS grade lumber is usually required for hardwood trim materials that have a natural or stained finish. Select grade lumber is lumber with one face graded FAS and the other face graded No. 1 Common.

No. 1 Common is standard furniture-grade lumber. No. 1 Common lumber is a minimum of 3″ wide and 4′ long. No. 2 Common lumber is the standard lumber grade for cabinets and millwork and is the same size as No. 1 Common lumber but yields a minimum of 50% clear-face cuttings a minimum of 3″ wide and 2′ long. No. 3 Common lumber has similar dimensions to No. 1 and No. 2 Common lumber, with minimum yields of 25% clear-face cuttings. No. 3 Common lumber is a utility-grade hardwood used for crates, pallets, and other rough applications.

Cabinetry. A *cabinet* is an enclosure fitted with any combination of shelves, drawers, and doors and is usually used for storage. Prefabricated or custom-built cabinets may be specified for a particular job. The specifications and drawings may indicate the use of standard cabinets available in stock sizes. The architect may also specify custom-built and finished cabinets. Cabinet sizes are shown on plan, elevation, and detail drawings using a width/height/depth notation. For example, a notation of 24/36/12 indicates a 24″ wide, 36″ high, and 12″ deep cabinet.

Standard prefabricated cabinets are commonly constructed of particleboard and plywood with a wood or plastic laminate veneer. Custom-built cabinets are commonly made of particleboard, plywood, wood veneer, plastic laminate, and hardwood or softwood materials. Architects provide drawings detailing the various materials and design for custom-built cabinets.

Millwork. Millwork is finished wood materials or parts, such as moldings, jambs, and frames completed in a mill or manufacturing plant. Millwork is cut and molded from hardwood or softwood and is available unfinished, primed, or prefinished. Millwork may also be formed from composite wood materials that are wrapped with paper or vinyl during the manufacturing process.

Standard sizes and shapes are used by mills when cutting and shaping millwork. Moldings are available in a variety of standard sizes and shapes. **See Figure 9-6.** Softwood molding with a primed finish is typically specified for applications where millwork is to be painted. Hardwood molding is usually specified for applications where millwork is to remain a natural wood finish.

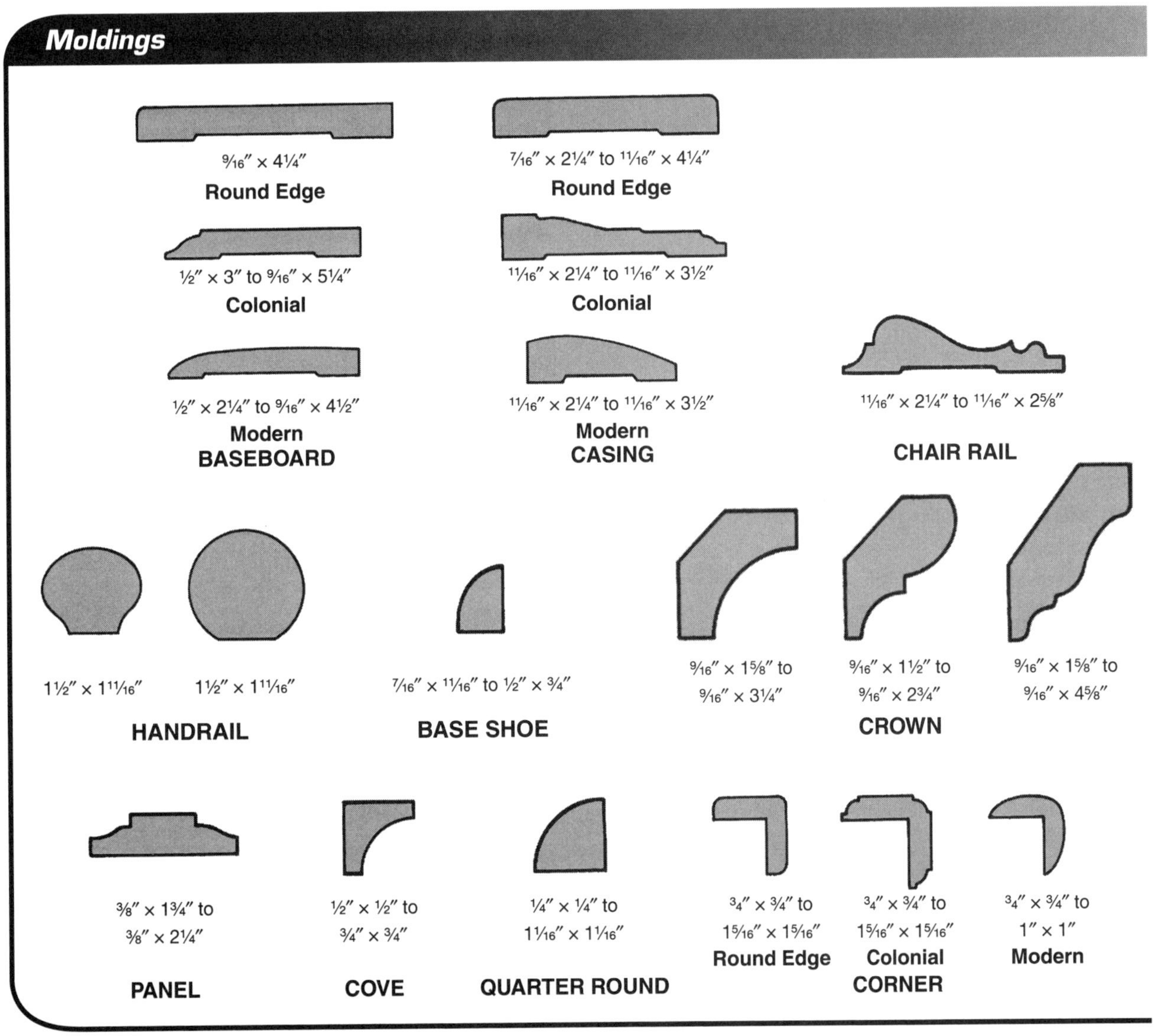

Figure 9-6. Millwork, such as moldings, is available from a variety of suppliers.

Paneling. Wood paneling is available in a wide variety of prefinished and unfinished styles and types of wood. Wood paneling is formed from ¼″ or ¾″ plywood panels covered with a hardwood veneer such as birch, oak, walnut, or mahogany. The surface may be plain or textured and prefinished or unfinished. Panels that simulate wood paneling are manufactured from fiberboard panels with a thin vinyl veneer that has the appearance of a finished wood grain.

Stairway Components. A large number of specialized stairway components are included in a completed staircase. **See Figure 9-7.** A *stringer,* also known as a carriage, is the support for a stairway. A *tread* is the horizontal surface of a step of a stairway. A *riser* is the piece forming the vertical face of a step. Structural wood members, typically 2 × 10s or 2 × 12s, are used for stringers to support the treads and risers.

For rough stairways, treads are constructed of 2 × 10s or 2 × 12s. For finished stairways, 1″ or greater thickness of oak or other hardwood is used for the treads. Risers and open and closed stringers are made of ¾″ thick softwood or hardwood lumber, depending on the finish. An *open stringer* is a stringer that has been cut out to support the treads on the open side of a stairway. A *closed stringer* is a stringer installed at the meeting of a stairway and wall.

Wood Stairway Components

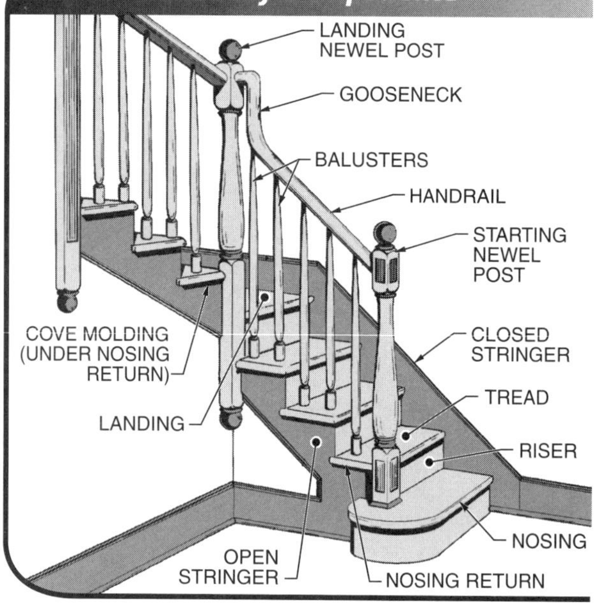

Figure 9-7. Wood stairways are made from a variety of components, many of which are provided precut from woodworking mills.

A *baluster* is the upright member that extends between the handrail and the treads. Balusters support the handrail and may be made from softwood or hardwood, depending on the finish. A *handrail* is a support member that is grasped by the hand for support when using a stairway. Handrails for rough stairways are constructed with 2″ thick lumber, and milled hardwood is used for finished stairs.

Special components such as goosenecks, starting and landing newel posts, and cove moldings are shown on the architectural detail drawings of wood stairs. A *gooseneck* is a curved section of a handrail. A *starting newel post* is the main post supporting the handrail at the bottom of a stairway. A *landing newel post* is the main post supporting the handrail at a landing. A *landing* is the platform that breaks the stairway flight from one floor to another. *Cove molding* is the trim material often used to cover the joint between the tread and the stringer and the joint between the tread and the top of the riser.

Plastics and Composites

Various building components may be made from plastic and composite material. The components are typically used in applications where decay resistance and weatherproof qualities are required, such as exterior decks and fences. Plastic is also used in cabinet construction as a surfacing material for horizontal and vertical surfaces.

Structural. Structural plastics and composites are used for exterior light-load-bearing applications. Plastics and composites shaped to common lumber sizes may also be used for fencing or other decorative exterior applications, such as trellises. Plastic and composite members are available in sizes similar to nominal size structural wood framing members. Plastics and composites are warp-, split-, and weather-resistant and are commonly fastened together with screws.

Laminates. *Plastic laminate* is sheet material composed of multiple layers of plastic and resins bonded together under intense heat and pressure. Plastic laminate is commonly referred to as high-pressure or low-pressure laminate. High-pressure plastic laminate, also known as horizontal-grade laminate, is used where heavy wear qualities are required, such as with countertops. Low-pressure plastic laminate is used for vertical applications, such as cabinet sides or where post-forming of the laminate is required. Plastic laminate is available in sheets of 1/32″ or 1/16″ thickness.

Solid-core laminates maintain their surface color throughout the entire thickness of the laminate. Plastic laminate is commonly applied as a finish material with contact cement to particleboard. Plastic laminate is available in widths from 16″ to 60″ and lengths of 8′ to 12′. Plastic laminate is available in many patterns and designs.

WOOD CONSTRUCTION METHODS

Wood is used for structural and finish construction. When used for structural applications, wood members are used to support all imposed loads of a structure and are typically concealed. When used for finish applications, the wood members are exposed.

Southern Forest Products Association

Figure 9-8. Platform framing is the most common construction method used for one-story and multistory buildings.

Structural Wood Construction

Structural wood construction includes frame construction and the use of trusses or laminated members. In frame construction, individual wood members are assembled and fastened into structural frames on the job site. Trusses and laminated members are engineered to span various distances and support specific loads. They are commonly built off-site, transported to the job site, and set in place. Frame construction includes platform, balloon, and timber framing. Each frame construction method requires different wood framing members used in different applications.

Platform Framing. *Platform framing* is a framing method in which single or multistory buildings are built one story at a time, with work proceeding in consecutive layers or platforms. **See Figure 9-8.** Wood members are used for joists, headers, and plates. A *joist* is a horizontal structural member that supports the load of a floor or ceiling. A *header* is a horizontal framing member placed at the top of a window or door opening. A *plate* is a horizontal support member at the top and bottom of a framed wall to which studs are attached.

Balloon framing is experiencing a resurgence in popularity due to the availability of long lengths of engineered lumber.

Vertical wood members include studs and corner posts. A *stud* is a vertical support member that extends from the bottom to the top plates in a framed wall. Studs are spaced at regular intervals to support imposed loads and provide a surface for finish wall material. A *corner post* is a vertical support member consisting of studs or studs and blocks placed at the outside and inside corners of a framed building.

Blocking is wood fastened between structural members to strengthen the joint, provide structural support, or block the passage of air. Blocking, permanent and temporary bracing, and backing are additional uses of wood in platform framing. Wood sheathing is attached to the frames to add lateral strength. Platform framing is the most common framing method. A header is placed over the top of a wall opening to distribute the load to either side of the opening. A *rough sill* is a horizontal framing member placed under window openings to support the window. A *cripple stud* is a short wall stud placed between the header and the top plate or between the rough sill and the bottom plate.

Balloon Framing. *Balloon framing* is a framing method in which individual studs extend from the sill plate of the first story to the top plate of the upper story. The primary difference between balloon framing and platform framing is that the studs for multistory balloon-framed load-bearing walls extend the full height of the walls from foundation to roof.

Wood members used in balloon framing are similar to those used in platform framing, with only the lengths of the members being different. Balloon framing is used in some applications to minimize the effects of lumber shrinkage and to create open-framed areas in multistory buildings. Engineered lumber is commonly used for balloon framing.

Timber Framing. *Timber framing* is a framing method in which large wood members are used to form large open areas. **See Figure 9-9.** Large glulam or timber members are spaced farther apart than in other wood framing systems to create these open areas. Imposed loads are conveyed to the posts with bents. A *bent* is a structural, interconnected system of timbers contained in a wall. Areas between timber framing members are fitted with glass or covered with wood planking of sufficient thickness to span between posts and beams with minimal deflection.

Classic Post & Beam

Figure 9-9. Timber framing uses large wood members to form large open areas.

The American Institute of Timber Construction (AITC) is the national technical trade association of the structural glued laminated (glulam) industry. Information about AITC and its constituency can be found at *www.aitc-glulam.org*.

Finish Wood Construction

Finish wood construction is composed of items made of wood or wood products and applied to the interior or exterior surfaces of walls, partitions, floors, or ceilings to provide decorative or functional finishes. Many components of finish wood construction are precut and prefinished prior to delivery to the job site. Finish wood components are typically installed near the completion of the work when climatic conditions or tradesworkers cannot damage the finished materials.

Due to the moisture content of some finished wood, certain installations require finish wood materials to be delivered with sufficient time for the wood to adjust to the interior climatic conditions of the final installation space prior to being cut and fitted into position. Finish wood construction includes cabinetry, millwork, paneling, and stairway components.

Cabinet units are commonly purchased or prefabricated at a cabinet shop and delivered to the job site ready for installation. Some custom-built cabinets may be built in-place. Countertops may be built of particleboard covered with plastic laminate, veneer, ceramic tile, stone, or concrete, or made of solid surface material. Dimensions and finishes for cabinets are based on standard sizes or custom sizes determined from the specifications, details, interior elevations, job site measurements, and architectural notes.

Millwork, including baseboard, crown, chair rail, and other specialized moldings, is measured, cut, and fitted in place at the job site. Wood paneling is installed by fastening it to structural members or special framing. Dimensions for finished wood stairways are shown on the prints but are commonly measured at the job site when structural work is completed. Stairway components are commonly fabricated at a woodworking shop, shipped to the job site, and installed. Two or three stringers are typically required for a stairway. Additional stringers may be required on very wide stairways.

QUANTITY TAKEOFF

Methods for specifying lumber quantities vary according to the use of the lumber. For example, structural wood framing members for platform and balloon framing are ordered according to the individual number of pieces needed and are priced based on board feet.

Timber components are specified according to each piece and dimension. Sheathing materials are based on the number of sheathing panels commonly calculated in square feet required to cover a specified area. Millwork quantities are expressed in linear feet. Cabinetry is noted by the dimensions, number of doors, number of drawers, and finish of each cabinet. Countertop calculations are based on the countertop area in square feet. Wood paneling is calculated as the number of sheets.

Structural Wood Members

Platform and balloon framing wood members include sill plates, floor joists, floor sheathing, wall plates, wall studs, headers, bracing and blocking materials, wall sheathing, ceiling joists, roof rafters, roof trusses, and roof sheathing. Timber framing members include glulam members, timbers, and planking.

The species, grade, stress rating, and moisture content of lumber to be used are noted in the specifications. **See Figure 9-10.** Lumber measurements are stated as thickness, width, and length. Structural lumber length is based on multiples of 2′, typically beginning at 8′ in length and extending to 16′ or 18′. Structural lumber length is always rounded up to the closest multiple of 2′. Lumber quantities are often stated in board feet when calculating prices for large construction projects.

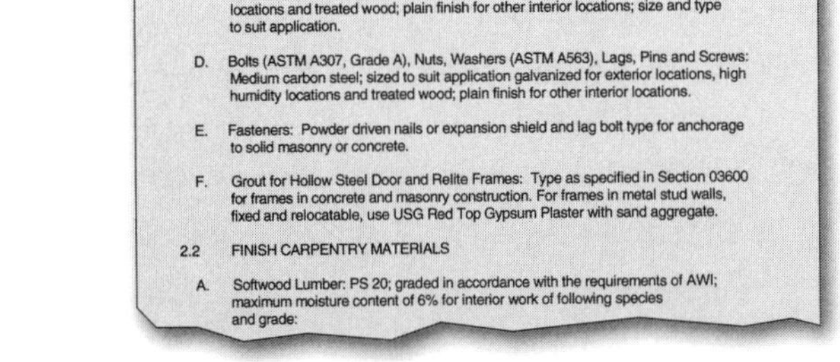

2. PART 2 PRODUCTS

2.1 ROUGH CARPENTRY MATERIALS

A. Lumber: PS 20; graded in accordance with established grading rules; maximum content of 19%; of following species and grades:

1. In-wall Blocking: Stress Group A or B; construction grade; fire-retardant treated.
2. All other Uses (nailers, roofing blocking, window blocking): Stress Group A; construction grade; treated with wood preservative for water resistance.

B. Douglas Fir Plywood: Sheathing grade.

C. Nails and Powder Driven Fasteners: Galvanized for exterior locations, high humidity locations and treated wood; plain finish for other interior locations; size and type to suit application.

D. Bolts (ASTM A307, Grade A), Nuts, Washers (ASTM A563), Lags, Pins and Screws: Medium carbon steel; sized to suit application galvanized for exterior locations, high humidity locations and treated wood; plain finish for other interior locations.

E. Fasteners: Powder driven nails or expansion shield and lag bolt type for anchorage to solid masonry or concrete.

F. Grout for Hollow Steel Door and Relite Frames: Type as specified in Section 03600 for frames in concrete and masonry construction. For frames in metal stud walls, fixed and relocatable, use USG Red Top Gypsum Plaster with sand aggregate.

2.2 FINISH CARPENTRY MATERIALS

A. Softwood Lumber: PS 20; graded in accordance with the requirements of AWI; maximum moisture content of 6% for interior work of following species and grade:

Figure 9-10. Specifications provide information concerning the species, grade, stress rating, and moisture content of lumber.

Southern Forest Products Association

Lumber is graded and sorted according to thickness, width, and length as it moves along a conveyor belt at the sawmill.

A *board foot (bf)* is a unit of measure based on the volume of a piece measuring 1″ thick by 12″ wide by 12″ long (144 cu in). Board feet is calculated by multiplying the thickness (in inches) by the width (in inches) by the length (in feet), and then dividing this total by 12. For example, a board that measures 2″ × 4″ × 16′ contains 10.66 bf ([2″ × 4″ × 16′] ÷ 12 = 10.66 bf). Lumber prices are often stated in cost per board foot or thousand board feet (Mbf).

Building dimensions are taken from the architectural drawings and used to determine the number of framing members. Depending on the estimating method or estimating program used, individual calculations may be necessary to determine the number of framing members, or the estimating program may automatically determine the number of framing members needed based on spacing and dimension information.

When using a spreadsheet, the number of framing members is entered into the proper spreadsheet cell. When using an estimating program, the number of framing members is entered into the takeoff quantity column on the row corresponding to the applicable framing member. A BIM model may calculate these quantities automatically based on the integrated model. In an estimating program, material and labor costs are calculated electronically based on information retrieved from a database containing cost information.

Sill Plates. A *sill plate* is a wood member, usually a pressure-treated 2 × 6, laid flat and fastened to the top of a foundation wall to provide a nailing base for floor joists or studs. The quantity of sill plates required is calculated by determining the linear feet of the foundation walls. The amount of linear feet is determined from the foundation and floor plans. For example, an L-shaped foundation wall measuring 23′-11″ × 32′-3″ requires 56′-2″ (56.17′) of sill plates. **See Figure 9-11.**

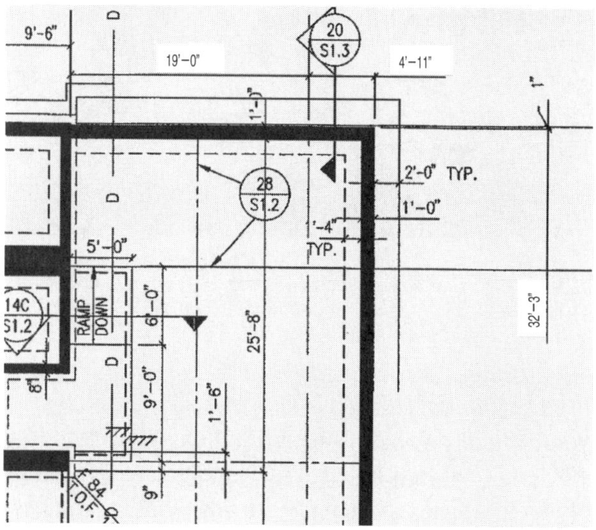

Figure 9-11. The quantity of sill plates needed (in linear feet) is determined from foundation and floor plans.

The nominal size of the sill plates is shown with an architectural note or detail drawings. An architectural note may indicate that 2 × 6s are used for sill plates. The number of board feet is determined by multiplying 2″ × 4″ × 56.17′ and dividing by 12 ([2″ × 4″ × 56.17′] ÷ 12 = 37.45 bf). The total number of board feet is multiplied by the cost of the lumber per board foot to determine the total cost of the sill plates.

Other costs for sill plates include labor for installation and fastening to the foundation system and any special treatment or insulation that may be specified in the construction documents. Labor costs are available from company historical records or standard industry data.

Floor Joists. The type and number of floor joists required is calculated by determining the header joist length and on-center (OC) spacing from the floor plan, and the nominal size is determined from the specifications, from an architectural note on the floor plan, or from details. Floor joist length depends on the distance between supporting structural members such as the sill plates and beams. Sufficient length is allowed for a certain amount of overlap where required. For example, a 14′ floor joist is used when the distance between the outside of a sill plate and the center of the supporting beam is 13′-4″.

The number of floor joists required is calculated based on the header joist length and the on-center spacing of the joists. **See Figure 9-12.** A multiplier is used to determine the actual number of joists required based on the on-center spacing. The multiplier is 1 when joists are spaced 1′-0″ OC, 0.75 when joists are spaced 16″ OC, 0.625 when joists are spaced 19.2″ OC, or 0.5 when joists are spaced 24″ OC. After the header joist length is multiplied by the appropriate multiplier, an additional joist is added as an end joist.

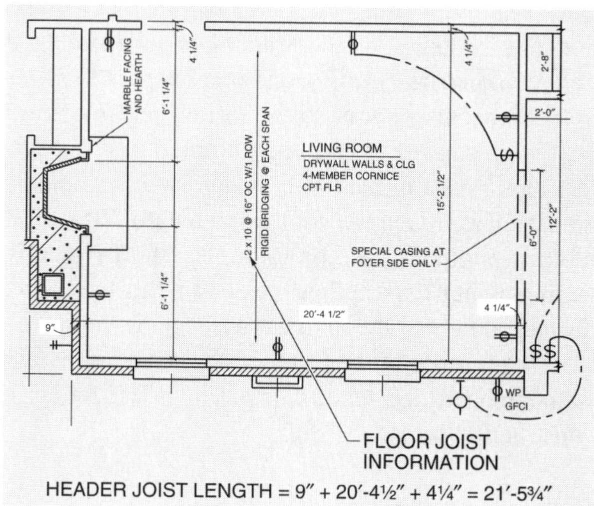

Figure 9-12. Wood floor joist calculations are determined from information provided on floor plans.

Determining the Number of Joists

Determine the number of joists required when the header joist length is 21′-5¾″ and the joists are spaced 16″ OC.

1. **Convert the header joist length to its decimal foot equivalent.**

   21′-5¾″ = 21.479′

2. **Determine the number of joists required.**

   number of joists = (*header joist length* × *multiplier*) + 1

   number of joists = (21.479′ × 0.75) + 1

   number of joists = 16.11 + 1

   ***number of joists* = 17.11; rounded to 18 joists**

Additional floor joists are added or subtracted around stairway openings, the perimeter of the structure, and at cantilevers where joists may be doubled to provide additional structural support. Additional joist material is also included for blocking between the joists. The number of board feet of floor joists is determined, and the total board foot value is multiplied by the cost of the particular floor joist lumber per board foot to determine the material cost.

Labor costs for joist installation include layout, placing and fastening of joists, and installation of joist hangers where required. Openings in floor structures, such as stairway and fireplace openings, may require additional labor costs for framing. Labor costs are available from company historical records or standard industry data.

Girders and Beams. A *girder* is a large horizontal structural member that supports loads at isolated points along its length. A *beam* is a horizontal structural member that is used to support loads over an opening. Dimensions for wood girders and beams are determined from floor plans and details. The lumber and design of any pre-engineered and trussed components are shown in the details and specifications. The cost per unit for prefabricated members is obtained from the supplier and is based on the engineering specifications. Wood girders and beams built and fabricated on the job site are priced according to the specified lumber dimensions and grade.

Girders, beams, and pre-engineered wood members are estimated as individual items. When large wood members are specified, the estimator should coordinate construction costs for lifting equipment, such as cranes, with the project field management. Where lifting equipment is required, additional equipment and labor costs are added to the overall construction estimate. These costs are available from company historical records or standard cost data.

Wall Framing Members. Wall framing members include the individual wood pieces used to provide structural support and partitioning. Wall framing members include plates, studs, headers, rough sills, blocking, bracing, and cripple studs. Material estimating for each wall framing member requires a different method.

Plates, studs, rough sills, blocking, and cripple studs are usually the same width. The linear feet of wall plates and lumber size are determined from the wall length dimensions on the floor plan and notes in the details or specifications. The required linear feet of wall plates is determined by adding the wall lengths and multiplying by 2 for walls with single top plates, or by 3 for double top plates.

For example, a rectangular building measuring 24′-6″ × 13′-0″ (24.5′ × 13′) with no interior partitions has a total of 75 lf of walls ([24.5′ × 2] + [13′ × 2] = 75′). If single top plates are specified, the total linear feet of plate material is 150′ (75′ × 2 = 150′). If double top plates are specified, the linear feet of wall is multiplied by 3 for a total of 225′ (75′ × 3 = 225′). Plate material costs are typically based on board feet.

Material quantities for window and door headers are determined by adding the opening width, 2″ for jamb installation, and 3″ to allow for 1½″ of bearing on each end of the header. This value is multiplied by 2 for double headers. For example, a 3′-0″ wide swinging door will require a 3′-5″ header (3′-0″ opening + 2″ jamb installation + 3″ of bearing surface = 3′-5″). This value is multiplied by 2 for a double header (3′-5″ × 2 = 6′-10″). An 8′ header is ordered for a built-up double header. A note on the floor plan, elevations, or specifications indicates the lumber grade, type, thickness, and width of header material. This amount is entered into a ledger sheet, spreadsheet, or estimating program in the appropriate location.

The method for estimating the number of wall studs is similar to the method used for taking off joists. The number of studs is based on the wall plate length and stud spacing. The same multipliers used for joist spacing are used for stud spacing (16″ OC = 0.75 multiplier; 19.2″ OC = 0.625 multiplier; 24″ OC = 0.5 multiplier).

After the wall-plate length is multiplied by the appropriate multiplier, one additional stud is added to the calculated amount to end the wall framing at the corner. As with other structural wood members, studs are priced by board foot. Other elements to consider when determining the number of studs required per wall include additional studs required at intersecting walls and studs that may be omitted at door or window openings.

> Diagonal metal or wood let-in braces may be required for improved lateral wall strength. The estimator should make sure to include braces in the quantity takeoff for wall framing members.

Determining the Number of Wall Studs

Determine the number of studs required when the wall plate length is 75′-0″ and the studs are spaced 19.2″ OC.

1. **Convert the wall plate length to its decimal foot equivalent.**

   75′-0″ = 75.0′

2. **Determine the number of studs required.**

   number of studs = (*wall plate length* × *multiplier*) + 1

   number of studs = (75.0′ × 0.625) + 1

   number of studs = 46.88 + 1

   ***number of studs* = 47.88; rounded to 48 studs**

Balloon framing requires a 1 × 4 ribbon board at the second-floor level. A *ribbon board* is a supporting ledger applied horizontally across studs to support the end of joists. To determine the length of the ribbon board, the dimensions of the two outside walls that support the ends of the joists are added together.

Additional wall framing members are added for rough sills, blocking, header framing, intersecting wall framing, and bracing. These additional members are needed based on the number of door and window openings and other framing requirements. An estimator should check these structural items when determining the total number of wood framing members entered into the quantity takeoff.

When using most estimating programs, calculations for an entire wall assembly may be performed from a single entry. An estimator can enter information regarding the wall construction into a database, including information on plates, studs, blocking, bracing, gypsum board, molding, and paint. **See Figure 9-13.** The database is linked to the spreadsheet in the estimating program. The estimator can then make a single entry to take off all material for a certain number of linear feet of standard wall including wood framing members, surface finish materials, fasteners, and labor. The number of linear feet of wall can be calculated traditionally, imported from the BIM model, or digitized into the estimating program.

Labor costs for wall framing are based on the entire structure as a unit, including plates, studs, headers, blocking, and bracing. Labor rates for wall framing are commonly based on the length of wall to be constructed and the height of the wall. Company historical records or standard industry labor data can be utilized to determine these costs.

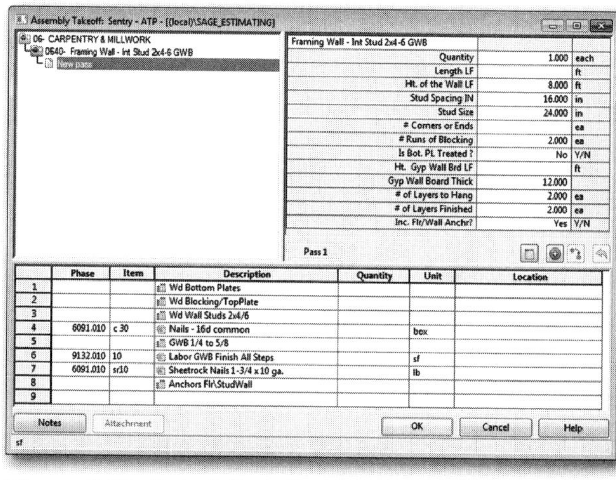

Figure 9-13. Calculations for entire wall assemblies may be made from a single entry when using most estimating programs.

Ceiling Joists and Trusses. Ceiling joist length is determined from the location of bearing walls on the floor plan. The quantity of ceiling joists required is based on the distance between the bearing walls and on-center spacing, similar to determining the quantity of floor joists and wall studs. The ceiling joist quantity is entered into a ledger sheet, spreadsheet, or estimating program. Material costs for ceiling joists are based on board foot, and labor calculations are similar to those for floor joists.

The number of roof trusses required is also based on the distance between the bearing walls and the on-center spacing of the trusses. For example, a building with 24′-6″ between bearing walls and roof trusses spaced 2′-0″ OC will require 13 roof trusses (24.5′ × 0.5 = 12.25 trusses; rounded up to 13 trusses). Truss manufacturers provide unit costs for trusses based on the style and size of the truss or trusses required. Estimators should coordinate construction of the trusses with project field management to determine any related lifting equipment requirements and labor costs.

Sheathing Panels

The number of sheathing panels required is based on the area of the floor, wall, or roof to be covered. Deductions in area may be made for large openings such as stairways or windows. Each 4′ × 8′ sheathing panel covers 32 sq ft. When using a ledger sheet, the number of sheathing panels required for a given application is determined by dividing the gross area (in square feet) by 32. Approximately 7% is added for waste to determine the total number of sheathing panels.

Floor Sheathing. For floors, the area to be covered with sheathing panels is calculated based on floor-plan dimensions. The type and thickness of sheathing panel to be installed is indicated on the floor plan with a note or shown in details and specifications.

Materials are priced on a per panel cost. Labor costs include fastening of flooring materials and may involve the use of adhesive to fasten the panels to the floor joists. Labor costs for installation are based on company historical data or standard industry cost data.

Determining the Number of Floor Sheathing Panels

Determine the number of 4′ × 8′ sheathing panels required to cover a floor measuring 60′ × 250′.

1. **Determine the area of the floor to be covered with sheathing.**

   area = *length* × *width*

   area = 60′ × 250′

   area = 15,000 sq ft

2. **Determine the number of sheathing panels (without waste).**

   number of sheathing panels = *area* ÷ 32

   number of sheathing panels = 15,000 sq ft ÷ 32

   number of sheathing panels = 468.75 panels

3. **Determine the total number of sheathing panels.**

   total number of sheathing panels = *number of sheathing panels* × *waste factor*

   total number of sheathing panels = 468.75 × 1.07

   ***total number of sheathing panels* = 501.56; round to 502 sheathing panels**

Wall Sheathing. A similar method is also used to calculate the number of wall sheathing panels. An estimator refers to the floor plan and elevations and determines the wall area and the area of the wall openings, such as door and window openings, and calculates the number of sheathing panels required. Materials are priced on a per panel basis. Labor costs for installation are based on company historical data or standard industry cost data.

Determining the Number of Wall Sheathing Panels

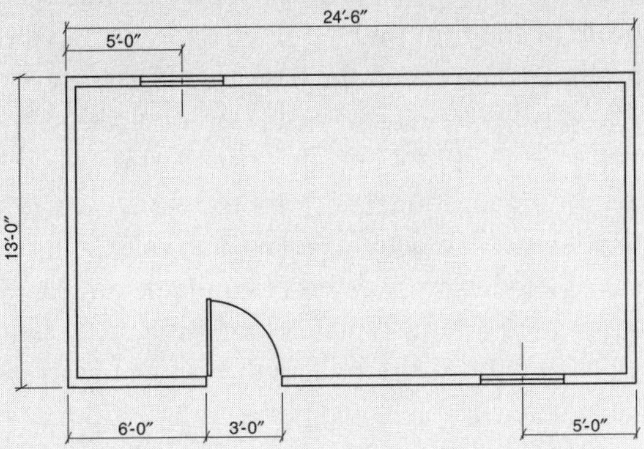

Determine the number of sheathing panels required to cover the 8′-0″ high walls of a building measuring 24′-6″ × 13′-0″. A door opening measuring 3′-0″ × 6′-8″ and two window openings measuring 2′-6″ × 3′-0″ each are placed in the walls.

1. **Determine the gross area of the walls.**

   gross area = [(24.5′ × 8′) × 2] + [(13′ × 8′) × 2]

   gross area = [196 sq ft × 2] + [104 sq ft × 2]

   gross area = 392 sq ft + 208 sq ft

   gross area = 600 sq ft

2. **Determine the area of the openings.**

   area of openings = (*opening 1 width* × *opening 1 length*) + [(*opening 2 width* × *opening 2 length*) × 2]

   area of openings = (3′ × 6.667′) + [(2.5′ × 3′) × 2]

   area of openings = 20 sq ft + 15 sq ft

   area of openings = 35 sq ft

3. **Determine the net area of the walls.**

   net area = *gross area* – *area of openings*

   net area = 600 sq ft – 35 sq ft

   net area = 565 sq ft

4. **Determine the number of sheathing panels (without waste).**

   number of sheathing panels = *net area* ÷ 32

   number of sheathing panels = 565 sq ft ÷ 32

   number of sheathing panels = 17.66 panels

5. **Determine the total number of sheathing panels.**

   total number of sheathing panels = *number of sheathing panels* × *waste factor*

   total number of sheathing panels = 17.66 × 1.07

   ***total number of sheathing panels* = 18.90; round to 19 sheathing panels**

Roof Sheathing. Roof sheathing calculations are based on the roof and rafter lengths, which are based on the pitch of the roof. Roof length is indicated on the floor plans. Rafter length is determined from the run of the roof and the pitch shown on exterior elevations. The run is the horizontal distance from the roof ridge to the outer edge of the bearing wall. Steeply pitched roofs result in longer rafter lengths and require additional sheathing panels.

Rafter length is determined prior to calculating roof area. Rafter length is determined mathematically or from printed tables. When determining rafter length mathematically, the rafter length per foot of run must first be determined. The rafter length per foot of run is equal to the square root of the sum of the squares of the rafter run and rise.

Determining the Length of Common Rafters

Determine the total length of the common rafters for a roof with a total run of 8′-0″ and a 7″ unit rise.

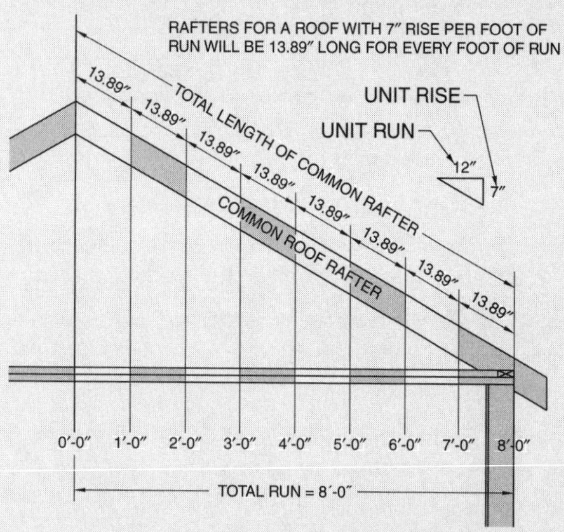

1. Determine the rafter length per foot of run.

rafter length per foot of run = $\sqrt{\textit{unit run}^2 + \textit{unit rise}^2}$

rafter length per foot of run = $\sqrt{12^2 + 7^2}$

rafter length per foot of run = $\sqrt{144 + 49}$

rafter length per foot of run = $\sqrt{193}$

rafter length per foot of run = 13.89″

2. Determine the total length of the common rafters.

total length = *total run* × *rafter length per foot of run*

total length = 8.0′ × 13.89″

***total length* = 111.12″ = 111⅛″ = 9′-3⅛″**

Rafter length multipliers become greater as the unit run becomes greater and the roof pitch becomes steeper. An estimator should consult typical wall sections or details to determine overhang distance to be added to rafter and sheathing totals. In certain estimating programs, formulas may be established for rafter items that will automatically calculate rafter length based on total run and rafter length per foot of run. The area covered by roof sheathing panels is determined by multiplying the rafter length by the roof length. The area is divided by 32 to determine the number of 4′ × 8′ sheathing panels.

Determining the Number of Roof Sheathing Panels

Determine the number of 4′ × 8′ sheathing panels required to cover the gable roof of a building measuring 24′-6″ long by 13′-0″ wide. The roof has a 3:12 pitch.

1. Determine the rafter length per foot of run.

rafter length per foot of run = $\sqrt{\textit{unit run}^2 + \textit{unit rise}^2}$

rafter length per foot of run = $\sqrt{12^2 + 3^2}$

rafter length per foot of run = $\sqrt{144 + 9}$

rafter length per foot of run = $\sqrt{153}$

rafter length per foot of run = 12.37″

2. Determine the total length of the common rafters.

total rafter length = (*total run* × *rafter length per foot of run*) ÷ 12

total rafter length = (6.5′ × 12.37″) ÷ 12

total rafter length = 80.41 ÷ 12

total rafter length = 6.7′

3. Determine the roof area.

roof area = (*roof length* × *total rafter length*) × 2

roof area = (24.5′ × 6.7′) × 2

roof area = 164.15 × 2

roof area = 328.3 sq ft

4. Determine the number of sheathing panels (without waste).

number of sheathing panels = *roof area* ÷ 32

number of sheathing panels = 328.3 sq ft ÷ 32

number of sheathing panels = 10.26 panels

5. Determine the total number of sheathing panels.

total number of sheathing panels = *number of sheathing panels* × *waste factor*

total number of sheathing panels = 10.26 × 1.05

***total number of sheathing panels* = 10.77; round to 11 panels**

In some estimating programs, a formula may be entered that will automatically calculate the sheathing amount based on pitch and rafter length calculations. The estimator may use rafter length information based on various pitches in a standardized formula. Each time the roof sheathing item is selected from an item listing, the program requests specific information concerning roof pitch, length, and width. The quantity of roof sheathing required is automatically calculated and entered into the takeoff quantity column corresponding to the item.

Roof sheathing is priced on a per panel basis. Labor costs for installation are based on company historical data or standard industry cost data.

Finish Wood Members

Finish wood members include cabinetry, millwork, paneling, and stairway components. Quantity takeoff methods for finish wood members include linear feet calculations and piece counts on the material to be determined. Subcontractors may submit bids for a large share of this work. Estimators rely primarily on interior elevations, floor plans, schedules, and the specifications to take off finish wood member quantities.

Cabinetry. Interior elevations and the related schedules provide location, dimension, and finish information for cabinetry and casework. **See Figure 9-14.** As BIM, computer-aided design (CAD), and computer-aided manufacturing (CAM) become more prevalent in prefabricated cabinet construction, cabinet information from the construction documents can be electronically transferred to cabinet manufacturing equipment in an increasing number of applications. In these instances, both a cutting schedule and a materials schedule may be generated by the computer, thereby automating much of the cabinet estimator's work. Estimators monitor the accuracy of these systems related to materials prices, labor requirements, and any applicable transportation and installation costs.

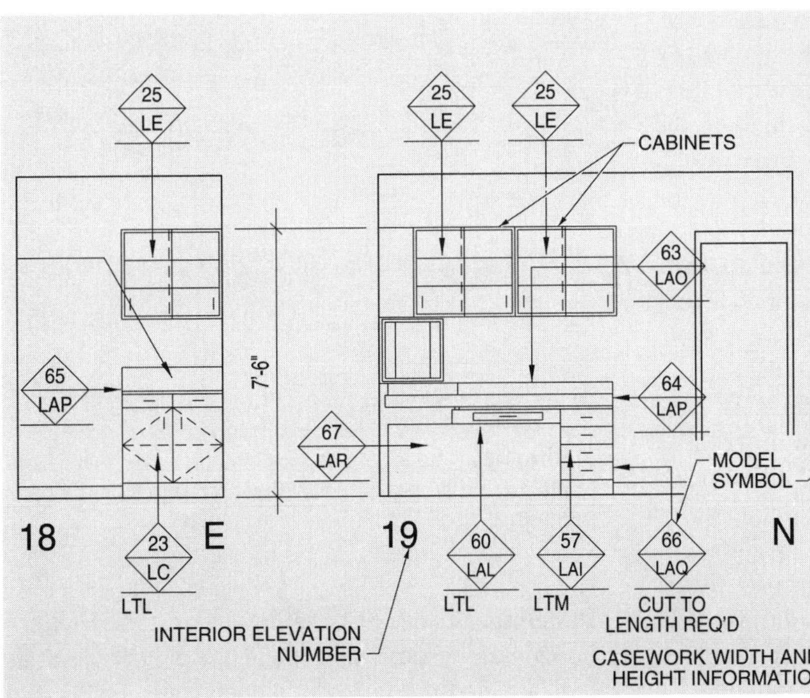

NOTE:
SEE SHEET A10.2 FOR ADDITIONAL LABORATORY CASEWORK SCHEDULE

LABORATORY CASEWORK SCHEDULE, ALTERNATE #1 (12301)						
MODEL SYMBOL	MODEL #	NOTE	SIZE SYMBOL	WIDTH	HEIGHT	DEPTH
54	91-838	PEDESTAL	LAF	35	5	9
55	97-640Q	APRON	LAG	47	4 3/4	22
56	90-600	LEGS	LAH	2	28 9/16	2
57	90-150Q	BASE	LAI	18 5/16	28 3/4	22
58	91-389-3	CORNER LEDGE	LAJ	32 5/16	4 3/4	3
59	97-690Q	CORNER APRON	LAK	24 3/8	4 3/4	24 3/8
60	97-630Q	APRON W/ DRAWER	LAL	35	4 3/4	22
61	97-625Q	APRON W/ DRAWER	LAM	30	4 3/4	22
62	91-381	LEDGE	LAN	48	4 3/4	3
63	91-382	LEDGE	LAO	96	4 3/4	3
64	91-364R	R. LEDGE END FILLER	LAP	3	9	-
65	91-364L	L. LEDGE END FILLER	LAP	3	9	-
66	91-347	BASE CAB END FILLER	LAQ	3	35 3/8	-
67	91-340	KNEE SPACE FILLER PANEL	LAR	24 [illegible]/4 -47	23 5/8	-
68	91-164	PEG BOARD DRAIN RACK	LAS	24	30	-

COUNTERTOP SCHEDULE

LTA	24"D, 4"H SPLASH, 3" LEDGE, 97-70B CORNER TOP
LTK	24"D X 47"L, 4"H SPLASH, NO LEDGE, 94-84B-U010
LTL	24"D X 35"L, 4"H SPLASH, 3" LEDGE, 94-84B-U010
LTM	24"D X 18"L, 4"H SPLASH, 3" LEDGE, 94-84B-U010
LTN	24"D X 30"L, 4"H SPLASH, 3" LEDGE, 94-84B-U010

Figure 9-14. Interior elevations and related schedules provide location, dimension, and finish information for cabinetry and casework.

Architectural notes and codes on interior elevations relate to a cabinetry and casework schedule that indicates the width, height, and depth of each cabinet. The construction documents contain a schedule for each cabinet for use by the cabinet shop. A price for these cabinets is commonly obtained from a cabinet shop or subcontractor and entered into the general contractor bid.

For job-built cabinets, materials calculated include particleboard, plywood, and exterior finish material. Labor costs for cabinet construction and installation are also included in this portion of the estimate, with information obtained from company historical data or standard industry data.

Millwork and finish flooring materials should be allowed to acclimate to environmental conditions before installation and should not be subjected to extreme dryness, excessive humidity, sudden changes in temperature, or direct sunlight.

Millwork. Millwork is estimated based on linear feet. Trim molding quantities for items such as baseboard, casing, chair rail, and crown molding are estimated by the linear foot. The total linear feet of these moldings is determined from wall length dimensions on the floor plan. A tabulation is made for each type of molding required. A cost-per-foot method that includes the material and labor costs is commonly used.

Baseboard molding is calculated in linear feet and is based on the total perimeter of the room. Deductions are made where doors or large openings exist. The perimeter of all rooms is added and the amount of baseboard is determined. For example, a 24′-6″ × 13′-0″ room (24.5′ × 13′) has 75 lf of wall ([24.5′ × 2] + [13′ × 2] = 75 lf). A minimum of 75′-0″ of baseboard, chair rail, and crown molding is required for the room (with no allowances for doors, windows, or waste). Some additions may be necessary for fitting at the corners and for splicing based on available material lengths. In applications where special trim moldings are noted, they are estimated individually for length and type.

Casing length depends on the width and height of each door or window opening. For a 3′-0″ × 6′-8″ door, casing is installed on each side (6′-8″ × 2 = 13′-4″). A 7′-0″ length is needed along each side to allow for mitering and fitting at each top corner. A 4′-0″ piece of casing is required across the top of the door opening to allow for mitering and fitting (depending on the width of the casing).

The linear feet of each type of molding is taken off from the drawings and entered into the proper row and column on a ledger sheet or spreadsheet. Print takeoff or a digitizer may be used to enter linear foot information into the proper item column and row, or cell when using an estimating program. **See Figure 9-15.**

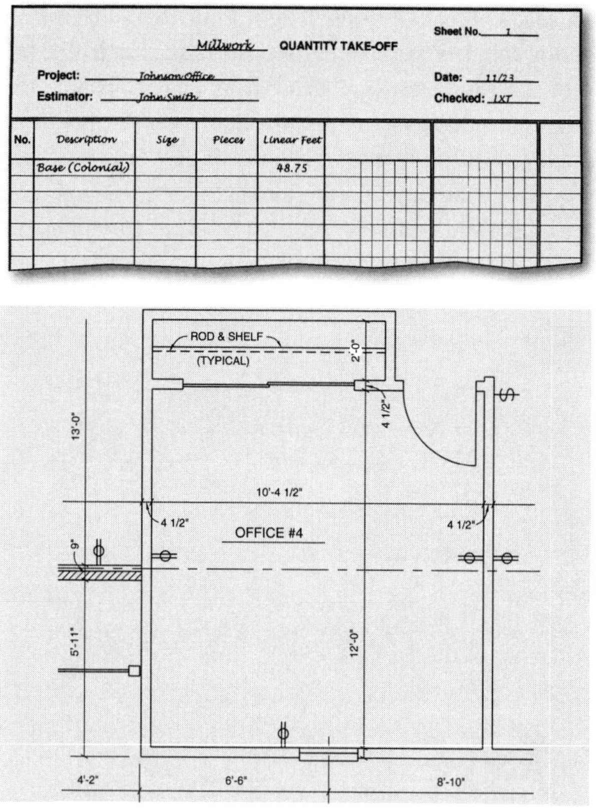

Figure 9-15. The length of moldings (in linear feet) is entered into the proper item information cell in an estimating program.

Paneling. Sheets of paneling are installed with a horizontal, vertical, or angular grain orientation as indicated in the construction documents. Each sheet of paneling covers a 4′-0″ width when sheets are installed vertically and an 8′-0″ width when sheets are installed horizontally. When paneling is to be installed vertically in rooms with 8′-0″ ceilings or lower, the perimeter measurement of the room (in linear feet) is divided by 4 to calculate the number of sheets of paneling needed.

For example, the perimeter of a rectangular room measuring 14′-0″ × 52′-0″ with 8′-0″ high ceilings is 132 lf ([14′ × 2] + [52′ × 2] = 132 lf). To determine the number of sheets of paneling needed, the perimeter measurement is divided by 4 (132 lf ÷ 4 = 33 sheets). Five percent waste is added, resulting in 34.65 sheets (33 sheets × 1.05 = 34.65 sheets), which is rounded up to 35 sheets.

Cost per sheet is based on the type and grade of paneling as quoted by the supplier. Allowances may be made for large door openings. Labor installation costs are based on company historical data or standard industry data. Other potential items to consider in paneling installation include fasteners, adhesives, or finishing requirements.

Stairway Components. Estimating stairway components requires information concerning finishes, the number of treads and risers, balusters, handrails, newel posts, and stringers and is obtained from interior elevations, floor plans, and specifications. **See Figure 9-16.** Rough stairways are not always shown in detail on the prints. An estimator must calculate stringer length and tread width before performing a quantity takeoff. The estimator must also calculate the rise and run of the stairs without details or interior elevations.

Variables in stairway calculations involve the total rise, total run, type of wood to be installed, and final finish. The actual number of treads, risers, balusters, goosenecks, and newel posts are counted individually from the architectural drawings when possible.

> Both hardwood and softwood are used for stairway construction. Softwood is typically used for structural members, such as the stringers, or when the wood is to be painted.

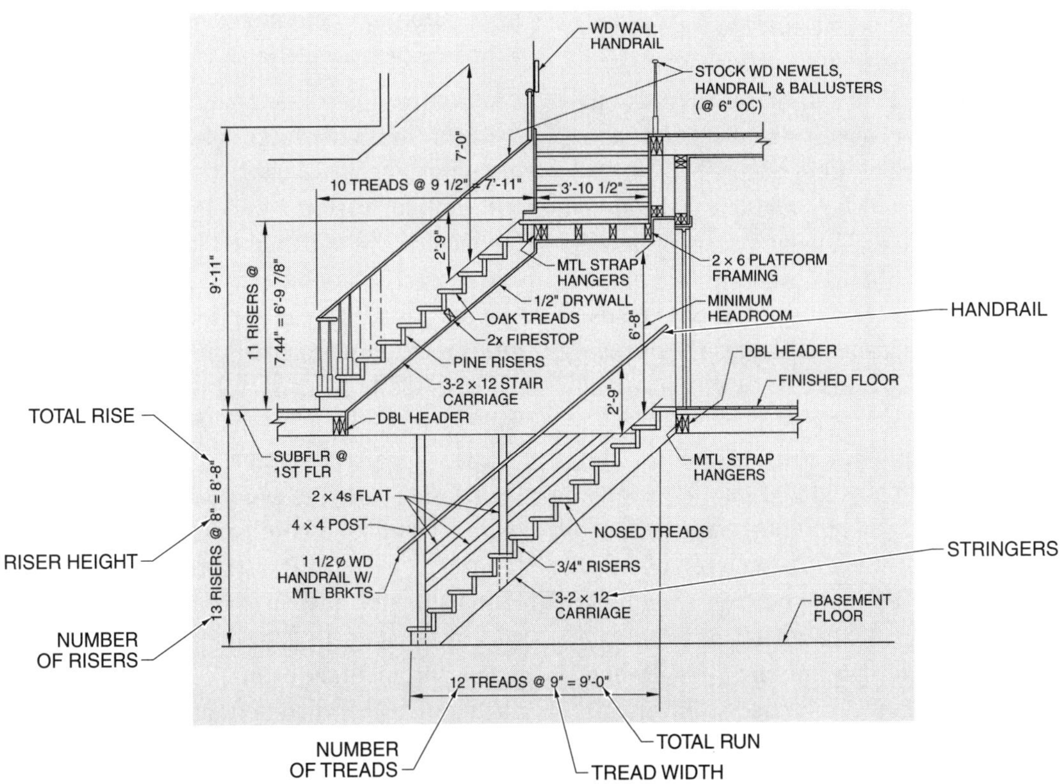

Figure 9-16. Interior elevations drawings and floor plans show finish wood stairway components and dimensions.

The linear feet of handrail and stringers is based on the hypotenuse of a triangle with the two legs being the total run and the total rise of the stair. The length of the hypotenuse is equal to the square root of the sum of the total run squared and the total rise squared ($h = \sqrt{run^2 + rise^2}$). For example, a stairway with a total run of 6′-0″ and a total rise of 4′-0″ requires stringers and handrails approximately 7′-3″ long ($\sqrt{6^2 + 4^2} = 7.21' = 7'\text{-}3''$). To allow for cutting and fitting, 8′ of material is purchased for each stringer or handrail required.

In some cases, the tread width and riser height is not provided on the prints. To find the tread and riser height and the number of risers, the total floor-to-floor distance is divided by the desired riser height (commonly 7½″). If the division does not result in a whole number, the closest whole number should be chosen, and the total floor-to-floor distance should be divided by the whole number. The result of the division is the actual unit rise.

The number of treads is always one less than the number of risers. The tread width is found by dividing the total run of the stairway by the number of treads in the stairway. In some situations, the total run of the stairway is limited. Using the total run as a limit, the same process is used as for calculating riser height. When the total run of the stairs is not limited, the tread width is determined using a ratio established by the local building code.

Stairway components, such as balusters and goosenecks, are counted individually and priced on a per piece basis. Material costs for handrails are based on linear feet, and costs for stringers, rough treads, and rough risers are based on board feet. The quantities determined are entered into the correct row and column on a ledger sheet or spreadsheet.

When using certain estimating programs, the quantities for each item are entered into the appropriate item takeoff cell. In addition, some estimating programs allow stairways to be entered into the database by the estimator as a common assembly. Formulas may be entered and stored for stairway components to assist in determining stringer and handrail lengths and number of risers and treads based on the total rise and run. Labor costs for fabrication of stair components and installation at the job site are based on company historical data or standard industry information and vary depending on whether the stair installation is finished or rough.

Finished stairway construction is a portion of the work that is commonly performed by a specialty contractor. This is an item that may be subcontracted and included in the overall final bid as a single item by the general contractor.

Plastics and Composites

Structural plastics are commonly used for exterior deck areas and fencing components. High- and low-pressure plastic laminate sheets are applied over cabinet framing materials. Linear foot, per piece, and area calculations are required for plastics and composites.

Structural. Structural plastics and composites are shown on floor plans and elevations and are included in the specifications. Estimating for structural plastics and composites is performed in the same manner as structural wood members, with the number of pieces required depending on the spacing of studs, joists, and plates. Costs for structural plastics and composites are calculated based on a per piece cost or a cost-per-board-foot calculation. Labor costs for installation are determined in a manner similar to structural wood materials.

Laminates. Plastic laminate cost is determined by the area (in square feet) to be covered and the type of laminate material used. This applies to horizontal and vertical surfaces and high- and low-pressure laminates. Where patterned laminate is used (rather than a solid color), quantity calculations must ensure that the pattern can be applied continuously without turning or reorienting the direction or design of the pattern. Additional costs for waste can be anticipated when patterned laminate is specified.

Costs for plastic laminate vary depending on the thickness, quality, manufacturer, finish, possible requirement of solid-core laminate, and pattern. Costs for plastic laminate are expressed based on square foot. Estimators must be sure to include plastic laminate material costs, adhesives, and fabrication labor in this portion of the estimate. For some laminate applications such as countertops, specifications should be checked to determine if the estimate is to include a base material, such as a particular type of particleboard, and labor costs for installation at the job site.

SAGE TIMBERLINE ESTIMATING . . .

ASSEMBLY TAKEOFF

An *assembly* is a collection of items needed to complete a particular unit of work. For example, a floor framing assembly may include joists, bridging, subflooring, underlayment, and nails or screws. Assemblies allow the estimator to take off multiple items with a single operation and to obtain a cost per unit for a group of items.

Assemblies may be database assemblies or one-time assemblies. A *database assembly* is an assembly that is stored in a database and can be used in any estimate based on that database. A *one-time assembly* is an assembly that is stored with the estimate and cannot be used in other estimates.

1. Start a new estimate called **Jones** and link the estimate to the **Sample Commercial GC** database.
2. Open the **Assembly Takeoff** screen by picking the **Assembly Takeoff/Review** button in the toolbar. Double-click **06 Carpentry & Millwork and 0640- Frame Wall – Int. Stud 2×4-6 GWB**. The **Assembly Takeoff** screen consists of the assembly list pane, dimension pane, and item grid pane. The assembly list pane displays the assemblies listed in the database. Dimensions and quantities are entered and displayed in the dimension pane. The item grid pane displays items in the selected assembly and reflects the dimensions and quantities entered in the dimension pane.

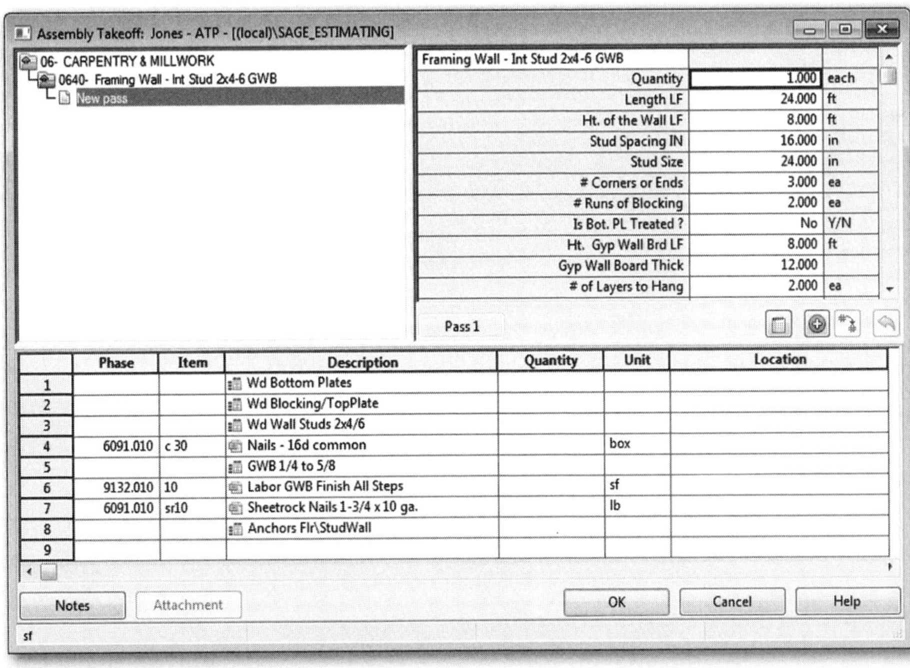

. . . SAGE TIMBERLINE ESTIMATING

3. Items in the selected assembly are displayed in the item grid pane, and the dimensions and quantities from the formulas assigned to those items are displayed in the dimension pane. Note that some quantities in the dimension pane prefill from the last time the assembly was used. Enter the dimensions for the assembly. Enter **1** for the **Quantity**, 24 for **Length LF**, **8** for **Ht. of the Wall LF**, **16** for **Stud Spacing IN**, **24** for **Stud Size**, **3** for **# Corners or Ends**, **2** for **# Runs of Blocking**, **No** for **Is Bot. PL Treated?**, **8** for **Ht. Gyp Wall Brd LF**, **12** for **Gyp Wall Board Thick**, and **2** for **# of Layers to Hang**.
4. Pick the **Add Pass** button to make the calculations and send the items to the item grid pane.
5. Review the prints and the items in the item grid pane to ensure a comprehensive estimate. Pick the **OK** button at the bottom of the **Assembly Takeoff** screen to send the items to the spreadsheet. The items taken off can be viewed by picking the **Assembly** tab at the bottom of the spreadsheet to sort the spreadsheet into assembly order.
6. Close the estimate.

Jones - ATP - [(local)\SAGE_ESTIMATING]

Assembly	Phase	Description	Takeoff Quantity	Labor Cost/Unit	Labor Price	Labor Amount	Material Price
0640-		Framing Wall - Int Stud 2x4-6 GWB					
	6110.010	Plates 2x4 RL	25.20 lf	0.56 /lf	20.00 /hr	14	480.00 /mbf
	6117.020	Blocking 2 x 4 R.L.	100.80 lf	0.56 /lf	20.00 /hr	56	480.00 /mbf
	6112.010	Studs 2x4x 8	27.00 ea	4.28 /ea	20.00 /hr	116	480.00 /mbf
	6091.010	Nails - 16d common	1.00 box	-	-	-	60.00 /box
	9131.010	GWB 1/2" All Size Regular	384.00 sf	0.50 /sf	20.00 /hr	192	0.30 /sf
	9132.010	Labor GWB Finish All Steps	384.00 sf	0.25 /sf	20.00 /hr	96	-
	6091.010	Sheetrock Nails 1-3/4 x 10 ga.	7.68 lb	-	-	-	6.00 /lb
	6090.010	Shot Anchors	12.00 ea	0.40 /ea	20.00 /hr	5	0.30 /ea

Phase/Item | Takeoff Order | Assembly | Location/Phase

Quick Quiz®

Quick Quiz®

Refer to the CD-ROM for the Quick Quiz® questions related to chapter content.

Key Terms

Illustrated Glossary

- balloon framing
- baluster
- beam
- board foot (bf)
- cabinet
- closed stringer
- column
- common board
- cross-lamination
- dead load
- dimension lumber
- finish wood member
- glulam timber
- grading
- handrail
- hardwood
- header
- joist
- landing
- live load
- lumber
- medium-density fiberboard (MDF)
- millwork
- nominal size
- open stringer
- oriented strand board (OSB)
- parallel-strand lumber (PSL)
- particleboard
- plastic laminate
- platform framing
- plywood
- riser
- rough sill
- sill plate
- softwood
- stringer
- stud
- timber
- timber framing
- tread
- truss

Web Links

Web Links

American Institute of Timber Construction
www.aitc-glulam.org

American Wood Council
www.awc.org

APA—The Engineered Wood Association
www.apawood.org

Architectural Woodwork Institute
www.awinet.org

California Redwood Association
www.calredwood.org

Canadian Wood Council
www.cwc.ca

Composite Panel Association
www.pbmdf.com

Hardwood Plywood and Veneer Association
www.hpva.org

National Hardwood Lumber Association
www.nhla.com

Powder-Actuated Tool Manufacturers Institute
www.patmi.org

Southern Forest Products Association
www.sfpa.org

Southern Pine Council
www.southernpine.com

Truss Plate Institute
www.tpinst.org

Truss Plate Institute of Canada
www.tpic.ca

United Brotherhood of Carpenters and Joiners of America
www.carpenters.org

Western Wood Products Association
www.wwpa.org

Wood Moulding and Millwork Producers Association
www.wmmpa.com

Wood Products Manufacturers Association
www.wpma.org

Wood Truss Council of America
www.sbcindustry.com

Woods, Plastics, and Composites 9

Review Questions

Estimating

______________ 1. A(n) ___ wood member is a wood component that provides support for live and dead loads.

______________ 2. ___ is finished wood material or parts completed in a mill or fabrication facility.

______________ 3. A(n) ___ load is the total of all the dynamic loads a structure is capable of supporting including natural forces such as wind loads, people, and any other imposed loads.

______________ 4. A(n) ___ load is a permanent stationary load composed of all construction materials, fixtures, and equipment permanently attached to a structure.

______________ 5. ___ is the classification of various pieces of wood according to their quality and structural integrity.

______________ 6. ___ is sawn and sized lengths of wood used in construction.

T F 7. Dimension lumber is lumber that has a nominal size from 1 × 4 up to 4 × 6.

T F 8. Timber is a heavy lumber that has a nominal size of 5 × 5 or larger.

______________ 9. A(n) ___ is a horizontal structural member used to support a load over an opening.

______________ 10. A(n) ___ is a vertical structural member used to support axial compressive loads.

______________ 11. A(n) ___ is a manufactured roof or floor support member with components commonly placed in a triangular arrangement.

______________ 12. ___ is an engineered wood panel product made of wood layers glued and pressed together under intense heat and pressure.

______________ 13. ___ is an engineered wood panel product in which wood strands are mechanically oriented and bonded with phenolic resin under heat and pressure.

______________ 14. ___ is an engineered lumber product composed of wood laminations (lams) that are bonded together with adhesives.

______________ 15. ___ is an engineered wood panel product constructed of wood particles and flakes that are bonded together with a synthetic resin.

______________ 16. Medium-density ___ is an engineered wood panel product manufactured from fine wood fibers mixed with binders and formed with heat and pressure.

______________ 17. A(n) ___ wood member is a decorative and nonstructural wood component in a structure.

T F 18. A cabinet is an enclosure fitted with any combination of shelves, drawers, and doors and is usually used for storage.

T F **19.** Structural plastics and composites are used for interior light-load-bearing applications.

______________ **20.** Plastic ___ is sheet material composed of multiple layers of plastic and resins bonded together under intense heat and pressure.

______________ **21.** A(n) ___ is a horizontal framing member placed at the top of a window or door opening.

______________ **22.** A(n) ___ is a horizontal support member at the top and bottom of a framed wall to which studs are attached.

______________ **23.** A(n) ___ is a vertical support member that extends from the bottom to the top plates in a framed wall.

______________ **24.** ___ framing is framing method in which large wood members are used to form large open areas.

Short Answer

1. Describe how nominal size differs from actual size and explain the cause of the difference.

____

____

____

____

2. Identify the wood members that may be included in a takeoff for a wall.

____

____

____

3. When estimating the number of studs and joists, one additional joist is added to the calculations before determining waste. What is the purpose of this stud or joist?

____

____

____

____

4. Refer to the International Building Code and cite the minimum and maximum requirements for treads and risers.

____

____

____

____

Woods, Plastics, and Composites 9

Activities

Activity 9-1—Ledger Sheet Activity

Refer to Print 9-1 and Quantity Sheet No. 9-1. Take off the linear feet of interior wall plates, number of studs (spaced 16″ OC), and number of 4′ × 8′ sheets of OSB floor sheathing for the Biotech Study. Assume a double top and a single bottom plate. Round all plates, studs, and sheets to the next highest whole unit. Add one stud at each wall intersection. Do not make allowances for door openings.

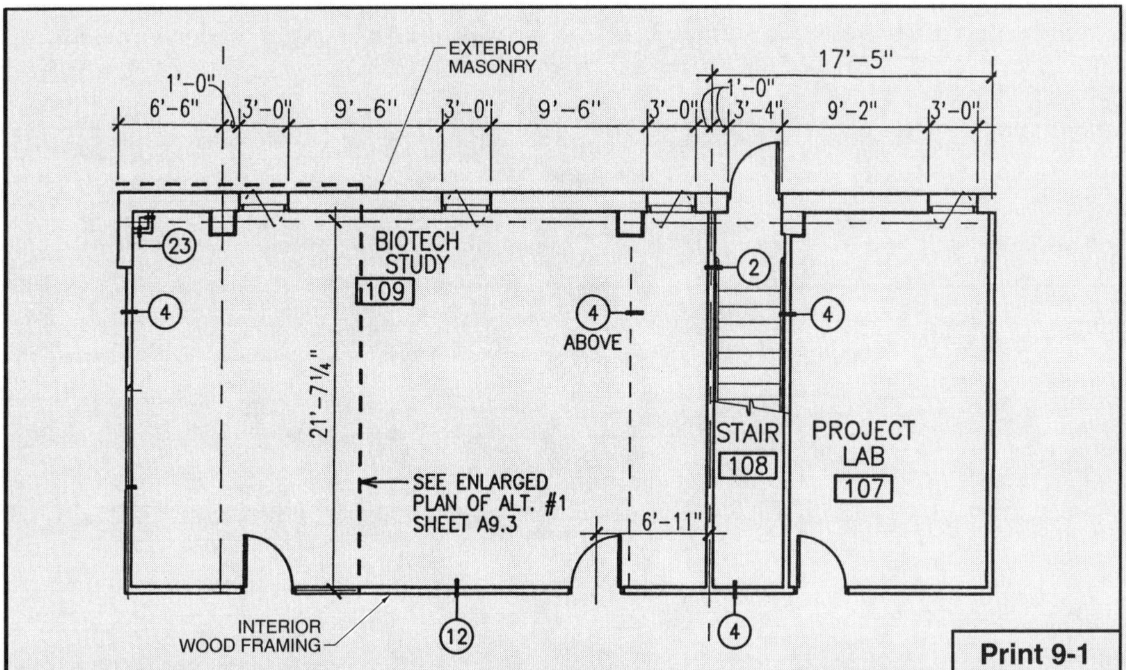

Print 9-1

QUANTITY SHEET

Project: ______________________ Sheet No. 9-1

Estimator: ______________________ Date: ______________________

Checked: ______________________

No.	Description	Dimensions L	W			Unit		Unit		Unit		Unit
	Biotech study											
	Interior wall plates											
	Studs											
	OSB floor sheathing											

Activity 9-2—Spreadsheet Activity

Refer to Print 9-2 and Quantity Spreadsheet No. 9-2 on the CD-ROM. Take off the total number of studs spaced 2′ OC, 4′ × 8′ sheets of OSB floor sheathing, and the linear feet of plates for Office 127, Project Lab 128, and Project Lab 129. Assume a double top and a single bottom plate. Round all studs, sheets, and plates to the next highest whole unit. Add one stud at each wall intersection. Do not make allowances for door openings.

Activity 9-3—Estimating Wood Products

Refer to Print 9-3 and Quantity Sheet No. 9-3. Take off the linear feet of 2 × 3, 2 × 4, and 2 × 6 interior wall plates, number of studs (spaced 16″ OC), and the linear feet of base molding required for the interior walls of Rooms 101A, 102, 103, and 104 (including the adjoining closet). In addition, calculate the number of 4 × 8 sheets of ⅝″ plywood sheathing required. Assume all walls are constructed with 2 × 4 studs with double top plates and single bottom plates unless otherwise noted. Wall height is 8′-0″. Do not make allowances for door openings.

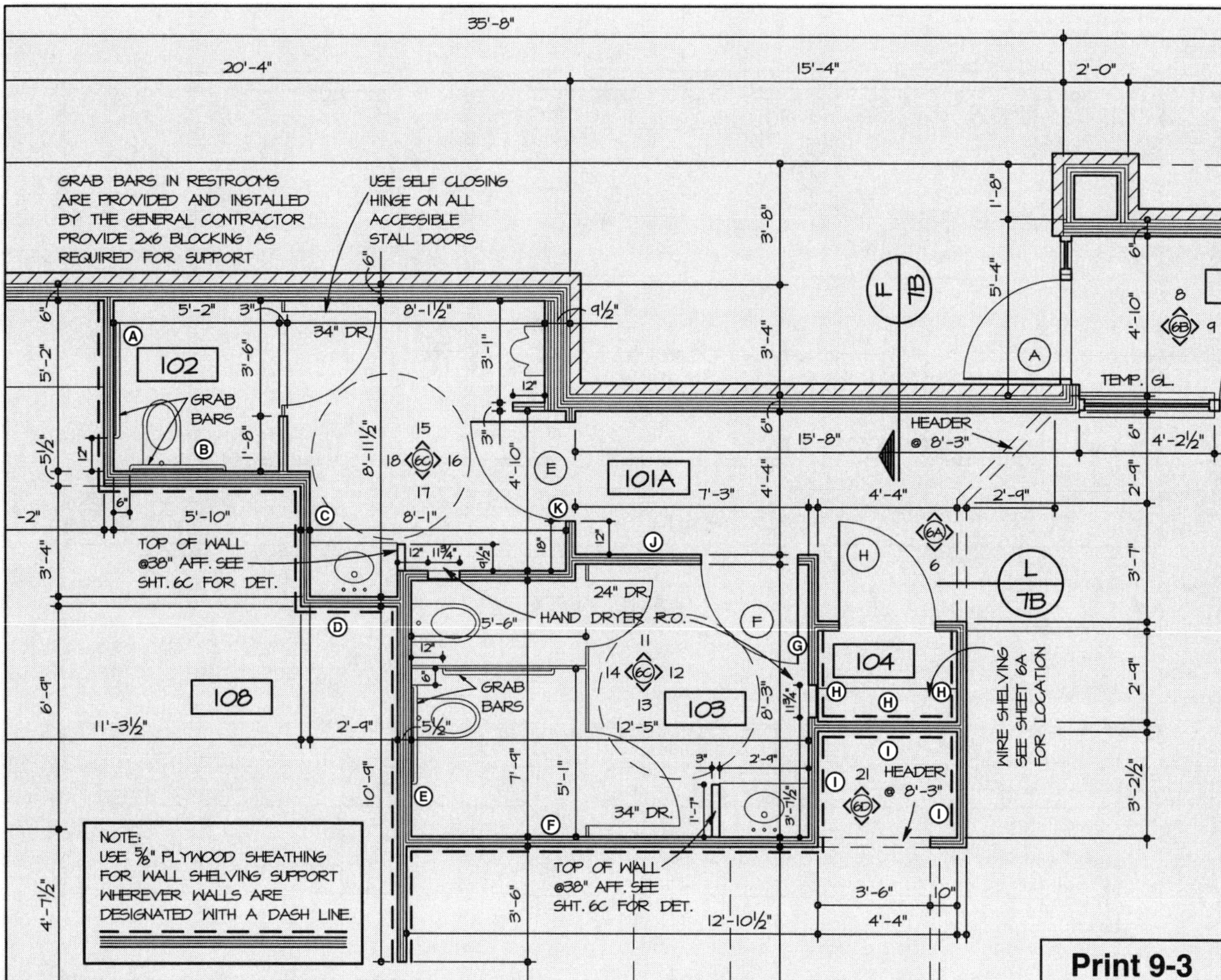

Print 9-3

ESTIMATE SUMMARY SHEET

Sheet No. 9-3
Project: ______
Date: ______
Estimator: ______
Checked: ______

Wall Length (lf)	2 × 3	2 × 4	2 × 6
a			
b			
c			
d			
e			
f			
g			
h			
i			
j			
k			
Total wall length (lf)			
Total length of plates (lf)			
Total number of studs			
Total length of studs (lf)			
Total lumber length (lf)			
Linear feet of ⅝″ sheathing			
a			
b			
c			
d			
e (return)			
f			
h			
i			
Total linear feet of ⅝″ sheathing			
Total number of 4 × 8 sheets			

Activity 9-4—Sage Timberline Estimating Activity

Create a new estimate and name it Activity 9-4. Link the new estimate to the **Sample Commercial GC** database. Take off items **6110.010 100 plates 2 × 6 RL**, **6110.020 100 plates 2 × 6 RL**, **6112.010 20 2 × 4 × 8 Studs**, and **6117.020 90 2 × 4 RL Blocking Misc.** for a 12′ × 24′ room. Add one stud at each intersection of walls. The studs and joints are 16″ OC. Print a standard estimate report.

Thermal and Moisture Protection 10

Key Concepts

- The method used for quantity takeoff and estimating of roofing materials depends on the type of material to be applied.
- Roofing material quantities are based on the roof area (in square feet).
- Quantities of exterior wall covering materials are based on the coverage area (in square feet). Deductions are made for large wall openings such as doors and windows.
- Estimators must refer to specifications and details to determine the required R values and types of insulation, water vapor materials, and fire protection coatings needed in each area of a building.
- Estimators use different takeoff methods depending on the insulation form used.

Introduction

Division 07 of the CSI MasterFormat™ includes information concerning thermal protection, moisture protection, roof covering materials, siding, fire and smoke protection, and joint sealants. Information related to exterior finish materials is provided on exterior elevations, specifications, details, mechanical plans (for rooftop units, vents, and exhausts), roof plans, and floor plans.

Thermal systems are designed to control temperatures that affect the comfort of building occupants, deter condensation, and reduce heat transmission to improve energy use within the structure. Thermal system materials are installed after

structural and framing members are in place. Estimators refer to specifications and details to determine the types of insulation, water vapor materials, and fire protection coatings to be installed in each area.

EXTERIOR FINISH MATERIALS AND METHODS

Exterior finish material information for takeoff and estimating is indicated on exterior elevations, specifications, and floor plans. **See Figure 10-1.** Division 07 of the CSI MasterFormat™ includes thermal protection, moisture protection, roof covering materials, siding and trim members, fire and smoke protection, and joint sealants. For example, title 07 22 16 of the CSI MasterFormat™ includes information on roof board insulation materials.

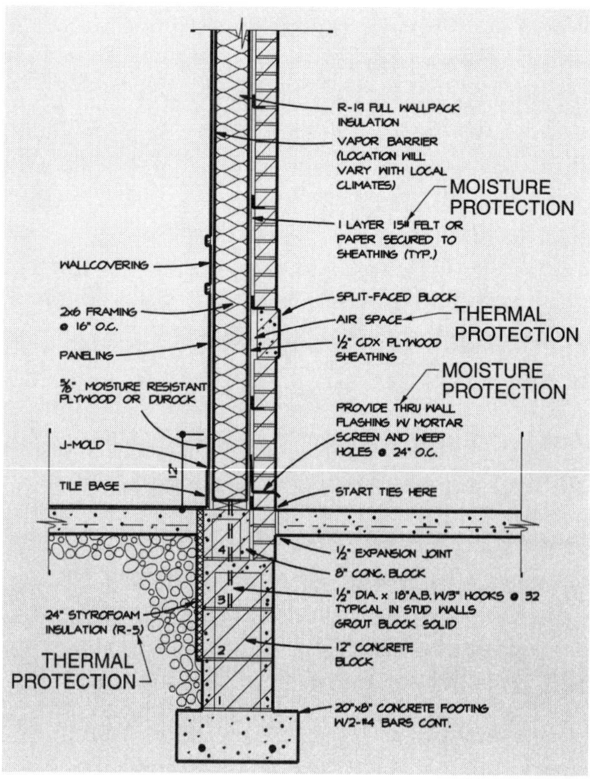

Figure 10-1. Exterior finish materials include thermal protection and moisture protection.

Roofing

A *roof* is the covering for the top exterior surface of a building or structure. Roofing includes all material placed on top of a building or structure to provide protection from environmental elements such as wind, rain, and snow. Roofing materials are selected for a particular application based on the appearance, cost, roof pitch, temperature variation, and local climatic conditions, such as annual rainfall. Roofing may be bituminous, elastomeric, or metal and consists of shingles or tiles, roof pavers, and other roof accessories. Sustainable roof systems may include natural plant and other materials that are designed to reduce energy consumption.

Bituminous Roofing. *Bituminous roofing,* commonly known as built-up or hot tar roofing, is roofing composed of layers (plies) of asphalt-impregnated felt or fiberglass material that is fastened to the roof deck and mopped with hot tar to create a waterproof surface. Rigid insulation may be applied to the top of the roof deck and covered with building paper prior to the roofing application. Gravel, slag, or a mineral top sheet is applied to the roof surface to finish the application and provide protection for the felt or tar layers.

Elastomeric Roofing. *Elastomeric roofing* is roofing made of a pliable synthetic polymer. An elastomeric roofing system is composed of large flexible sheets that are laid in place and sealed at the joints. **See Figure 10-2.** The sheets are made of chlorinated polyethylene (CPE), ethylene propylene diene monomer (EPDM), polyvinyl chloride (PVC), or other chemical compounds.

A vapor barrier and rigid roof insulation are installed on the surface of the roof deck. Elastomeric sheets are rolled out across the entire surface of the roof. Joints are sealed with a solvent that joins the sheets into a single unit. The entire surface may be covered with gravel or pavers after all joints are sealed.

Figure 10-2. Elastomeric roofing is made of pliable synthetic polymer sheets that are laid in place and sealed at the joints.

Metal Roofing. *Metal roofing* is roofing made of steel, aluminum, copper, or various metal alloys. Metal roofing materials may be formed into shingles, corrugated sheets, or sheet-metal strips. Rigid insulation may be applied to the top of the roof deck and covered with building paper.

Building paper is felt material saturated with tar to form a waterproof sheet. Sheets of prefinished or ornamental metal are set in place with the long dimension placed parallel to rafters and extending from the eaves to the ridge board. Seams along the sides of the metal sheets are fastened together using flat, ribbed, or standing seams. **See Figure 10-3.** Metal roof sheets are fastened to the roof deck with clips or self-tapping screws. Overlapped joints are coated with a waterproof joint sealant.

Shingles. A *shingle* is a thin piece of wood, asphalt-saturated felt, fiberglass, lightweight concrete, or other material applied to the surface of a roof or wall to provide a waterproof covering. A *composition shingle* is an asphalt or fiberglass shingle coated with a layer of fine mineral gravel. Composition shingles are available in a wide range of styles, colors, and thicknesses. Composition shingles include three-tab, laminated, and architectural shingles. The quality of composition shingles is determined by weight. Heavyweight shingles are of higher quality with additional weather resistance and longer life expectancy than lightweight shingles.

Solar shingles have a similar pattern, texture, and size as traditional composition shingles, but they are also a source of electricity. Electrical lead wires extend from the underside of each shingle and pass through the roof decking material to allow interior roof space connections.

A *shake* is a hand-split wood roofing material. *Exposure* is the amount a shingle or shake is visible after installation. Exposure of wood shingles or shakes varies with width and thickness and with roof slope.

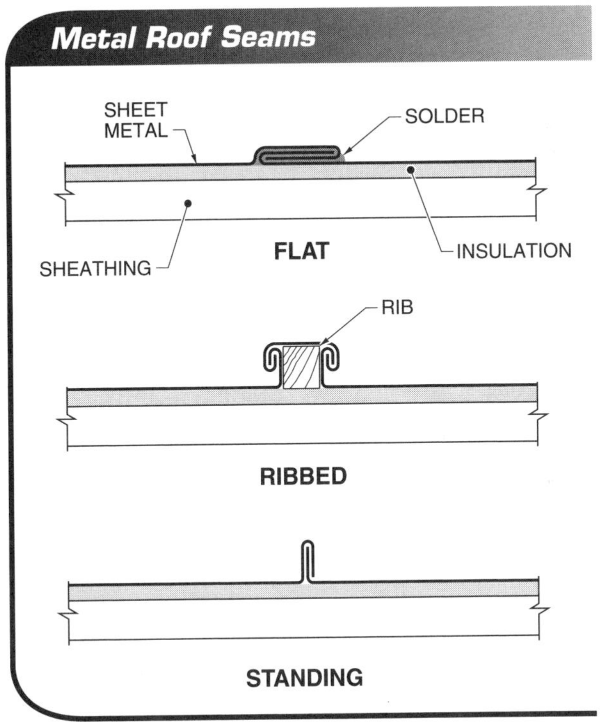

Figure 10-3. Sheets of metal roofing may be joined with flat, ribbed, or standing seams.

Tiles. *Roofing tile* is clay, slate, or lightweight concrete material installed as weather-protective finish on a roof. Roofing tile is manufactured in several different shapes. **See Figure 10-4.** Roofing tiles are usually heavier than other roofing materials and require stronger roof support systems. Roofing tiles are laid in successive rows from the eaves toward the ridge board, with each tile overlapping and covering a portion of the previous row.

Roofing Tile Shapes

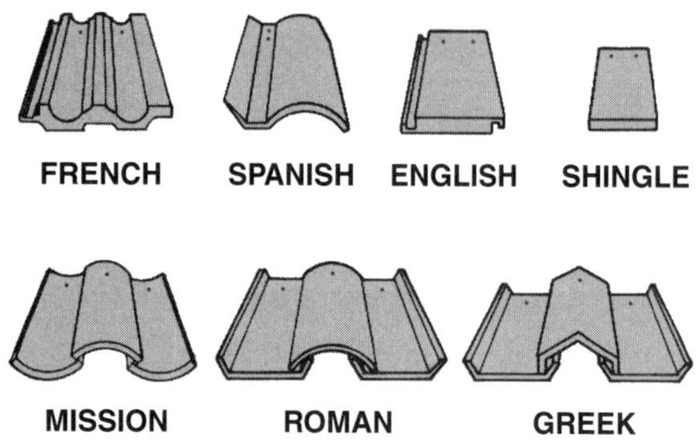

Figure 10-4. Roofing tile is made of clay, slate, or lightweight concrete material and is available in a variety of designs.

Roof Pavers. A *roof paver* is a flat, precast concrete unit that is set in place on a roof deck to provide a walking surface and to protect waterproof roofing materials. Roof pavers are placed on the top of roofs with support devices placed between the pavers and the roof surface that are designed to support the pavers and prevent roof punctures and leaks. Roof paver locations are occasionally shown on roof plans or mechanical plans for rooftop units.

Roof Accessories. Roof accessories include vents to provide airflow and remove smoke, hatches to allow access to roof surfaces, and curbs around the perimeter of roof surfaces. Specifications and roof plans indicate the placement and types of roof accessories required. Items such as gutters and downspouts are included in title 07 71 23 of the specifications.

Sustainable Roof Systems. Installations of planting media and plant materials on roof systems are used to create sustainable, or "green," roofs. Rigid insulation is installed on top of the roof decking material and covered with one or more layers of elastomeric or rubberized roofing material to create a waterproof membrane. **See Figure 10-5.** A drain layer is installed on top of the waterproof membrane to allow for water runoff. Planting media and plant materials cover the top of the roof surface. Sustainable roof systems provide some insulation advantages, reduce the amount of stormwater runoff, and create a quieter roof covering.

Green Roof Layers

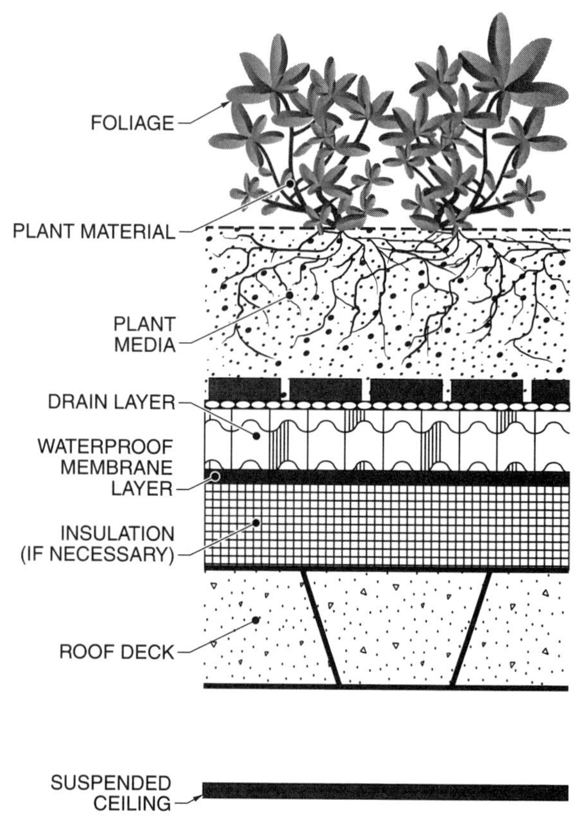

Figure 10-5. The selection of plants and materials for the supporting layers is based on the local climatic conditions and geographic regions.

Walls

Building walls may be faced with a wide variety of materials, including exposed concrete, brick, stone, and glass. Exterior wall finish materials are included in various sections of the specifications and are referenced under a variety of titles in the CSI MasterFormat™.

Exterior wall finish materials included in Division 07 include exterior insulation and finish systems (07 24 00), sheet metal (07 42 13), and various siding materials including steel (07 46 19), hardboard (07 46 26), fiber cement (07 46 46), vinyl (07 46 33), and wood (07 46 23). Fiber cement siding is similar in manufacture and design to precast concrete panels but is smaller and lighter. Insulating building wrap is commonly applied to the exterior face of a wall prior to installing many types of siding.

A variety of trim members are also applied in conjunction with each type of exterior wall finish. Estimators should be knowledgeable about all components of the specified exterior finish system. Wood, metal, injected foam products, plastic, lightweight concrete, and other trim members are important components of an exterior wall finish system and help create a weatherproof building.

Exterior Insulation and Finish Systems. An *exterior insulation and finish system (EIFS)* is an exterior finish system composed of a layer of exterior sheathing, insulation board, reinforcing mesh, a base coat of acrylic copolymers, and a textured finish. **See Figure 10-6.** Insulation board is made of extruded or molded expanded polystyrene and provides thermal insulation and flexibility in the plaster structure to minimize cracking.

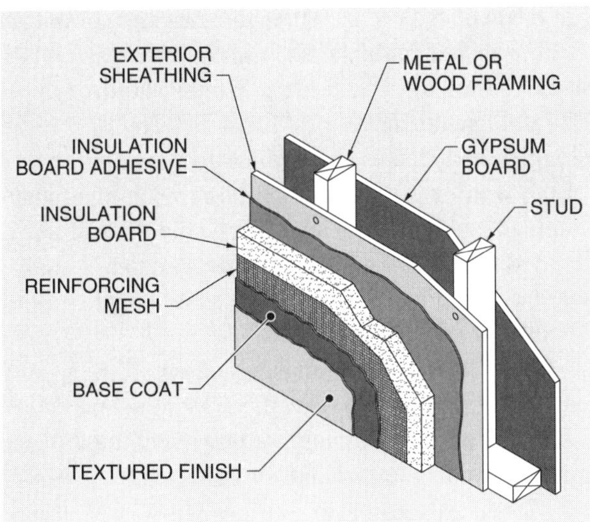

Figure 10-6. An exterior insulation and finish system (EIFS) is composed of a layer of exterior sheathing, insulation board, reinforcing mesh, a base coat, and a textured finish.

A base coat approximately ¼″ thick containing acrylic copolymers and portland cement is troweled onto the surface of the insulation board and reinforced with one or two layers of an open-weave glass fiber reinforcing mesh. After the base coat has set, the textured finish of acrylic resins is troweled or sprayed onto the structure to create the desired finish. As with all wall siding systems, care is taken at openings to fully waterproof this assembly with the proper trim assemblies and eliminate water penetration behind the finish surface.

Sheet Metal. A common exterior wall finish for small metal buildings is sheet-metal siding panels. Erection plans and elevations indicate the types and applications for sheet-metal siding panels. **See Figure 10-7.** Information on the erection plans and elevations includes the direction of the application, the types of finish trim members at the roofline and corners, and possibly the manufacturer name and identifying code numbers and colors for the panels.

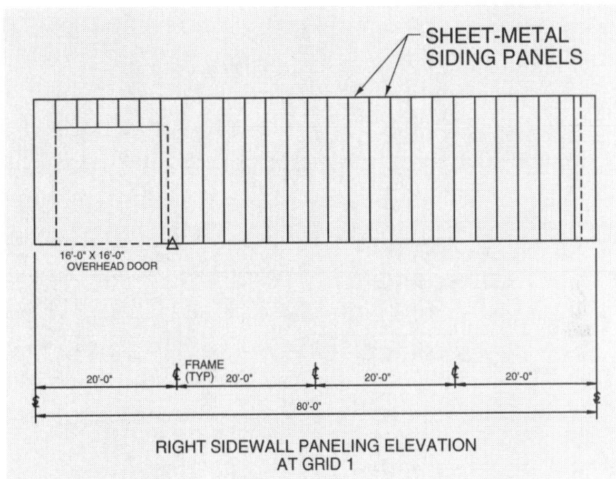

Figure 10-7. Sheet-metal siding panels are shown on erection plans and exterior elevations.

Sheet-metal wall panels are attached to purlins or girts with self-tapping and self-sealing screws. A *purlin* is a horizontal support member that spans across adjacent roof rafters or between beams, columns, or joists to carry intermediate loads, such as the roof deck or wall panels. A *girt* is a horizontal bracing member placed around the perimeter of a structure. Exterior wall panels may also be used for soffit and canopy coverings.

Siding. Wood siding is available in long, narrow pieces of tempered hardboard or solid wood, 4′ × 8′ siding panels, or wood shingles or shakes. Tempered hardboard siding is 4″ to 12″ wide, ½″ thick, and is available in smooth or textured finishes. Unlike solid lumber, tempered hardboard siding is not susceptible to warping and twisting. Tempered hardboard siding may be applied horizontally or at an angle.

Solid wood siding includes bevel, shiplap, and clapboard siding. **See Figure 10-8.** Bevel siding has a tapered cross-section. Bevel siding is 8″ to 12″ wide and tapered in thickness from approximately ¾″ at the bottom to ⅛″ at the top. It may be applied horizontally or at an angle.

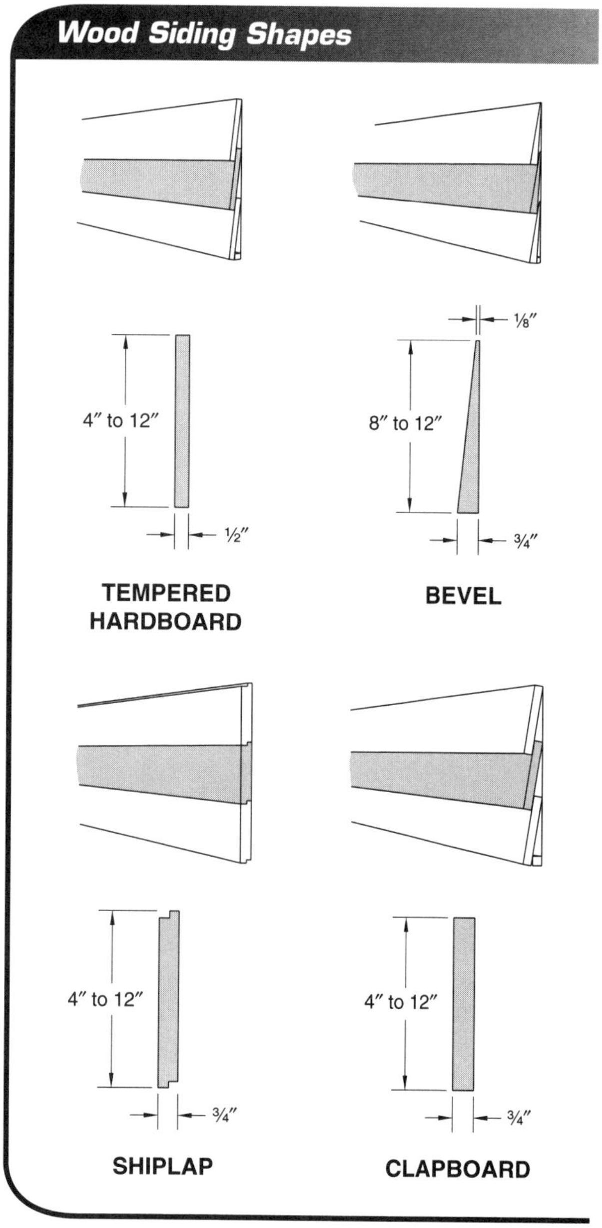

Figure 10-8. Wood siding is available in tempered hardboard and solid wood. Solid wood siding includes bevel, shiplap, and clapboard siding.

Shiplap siding has rabbeted edges that allow for a fitted overlap. It is designed to allow for the overlapping of pieces without increasing the thickness of the wall covering. Shiplap siding is 4″ to 12″ wide and ¾″ thick. It may be applied horizontally, vertically, or at an angle.

Clapboard siding has a consistent thickness and square edges. Clapboard siding is 4″ to 12″ wide and ¾″ thick. It may be applied horizontally or at an angle.

Wood siding applied horizontally is attached to exterior walls by starting at the bottom of the wall and proceeding to the top. Each row overlaps the previous row by approximately 1″. Wood siding may also be applied vertically with battens covering the joints or at an angle for a decorative appearance. Red cedar and redwood are the most common species of wood used for solid wood siding because they are more weather- and decay-resistant than other species.

Sheet siding is commonly manufactured in 4′ × 8′ panels that are ⅜″ to ⅝″ thick. The panels are made of tempered hardboard or plywood with a special grade of adhesive binder and face veneer. Panels are flat or patterned to simulate wood grain, reverse board-and-batten, stucco, and shiplap siding. A face veneer of cedar or other water-resistant wood may be applied to the plywood panels.

Wood shingles and shakes for vertical applications are made of red cedar. Wood shingles and shakes are flat or tapered in thickness from approximately 1¼″ at the bottom to ½″ at the top. Widths vary from 3″ to 14″ and lengths include 16″, 18″, 24″, and 32″. Wood shingles and shakes are applied to the exterior face of the wall over a layer of building wrap or asphalt-impregnated building paper. Each shingle or shake is applied separately, with successive shingles or shakes overlapping previous shingles or shakes to close horizontal and vertical joints.

Metal and vinyl exterior siding provides the appearance of horizontal lap or vertical wood siding. Metal siding is made of aluminum or prepainted steel. Metal and vinyl siding are available with smooth finishes or woodgrain textures and in a variety of colors. Metal and vinyl siding panels are approximately 1⁄16″ thick, 8″, 10″, or 12″ wide, and 12′-6″ long. Each panel may have one or more horizontal offsets in the face to make the panel appear as several smaller pieces. **See Figure 10-9.**

Fiber cement siding is a highly durable siding material made of lightweight concrete materials. Fiber cement siding components are manufactured in a wide variety of shapes, sizes, and colors and are designed to give the appearance of wood lap siding or other siding types.

> Special fasteners may be required for certain materials. The specifications should be checked to ensure the proper fasteners are included in the estimate.

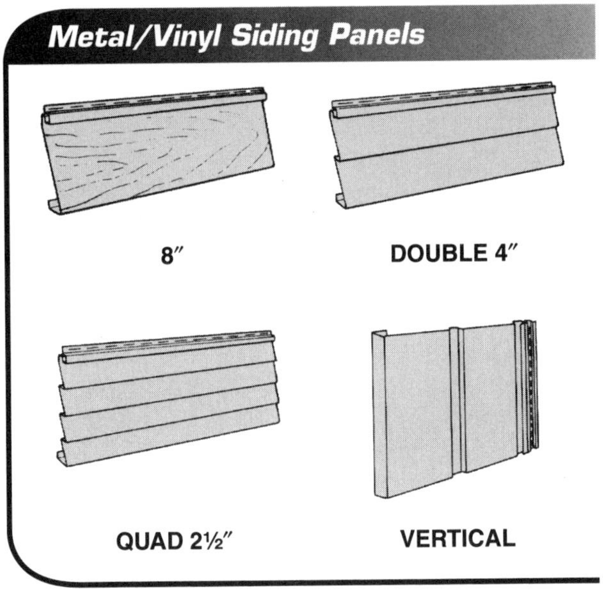

Figure 10-9. Interlocking siding panels made of metal or vinyl are used for exterior wall finishes.

EXTERIOR FINISH QUANTITY TAKEOFF

The primary sources of information for the estimator concerning exterior finish system materials include exterior elevations; specifications; details; mechanical plans for rooftop units, vents, and exhausts; and roof and floor plans. Exterior elevations provide information concerning roofing material and siding types. Specifications provide detailed information concerning architectural requirements for all building materials and any manufacturer references for items such as trim members, fireproofing materials, and sealants.

Details show applications of flashing, roof accessories, and sustainable roof structures. Mechanical plans show roof vents, roof vent flashing materials, and other exterior penetration finishes. Roof and floor plans may provide additional quantity information for exterior wall finishes. Most exterior finish material quantities, including roofing and siding materials, are determined by the area of coverage required (in square feet).

Roofing Takeoff

The method used for quantity takeoff and estimating of roofing materials depends on the type of material to be applied. Bituminous, elastomeric, composition shingles, wood shakes or shingles, and sheet metal each require slightly different calculations. For all roofs, estimators should take into account roof accessibility and job scheduling. In some geographic locations, climatic conditions may cause roof labor application costs to vary depending on the time of year. The estimator should also note the pitch of the roof and the overall height of the building. For steep-sloped roofs, additional fall protection equipment and application time may be required.

Roofing material quantities are based on the roof area (in square feet) or number of pieces (for sheet-metal roofs). **See Figure 10-10.** The roof area or number of pieces is entered in the proper location on the estimate sheet or spreadsheet. When using some estimating programs, roof length and width dimensions may be entered and the amounts of roofing materials calculated automatically. Costs can be based on historical data, either by individual item or by overall construction assemblies. Individual items, such as roof accessories, are counted by the estimator and entered into the proper ledger sheet, spreadsheet, or estimating program cell.

Figure 10-10. Quantities for roofing materials, such as shingles and underlayment, are based on the roof area.

Bituminous and Elastomeric. Bituminous and elastomeric roofing, commonly used for flat or slightly pitched roofs and for sustainable roofs, is taken off based on the roof area to be covered. The number of roof accessories and type of edge treatment can affect labor costs. **See Figure 10-11.** Additional installation cost allowances may be required for buildings with several roof openings and irregular or lengthy edge treatments that require additional labor time for proper sealing and flashing. Installation costs per square foot of roof are obtained from industry standard information or company historical data.

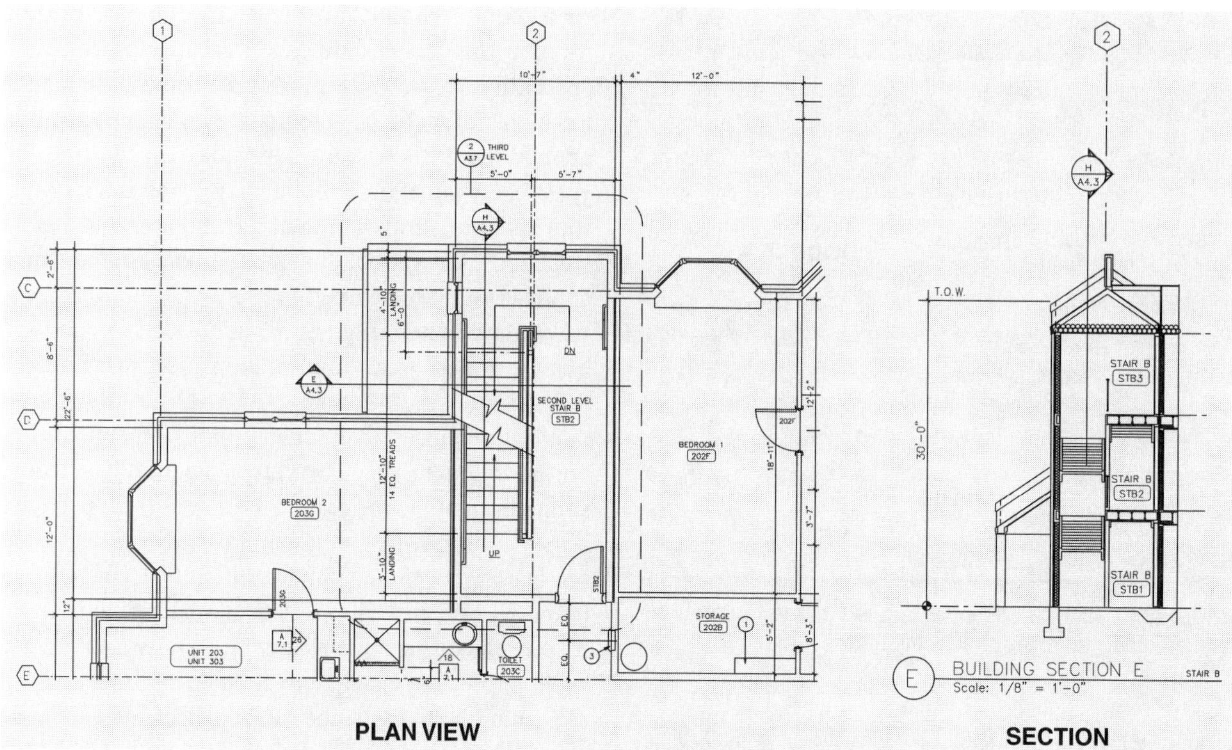

Figure 10-11. Plan views and sections are used to determine roof area.

Sheet Metal. An estimator must determine the width of the roofing sheets, the types of seams used along the edges, the types of sheets specified by the architect, and the total roof area (in square feet) to calculate the number of sheets required. The length of run of each piece of sheet-metal roofing must be compared to the length of roofing materials available to determine the number of seams necessary. The total length of the roof is divided by the width of a sheet-metal roof sheet to determine the number of sheets and the number of seams needed. **See Figure 10-12.** Costs of installation per square foot of roof coverage are obtained from industry standard information or company historical data.

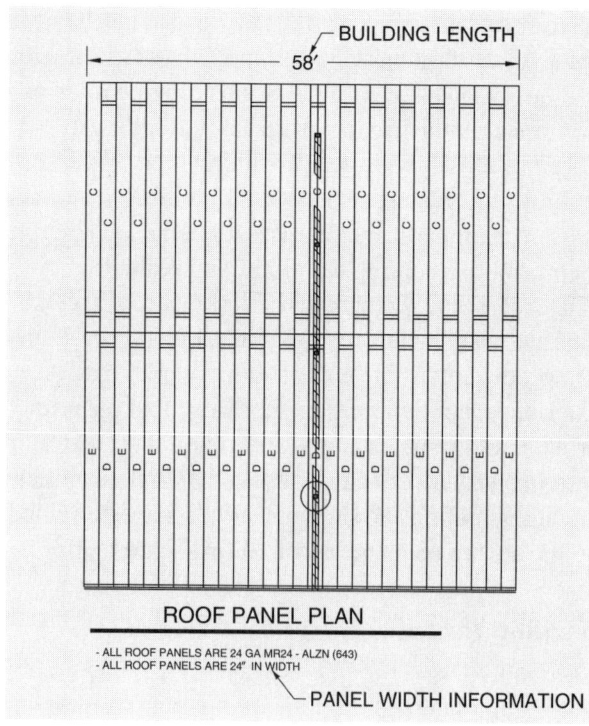

Figure 10-12. Metal roof panel quantities are based on the overall building length and the panel width.

Shingles and Tiles. Shingle and tile quantities are based on the number of squares of shingles or tiles required. A square provides 100 sq ft of coverage. Composition roof shingles are commonly packaged in bundles containing 33 sq ft. Three bundles of composition roof shingles or rolls of roll roofing material cover one square, or 100 sq ft. Each bundle of composition roof shingles covers 33 lf of ridge.

The area of the roof must be calculated before determining the number of squares of roofing material required. For sloped roofs, the roof length (in feet) is multiplied by the total rafter length (in feet) to determine the roof area (in square feet). Total rafter length is determined from floor plan dimensions and slope information on exterior elevations. The roof area is divided by 100 to determine the number of squares. A 10% waste factor is added for composition roof shingles; a 20% waste factor is added for slate shingles; and an 18% waste factor is added for tile. The lengths of any overhangs, ridges, valleys, and hips must also be included for an accurate estimate.

Determining Quantities of Roof Shingles

Determine the number of squares of composition shingles required to cover the gable roof of a building measuring 39′-0″ long with a total rafter length of 12′-0″.

1. Determine the roof area.

roof area = (*roof length* × *total rafter length*) × 2

roof area = (39′ × 12′) × 2

roof area = 936 sq ft

2. Determine the number of squares (without waste).

number of squares = *roof area* ÷ 100

number of squares = 936 sq ft ÷ 100

number of squares = 9.36 squares

3. Determine the total number of squares required.

total number of squares = *number of squares* × *waste factor*

total number of squares = 9.36 squares × 1.10

***total number of squares* = 10.3 squares; round to 11 squares**

An additional 2 bundles of shingles are added to cover the 39′ of ridge. Eleven squares plus 2 bundles of roof shingles are ordered. For other roofing materials, a similar amount of ridge material is added.

Quantities of wood shingles and shakes are usually expressed in squares or in bundles. Four bundles of wood shingles or shakes make up one square (100 sq ft). The minimum roof slope for wood shingles and shakes is a 4″ rise per foot of run. For wood shingles on a slope of 4 to 12 or steeper, the standard exposures are 5″ for 16″ shingles, 5½″ for 18″ shingles, and 7½″ for 24″ shingles. The standard exposure for wood shakes is 7½″ for 18″ shakes, 10″ for 24″ shakes, and 13″ for 32″ shakes.

To estimate the proper quantity of wood shingles or shakes, the pitch of the roof and the required shingle exposure should be determined first. In areas with severe climatic conditions, up to three layers of wood shingles or shakes may be required. To take off the amount of wood shingles or shakes for various roof pitches, an estimator multiplies the appropriate exposure factor by the roof area to be covered. **See Figure 10-13.**

WOOD SHINGLE TAKEOFF	
Roof Pitch	**Exposure Factor**
4:12	1.05
5:12	1.085
6:12	1.12
8:12	1.20

Figure 10-13. Wood shingle exposure factor increases with roof pitch.

After the quantity of wood shingles or shakes is determined, allowances are made for starter courses and double courses at eaves. One square of hand-split shakes provides approximately 120 lf of starter course. Approximately one square of cedar shingles and two squares of hand-split shakes are allowed for every 100 lf of valley. At standard exposure, it takes approximately 2½ lb of nails per square of cedar shingles and hand-split shakes. Installation costs for roof shingles vary with the type of material applied. Installation costs are calculated based on company historical data or standard industry information.

Roof Pavers and Roof Accessories. Roof plans provide information concerning the area of coverage (in square feet) for roof pavers. Estimators use length and width calculations to determine the number of pavers required for an area of the entire roof. Roof accessories are calculated on an individual basis. The quantity of each roof accessory is determined from roof plans, specifications, detail drawings, or a schedule provided by the architect. Installation costs are based on standard labor tables or company historical data.

Sustainable Roofs. Estimators take off for sustainable roofs based on square foot calculations similar to other flat or slightly sloping roof surfaces. Additional calculations are made to include quantities and labor costs for drainage materials, planting media, and surface plants as shown on detail drawings or indicated in the specifications.

Wall Takeoff

Quantities of exterior wall covering materials are based on the coverage area (in square feet). Deductions are made for large wall openings such as doors and windows. Trim members included in the material takeoff depend on the type of exterior wall covering and the architectural design. Wall height and accessibility are key elements in accurately pricing the labor. Provisions may need to be made for the rental, erection, and dismantling of scaffolding or other worker lifts and for safety restraints required for work crews working on buildings several stories high.

Exterior Insulation and Finish Systems. Exterior insulation and finish system (EIFS) wall coverings include takeoff of insulation board, reinforcing mesh, and various coats of cementitious materials. These materials are taken off according to the square feet of coverage required, resulting in the number of required sheets of insulation board, rolls of reinforcing mesh, and gallons of coating materials. Trim members for edge and opening treatments are taken off by the linear foot.

Roof pavers are set in place on a roof deck to provide a walking surface and to protect waterproof roofing materials.

Architects may use EIFS materials for the creation of elaborate geometric designs. Estimators should include additional labor costs when complex layout and execution of designs are required. Some estimating programs may allow for a quick takeoff of EIFS materials and labor, where the program leads the estimator through a series of dimension questions that connect to a database that determines labor and material quantities. **See Figure 10-14.** A final review of all quantities should be performed to ensure all job-site conditions have been considered.

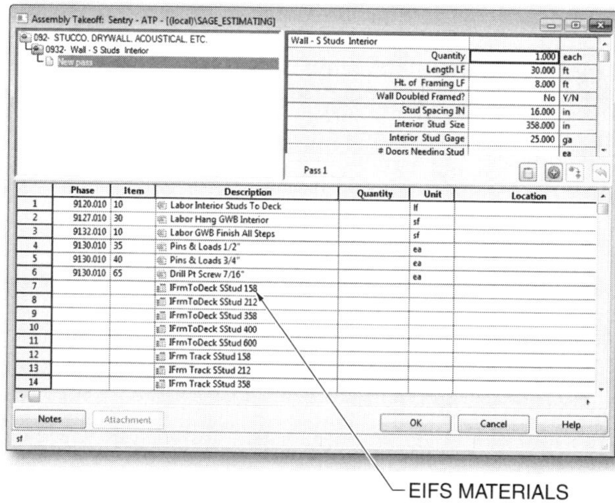

Figure 10-14. The assembly takeoff function in an estimating program allows for a quick takeoff of EIFS materials and labor.

Sheet Metal. As with sheet-metal roofing, the length of available metal siding panels is compared to the required wall height. The amount of linear feet of wall to be covered is divided by the width of each metal panel to determine the number of panels required. Metal trim members or anchor clips may also be required by the manufacturer. **See Figure 10-15.** Specifications and manufacturer notes are checked to ensure all items and components are included for metal siding panel installation.

Siding. All siding types and styles appear on exterior elevations with various symbols and architectural notes. Manufacturer codes and identification numbers may appear on the elevations and in the specifications. The area of a wall covered with siding is calculated in square feet and divided by 100 to determine the number of squares of material required.

Sheet Metal Siding Panels

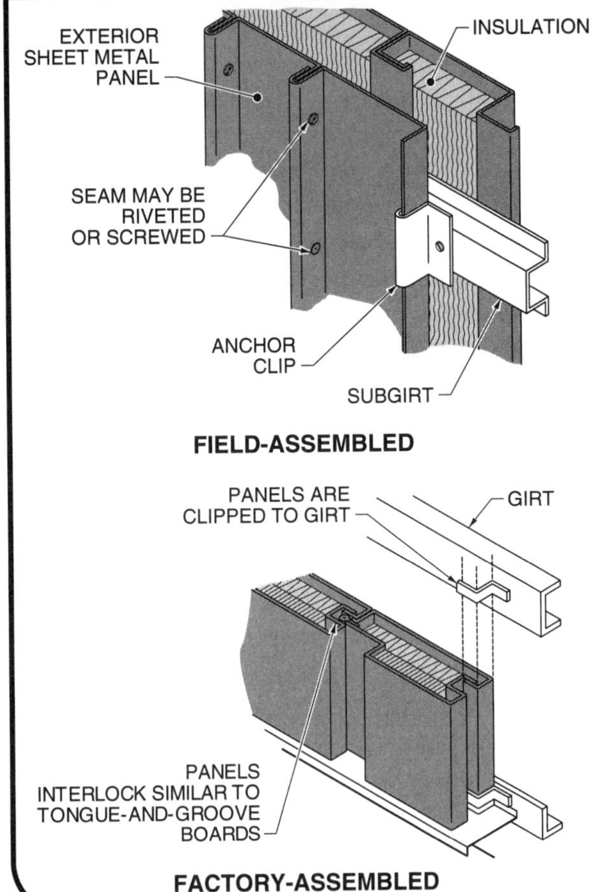

Figure 10-15. Sheet-metal siding panels may be field-assembled using rivets or screws or factory-assembled using interlocking panels.

Determining Quantities of Siding

Determine the number of squares of siding required to cover an exterior wall measuring 35′-0″ long by 9′-0″ high.

1. Determine the wall area.

wall area = *wall length* × *wall height*

wall area = 35′ × 9′

wall area = 315 sq ft

2. Determine the number of squares (without waste).

number of squares = *wall area* ÷ 100

number of squares = 315 ÷ 100

number of squares = 3.15 squares

3. Determine the total number of squares required.

total number of squares = *number of squares* × *waste factor*

total number of squares = 3.15 squares × 1.05

***total number of squares* = 3.31 squares; rounded up to 4 squares**

Wood siding is packaged and delivered to a job site in bundles containing one square each. Metal siding is bundled in one-square packages with expanded foam backing board, or in two-square packages without foam backing board. Vinyl siding bundles contain two squares of siding. Fiber cement siding is ordered by the square. In addition to the siding, the linear feet of each type of trim member and soffit material is determined from the elevations. Unique exterior components, such as injected-foam trim members or design accents, are taken off individually.

Reducing the air infiltration from the outside to the inside of a building can significantly reduce the convective heat transfer rate. This increases the effectiveness of insulation placed in the walls and ceiling of buildings.

THERMAL SYSTEM MATERIALS AND METHODS

Thermal systems are generally designed to perform within ambient temperature limits. Thermal systems are designed to control temperatures that affect the comfort of the building occupants, deter condensation, and reduce heat transmission to improve energy use within the structure. Thermal systems may also be used to add structural strength, support surface finishes, and reduce water vapor and noise transmission.

Insulation

Insulation is material used as a barrier to inhibit thermal and sound transmission. The North American Insulation Manufacturers Association (NAIMA) has adopted a uniform method of rating the effectiveness of all thermal system insulation when installed according to manufacturer instructions. The resistance of a specific thickness and type of insulation is indicated by R value.

R value is the unit of measure for resistance to heat flow. Higher R values indicate higher heat-flow resistance. For example, an insulating material with an R value of 12 (R-12) offers three-fourths as much resistance as an insulating material with an R value of 16 (R-16). The resistance of any thickness of material is equal to its resistivity per inch multiplied by its total thickness.

Materials. Insulating materials are categorized based on their structure and form. Insulation structures include cellular, granular, and fibrous. Cellular insulation is composed of small individual cells separated from each other. Cellular insulation materials include polystyrene and polyurethane. Granular insulation is composed of small nodules that contain voids or hollow spaces. Granular insulation materials include vermiculite, perlite, and cellulose. Fibrous insulation is composed of small-diameter fibers. Fibrous insulation materials include rock wool, slag wool, and glass (fiberglass).

Each of these materials has its own density, R value, water vapor permeability, and dimensional stability properties. **See Figure 10-16.** Insulation is manufactured in various forms for specific uses including loose fill; flexible blankets and sheets; semirigid blankets and batts; rigid board, blocks, and sheets; tapes; spray-on fibers and cements; and foams.

Applications. The type and amount of thermal insulation used depends on the geographic location and intended use of the building. For example, a parking garage has very different insulation requirements than a medical facility. Architects and engineers use standard tables and charts to determine minimum R values for common building usage in various zones of the United States. **See Figure 10-17.**

Water Vapor and Fire Protection

Waterproofing is the treatment of a material that makes it impervious to water. Waterproofing prevents the passage of water through the walls and floors of a building. *Dampproofing* is the treatment of a material that makes it moisture-resistant. Dampproofing prevents the passage of moisture, but does not prevent the effects of hydrostatic pressure. *Hydrostatic pressure* is pressure exerted by a fluid at rest.

The choice of the proper waterproofing or dampproofing system depends on the hydrostatic conditions. Applications for waterproofing and dampproofing include below-grade walls; under floors; under the walking surfaces of roofs, walls, and floors above grade; balconies; concrete canopies; pools; and around floor and roof drains. Waterproofing and dampproofing systems include membranes, hydrolithic coatings, concrete admixtures, bentonite materials, and sheet metal.

INSULATION MATERIAL PROPERTIES

Material	Density*	Resistance†	Water Vapor Permeability‡	Dimensional Stability
Polystyrene	0.8 to 2.0	3.8 to 4.4	1.2 to 3.0	None
Polyurethane	2.0	5.8 to 6.2§	2.0 to 3.0	0% to 12% change
Vermiculite	4.0 to 10.0	2.4 to 3.0	High	None
Perlite	2.0 to 11.0	2.5 to 3.7	High	None
Cellulose	2.2 to 3.0	3.2 to 3.7	High	Settles 0% to 20%
Rock or slag wool	1.5 to 2.5	3.2 to 3.7	100	None
Fiberglass	0.6 to 1.0	3.16	100	None

* in lb/cu ft
† R value
‡ in perms per in.
§ aged unfaced or spray applied

Figure 10-16. Insulation materials are selected for an application based on their density, resistance, water-vapor permeability, and dimensional stability.

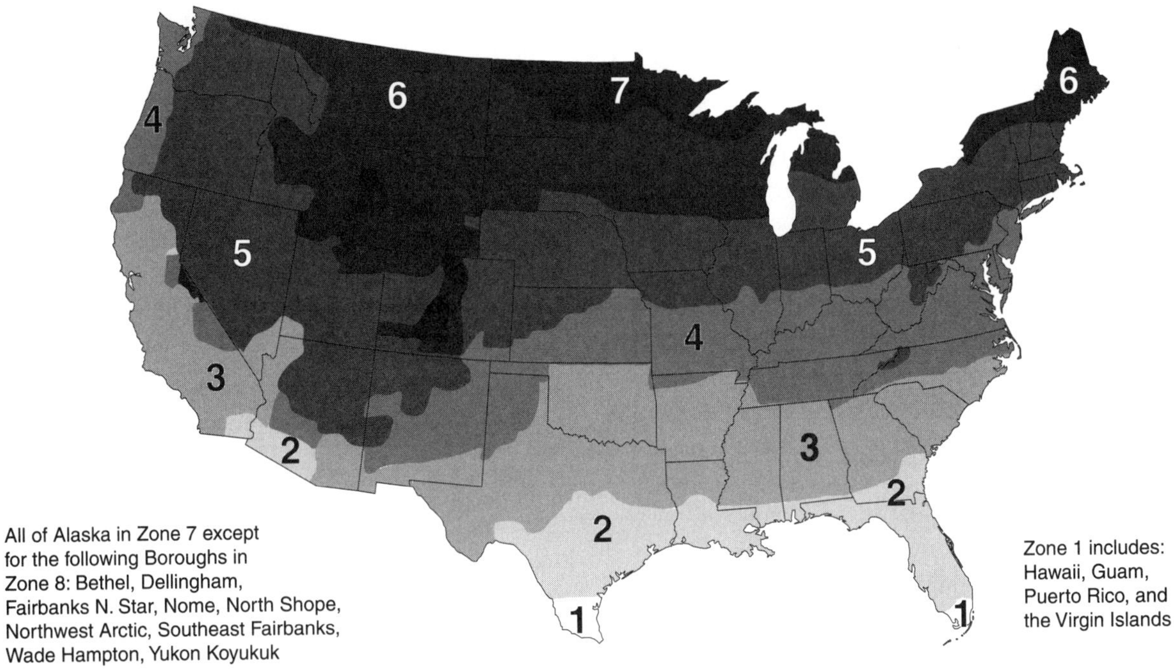

RECOMMENDED MINIMUM INSULATION R VALUES					
Zone	Attic	Cathedral Ceiling	Wall		Floor
			Cavity	Insulation Sheathing	
1	R30 to R49	R22 to R15	R13 to R15	None	R13
2	R30 to R60	R22 to R38	R13 to R15	None	R13, R19–R25
3	R30 to R60	R22 to R38	R13 to R15	None	R25
4	R38 to R60	R30 to R38	R13 to R15	R2.5 to R6	R25 to R30
5	R38 to R60	R30 to R38	R13 to R15	R2.5 to R6	R25 to R30
6	R49 to R60	R30 to R60	R13 to R21	R5 to R6	R25 to R30
7	R49 to R60	R30 to R60	R13 to R21	R5 to R6	R25 to R30
8	R49 to R60	R30 to R60	R13 to R21	R5 to R6	R25 to R30

Figure 10-17. Recommended minimum insulation R values relate to climatic conditions in specific geographic zones.

Membranes include built-up bituminous layers and rubber or PVC sheets with sealed joints. Hydrolithic coatings include materials such as tar, plastics, plaster, or cement that is troweled or sprayed onto building surfaces. Concrete admixtures include various chemical compounds added to concrete during placement to create an impervious surface. Bentonite is a natural compound that expands when exposed to moisture, allowing it to seal joints when water is present. Sheet-metal materials include coated aluminum, various alloys, and galvanized sheets used for roof valleys and ridges, gutters, downspouts, and flashing.

Fire protection materials include a variety of chemical compounds that are sprayed, troweled, or caulked into place. Some panel materials, such as calcium silicate and slag fiber, are formed into sheets and applied as boards or panels to provide fire resistance. These materials resist high temperatures and retain their integrity by not allowing the flow of air, thus inhibiting the spread of smoke and fire.

Flashing. *Flashing* is material installed at the seam or joint of two building components to inhibit air, water, or fire passage. Types of flashing include base, cap, concealed, and exposed flashing. Flashing typically consists of job-formed and preformed sheet-metal materials. Base flashing is installed at the lowest meeting point of a vertical and horizontal surface. Base flashing is a continuation of roofing membranes that are turned upward at the edge. Cap flashing is wrapped around the top of a wall or other vertical projection to provide moisture protection. Cap flashing overlaps base flashing.

Joint Sealants. *Joint sealant* is material installed between two members to seal the seam between the members. Locations such as construction joints require an allowance to compensate for expansion and contraction due to variations in climate and differences in building materials. A solid substrate ensures a properly sealed joint.

A *substrate* is the underlying surface of a finish material. A primer may be applied to ensure proper adhesion of the substrate and the joint sealant. A joint filler or backer rod may be used to control the depth of the sealant and permit proper installation of the sealant. A bond breaker may also be applied to prevent adhesion of the sealant to surfaces, which would reduce the performance of the sealant.

Common elastomeric sealants include acrylic materials, polysulfide, polyurethane, and silicone. Acrylic materials are one-part solvent-release compounds that cure by evaporation of solvents. Polysulfide, polyurethane, and silicone are one- or two-part chemical compounds that cure by reaction with moisture or oxygen in the air. Variables in these materials include the percent of solids, the curing process and characteristics, primers, application temperature ranges, hardness, and setting time.

THERMAL SYSTEM QUANTITY TAKEOFF

Insulation, waterproofing, and dampproofing materials are installed after structural and framing members are in place. Insulation installation must be coordinated and scheduled with the project manager to ensure that there is accessibility to the required work areas and that installation is completed prior to the application of surface finish materials.

Estimators refer to specifications and details to determine required R values and the types of insulation, water vapor materials, and fire protection coatings needed in each area. Calculations are based on cubic feet, square feet, linear feet, or liquid measure, depending on the insulation and application. The square feet of coverage required are also applicable to thermal protection such as waterproofing and dampproofing. Linear feet calculations are used for various exterior trim members, flashing, and joint sealants.

Insulation Takeoff

Estimators use different takeoff methods depending on the insulation form. Insulation forms include rigid boards, blocks, and sheets, loose fill, flexible blankets and sheets, and foam.

Rigid insulation material quantities are calculated by the area to be covered, similar to floor or wall sheathing or siding. The length (in feet) is multiplied by the height or width (in feet) to calculate the area of coverage required (in square feet).

One type of loose fill insulation material is dry granules or fibers. Loose fill insulation is poured or blown into place. Loose fill insulation quantities are measured in cubic feet or cubic yards of material. Estimators determine the area to be covered (in feet) and the depth or thickness (in feet) to determine the quantity of loose fill insulation.

Determining the Volume of Loose Fill Insulation

Determine the volume (in cubic yards) of loose fill insulation required for a ceiling measuring 45′-0″ × 30′-0″ with insulation blown to a depth of 8″.

1. Determine the volume in cubic yards.

volume (in cubic yards) = *thickness (in feet)* × *width (in feet)* × *length (in feet)* ÷ 27

volume (in cubic yards) = 0.67 × 30′ × 45′ ÷ 27

volume (in cubic yards) = 904.5 ÷ 27

***volume (in cubic yards)* = 33.5 cu yd; rounded to 34 cu yd**

Quantities of flexible insulation for frame walls are expressed in square feet. Estimators take off the surface area to be insulated and deduct large openings such as windows and doors. Flexible insulation for walls is manufactured in blankets or batts that are 3½″, 5¼″, or 6½″ thick, or sheets that are 1″, 2″, 3″, or 4″ thick and designed to fit between studs. Each batt is precut to 24″, 48″, or 96″ long or may be obtained in blanket form in a continuous roll.

The wall area is calculated in the same manner as for sheathing, by multiplying the wall length by the wall height. When using some estimating programs, the installer may include flexible insulation material and installation in common wall unit assemblies.

For masonry walls, the takeoff process is the same as for frame walls. Flexible blanket or sheet insulation is often applied to the interior of masonry walls. Blanket insulation is stapled to furring strips or studs and insulation sheets are glued to the masonry walls to fill the space between the masonry units and the interior finish materials.

Foam insulation is purchased in liquid form and applied using compressed air or gas. Estimators should check manufacturer specifications concerning the amount of coverage or volume per gallon for various foam insulation materials. The total coverage area is divided by the coverage or volume per gallon to determine material quantities. Installation costs include labor, material transportation, and any pumping or placing equipment.

> Foam insulation is a chemical foam that is applied by through spraying. For new construction, the foam is sprayed into open wall cavities. The foam expands to fill the entire cavity and the excess is cut away with long blades.

Water Vapor and Fire Protection Takeoff

Foundation plans, floor plans, elevations, roof plans, specifications, and details are reviewed to check for locations of waterproofing, dampproofing, and fire protection materials. Special items that are taken off individually include smoke vents and joint gaskets that provide fire protection where pipes penetrate floors, roofs, and walls. Other waterproofing, dampproofing, and fire protection materials are taken off by the square or linear foot.

Floors. Concrete slab-on-grade floors commonly require waterproofing and dampproofing materials under the slab. The slab area is determined from dimensions on foundation plans and floor plans. Detail floor sections indicate the types of moisture protection and insulation used. **See Figure 10-18.**

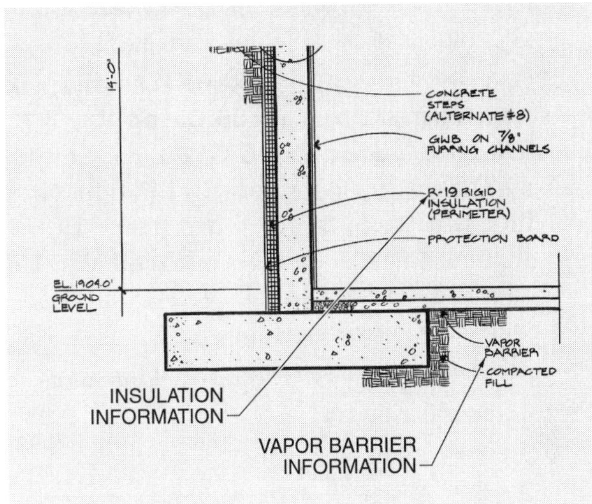

Figure 10-18. Floor sections are used to determine the moisture protection and insulation required.

For above-grade floors, insulation may occasionally be specified on the underside of the floor for sound or thermal protection. These materials are taken off, and cost estimates for materials and labor installation are calculated by the square foot. Joint sealant quantities and types of materials for joints in pavement or concrete floors are determined by the number of linear feet of joints to be sealed.

Roofs. Estimators use floor and roof plans along with details to take off other items such as flashing and sheet-metal roof specialties such as gutters, downspouts, and roof vents. Flashing material and installation costs are based on the number of linear feet of flashing required. Gutter and downspout quantities are also determined by the linear foot. Roof vents and other roof specialties are counted individually.

Walls. Waterproofing and dampproofing for below-grade foundation walls is shown on foundation plans and exterior wall sections. Membrane or rigid materials for this application are taken off based on the coverage area (in square feet). Cementitious or other waterproofing coatings are taken off as square feet per gallon of material.

SAGE TIMBERLINE ESTIMATING . . .

ASSEMBLY ITEM ADDITION

In the Sage Timberline Estimating program, adding, deleting, or substituting items in an assembly may be necessary for unique applications. Any changes made to an assembly apply to the current estimate only and do not affect the assembly in the database.

1. Create a new estimate named **Parker**. Specify the database in the **Sample** Commercial GC folder for use with this estimate.
2. Open the **Assembly Takeoff** screen by picking the **Assembly Takeoff/Review** button in the toolbar. Double-click **06 Carpentry & Millwork** and then double click **0640- Frame Wall – Int. Stud 2×4-6 GWB**. Enter the following dimensions for the assembly. Enter **1** for the **Quantity**, 24 for **Length LF**, **8** for **Ht. of the Wall LF**, **16** for **Stud Spacing IN**, **24** for **Stud Size**, **3** for **# Corners or Ends**, **2** for **# Runs of Blocking**, **No** for **Is Bot. PL Treated?**, **8** for **Ht. Gyp Wall Brd LF**, **12** for **Gyp Wall Board Thick**, and **2** for **# of Layers to Hang**.
3. Pick the **Add Pass** button.
4. Display the item list by right-clicking in the assembly list pane and selecting **List Items** from the shortcut menu.

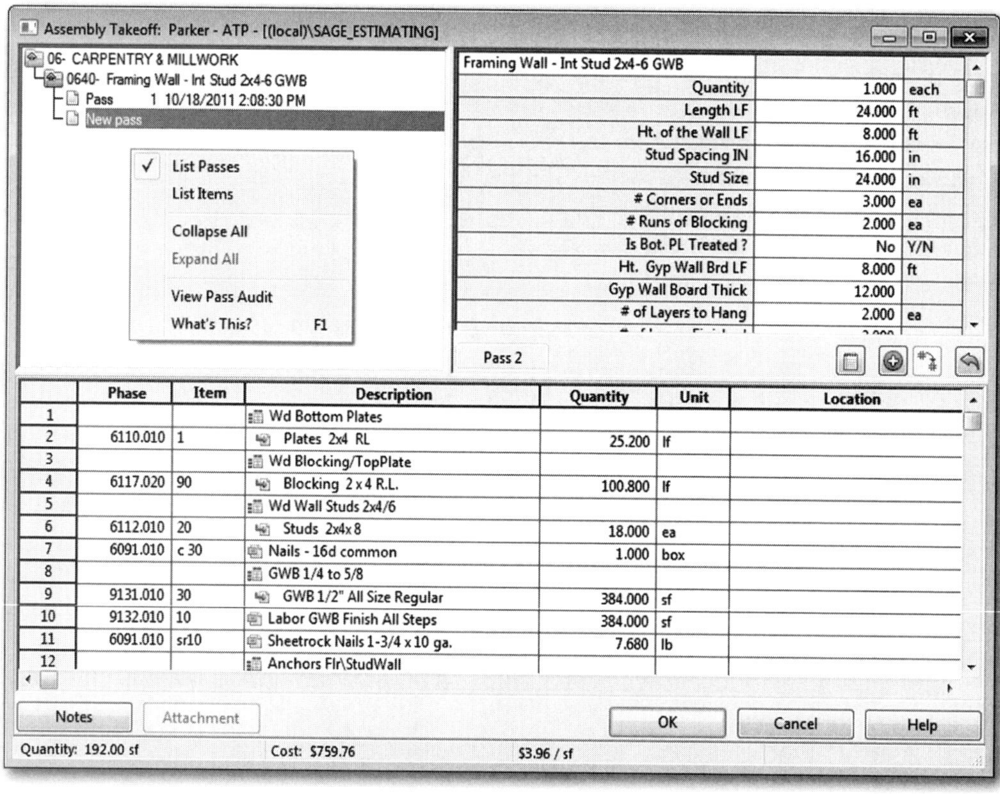

5. Add an item to the assembly by double-clicking **7000.000 Thermal & Moisture Prot.**, **7210.030 Insulation: Sound Blanket**, and **30 Sound Blanket 16″ × 3⅝″**.

. . . SAGE TIMBERLINE ESTIMATING

6. Assign a formula to the **Quantity** cell for the insulation by right-clicking the cell and selecting **Edit Formula...** from the shortcut menu. Navigate through the list and select **aSF SAB 16″** and pick **OK**.
7. Input the value **8′** for the assembly addition in the **Height LF** dialog box (the values may already be prefilled based on information previously input).

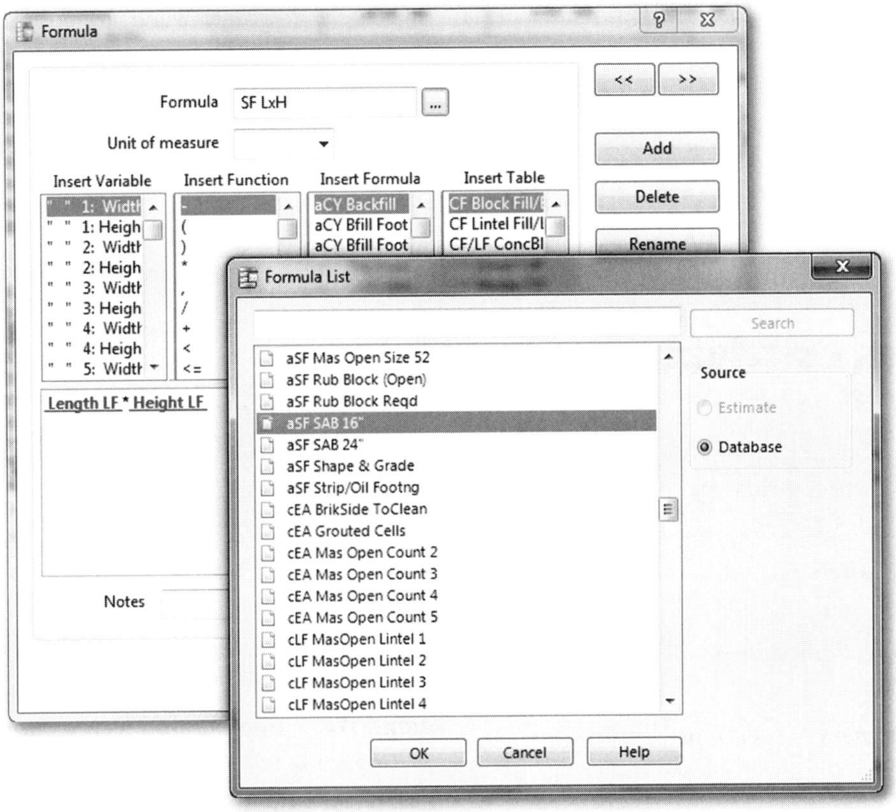

8. Send the calculated value to the item grid pane by picking the **Add to Qty** button.
9. Pick the **OK** button to close the **Assembly Takeoff** dialog box and send the takeoff information to the spreadsheet.

Quick Quiz®

Quick Quiz®

Refer to the CD-ROM for the Quick Quiz® questions related to chapter content.

Key Terms

Illustrated Glossary

- *bituminous roofing*
- *building paper*
- *composition shingle*
- *dampproofing*
- *elastomeric roofing*
- *exterior insulation and finish system (EIFS)*
- *flashing*
- *girt*
- *hydrostatic pressure*
- *insulation*
- *joint sealant*
- *metal roofing*
- *purlin*
- *roof*
- *roofing tile*
- *roof paver*
- *R value*
- *shake*
- *shingle*
- *substrate*
- *waterproofing*

Web Links

Web Links

Adhesive and Sealant Council, Inc.
www.ascouncil.org

Asphalt Roofing Manufacturers Association
www.asphaltroofing.org

Cedar Shake and Shingle Bureau
www.cedarbureau.org

Cellulose Insulation Manufacturers Association
www.cellulose.org

EIFS Industry Members Association
www.eima.com

Firestop Contractors International Association
www.fcia.org

National Insulation Association
www.insulation.org

National Roofing Contractors Association
www.nrca.net

National Tile Roofing Manufacturers Association
www.tileroofing.org

North American Insulation Manufacturers Association
www.naima.org

Polyisocyanurate Insulation Manufacturers Association
www.pima.org

Roof Coating Manufacturers Association
www.roofcoatings.org

Sealant, Waterproofing and Restoration Institute
www.swrionline.org

Sheet Metal Workers International Association
www.smwia.org

Single-Ply Roofing Institute
www.spri.org

Structural Insulated Panel Association
www.sips.org

Thermal and Moisture Protection 10

Review Questions

Estimating

______ 1. A(n) ___ is the covering for the top exterior surface of a building or structure.

______ 2. ___ is felt material saturated with tar to form a waterproof sheet.

______ 3. A(n) ___ is a thin piece of wood, asphalt-saturated felt, fiberglass, or other material applied to the surface of a roof or wall to provide a waterproof covering.

______ 4. A(n) ___ is a hand-split wood roofing material.

______ 5. Roofing ___ is clay, slate, or lightweight concrete material installed as a weather-protective finish on a roof.

______ 6. A(n) ___ is a flat, precast concrete unit that is set in place on a roof deck to provide a walking surface and to protect waterproof roofing materials.

______ 7. A(n) ___ system is an exterior finish system composed of a layer of exterior sheathing, insulation board, reinforcing mesh, a base coat of acrylic copolymers, and a textured finish.

T F 8. A purlin is a horizontal support member that spans across adjacent roof rafters.

T F 9. A girt is a horizontal bracing member placed around the perimeter of a structure.

T F 10. R value is the unit of measure for resistance to heat flow.

______ 11. ___ is material used as a barrier to inhibit thermal and sound transmission.

______ 12. ___ insulation is composed of small individual cells separated from each other.

______ 13. ___ is the treatment of a material that makes it impervious to water.

______ 14. ___ is the treatment of a material that makes it moisture-resistant.

______ 15. ___ pressure is pressure exerted by a fluid at rest.

______ 16. ___ is material installed at the seam or joint of two building components to inhibit air, water, or fire passage.

______ 17. ___ is the underlying surface of a finish material.

______ 18. ___ roofing is roofing made of a pliable synthetic polymer.

______ 19. ___ is the amount a shingle or shake is visible after installation.

Short Answer

1. Identify five types of drawings on a set of prints where information related to thermal and moisture protection is located.

2. Identify at least three factors that should be taken into account when estimating roofing projects.

3. Describe the difference between cellular, granular, and fibrous insulation.

Thermal and Moisture Protection 10

Activities

Activity 10-1—Ledger Sheet Activity

Refer to Print 10-1A, Print 10-1B, and Quantity Sheet No. 10-1. Take off the quantities of composition shingles required (in squares) including hips. Calculate the building roof perimeter as square to obtain the material estimate. Do not make deductions for the cupola or entrance gable.

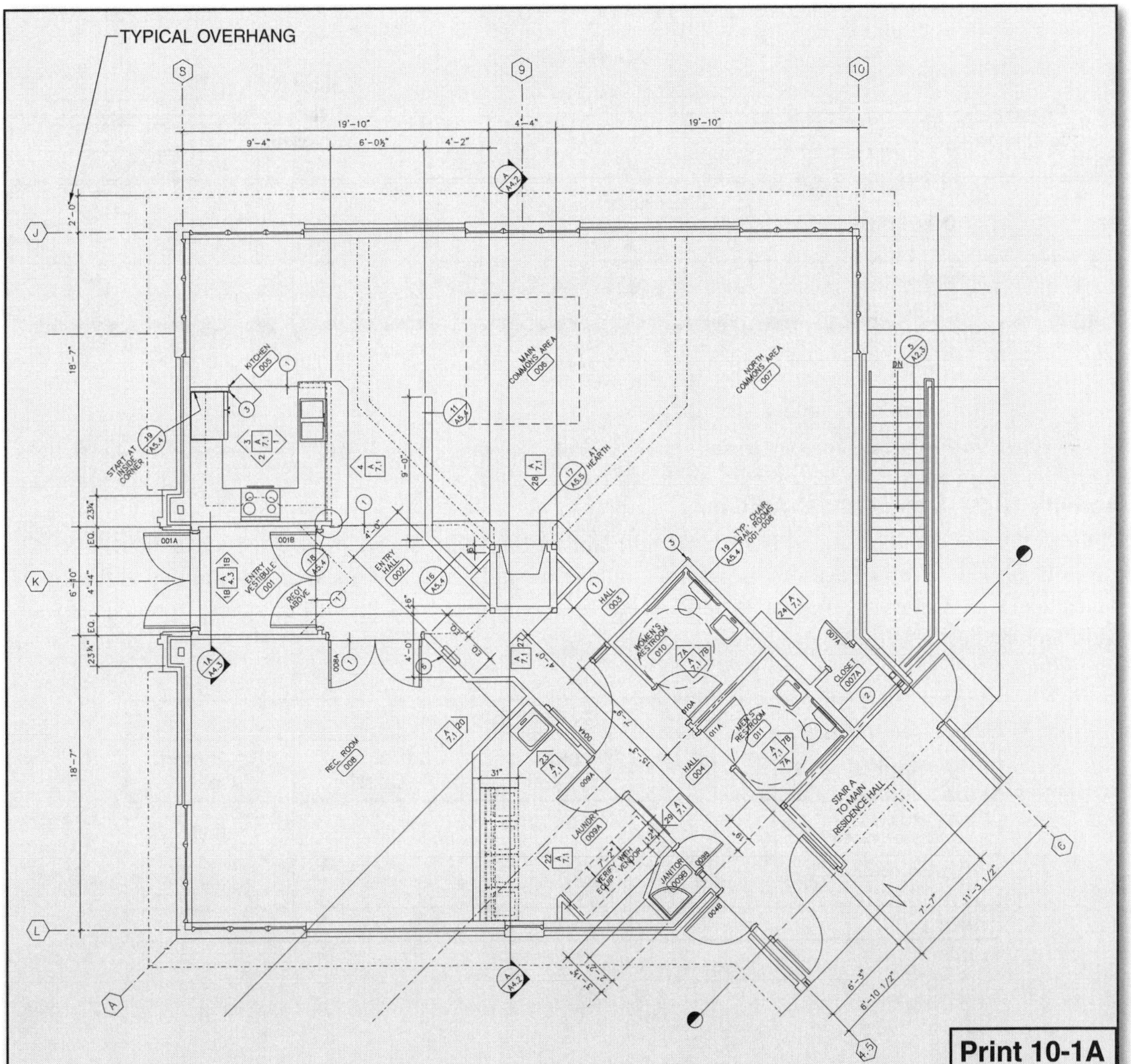

Print 10-1A

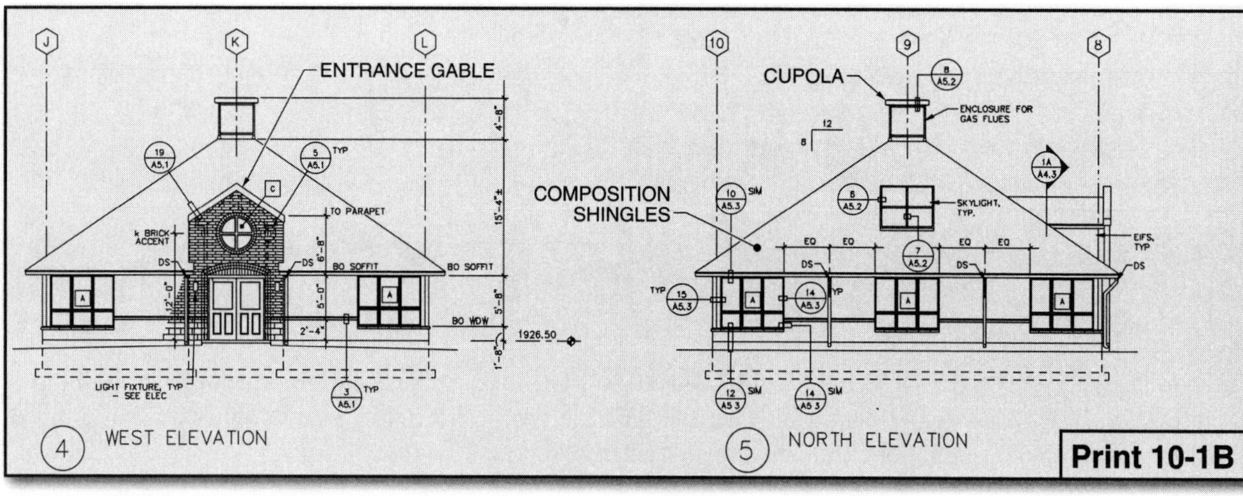

Print 10-1B

QUANTITY SHEET

Project: ____________________ Sheet No. 10-1

Estimator: ____________________ Date: ____________________

Checked: ____________________

No.	Description	Dimensions										
		L	W	D		Unit		Unit		Unit		Unit
	Composition shingles											

Activity 10-2—Spreadsheet Activity

Refer to the cost data below, and Print 10-2 and Estimate Summary Spreadsheet No. 10-2 on the CD-ROM. Take off the square feet of elastomeric roofing, number of walk pads, and linear feet of flashing at the top of the parapet wall. Determine the total material, labor, and equipment costs for the elastomeric roofing, the walk pads, and the flashing.

COST DATA				
Material	**Unit**	**Material Unit Cost***	**Labor Unit Cost***	**Equipment Cost***
Elastomeric roofing, 50 mils, reinforced	sq ft	3.08	0.81	0.15
Walk pads, concrete, 2″ thick	sq ft	1.53	1.64	—
Flashing	lf	0.36	0.61	—

* in $

Activity 10-3—Estimating Insulation

Refer to Print 10-3, Specifications 10-3, and Estimate Summary Spreadsheet No. 10-3 on the CD-ROM. Perform an item takeoff for insulation for the ceiling of the building. Use the perimeter of the building dimensions in your calculations and assume 6″ of insulation. Determine the volume of insulation required (in cubic yards), material cost, and labor cost for the ceiling. The material cost is $35.46 per cubic yard and the labor cost is $15.00 per cubic yard.

Activity 10-4—Estimating Finish Fascia

Refer to Print 10-4A, Print 10-4B, and Estimate Summary Spreadsheet No. 10-4 on the CD-ROM. Determine the length (in linear feet) of the "Series II" fascia system required for the front, left, and right sides of the building.

Openings 11

Key Concepts

- Items considered in takeoff for doors include frames, doors, and hardware.
- Labor costs for door installation are based on the average time required to install each type of door frame, hang various types of doors, and install the necessary hardware.
- Information required for window quantity takeoff includes the grade, thickness, and type of glazing to be used at each location, type of each window, and quantity and size of the glazing.
- Depending on the height of the building and the type of window system, scaffolding or lifting equipment costs may be incurred for window installation.

Introduction

Division 08 of the CSI MasterFormat™ includes information about openings, such as door and window materials and methods. Metal and wood door descriptions include information such as fire ratings, core materials, and finishes. Specialty door types include fire, access, coiling, folding and sliding, overhead, and entrance doors. Materials for window construction include various metals, woods, plastics, and glass. Window frames and glazing are commonly set in place after construction for structural members has been put in place.

DOORS

Information about door materials and methods is part of Division 08 of the CSI MasterFormat™ and includes items such as metal doors (08 11 00), wood doors (08 14 00), plastic doors (08 15 00), and entrances and storefronts (08 41 00). Metal and wood doors are described in the specifications with their fire rating, core materials, and finishes. Estimators review door information on floor plans, in specifications, and on elevations to determine the type of door and frame, door finish, and hardware.

Metal door finishes include primer and finish paint coatings. Wood door finishes include veneers of various species and qualities and the application of various trim and finish materials such as moldings, panels, and lights. The specifications or prints themselves commonly contain a door schedule providing detailed information about all doors on a construction project. **See Figure 11-1.**

Standard door sizes are expressed with the door width stated first, followed by the height and then the thickness. Door widths are sized in 2″ increments beginning at 1′-6″. Common swinging door heights are 6′-8″ and 7′-0″. The most common swinging and sliding door thicknesses are 1⅜″ and 1¾″. Swinging door information also indicates the hand of the door, which indicates the direction of door swing. Details are provided by the architect for special door applications with nonstandard widths, heights, and thicknesses. Common door types include flush, panel, louvered, view, and full view. **See Figure 11-2.** Flush doors contain no lights and are covered with metal or wood on both sides. A *light* is a pane of glass or translucent material in a door. Panel doors are composed of a series of rails and stiles that form frames for panels. Louvered doors contain a frame with vents that allow air passage. View doors contain an opening that is filled with a clear glass or plastic light. Full view doors are made primarily of a large pane of glass supported by a wood or metal frame.

> Temporary doors may need to be installed at building entrances to secure the building during most of the construction. This prevents damage to the permanent building doors while allowing access for construction materials.

DOOR SCHEDULE			
Number	**Unit**	**Type**	**Remarks**
1	2′-8″ × 6′-8″ × 1⅜″	fl hollow core	—
2	2′-8″ × 6′-8″ × 1⅜″	fl hollow core	—
3	3′-0″ × 6′-8″ × 1¾″	solid core	—
4	2′-8″ × 6′-8″ × 1⅜″	fl hollow core	—
5	2′-6″ × 6′-8″	bifold	2′-6″ opening
6	2′-6″ × 6′-8″ × 1⅜″	fl hollow core	—
7	2′-8″ × 6′-8″ × 1¾″	fl solid core	—
8	3′-0″ × 6′-8″	bifold	3′-0″ opening
9	3′-6″ × 6′-8″	bifold	3′-6″ opening
10	5′-0″ × 6′-8″	alum siding	5′-0″ opening
11	2′-4″ × 6′-8″ × 1⅜″	fl hollow core	—
12	2′-8″ × 6′-8″ × 1⅜″	fl hollow core	—
13	4′-0″ × 6′-8″	alum siding	12′-0″ unit
14	2′-8″ × 6′-8″	alum storm	—
15	2′-8″ × 6′-8″ × 1¾″	fl solid core	—
16	2′-6″ × 6′-8″	bifold	2′-6″ opening
17	2′-6″ × 6′-8″	alum siding	5′-0″ opening
18	2′-6″ × 6′-8″ × 1⅜″	fl hollow core	—
19/20	2′-6″ × 6′-8″	bifold	2′-6″ opening
21	2′-4″ × 6′-8″ × 1⅜″	fl hollow core	2′-4″ opening
22/23	2′-6″ × 6′-8″ × 1⅜″	fl hollow core	—
24	4′-0″ × 6′-8″	2 bifold	4′-0″ opening
25	4′-0″ × 6′-8″	2 bifold	4′-0″ opening
26	2′-6″ × 6′-8″	bifold	2′-6″ opening
27	2′-8″ × 3′-0″ × 1⅜″	fl hollow core	crawl space
28	16′-0″ × 7′-0″	ovhd 4-panel	cedar faced

Figure 11-1. The specifications or prints contain a door schedule that provides detailed information about doors to be installed on a construction project.

Full view doors consist of a large pane of glass surrounded by a wood or metal frame.

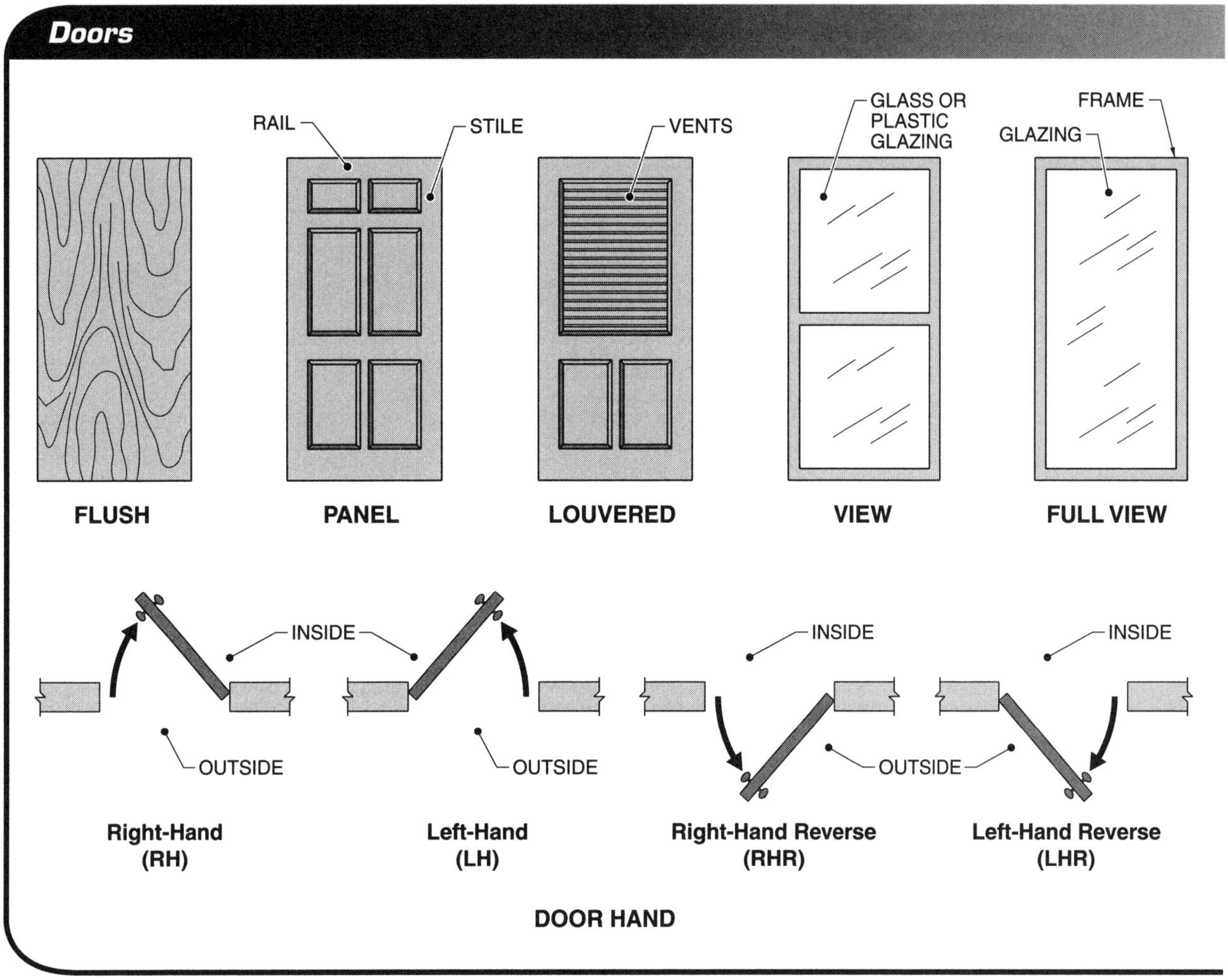

Figure 11-2. Common residential and commercial doors include flush, panel, louvered, view, and full view.

Metal Frames and Doors

Metal frames and doors are commonly referred to as hollow metal frames and doors. Specifications for hollow metal frames and doors are available from the National Association of Architectural Metal Manufacturers (NAAMM). Metal frames and doors are selected for commercial and residential applications based on frequency of use, traffic patterns, fire ratings, and climatic and temperature conditions.

Metal Frames. Hollow metal door frames are designed to receive a variety of doors, including metal and wood doors. Metal door frame gauge varies based on the design of the door to be supported and the required fire rating. A variety of metal door frame anchors are available. The type of anchor utilized for any application is based on the type and size of metal frame and the structure of the wall into which the frame is installed. **See Figure 11-3.** Metal door frame anchors may be screwed, snapped, or welded into position. Various metal door anchors are used to secure metal door frames to a variety of interior and exterior wall materials.

> The National Association of Architectural Metal Manufacturers (NAAMM) is composed of four divisions—the Architectural Metal Products Division, the Expanded Metal Manufacturers Association, the Hollow Metal Manufacturers Association, and the Metal Bar Grating Division. Manufacture of hollow metal doors and frames is covered under the jurisdiction of the Hollow Metal Manufacturers Association.

Hollow Metal Door Frame Anchors

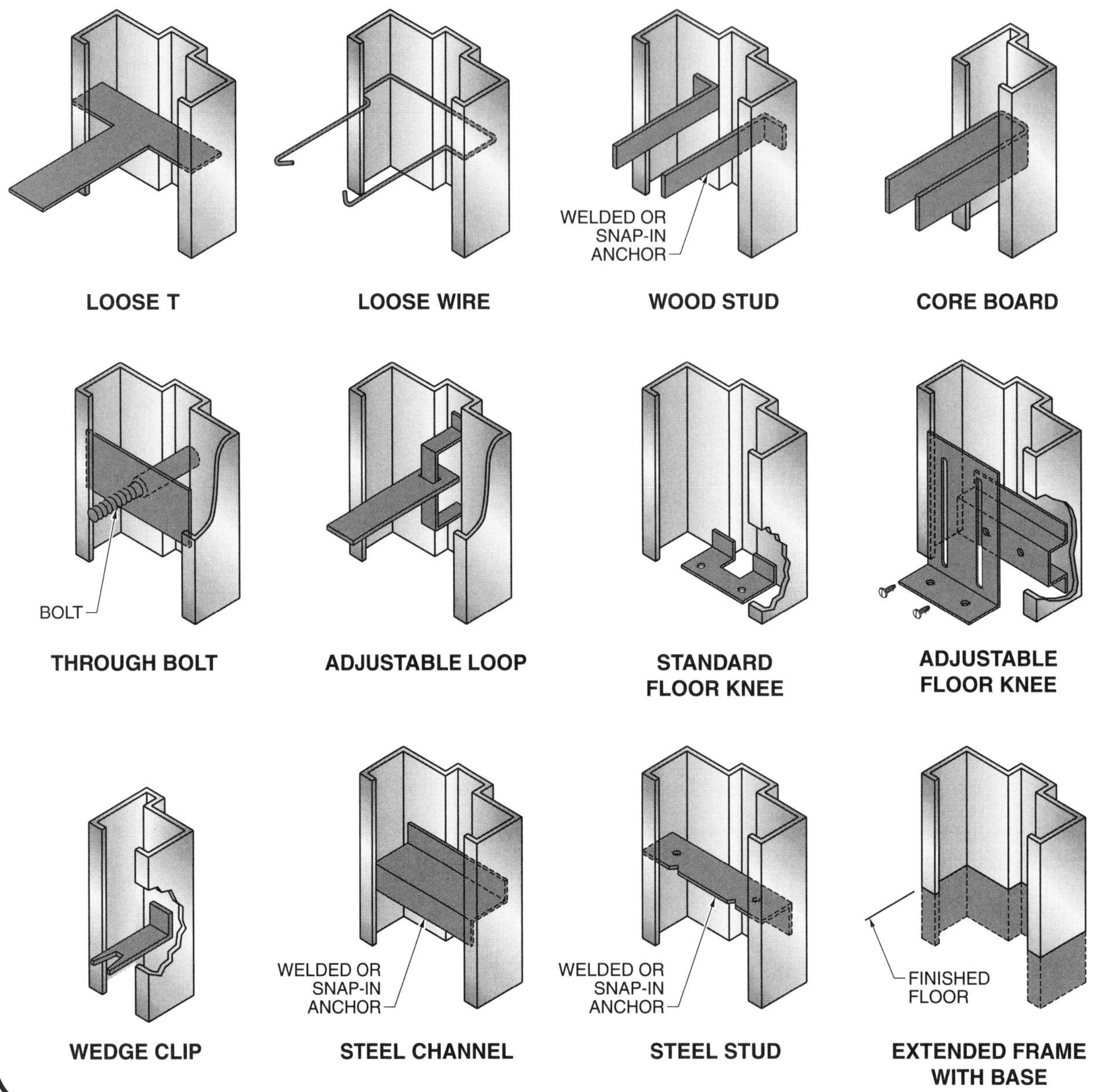

Figure 11-3. Metal door frames are fastened to the surrounding supporting structure using various types of anchors.

In addition to width, height, and door thickness, variables in metal door frame design include jamb depth, rabbet, soffit, face, stop depth, and backbend. **See Figure 11-4.** The open area between the metal frame and the supporting structure may need to be filled with grout or another material as specified by the architect to provide additional structural support, insulation, or fireproofing.

Metal Doors. Hollow metal doors are composed of a metal perimeter channel frame filled with a foam or fiber core and covered by a metal sheet on each side of the frame. The core has insulating and soundproofing properties. Minimum metal thicknesses are specified for metal door frames and faces depending on the door application and fire rating. **See Figure 11-5.**

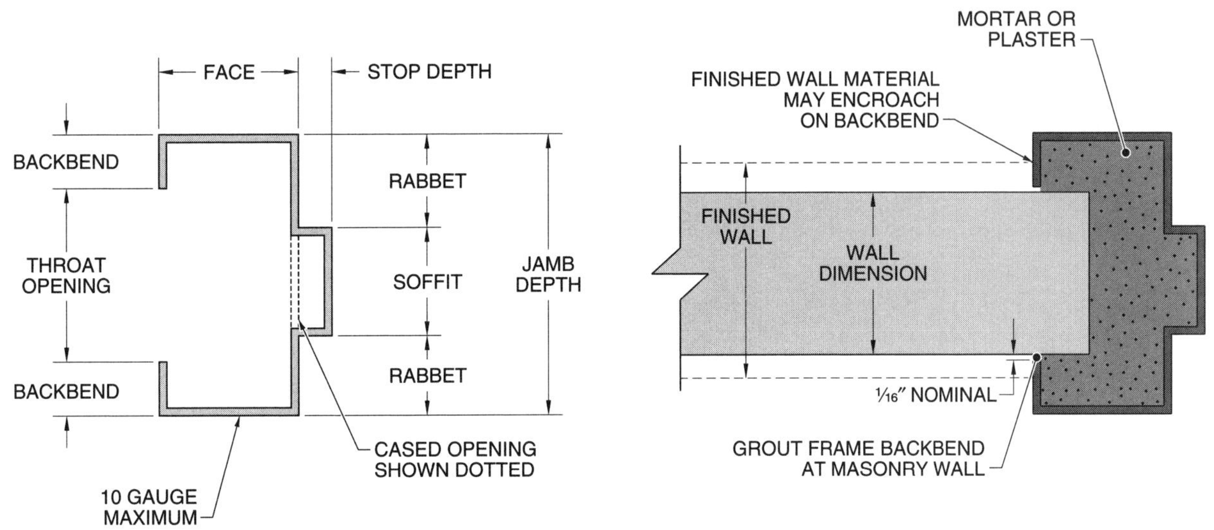

Figure 11-4. Variables in metal door frame design include jamb depth, rabbet, soffit, face, stop depth, and backbend.

STEEL DOOR MINIMUM METAL THICKNESS		
Commercial Doors		
Item	**Thickness***	**Equivalent Thickness†**
Door frames Surface-applied hardware reinforcement	16 16	0.0598 0.0598
Doors — Hollow steel construction Panels and stile	18	0.0478
Doors — composite construction Perimeter channel Surface sheets	18 22	0.0478 0.0299
Reinforcement Surface-applied hardware Lock and strike Hinge Flush bolt	16 16 10 16	0.0598 0.0598 0.1345 0.0598
Glass molding	20	0.0359
Glass muntins	22	0.0299
Interior Doors		
Item	**Thickness***	**Equivalent Thickness†**
Door frames, 1⅜" thick	18	0.0478
Door frames, 1¾" thick	16	0.0598
Stiles and panels	20	0.0359
Reinforcement Lock and strike Hinge Closer	16 11 14	0.0598 0.1196 0.0747

* in ga
† in in.

Figure 11-5. Minimum metal thicknesses are specified for commercial and interior metal door faces and frames.

For example, commercial metal doors require 16 ga or 18 ga metal. Institutional-use metal doors require 12 ga or 14 ga metal. Standard finish for metal doors is a primer applied at the factory prior to delivery to the job site. Metal doors may also receive a baked enamel finish, an applied finish such as vinyl cladding, a textured embossed finish such as stainless steel or aluminum, or a polished metal finish.

Wood Frames and Doors

As with metal frames and doors, traffic and climatic conditions determine the selection of wood frames and doors used in building construction. Wood frames and doors are most commonly used in residential and light commercial applications. They may be prefinished or finished at the job site. Wood frames and doors may be painted or finished with coatings that expose the wood grain.

Wood Frames. Wood frames are most common in residential and light commercial buildings. Two types of wood door frames are built-up jambs and split jambs. **See Figure 11-6.** Built-up jambs are made from ¾″ thick lumber, which is cut to the same width as the wall thickness, including the finish materials. Stops and casings are applied to the wood jamb.

Split jambs are made from molded wood members designed to interlock. The two interlocking jamb pieces may be slid together or apart to adjust the jamb depth according to slight variations in the wall width. Wood frames are also used for specialty doors, such as overhead doors and pocket doors. Fire-rated wood frames are also available.

Wood Doors. The general categories of wood doors include solid-core, hollow-core, and panel. Solid-core wood doors are made of a wood frame reinforced by a solid particleboard or staggered-block core. The frame and core are covered with veneer to create a solid wood door. Hollow-core doors are made of a wood frame with a honeycombed core. The wood frame and honeycombed core are covered with veneer to form a light-traffic door. Panel doors are made of solid lumber or built-up rails and stiles that frame flush or raised panels. Panel door designs are also available as hollow-core doors with specially molded veneer face panels.

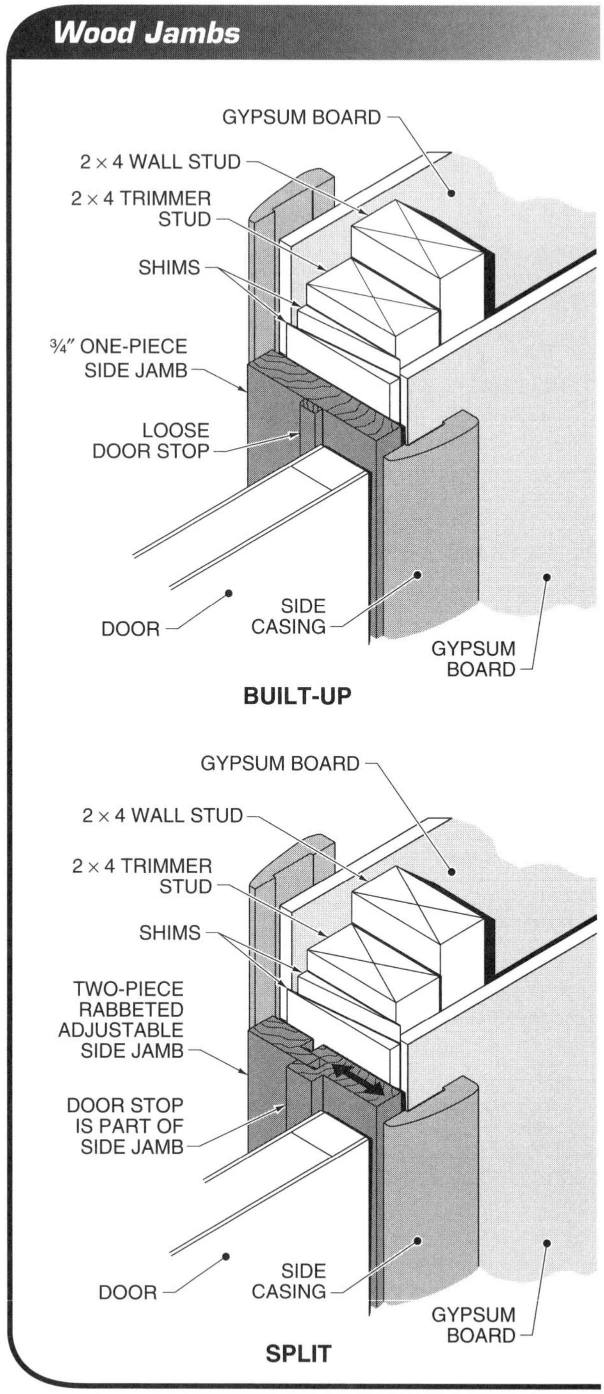

Figure 11-6. Built-up jambs are made from ¾″ thick lumber. Split jambs are made from molded wood members designed to interlock.

The panels in panel doors are designed to "float" between the stiles and rails to allow movement during changes in humidity.

Specialty Doors

Specialty doors include fire, access, coiling, folding and sliding, overhead, and entrance doors. Estimators obtain information concerning each door from floor plans, elevations, and door schedules. Specialty doors, such as revolving, coiling, and overhead doors, are commonly estimated by subcontractors.

Fire Doors. A *fire door* is a fire-resistant door and assembly, including the frame and hardware, commonly equipped with an automatic door closer. A *door closer* is a device that closes a door and controls the speed and closing action of the door. Fire doors are specially built and installed to inhibit the transmission of fire from one building area to another.

The fire rating classification of the wall into which a fire door is installed determines the required classification of the fire door. Steel fire doors are rated using the amount of time the door can withstand fire test conditions and include 3-hour (180 minute), 1½-hour (90 minute), 1-hour (60 minute), ¾-hour (45 minute), and ⅓-hour (20 minute) doors.

Three-hour doors are installed in walls separating buildings or in walls dividing one building into separate fire areas. One and one-half hour doors are installed in enclosures of vertical egress in a structure, such as stairways and elevators, or in exterior walls where the potential for severe fire exposure from the exterior exists. Three-quarter-hour doors are installed in hallways and room partitions or in exterior walls where the potential for light to moderate fire exposure from the exterior exists. One-third-hour doors are installed in corridors or hallways where smoke and draft control is required and the minimum wall rating is 1 hour.

Fire doors include composite wood and hollow metal. Composite wood fire doors have a minimum thickness of 1¾″. Hollow metal fire doors are formed of 20 ga or heavier steel. All fire doors and frames must display a label listing their fire rating. **See Figure 11-7.**

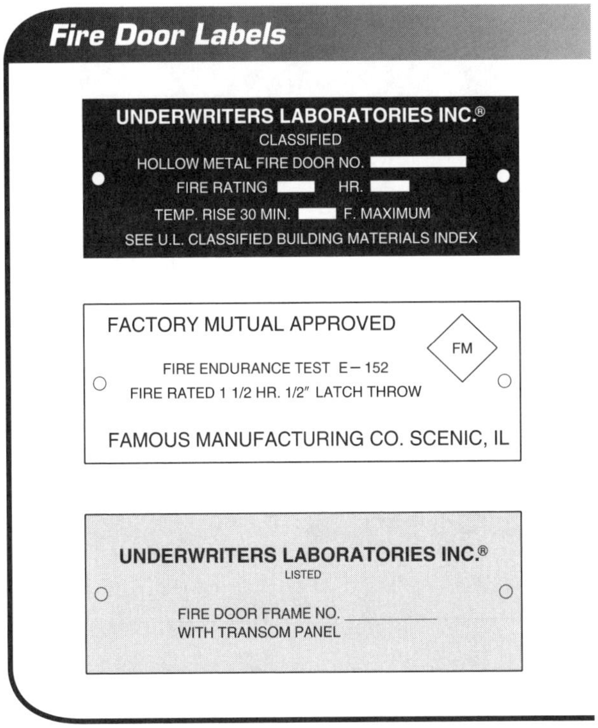

Figure 11-7. Fire doors and frames must display a label listing fire-rating information.

Access Doors. An *access door* is a door used to enclose an area that houses concealed equipment. Access doors include any special doors such as sidewalk, floor, and other small doors for access to mechanical and electrical equipment. Sidewalk doors are available in sizes from 2′-0″ to 3′-6″ in 6″ increments. Sidewalk doors are made of steel or aluminum ¼″ checker or diamond plate and can withstand 300 lb/sq ft of live load.

Floor doors are made in a manner similar to sidewalk doors with an allowance for floor covering that can match the surrounding floor area. Access doors are commonly made of metal and are premounted in a frame. Many different sizes and designs of access doors are available depending on the application.

Coiling Doors and Gates. A coiling door or gate is a door that moves vertically to coil around a steel rod. Interlocking strips of galvanized steel, stainless steel, aluminum, or grillwork are fabricated into coiling doors. **See Figure 11-8.** Tracks mounted on each side of the opening hold the coil in place. Coiling doors are available in many standard sizes varying in height up to 30′ and in width up to 33′.

Folding and Sliding Doors. A *folding door* is a door formed of panel sections joined with hinges along the vertical edges of the panels and supported by rollers in a horizontal upper track. Folding door tracks are suspended from the ceiling or roof. A series of rollers on the top of the folding door supports its movement. Folding door sizes vary from small sizes for residential installations to large sizes for meeting-room dividers.

Figure 11-8. A coiling door or gate moves vertically and wraps around a steel rod at the top.

A *sliding door* is a horizontal-moving door suspended on rollers that travel in a track that is fixed at the top of the opening or that moves on rollers mounted in the bottom of the door. Sliding doors are commonly constructed of wood or metal frames and large tempered glass lights. Common sliding door heights include 6′-8″, 8′-0″, and 10′-0″. Widths vary depending on the number of sliding panels. Wood, metal, and reinforced fabrics are used for folding and sliding doors.

Overhead Doors. An *overhead door* is a door that, when opened, is suspended in a track above the opening. Overhead doors consist of a series of horizontal panels joined with hinges, allowing the sections to fold as the door is raised and lowered. Overhead doors include fiberglass, steel, and wood panel doors. Panel and section dimensions and designs are set at the manufacturer to provide door dimensions in accordance with the design requirements.

Approximate heights range up to 20′ and widths up to 30′. Overhead doors are held in place and moved vertically by rollers attached to the inside of the door. The rollers move in a track mounted to the structural members on both sides of the door opening. **See Figure 11-9.** Springs facilitate raising and lowering the door. Overhead doors may use electric motors to raise and lower the door.

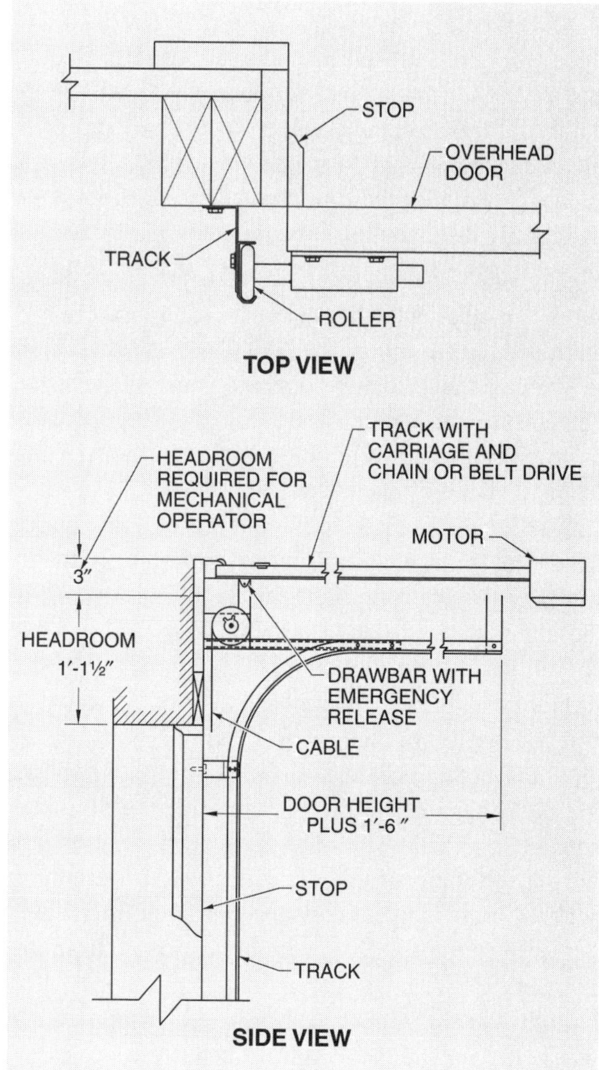

Figure 11-9. Overhead doors travel in vertical and horizontal tracks mounted to structural members on both sides of the door opening.

Entrance Doors. An *entrance door* is the primary means of ingress and egress for pedestrian traffic in a commercial building. Entrance doors used in commercial construction include metal-framed and all-glass entrances, automatic doors, and revolving doors. Metal-framed entrance door systems may be fabricated from aluminum, bronze, stainless steel, or other metal finishes. The metal frames commonly support tempered glass and use heavy-duty hardware due to the high traffic requirements. Standard sizes are available depending on the manufacturer.

Frameless glass doors are fitted with metal channels or brackets at the top and bottom. Pivot rods extend from the channel or bracket and fit into recessed hinges. Tempered glass is mandatory in these applications because it can withstand traffic and wind loads.

Standard revolving door dimensions vary from 6′-6″ to 7′-6″ in diameter in 2″ increments. Revolving door standard heights are 6′-8″ and 7′-0″. The circular enclosure for the door unit is commonly constructed of metal and glass, with an allowance of approximately 3¾″ for the rubber sweeps at the outside edges of the revolving door sections. A revolving door reduces air transmission, which has the effect of reducing energy consumption for heating and cooling.

Entrances for commercial businesses typically consist of paired swinging doors accompanied by lights to allow ample outdoor light into the building.

Door Hardware

Door hardware includes hinges, locksets, closers, and any other miscellaneous hardware such as panic devices, weatherstripping, and thresholds. The primary source of hardware information is the door schedule. Architects commonly use manufacturer-specific identification codes to specify hardware requirements at each door. Standard hardware finish codes are used for specifying door hardware. **See Figure 11-10.**

DOOR HARDWARE FINISH SYMBOLS	
Symbol	**Finish**
USP	Primed for painting
US 3	Bright brass
US 4	Dull brass
US 10	Dull bronze
US 10B	Dull bronze, oxidized, and oil rubbed
US 14	Nickel plated, bright
US 26	Chromium plated, bright
US 26D	Chromium plated, dull
US 27	Satin aluminum, lacquered
US 28	Satin aluminum, anodized
US 32	Stainless steel, polished
US 32D	Stainless steel, dull

Figure 11-10. Standard hardware finish codes are used to specify door hardware.

Hinges. A *hinge* is a pivoting hardware device that joins two surfaces or objects and allows them to swing around a pivot. Hinges allow for pivoting at the point of the door and jamb attachment. The metal, finish, fire rating, closing option, and door and jamb configuration are taken into account when selecting the proper hinge for an application.

Locksets. A *lockset* is the complete assembly of a bolt, knobs, escutcheon, and all mechanical components for securing a door and providing a means for opening. A lockset provides security, locking, and manual ability to open and close a door. As with other door hardware, a variety of designs are available based on finish, security requirements, and amount of traffic. **See Figure 11-11.** Deadbolt locksets are commonly used for coiling and overhead doors, with a deadbolt projecting into the door track. Sliding doors and folding doors normally incorporate a concealed lockset.

Door and hardware manufacturers offer certain accessories, such as weather seals or chain hoists, as standard equipment on some door models and as optional equipment on others.

Locksets

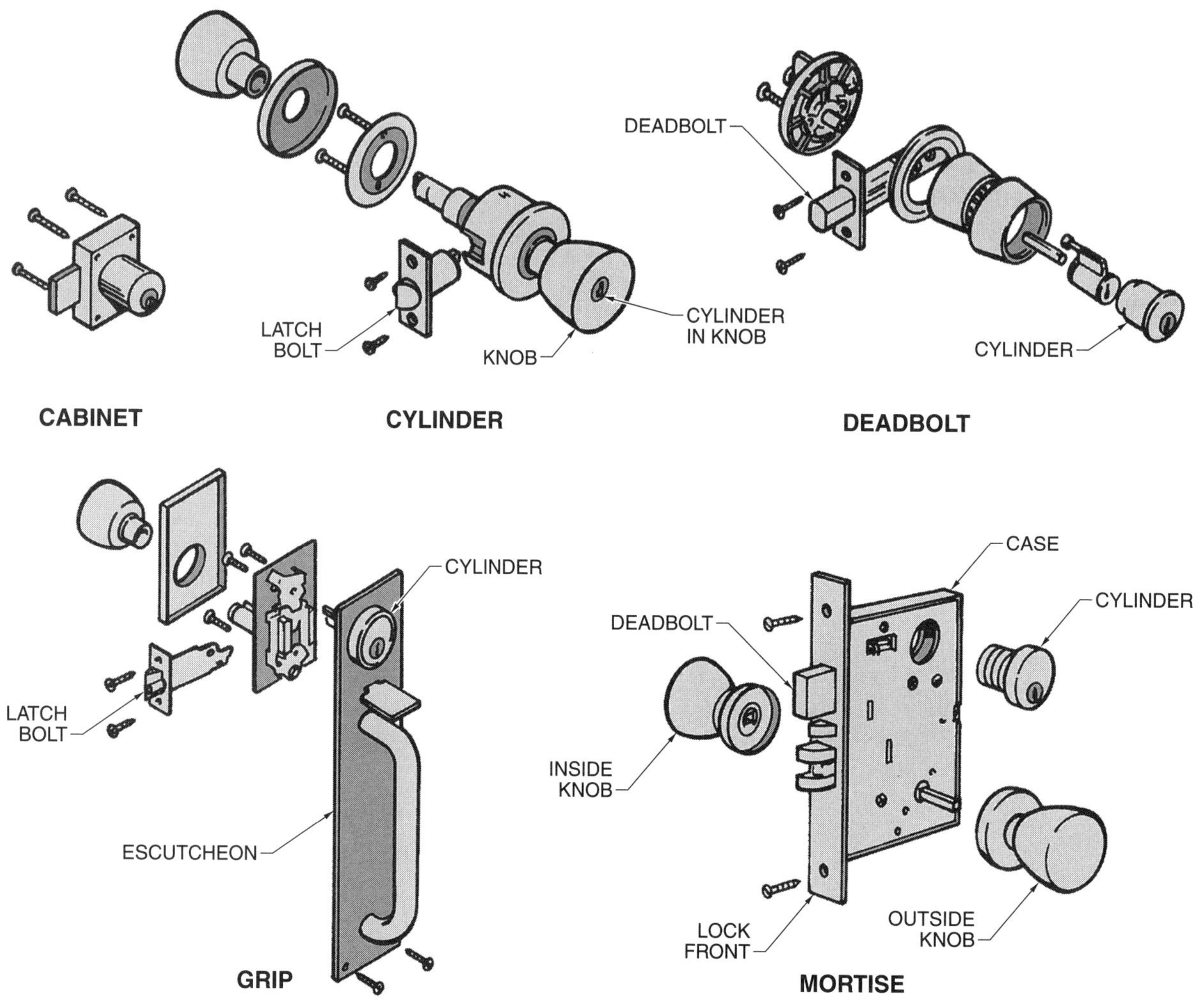

Figure 11-11. Lockset hardware varies depending on security requirements, durability, and finish.

Closers. A door closer closes a door and controls the speed and closing action of the door using a hydraulic or pneumatic device and pivots. The two primary categories of closers are concealed and surface-mounted. Concealed door closers may be mounted into the top or bottom of the door or top or bottom of the frame depending on the closer design. **See Figure 11-12.**

Some automatic door closers are activated by various detectors. These powered doors are controlled by electric or pneumatic sensing mates, photoelectric cells, motion detectors, or other methods and are commonly integrated with fire detection systems. Power-assisted opening and closing equipment is also required for applications where access for individuals with disabilities is required.

Miscellaneous Hardware. Additional door hardware includes panic devices, handles, push plates, and doorstops and holders. A panic device allows for quick exit from public areas in case of emergency. Panic devices include rim, mortise, exposed vertical rod, and concealed vertical rod. Rim panic devices are mounted on the surface of the door without any fitting into the door itself. Mortise panic devices allow for the bolt to project from the edge of the door in a manner similar to a cylinder or mortise lockset. Rod panic devices use a mechanism with latches at the top and bottom of the door fastened to vertical rods. **See Figure 11-13.**

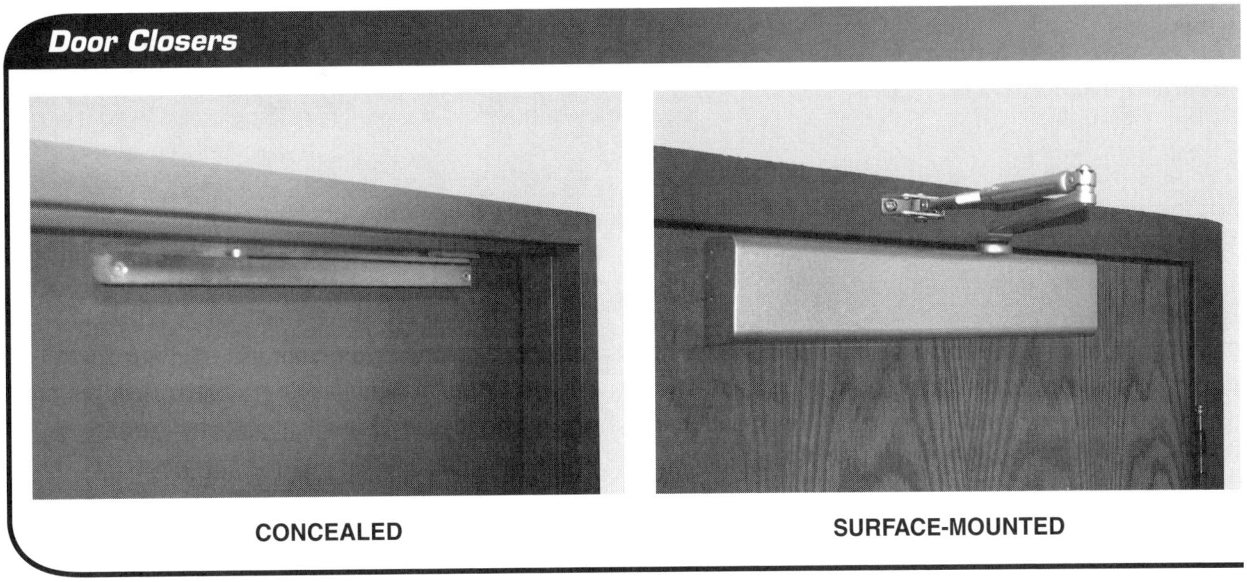

Figure 11-12. A door closer is a device that closes a door and controls the speed and closing action of the door.

Figure 11-13. Panic devices allow for quick exit from public areas in case of emergency.

DOOR AND FRAME CONSTRUCTION METHODS

Door frames are installed as the first component of door construction. Sizes and types of frames are determined from door schedule specifications. After frames are set plumb and level, doors are then set and hung in place. Hardware is installed as jambs are set and doors are hung according to manufacturer instructions. Latching hardware is installed to complete the door construction.

Frame Construction

Door frame construction begins with checking the rough opening to ensure the proper size opening is provided for the door and frame. Allowances are made in the wall opening construction for the door, the frame materials, the trim and finish, and any open space required for opening the door. Rough opening allowances vary depending on the frame.

Swinging Door Frames. Metal and wood frames for swinging doors are set in place using different methods, depending on the wall materials and methods. For concrete walls, frames and jambs may be set into formwork prior to concrete placement. Care is taken to ensure the frame is set at the proper location and set plumb and level and securely braced to prevent movement during concrete placement. Blockouts may be installed in concrete forms to allow for installation of doorjambs after the concrete has set and the forms have been removed.

For masonry walls, door frames are set in place and braced plumb and level prior to laying masonry. Masonry members are set around the frames with proper anchors set in mortar joints. For steel frame and wood frame walls, door frames are set after framing is in place.

Folding, Coiling, and Overhead Door Frames. Frames for folding, coiling, and overhead doors are sized based on allowances for door size and supporting hardware. Folding doors often need no additional framing material. Coiling and overhead door frames provide additional support at the header for the weight of the door and opening hardware. The side jambs for coiling and overhead doors are set in place to create a tight seal at the door sides and stops.

Door Construction

Doors may be prehung or hung at the job site. Prehung doors are set into jambs at the manufacturing facility. Hinges are fastened to the door and jamb, and temporary bracing is attached to secure the door in place during shipping. The prehung unit is set into the wall opening and adjustments are made to the jamb position with the door in place. Doors hung on the job site are set into preset jambs. Doors are fitted into the jambs with hanging hardware attached. Stops and finish materials are fastened to the doors and jambs to complete the door hanging.

Hardware. Standard allowances are made in metal doors and jambs for hinges and for lockset and hardware installation. Metal doors are delivered prefabricated. Lockset holes may be predrilled or drilled at the job site. Holes for other hardware such as closers and panic bars are drilled and tapped at the job site. Wood doors may be premilled for hinges and locksets or drilled and fitted at the job site.

Door hardware can add considerable cost to each door. Door hardware requirements are determined by reviewing the door schedule and specifications for the project. The door schedule may be included with the prints or specifications.

DOOR QUANTITY TAKEOFF

Items considered in takeoff for doors include door frames, doors, and hardware. Estimators should prepare a schedule or worksheet indicating the location of each door and frame by floor or building area. As with other estimating tasks, a systematic method should be used. As takeoff proceeds, drawings and schedules should be marked to indicate that all door openings, including frame, door, and hardware, have been included in the quantity takeoff total. BIM systems may automatically provide these door quantity totals.

Baldwin

Door assemblies may include sidelights, which must be included in the estimate.

Door, Frame, and Hardware Takeoff

Each type of door and frame is individually counted and totaled for each floor plan. Drawing symbols and architectural notes describe each door and frame. **See Figure 11-14.** The estimator cross-references the floor plans to the door schedule to further ensure an accurate count of the door types. Interior elevations may be crossreferenced with floor plans during quantity takeoff to ensure the proper types and sizes of doors are specified.

When using a ledger sheet, an estimator creates a chart for each area of the structure. Columns are used for frames, doors, and hardware. Totals are entered in each row to achieve a final total. Calculations for costs for frames, doors, hardware, and labor are made and totaled. Estimators should work with project managers on the job site and with suppliers to determine transportation or storage costs for doors, frames, and hardware.

Spreadsheets allow information to be entered in the same format as ledger sheets with totals calculated according to estimator-established formulas. Various estimating programs allow for assembly takeoff of frames, doors, and hardware as a single unit, with the appropriate entries made by entering one item assembly and labor costs calculated based on a per unit method. **See Figure 11-15.**

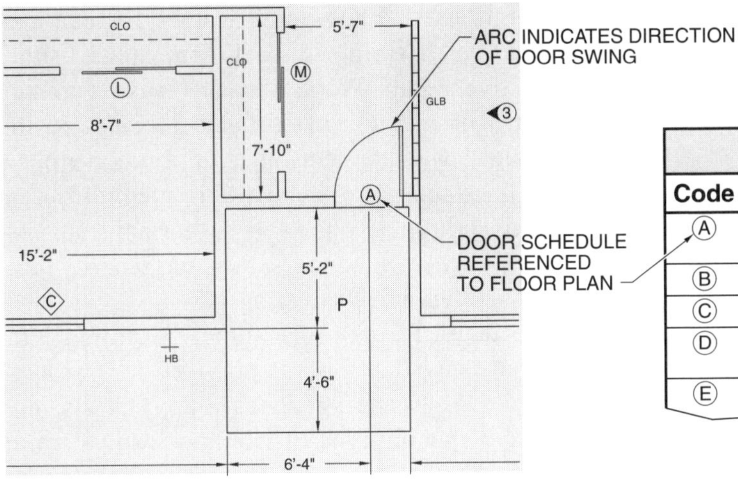

DOOR SCHEDULE				
Code	**Quantity**	**Size**	**Swing**	**Remarks**
Ⓐ	1	$3^0 \times 6^8$	RH	THERMA-TRU MODEL # CC45
Ⓑ	1	$3^0 \times 6^8$	LH	STEEL
Ⓒ	1	$3^0 \times 6^8$	RH	STEEL
Ⓓ	1	$16^0 \times 8^8$	—	WAYNE DALTON MODEL # 226
Ⓔ	1	$6^0 \times 6^8$	SLIDING	HURD MILLWORK MODEL # CHES-6

Figure 11-14. Estimators use floor plans and door schedules to perform accurate door, frame, and hardware takeoff.

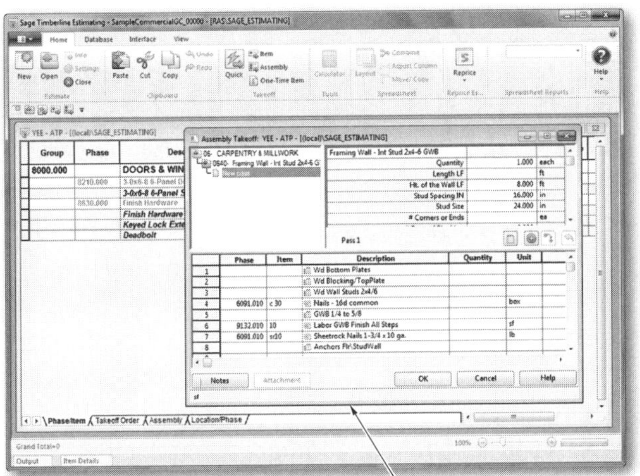

Figure 11-15. Estimating programs allow for assembly takeoff as a single item, with labor costs calculated based on a per unit method.

BIM systems using intelligent models provided by the architect and suppliers may provide door frame, door, and hardware quantities automatically. Estimators should review the totals provided to verify accuracy and ensure labor and installation costs are properly assigned to each door type and location.

Labor costs are determined based on the average time required to install each type of door frame, hang the various types of doors, and install the necessary hardware. Labor costs are based on standard industry sources or company historical data. A general contractor normally subcontracts specialty door installation, such as for overhead, revolving, and automatic doors.

Door Takeoff. Items considered by an estimator during door takeoff include door type, possible panel matching requirements, width, height, thickness, door hand (for swinging doors), hinge type and placement, locksets, door material, fire rating, any required specialty hardware, closers, and door inserts such as louvers or lights. The door schedule is the primary source of door requirements and specifications. Floor plans are the primary source of quantity information.

Frame Takeoff. Frame information considered during quantity takeoff includes door type, width, height, thickness, door hand (for swinging doors), sidelights or other openings included in the frame structure, hinge type and placement, lockset, frame material, and fire rating. The door schedule is the primary source of frame requirements and specifications. Floor plans are the primary source of quantity information.

Specialty Door Takeoff. Specifications, floor plans, elevations, and details provide specialty door information. **See Figure 11-16.** Specialty doors are taken off as unique, individual items. Manufacturers are consulted to obtain accurate detailed information to complete the estimate.

Hardware Takeoff. Takeoff of door hardware items is commonly performed by using specific manufacturer codes and designations. Quantities for hardware are related to the identification code for each door and frame. Items such as hinges, locksets, panic devices, thresholds, and closers are totaled according to the number of individual units needed for various doors. Door schedules and specifications are the primary source of door hardware detail information.

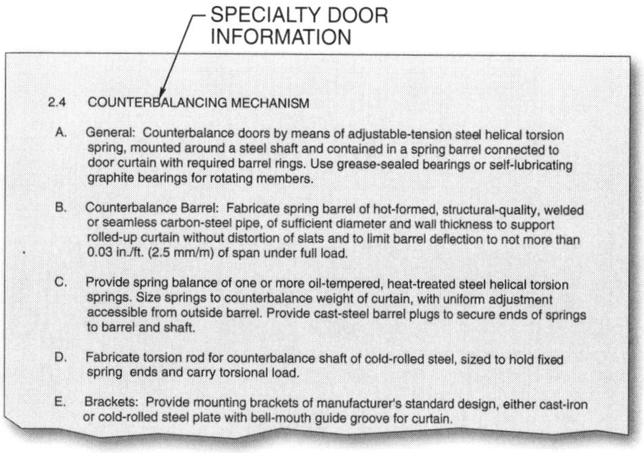

2.4 COUNTERBALANCING MECHANISM

A. General: Counterbalance doors by means of adjustable-tension steel helical torsion spring, mounted around a steel shaft and contained in a spring barrel connected to door curtain with required barrel rings. Use grease-sealed bearings or self-lubricating graphite bearings for rotating members.

B. Counterbalance Barrel: Fabricate spring barrel of hot-formed, structural-quality, welded or seamless carbon-steel pipe, of sufficient diameter and wall thickness to support rolled-up curtain without distortion of slats and to limit barrel deflection to not more than 0.03 in./ft. (2.5 mm/m) of span under full load.

C. Provide spring balance of one or more oil-tempered, heat-treated steel helical torsion springs. Size springs to counterbalance weight of curtain, with uniform adjustment accessible from outside barrel. Provide cast-steel barrel plugs to secure ends of springs to barrel and shaft.

D. Fabricate torsion rod for counterbalance shaft of cold-rolled steel, sized to hold fixed spring ends and carry torsional load.

E. Brackets: Provide mounting brackets of manufacturer's standard design, either cast-iron or cold-rolled steel plate with bell-mouth guide groove for curtain.

Figure 11-16. Specialty door information is detailed in the specifications, floor plans, elevations, and details.

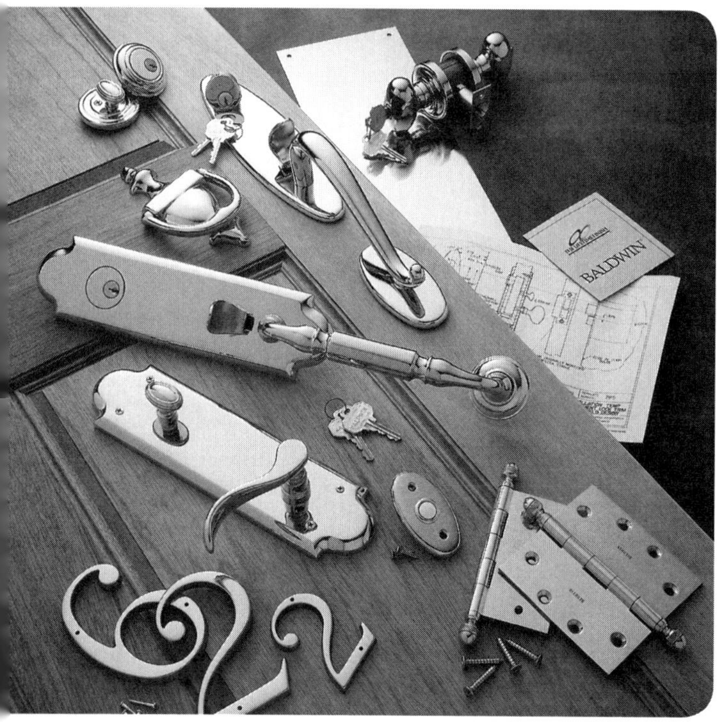

Baldwin

A variety of door hardware is typically required for each door.

WINDOWS

Information about window materials and methods are part of Division 08 of the CSI MasterFormat™ and include items such as windows (08 50 00), skylights (08 60 00), hardware (08 70 00), glazing (08 80 00), and curtain walls (08 44 00). Materials used for window construction include various metals, woods, plastics, and glass.

Metals used for window frames include aluminum, steel, stainless steel, bronze, and other decorative metals. Wood windows may be painted or clad with metal or plastic to improve corrosion resistance, weather resistance, and other properties. Glass sheets are available in many different designs and types. Wind loads, thermal transmission, privacy, appearance, safety, and security must be considered when selecting glass.

The majority of information concerning windows and installed glass is provided in a window schedule or in the specifications. Details may refer to a specific type of window or glass for an identified area. Exterior elevations and floor plans provide dimensional information for the sizes of windows and glass panels.

> Specialty hardware may be required to meet the needs of specialty doors. The specifications should be checked for the requirements.

Most windows are available in steel, aluminum, alloy, or wood sash. Many window manufacturers produce windows of varying quality depending on the weight of the sash, glass type, and other factors. Most building codes include restrictions on windows concerning minimum light, ventilation, wind load resistance, and egress requirements for various types of construction.

Elements considered during window design include air and water resistance, ventilation requirements, insulation, light transmission, visual and acoustical separation, safety, access, ease of operation, security, and maintenance. Light transmission and insulation capacity of windows affect the sustainable construction ratings of a structure. Architects and designers review these elements when selecting the window to be installed in each opening. Common windows include fixed, single- or double-hung, sliding, casement, hopper, awning, bay, bow, jalousie, and skylight. **See Figure 11-17.**

The two general categories of windows are fixed and movable. Window units may be composed of fixed and/or movable windows. Fixed windows are set in a frame and the entire window unit is stationary. Movable windows, as the name implies, can be opened and closed.

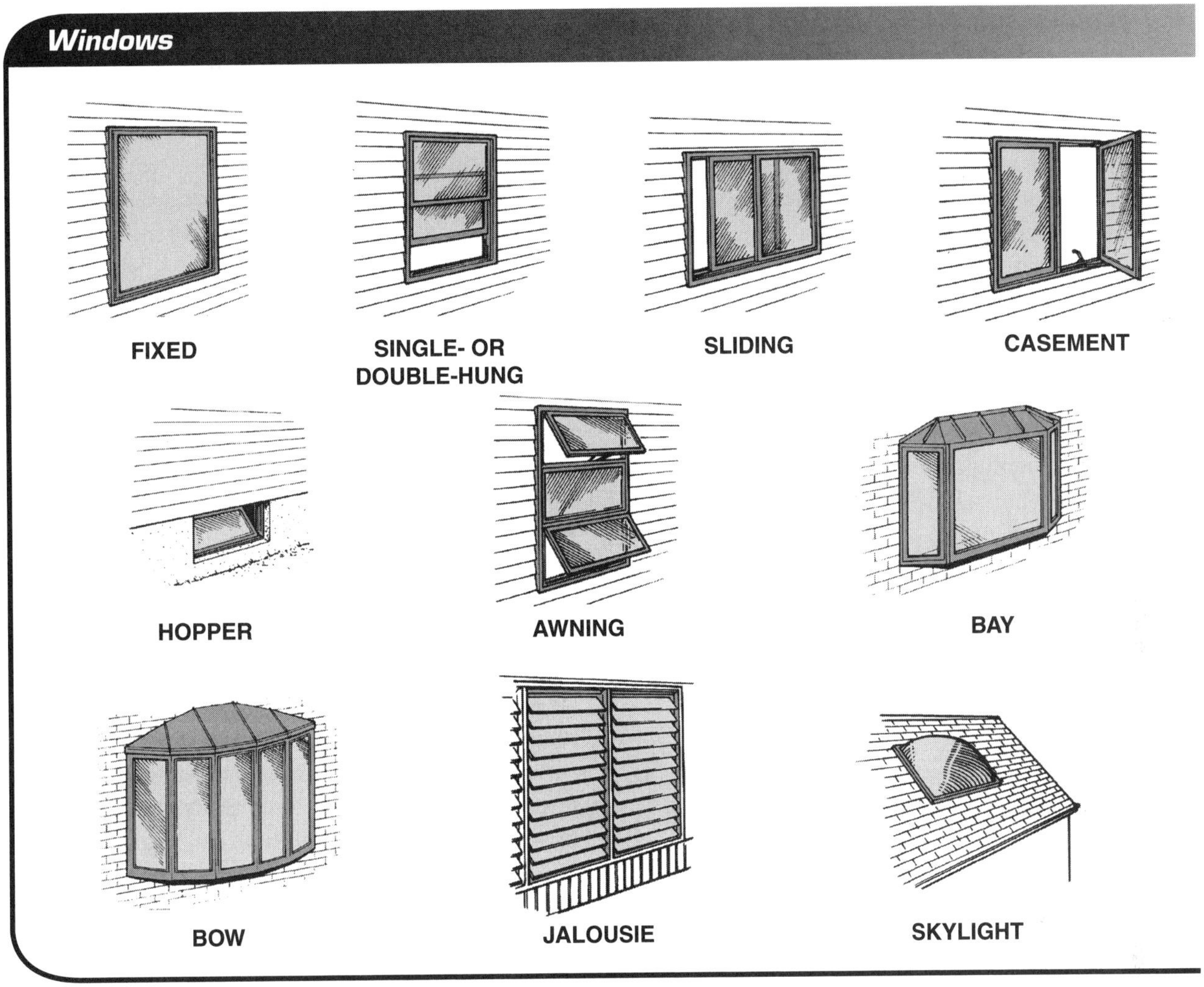

Figure 11-17. Common windows include fixed, single- or double-hung, sliding, casement, hopper, awning, bay, bow, jalousie, and skylight.

Fixed Windows

Fixed windows are made of solid pieces of glass set in a frame with wood, metal, plastic, or composite trim holding the glass in place. Fixed glass windows may contain a double layer of glass with a vacuum or various types of inert gases between the glass panels. The vacuum or gas increases insulation qualities and R factors. Fixed glass windows are commonly used in commercial buildings for security and safety reasons. **See Figure 11-18.** Glass for fixed windows may be clear, reflective, or tinted. Fixed glass panels are commonly applied in curtain wall systems. Bay and bow windows are made with fixed glass panels and possibly movable panels.

Andersen Windows, Inc.

Figure 11-18. Fixed windows in commercial buildings provide security and safety.

Movable Windows

Movable windows are windows that may be opened and closed. Single-hung windows have one fixed window panel (usually the top window) and one vertical sliding window panel. Springs or weights are attached to the windows to facilitate opening and closing the windows. Double-hung windows consist of two window panels that slide vertically. The upper or lower half of the window may be opened.

Sliding windows move in a horizontal direction. Rollers are fastened to the bottom of the sliding window frame to allow the window to move easily. Casement windows are hinged at one side and open similar to a swinging door. Casement windows open and close by means of a roto-operator built into the jamb and attached to the window frame.

Hopper windows consist of a single sash hinged at the bottom of the frame. Awning windows are made of a single sash hinged at the top of the frame. Jalousie windows are made of narrow pieces of glass mounted horizontally in a frame and are opened and closed by a roto-operator. A *skylight* is a roof opening covered with single, double, or triple layers of glass or plastic for insulating purposes and may be fixed or movable.

Glazing

Glazing is the transparent or translucent material installed in a window opening. Glazing includes sheet (window) glass, float glass, plate glass, and a variety of other types of glass that are used less frequently.

Sheet glass is clear or translucent glass manufactured in continuous long, flat pieces that are cut to desired sizes and shapes. Sheet glass is used in residential and industrial applications. *Float glass* is high-quality glass used in architectural and specialty applications. *Plate glass* is polished glass manufactured in large sheets. Plate glass is thicker and of higher quality than sheet glass. Plate glass may be bent to various curves when required. Mastics, caulks, sealants, rubber or plastic seals, and metal or wood stops secure glass into frames.

Plastics. High-strength plastics may be installed as glazing where high resistance to shattering and cracking is required and glass materials may not be sufficient to meet the architectural requirements. Acrylic plastic and polycarbonate sheets are shatter- and crack-resistant thermoplastics. Plastic glazing surfaces may be susceptible to scratching, however.

Acrylic materials have higher weather resistance than polycarbonates. Plastic glazing may also be bent to various curves where required. Plastic glazing is available in sizes up to 10′ × 14′ and in a wide range of thicknesses. Common applications for plastic glazing include skylights, domes, protective shields, and bullet-resistant applications.

WINDOW CONSTRUCTION METHODS

Window frames and glazing are commonly set in place after structural construction is completed. Opening sizes for various standard windows are determined from architectural and manufacturer specifications. Structural openings for window frames are slightly oversized to allow for leveling and plumbing. Sealants and trim are applied after frames and glazing are set in place. Depending on the type of window and glazing, frames and windows may be installed from either inside or outside of the structure. For some applications, scaffolding or lift equipment may be required for safe and efficient window installation.

Small, prefabricated window units are glazed at the mill and shipped to the job site ready for installation. For commercial buildings with large metal frames and glass panels, windows are glazed on-site when construction is near completion to avoid breakage.

Frame Construction

Frames may be delivered to the job site with glazing installed or may be glazed after being set in place. Small window frames are commonly manufactured to be set as a single unit and fastened in place by leveling and securing the frame to the surrounding structural members. Frames, clips, or various fastening materials for curtain walls and large sheet glass are set in place and leveled to receive glass or plastic.

Sealants, mullions, stops, and finish trim members are applied to secure the glass in the frame. Sealants may be wet, such as oil-based glazing compound, two-part rubber-base compound, one-part elastic compound, polybutene, polyvinyl chloride, or butyl materials, or dry, such as neoprene or polyvinyl chloride gaskets.

Skylights. Skylights may be manufactured with domes set into frames or they may be made of glass panels set into metal frames. Manufactured skylights are set onto preframed openings in roof systems. Skylights are set onto the roof or mounted slightly above the roof on a curb to prevent leakage. **See Figure 11-19.**

Pella Corporation

Figure 11-19. Skylights are set onto the roof or mounted slightly above the roof.

Curtain Walls. A *curtain wall* is a non-load-bearing prefabricated panel suspended on or fastened to structural members. Insulated metal panels and/or glass are set in a frame that is attached to structural members. Exterior walls made up of glass panels set in metal frames may also be referred to as curtain walls.

Andersen Windows, Inc.

Glass curtain walls surrounding entrance doors are fastened to structural members for support.

Curtain walls include custom, commercial, and industrial. **See Figure 11-20.** Custom curtain walls are designed for specific construction projects. Commercial curtain walls are made of standard-size parts and materials. Industrial curtain walls use ribbed or preformed metal sheets in standard sizes along with standard-size metal sashes.

Curtain wall systems take into account the expansion and contraction properties of all the materials in the curtain wall panels. Curtain wall systems are fastened to structural members with welded clips or inserts. Curtain wall panel locations are shown on elevations. Curtain wall manufacturers may provide shop drawings that identify each panel and location by code letters and numbers.

A *mullion* is a vertical dividing member between two window units. Mullions several stories in length can be installed to cap the joints between curtain wall panels. Mullions and curtain wall frames are made of aluminum or other metals that are treated to be corrosion- and weather-resistant. Methods for sealing joints between curtain wall panels are shown on the drawing details.

> Spandrel glass may be installed in curtain walls. Spandrel glass can be single-piece, laminated, or insulating glass. Spandrel glass can be made opaque through the use of opacifiers (film/paint or ceramic frit) applied to an unexposed surface, or through "shadow box" construction by providing a dark enclosed space behind the spandrel glass.

WINDOW QUANTITY TAKEOFF

Information required for window quantity takeoff includes the grade, thickness, and type of glazing to be used at each location, type of each window, and quantity and size of the glazing. Estimators should systematically prepare a schedule or work code indicating the location of each window by floor or building area. Manufacturer identification codes may be used for window information quantity takeoff. As with other estimating tasks, plans should be marked to indicate that a window has been included in the takeoff totals to avoid missing a window or duplicating an item. BIM systems may automatically provide these window quantity totals.

Curtain Walls

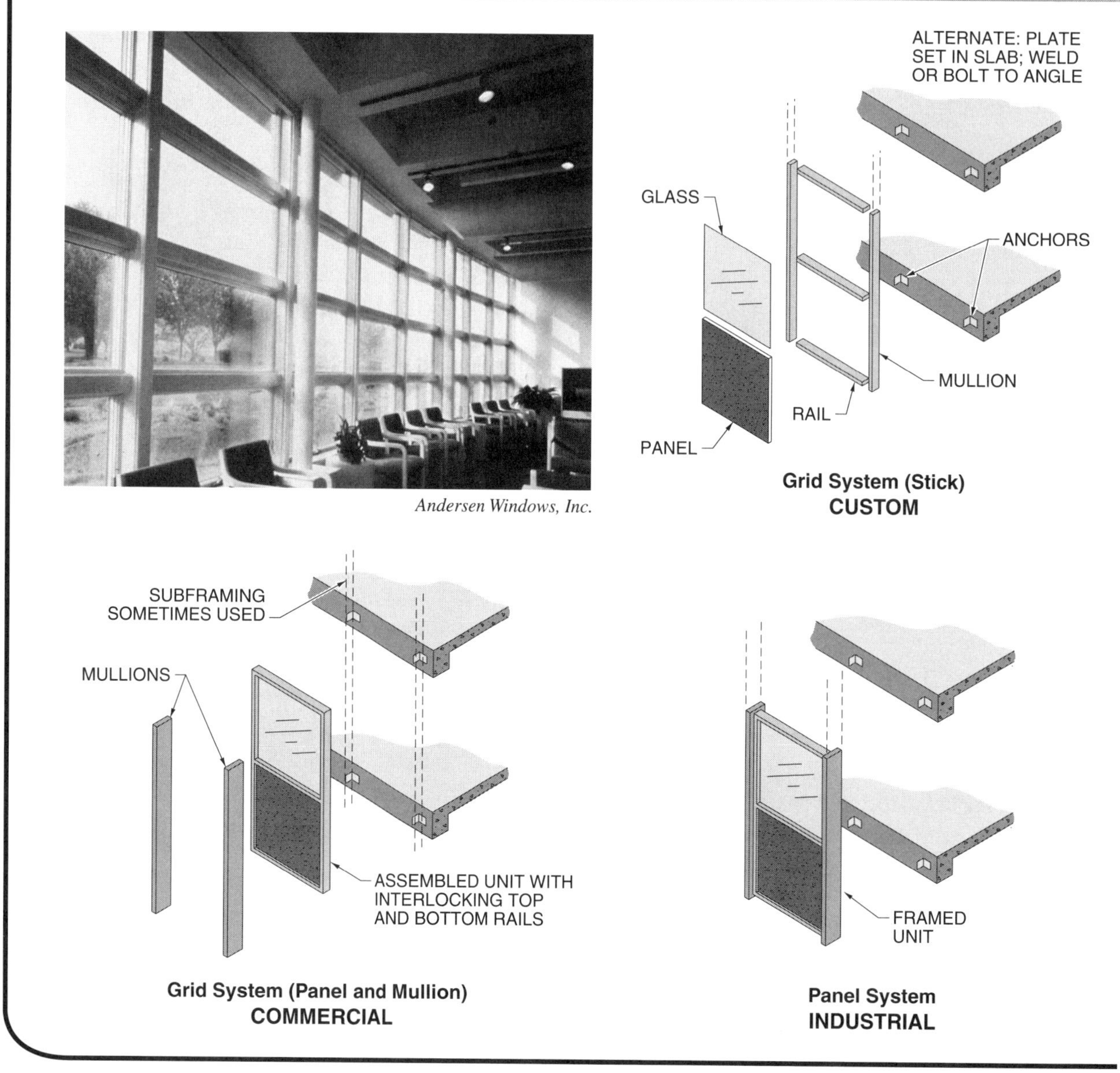

Figure 11-20. Curtain walls are composed of metal frames holding glass and other panels in place.

Manufactured Unit Takeoff

Each window and frame is individually counted and totaled on each floor plan. Similar windows and frames are grouped by type. In addition to information provided on floor plans and elevations, a window schedule and other related information may be provided in the specifications. **See Figure 11-21.** Manufacturer identification codes that denote the frame and glass type may be listed in the schedule.

When using a ledger sheet to calculate windows and frames, an estimator creates a table for each area of the structure. Columns are established for each type of window. Totals are entered in each row and added to determine a final total. Additional columns for material costs and labor are added and a total cost is determined. Depending on the height of the building and the type of window system, scaffolding or lifting equipment costs may be incurred.

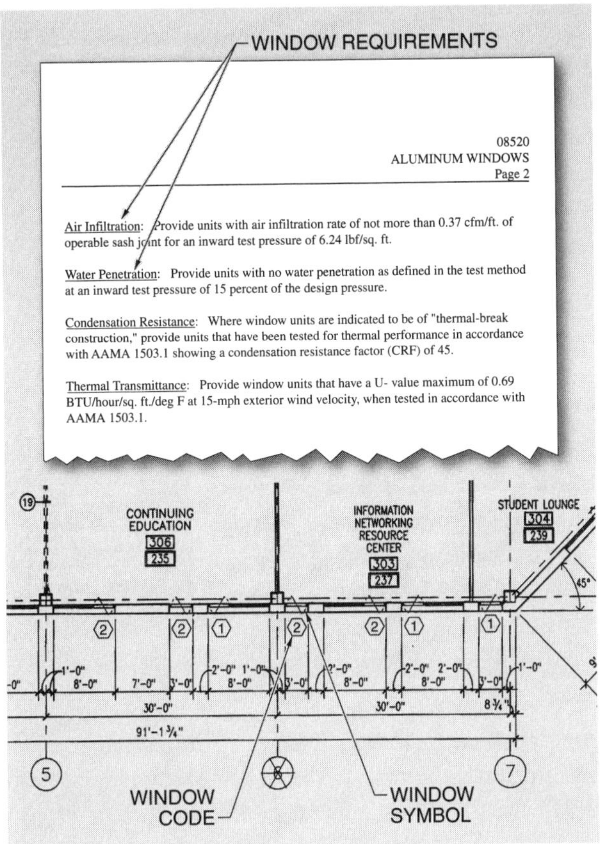

Figure 11-21. Cross-referencing windows shown on floor plans with the specifications provides complete window takeoff information.

Pella Corporation

Skylights should be included as an individual item in an estimate.

When using a spreadsheet, columns are established for each type of window, similar to ledger sheets. The spreadsheet calculates the totals as determined by formulas entered by the estimator. When using certain estimating programs, an estimator may enter standard windows into a database for repeated use. **See Figure 11-22.**

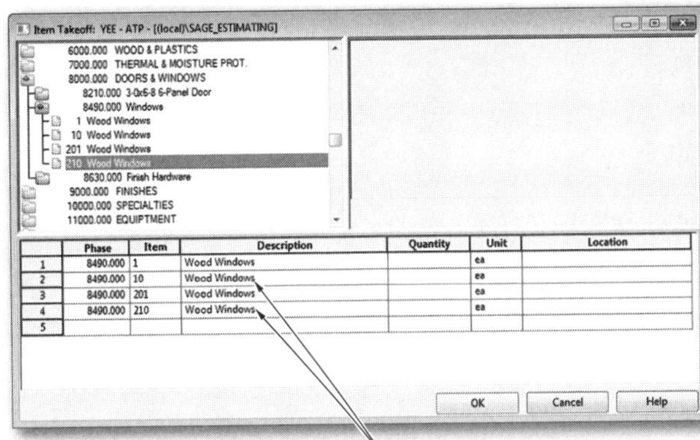

Figure 11-22. An estimator may enter standard windows into the item database for repeated use in an estimating program.

BIM systems using intelligent models provided by the architect and suppliers may provide individual window frame and type quantities automatically. Estimators should review the totals provided to verify accuracy and ensure labor and installation costs are properly assigned to each window installation.

Labor costs are based on a fixed cost per unit method for manufactured window frame installation. Curtain wall installation labor costs are determined based on variables such as the height of the installation, glass type and size, and trim members. Labor costs are based on standard industry sources or company historical data. A general contractor commonly subcontracts curtain wall systems and large fixed-glass panel installations.

Sizing. Window size is determined based on information included on exterior elevations, wall sections, and floor plans. **See Figure 11-23.** Depending on the construction method used and the structural supporting members, the window supplier may take on-site measurements to verify the opening dimensions.

The size of small windows may be expressed on elevations by the size of the glazing. Window sizes are expressed in inches, with the width given first followed by the height. The two dimensions are separated by a slash. For example, a window with 32″ wide by 28″ high glazing is noted on elevations as 32/28.

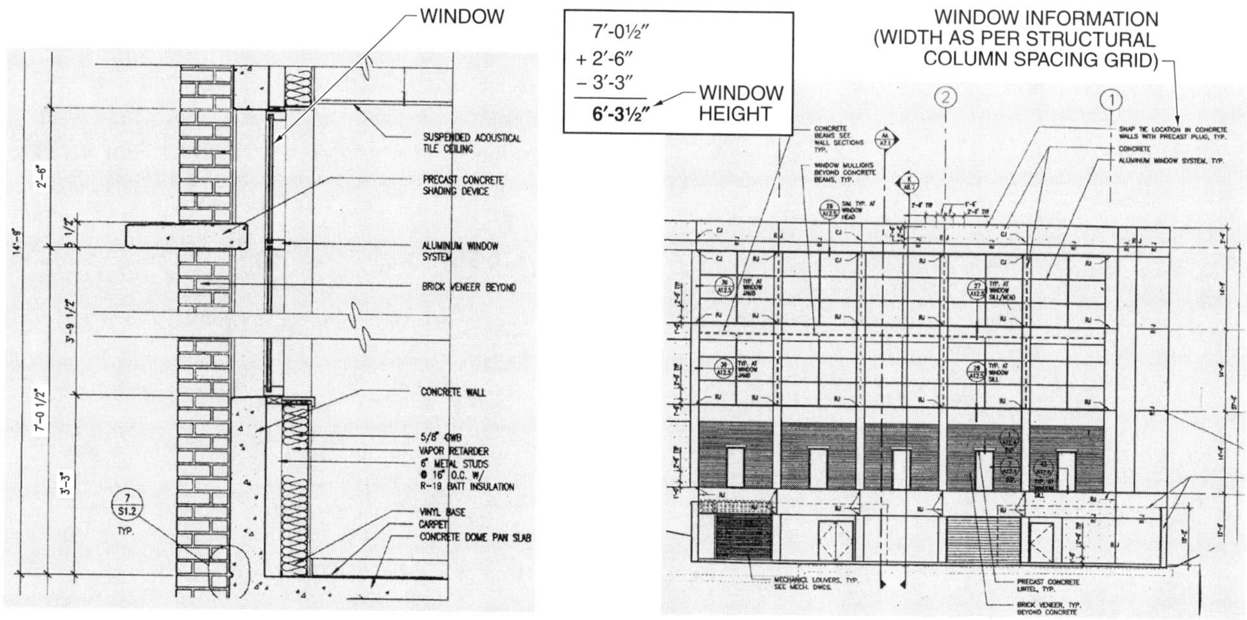

Figure 11-23. Window size is noted on exterior elevations and wall sections.

Curtain Wall Takeoff

Quantity takeoff for curtain walls includes frame and sash materials, anchoring systems, and facing materials such as glass, metal panels, plastic materials, precast panels, or other architectural panels. Scaffolding or lift equipment may be a consideration in curtain wall installation depending on job-site conditions and the curtain wall system being installed. Heavy curtain wall panels may require a crane or other lifting equipment to lift and hold panels in place during installation.

Curtain Wall Frame Takeoff. Estimators total the number of frame components for each section of curtain wall. The frame components include vertical and horizontal mullions, hardware required to fasten the system to structural members and at joints where frame members meet sash members, and flashing. Curtain wall frame members may be available in standard sizes from the manufacturer as specified by the architect or may need to be custom fabricated for the specific job. The quantity of each component is entered into a ledger sheet, spreadsheet, or estimating program.

Anchors are designed to provide for some adjustment both vertically and horizontally. Anchoring calculations include the number and types of anchors and the structural members to which the curtain wall members are fastened. Shimming of anchors may be required depending on the quality of the structural surfaces. Estimators also total the linear feet of trim and sealant members required, such as stops, frames, and gaskets.

Labor calculations for installation of curtain wall frames are based on company historical data or standard labor rate information. Items that may require additions or adjustments to these costs include job-site accessibility, lifting equipment or scaffolding requirements, and safety protection for installation at heights.

Curtain Wall Panel Takeoff. The size and type of each curtain wall panel member is entered into a ledger sheet, spreadsheet, or estimating program. Each panel should be identified as to the type of material (such as glass or metal), size, and installation location. Manufacturers of curtain wall systems use many standard panel sizes.

Some standard glass sizes are noted in standard or officially published lists, such as those published by the National Glass Association. Plate glass and mirrors are estimated by the square foot. The allowance for breakage is usually between 3% and 6%. Labor installation cost considerations are similar to those for the framing system and normally included as one cost for installation of framing and panels.

SAGE TIMBERLINE ESTIMATING . . .

CREATING AN ASSEMBLY

An assembly is created in the Sage Timberline Estimating program by taking off several items and saving these items as an assembly.

1. Start a new estimate called **Wyee & Associates** and link the estimate to the **Sample Commercial GC** database.
2. Pick the **Database** tab and pick **Phases** in the Phase/Item Group.
3. Start a new Phase. Pick the **Add** button.

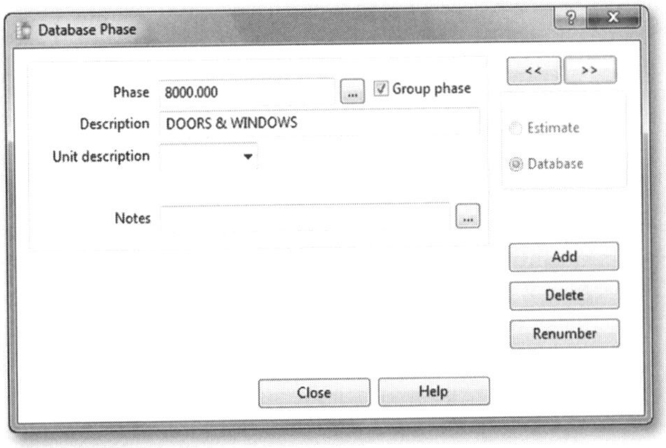

4. Enter **8210.000** in the **Phase** cell, **3-0x6-8 6-Panel Door** in the **Description** cell, and **ea** from the menu in the **Unit description** cell. Pick **OK** to accept and **Close** to close. Repeat to create a second phase. Enter **8630.000**, in the **Phase** cell, **Finish Hardware**, in the **Description** cell, and **ls** from the menu in the **Unit description** cell. Pick **OK** to accept and **Close** to close.

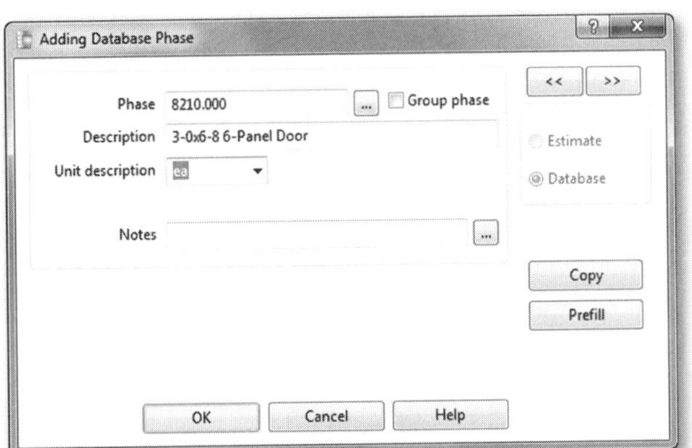

. . . SAGE TIMBERLINE ESTIMATING . . .

5. Add a new item. Pick the **Items** tab in the Phase/Item Group. Pick the **Add** button.

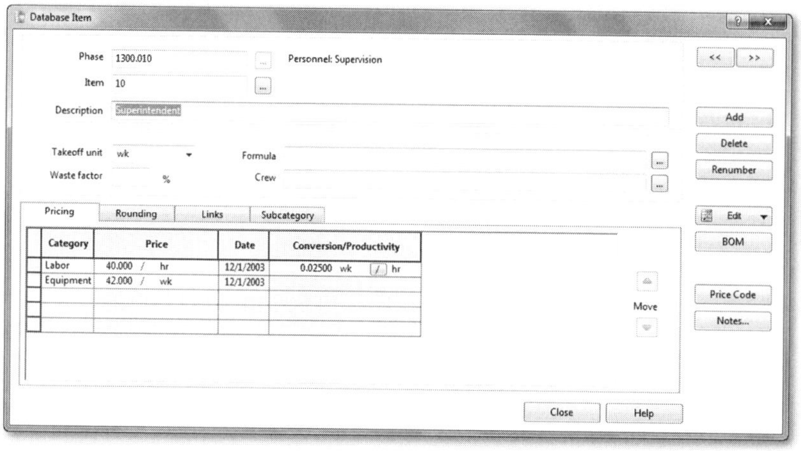

6. Enter **8210.000** in the **Phase** cell, **1130** in the **Item** cell, **3-0x6-8 6-Panel Steel Door w/SL** in the **Description** cell, and **ea** from the menu in the **Takeoff unit** cell. In the **Pricing Category**, select **Labor** in the first cell and enter **20.00** for the price, then select **Material** in the second cell and enter **120.00** for the price. Pick **OK** to accept.

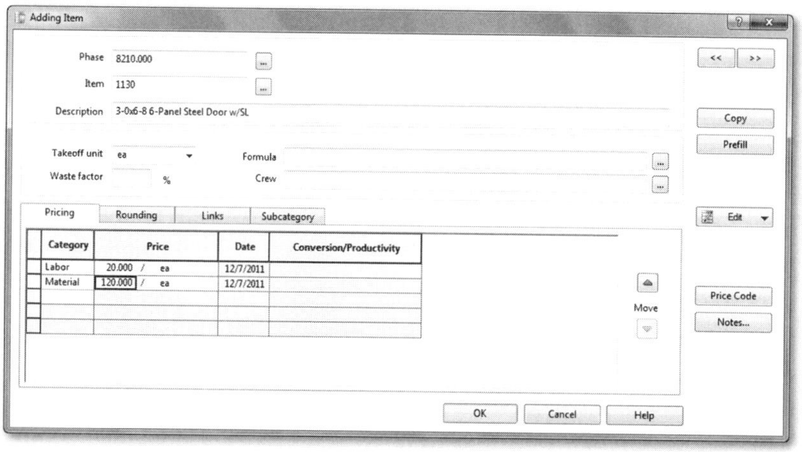

7. Repeat to add a second item. Pick the **Add** button. Enter **8630.000** in the **Phase** cell, **10** in the **Item** cell, **Finish Hardware** in the **Description** cell, and **ls** from the menu in the **Takeoff unit** cell. In the **Pricing Category**, select **Labor** in the first cell and enter **20.00** for the price, then select **Material** in the second cell and enter **20.00** for the price. Pick **OK** to accept.
8. Repeat to add a third item. Pick the **Add** button. Enter **8630.000** in the **Phase** cell, **21** in the **Item** cell, **Keyed Lock Exterior** in the **Description** cell, and **ea** from the menu in the **Takeoff unit** cell. In the **Pricing Category**, select **Labor** in the first cell and enter **20.00** for the price, then select **Material** in the second cell and enter **40.00** for the price. Pick **OK** to accept.
9. Repeat to add a fourth item. Pick the **Add** button. Enter **8630.000** in the **Phase** cell, **30** in the **Item** cell, **Deadbolt** in the **Description** cell, and **ea** from the menu in the **Takeoff unit** cell. In the **Pricing Category**, select **Labor** in the first cell and enter **20.00** for thc price, then select **Material** in the second cell and enter **20.00** for the price. Pick **OK** to accept and **Close** to close.

. . . SAGE TIMBERLINE ESTIMATING

10. Pick the **Home** task, then pick the **Item** task in the takeoff group.
11. The **Item Takeoff** window is now open. Double-click on **8000.000 DOORS & WINDOWS**. Then double-click on **8210.000 3-0x6-8 6-Panel Steel Door** and **8630.000 Finish Hardware**. Then double-click on **1130 3-0x8-6 6-Panel Steel Door w/SL**, **10 Finish Hardware**, **21 Keyed Lock Exterior**, and **30 Deadbolt**.

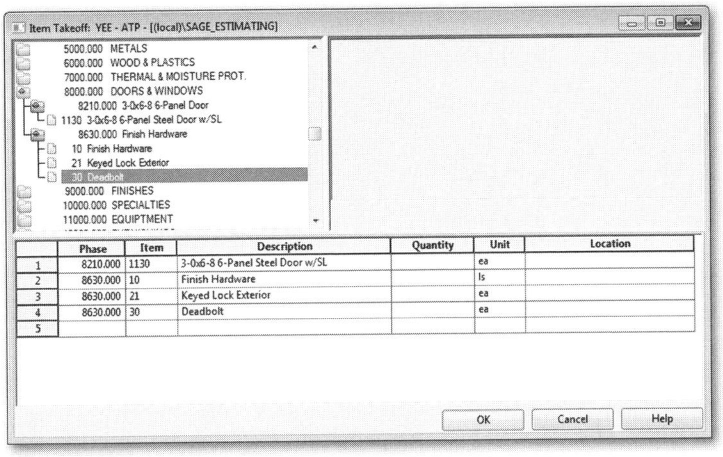

12. Pick the **Database** tab, then pick the **Update Database** task in the takeoff group. Select **Save as Assembly** from the menu.
13. The **Adding Assembly Windows** window is now open. Enter **10** in the **Assembly** cell, **6-Panel Steel Door Assembly** in the **Description** cell, **ea** as the assembly unit. Pick **OK** to accept and **Close** to close.

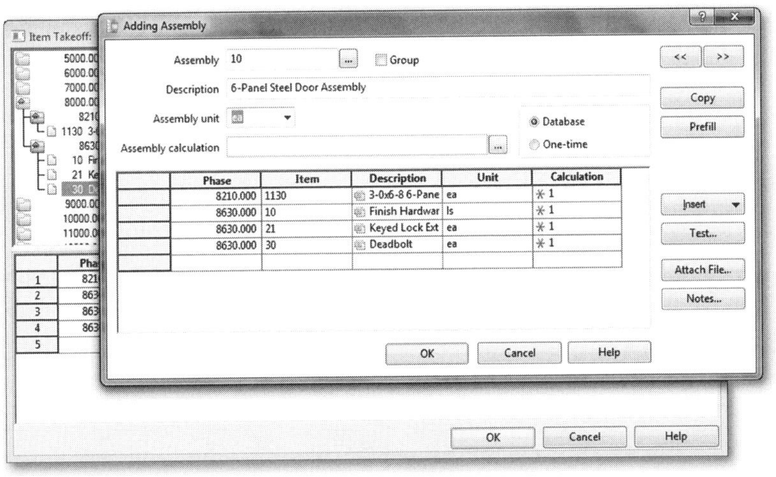

14. Pick **Assembly** in the **Takeoff/Review** task button to see the estimate name added to the **Assembly** list. Double-click on **Wyee & Associates** to view the new assembly. Double-click the assembly name to perform an assembly takeoff.

Quick Quiz®

Quick Quiz®

Refer to the CD-ROM for the Quick Quiz® questions related to chapter content.

Key Terms

Illustrated Glossary

- *access door*
- *curtain wall*
- *door closer*
- *entrance door*
- *fire door*
- *float glass*
- *folding door*
- *glazing*
- *hinge*
- *light*
- *lockset*
- *mullion*
- *overhead door*
- *plate glass*
- *sheet glass*
- *skylight*
- *sliding door*

Web Links

Web Links

American Architectural Manufacturers Association
www.aamanet.org

American Association of Automatic Door Manufacturers
www.aaadm.com

American Hardware Manufacturers Association
www.ahma.org

Builders Hardware Manufacturers Association
www.buildershardware.com

Door and Access Systems Manufacturers Association
www.dasma.com

Door and Hardware Institute
www.dhi.org

Glass Association of North America
www.glasswebsite.com

International Door Association
www.doors.org

National Association of Architectural Metal Manufacturers
www.naamm.org

National Fenestration Rating Council
www.nfrc.org

National Glass Association
www.glass.org

Steel Door Institute
www.steeldoor.org

Steel Window Institute
www.steelwindows.com

Window and Door Manufacturers Association
www.wdma.com

Openings 11

Review Questions

Estimating

______ **1.** Information related to door and window materials and methods is included in Division ___ of the CSI MasterFormat™.

______ **2.** A(n) ___ is a pane of glass or translucent material in a door.

______ **3.** A(n) ___ door is a fire-resistant door and assembly, including the frame and hardware, commonly equipped with an automatic door closer.

______ **4.** A(n) door ___ is a device that closes a door and controls the speed and closing action of the door.

______ **5.** A(n) ___ door is used to enclose an area that houses concealed equipment.

A. access
B. exit
C. barrier
D. entrance

______ **6.** A(n) ___ gate moves vertically to coil around a steel rod at the top.

______ **7.** A(n) ___ door is formed with panel sections joined with hinges along the vertical edges of the panels and supported by rollers in a horizontal upper track.

______ **8.** A(n) ___ door is a horizontal-moving door suspended on rollers that travel in a track that is fixed at the top of the opening or that moves on rollers mounted in the bottom of the door.

______ **9.** A(n) ___ door is a door which, when opened, is suspended in a horizontal track above the opening.

T F **10.** A hinge is pivoting hardware that joins two surfaces or objects and allows them to swing around a pivot.

______ **11.** A(n) ___ is the complete assembly of a bolt, knobs, escutcheon, and all mechanical components for securing a door and providing means for opening.

______ **12.** A(n) ___ is a roof opening covered with glass or plastic designed to let in light.

______ **13.** ___ is the transparent or translucent material installed in a window opening.

______ **14.** ___ glass is clear or translucent glass manufactured in continuous long, flat pieces that are cut to desired sizes and shapes.

______ **15.** A(n) ___ wall is a nonload-bearing prefabricated panel suspended on or fastened to structural members.

Short Answer

1. Describe one estimating method that can be implemented to ensure that all doors, door frames, and hardware are included in an estimate.

2. Identify three factors that are used to determine labor costs for door installation.

3. List the information required for window quantity takeoff.

4. Explain why the height at which windows are installed may impact the overall cost for window installation.

CONSTRUCTION ESTIMATING

Openings 11

Activities

Activity 11-1—Ledger Sheet Activity

Refer to the Door Schedule and Quantity Sheet No. 11-1. Perform a quantity takeoff of the doors based on the door schedule. List each door by type and the quantity required.

DOOR SCHEDULE			
Number	**Type**	**Size**	**Material**
101	Flush solid core	3′-0″ × 7′-0″	Oak
102	Panel	3′-0″ × 7′-0″	STL
103	Flush	3′-0″ × 7′-0″	STL
104	Flush solid core	3′-0″ × 6′-8″	Oak
105	Flush	4′-0″ × 7′-0″	STL
106	Flush solid core	3′-0″ × 7′-0″	Oak
107	Panel	3′-0″ × 6′-8″	STL
108	Panel	3′-0″ × 6′-8″	STL
109	Panel	3′-0″ × 7′-0″	STL
110	Panel	3′-0″ × 7′-0″	STL
111	Flush	4′-0″ × 7′-0″	STL
112	Flush solid core	3′-0″ × 6′-8″	Oak
113	Flush solid core	3′-0″ × 6′-8″	Oak
114	Flush solid core	3′-0″ × 7′-0″	Oak
115	Panel	3′-0″ × 6′-8″	STL

QUANTITY SHEET

Project: ____________________ **Sheet No.** 11-1

Estimator: ____________________ **Date:** ____________________

Checked: ____________________

No.	Description	Dimensions										
		W	H			Unit		Unit		Unit		Unit
	Doors											

Activity 11-2—Door Takeoff I

Refer to Print 11-2 and Estimate Summary Spreadsheet No. 11-2 on the CD-ROM. Take off the doors, frames, and hinges.

Activity 11-3—Door Takeoff II

Refer to the cost data below and Estimate Summary Spreadsheet No. 11-3 on the CD-ROM. Take off four 3′-0″ × 7′-0″ commercial steel doors, frames, and associated hardware. The hardware includes hinges, locksets, pull handles, and push bars. Each door requires 1½ pairs of hinge material. In addition, perform a takeoff of twelve 3′-0″ × 3′-0″ aluminum casement windows. Determine the total material and labor cost for the 4 doors and 12 windows.

COST DATA			
Material	**Unit**	**Material Unit Cost***	**Labor Unit Cost***
Steel doors, 3′-0″ x 7′-0″	ea	179.30	24.20
Frames, steel door, 3′-0″ x 7′-0″	ea	205.70	84.15
Hinges	pair	19.31	—
Locksets	ea	70.95	24.20
Pull handles and push bars	ea	115.50	21.84
Aluminum casement windows, 3′-0″ x 3′-0″	ea	247.50	53.90

* in $

Activity 11-4—Sage Timberline Estimating Activity

Create a new estimate and name it Activity 11-4. Create Database Phases **8280.000** with description **Glass Door** and **8630.000** with description **Finish Hardware**. Create Database Items **8280.000 80 8-0 × 6-8 Full Glass Alum Door**, **8630.000 10 Hardware Allowance**, and **8630.000 20 Finish Hardware**. Take off items **8280.000 80 8-0 × 6-8 Full Glass Alum Door**, **8630.000 10 Hardware Allowance**, and **8630.000 20 Finish Hardware** for 25 doors. Create an assembly called **08** with an appropriate description. The cost of finish hardware is $45.00/door. Print a standard estimate report.

CONSTRUCTION ESTIMATING

Finishes 12

Key Concepts

- Estimators determine the areas of walls and ceilings (in square feet) to begin quantity takeoffs for interior wall and ceiling finish materials.
- The area of flat ceilings is determined in the same manner as the area of flooring; sloping ceiling areas are calculated in the same manner as roof sheathing for a sloping roof.
- Components in the quantity takeoff for plaster finish areas include lath, plaster, and accessories such as trim members and control joints.
- The coverage area (in square feet) is used to determine tile and terrazzo material quantities.
- Tile moldings and trim members are taken off in linear feet.
- Material quantities to be taken off for suspended ceilings include hanger wire, clips or fasteners used to fasten the wire to the structural ceiling, main tees, cross tees, L-channel, splines, and ceiling finish materials.
- Wood flooring, such as interlocking plank flooring, is taken off by the square foot; parquet and block flooring is taken off by the square foot or by the number of pieces required for the area to be covered.
- The quantity of resilient flooring required is taken off by the square foot or by the square yard.
- Carpet is taken off by the square yard, with allowances made for carpet roll width, length, and pattern.
- Paint coverage is based on the number of square feet covered per quart or per gallon.

Introduction

Interior walls and ceilings are finished with a variety of gypsum and plaster surfaces. Tile and terrazzo provide durable and long-lasting finishes for interior areas. Ceiling finishes include

furred and suspended ceilings. For furred ceilings, furring is attached directly to structural members with ceiling tiles, gypsum board, or lath and plaster attached to the furring members. For suspended ceilings, wires, tees, cross tees, tiles, and other devices are hung from structural members to create a finish ceiling system. Commercial floor finishes include masonry, hardwood, resilient flooring, carpet, and specialized finishes such as asphalt composition floors. Wall-covering materials include paints, primers, stains, sealers, and wallpaper.

GYPSUM AND PLASTER MATERIALS AND METHODS

Interior and exterior walls and ceilings are finished with a variety of gypsum and plaster surfaces. Gypsum board is commonly used as a wall or ceiling finish material and is fastened to structural members. Plaster is composed of various cementitious materials and aggregate that form a plastic material when combined with water. Plaster hardens into a solid wall and ceiling finish. Plaster is held in place with various types of lath and may receive different surface finishes.

Gypsum Board

Gypsum board is the most common material used for interior wall and ceiling finish. Gypsum board, also known as drywall, is fastened to structural members with screws or nails. **See Figure 12-1.** A vapor barrier may be installed between the structural members and the gypsum board. The joints between the gypsum board panels are sealed with joint compound or paper tape or are framed with various metal trim members.

Gypsum Board Composition. Gypsum board sheets are made of a core of gypsum coated with layers of paper. Gypsum board is available in 4′ × 8′ sheets, but 12′ and 14′ lengths are more common for commercial applications because they minimize installation time. Thicknesses vary in ⅛″ increments, ranging from ¼″ to 1″. Special types of gypsum board are designed for fireproofing applications. Special paper finishes are applied during manufacturing to create waterproof grades. Gypsum board may have tapered, beveled, square, tongue-and-groove, rounded, or rounded-tapered edges.

Gypsum Board Accessories. Gypsum board accessories include various fasteners, corner beads, control joints, and finishing materials. Gypsum board is fastened to wood or metal structural members and furring strips with self-tapping drywall screws or nails. Where framing members are spaced at 16″ OC, screws are spaced at a maximum of 12″ OC on ceilings and a maximum of 16″ OC on walls. Where framing members are spaced 24″ OC, screws are spaced a maximum of 12″ OC on walls and ceilings.

Hilti, Inc.

Figure 12-1. Gypsum board sheets are applied to structural supporting members as a finish material for walls and ceilings.

In addition to the use of mechanical fasteners, gypsum board may be glued to framing members with adhesive. Metal corner beads, control joints, and end caps may be applied as noted on the details or in the specifications. Joints between sheets are finished with joint compound and paper tape. **See Figure 12-2.** Several coats of joint compound are applied and sanded after drying to create a smooth finish.

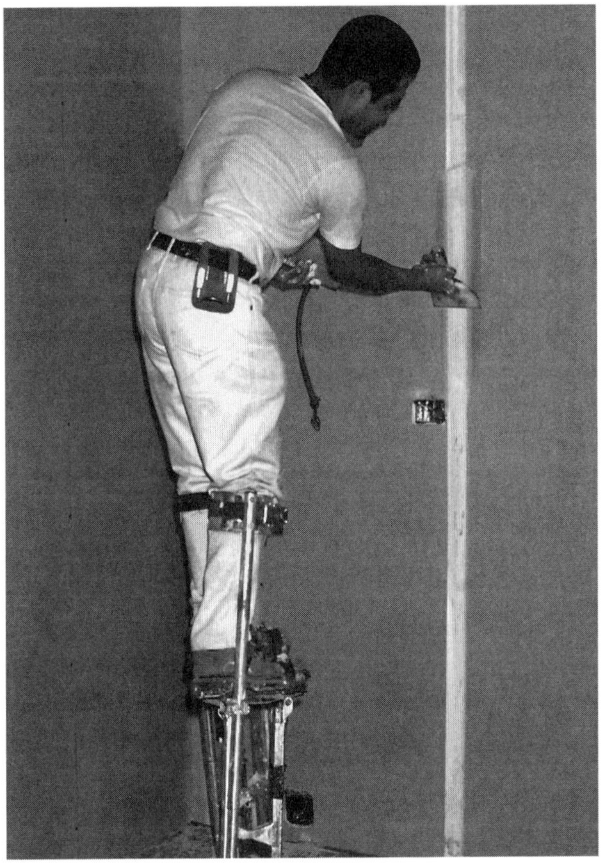

Figure 12-2. Fastener holes and joints between gypsum board sheets are filled with joint compound and sanded to a smooth finish.

Plaster Finishes

Interior and exterior building surfaces may be coated with various plaster finishes. The surface of the plaster is troweled smooth, swirled, or left with a rough finish as noted by the architect in the specifications or shown on the elevations. **See Figure 12-3.** Plaster wall finish materials include portland cement plaster on metal lath and thin-coat finishes applied directly to gypsum board. Control joints and trim pieces are installed to create a complete finish system.

Figure 12-3. Plaster provides a durable surface finish that may be smooth, swirled, or left with a rough finish as specified by the architect.

Lath. *Lath* is the backing fastened to structural members onto which the plaster is applied. Metal lath is tied, screwed, or clipped to structural members, such as metal studs or furring. Metal lath is available in sheets of diamond mesh or rib expanded metal. **See Figure 12-4.** Metal lath is available in 27″ wide by 8′-0″ long sheets. Metal lath sheets are made of painted or galvanized steel and may also be manufactured with a paper backing material for vapor resistance. Metal lath commonly contains recycled steel, thereby increasing the sustainability rating for a structure.

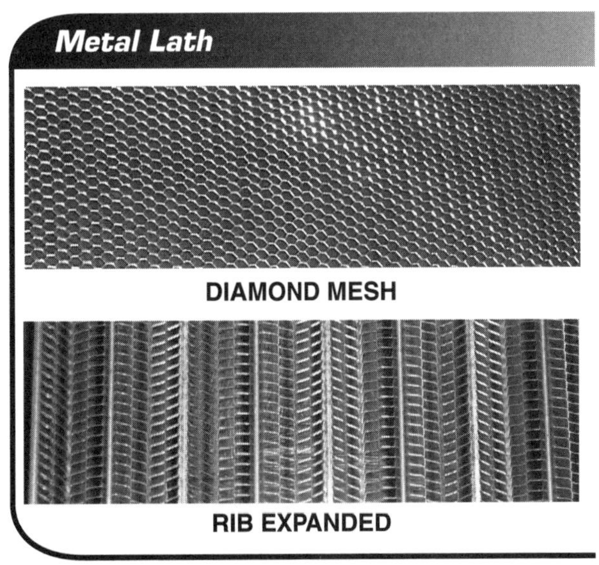

Figure 12-4. Metal lath is available in sheets of diamond mesh or rib expanded metal.

Plaster. *Plaster* is a mixture of portland cement, water, and sand used as an interior and exterior wall finish. Plaster may be applied in two or three thin coats until the desired thickness is achieved. A base (scratch) coat of plaster is applied to the lath and left with a rough finish. A second (brown) coat is applied over the surface of the base coat. A third coat is applied to add strength to the plaster finish. Keene's cement is used for applications where a very smooth finish is required. Keene's cement is a hard, dense-finish plaster.

Plaster Accessories. Grounds, corner beads, and control joints are installed as part of the lath-and-plaster finish system. **See Figure 12-5.** A *ground* is a wood or metal shape used as an edge for plaster material and a gauge for plaster thickness. A *corner bead* is a light-gauge, L-shaped galvanized metal or vinyl device used to cover outside corner joints in plaster walls. Corner beads are available in a wide variety of shapes and sizes depending on their application.

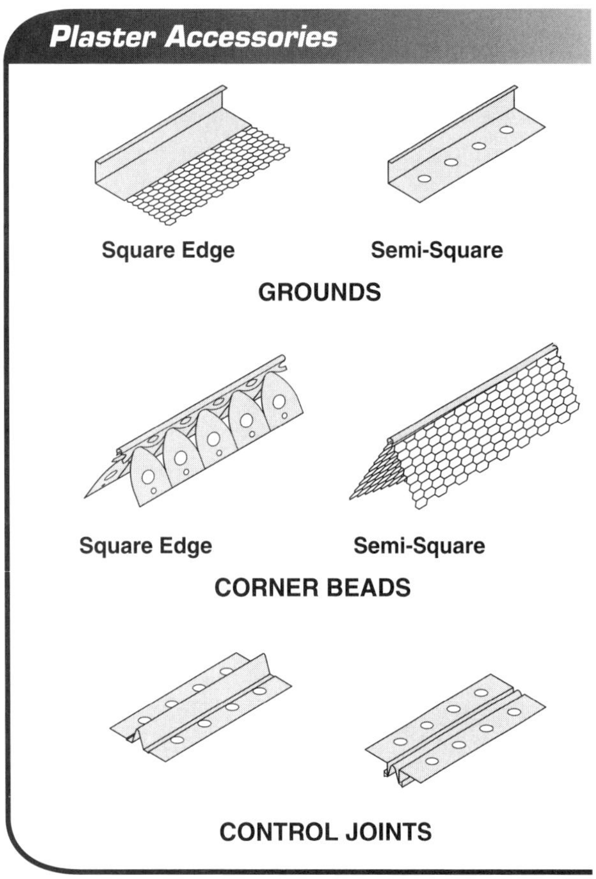

Figure 12-5. Grounds, corner beads, and control joints are accessories used with a lath-and-plaster finish system.

A *control joint* is a thin strip of perforated metal or vinyl applied in lath-and-plaster surfaces to relieve stress resulting from expansion and contraction in large ceiling and wall surfaces. As with corner beads, control joints are made from light-gauge metal and are available in many styles and shapes depending on the application.

Plaster accessories are made from a variety of materials including galvanized steel, zinc alloy, anodized aluminum, vinyl, polyvinyl chloride (PVC), and chlorinated polyvinyl chloride (CPVC). Anodized aluminum and vinyl lath accessories are used for interior work. Most lath accessories are installed before metal lath is installed and the metal lath is then butted or lapped over the flanges.

GYPSUM AND PLASTER QUANTITY TAKEOFF

Estimators determine the areas (in square feet) of walls and ceilings to begin calculation for interior wall and ceiling finish material quantities. Wall length and height are used to determine interior wall areas, with allowances made for larger or multiple door and window openings. The area of flat ceilings for gypsum board and plaster ceiling finishes is determined in the same manner as the area of flooring. Sloping ceiling areas are calculated in the same manner as roof sheathing for a sloping roof, with the sloped surface providing the actual ceiling dimensions. Quantity takeoffs for gypsum board and plaster finish are commonly performed by specialty contractors and are included as a bid submitted to the general contractor. BIM systems may provide some square foot quantity information.

Gypsum Board Takeoff

Gypsum board thickness and face finish are shown on details and room finish schedules. Takeoff is based on the area of coverage (in square feet). Estimators refer to the prints to determine the number of layers of gypsum board required and the direction of application. For some applications, two or more layers of gypsum board may be specified. For most commercial wall applications, gypsum board is applied with the long dimension of the sheet oriented vertically. Residential applications commonly use horizontal orientation.

Orientation of the gypsum board may affect the length of the sheets ordered. The area of coverage for each gypsum board sheet depends on the width and length of the sheets to be installed. Where possible, sheet lengths are matched to room sizes to prevent an excessive number of joints between sheets. The quantity of joint compound required is also based on area calculations. Trim members are taken off by the linear foot or by the piece.

For gypsum board finishes, ledger sheets and spreadsheets include columns for the square foot quantity in each room, type and number of gypsum board sheets required, linear feet of trim members, joint finishing materials, and labor. Estimating programs may perform calculations for gypsum board finishes using either the item or assembly method. Gypsum board finish materials may be included in a typical wall assembly for interior framed walls. **See Figure 12-6.**

Gypsum board may be applied to walls and ceilings.

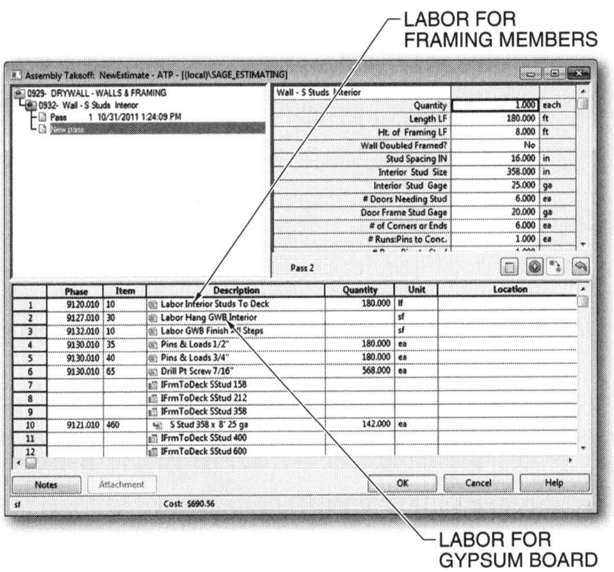

Figure 12-6. Gypsum board is commonly included in wall frame and finish assemblies in estimating programs.

Additional factors to consider when estimating gypsum board application costs may be the need for heat or weather protection during cold weather installation to allow for proper drying of the joint compound, and the moisture content of the gypsum board during finishing. Scaffolding or lift equipment may also be required for installations at heights.

Gypsum Board Sheet Takeoff. The number of sheets of gypsum board required for an application is determined by calculating the area of the wall or ceiling surface and dividing by the area of the gypsum board sheet to be installed. The wall area (in square feet) is determined by multiplying the wall length (in feet) by the wall height (in feet). Deductions are made for large door and window openings where significant amounts of gypsum board may be omitted.

The number of sheets of gypsum board for flat ceilings is determined in the same manner as sheathing for a subfloor. The floor area (in square feet) is calculated to determine the ceiling area. The number of sheets of gypsum board for sloped or cathedral ceilings is estimated using the same method as sheathing for roofs. Rafter length is multiplied by the length of the room to calculate the area of gypsum board.

When the number of required sheets is determined, a waste factor of 8% is added. When this calculation is completed, the length of the sheets to be ordered is verified to determine the best possible coverage and use of labor. For example, if an area is 13′-8″ long, ordering 14′ long sheets is more economical than ordering 12′ long sheets.

Determining the Number of Gypsum Board Sheets

Determine the number of 4′ × 8′ sheets of gypsum board required for the walls and ceiling of a room measuring 24′-6″ × 13′-0″. One wall has a 3′-0″ × 7′-0″ door and another wall has a 2′-6″ × 4′-6″ window.

1. Determine the total wall length.

total wall length = (*wall 1 length* × 2) + (*wall 2 length* × 2)

total wall length = (24.5′ × 2) + (13′ × 2)

total wall length = 75′

2. Determine the gross area of the walls.

gross wall area = *total wall length* × *wall height*

gross wall area = 75′ × 8′

gross wall area = 600 sq ft

3. Determine the area of the openings.

area of openings = (*opening 1 width* × *opening 1 length*) + (*opening 2 width* × *opening 2 length*)

area of openings = (3′ × 7′) + (2.5′ × 4.5′)

area of openings = 21 sq ft + 11.25 sq ft

area of openings = 32.25 sq ft

4. Determine the net area of the walls.

net wall area = *gross area* − *area of openings*

net wall area = 600 sq ft − 32.25 sq ft

net wall area = 567.75 sq ft

5. Determine the area of the ceiling.

ceiling area = *room width* × *room length*

ceiling area = 24.5′ × 13′

ceiling area = 318.5 sq ft

6. Determine the overall area of the surfaces to be covered.

room area = *net wall area* + *ceiling area*

room area = 567.75 sq ft + 318.5 sq ft

room area = 886.25 sq ft

7. Determine the number of gypsum board sheets (without waste).

number of sheets = *room area* ÷ *area of gypsum board sheet*

number of sheets = 886.25 ÷ 32

number of sheets = 27.7 sheets

8. Determine the total number of gypsum board sheets.

total number of sheets = *number of sheets* × *waste factor*

total number of sheets = 27.7 × 1.08

total number of sheets **= 29.9; round to 30 sheets**

Gypsum Board Accessories. Corner bead quantities are calculated by the linear foot. The number of outside corners requiring beads is determined, and the total length of bead is calculated for each corner. The linear feet of corner beads required for curved edges is determined using geometric formulas for linear curves.

For sheets of gypsum board, allow 10 lb of screws or nails per 1000 sq ft of gypsum board. The number of fasteners per sheet varies with the sheet size. Fastener requirements may be noted in the local building code or manufacturer specifications. Joint compound and paper tape for seams and corners are estimated based on formulas linked to the wall and ceiling area (in square feet). Manufacturer information is commonly used to perform these calculations. **See Figure 12-7.**

Plaster Finish Takeoff

Items in the quantity takeoff for plaster finish areas include lath, plaster, and accessories such as trim members and control joints. Plaster quantities are based on the area (in square yards) to be covered. Lath quantities are based on the area (in square feet) to be covered. Control joints are installed at locations shown on the prints or detailed in the specifications. Control joints are normally calculated by the piece or the number of linear feet. Costs for plaster materials and labor are typically based on the cost per square foot from standard industry sources or company historical data.

Hanger wire is used to suspend a plaster ceiling from the supporting members. To determine hanger wire length requirements, add the length of the hanger wire from the support to the ceiling and 12″ (6″ on each end to tie to the channels and supporting member). Multiply the number of wires by the wire length to determine the total linear feet required.

JOINT COMPOUND SPECIFICATIONS		
Product	**Container Size**	**Amount Required for 1000 sq ft of Coverage**
Sheetrock® Ready-Mixed Joint Compound	12 lb pail 42 lb pail 61.7 lb pail 48 lb carton 50 lb carton 61.7 lb carton	138 lb
Sheetrock® Lightweight All-Purpose Joint Compound (Plus 3)	1 gal pail 4.5 gal pail 4.5 gal carton 3.5 gal carton	9.4 gal
Sheetrock® Powder Joint Compound	25 lb bag	83 lb
Sheetrock® Lightweight All-Purpose Joint Compound (AP Lite)	20 lb bag	67 lb
Sheetrock® Setting-Type Joint Compound (Durabond)	25 lb bag	72 lb
Sheetrock® Setting-Type Joint Compound (Easy Sand)	18 lb bag	52 lb

Figure 12-7. Joint compound is available in many forms and standard quantities. Different types of joint compound have different applications and provide varying coverage per 1000 sq ft.

When using a ledger sheet or spreadsheet, the area (in square feet) of the surfaces to be plastered in each room is entered into the appropriate row and column. The ledger sheet or spreadsheet for each room or space should also include columns for lath, plaster, accessories, and labor. When using an estimating program, either the item or assembly takeoff method can be used to calculate lath-and-plaster materials. **See Figure 12-8.** Additional factors to consider for lath-and-plaster applications may be the provision of heat or weather protection during cold-weather installation. Scaffolding or lift equipment may also be required for installations at heights.

Plaster. Plaster quantities for plain surfaces, such as walls and ceilings, are calculated by the square yard (1 sq yd = 9 sq ft). Plaster quantities for beams, pilasters, or specialty work are calculated by the square foot. Estimators make allowances for wall openings by deducting the opening areas from the total wall area.

To determine the overall size of an area 1 cu yd of plaster will cover, the number of ⅛″ slices must be calculated. The ⅛″ slice is used as the basic unit of calculation because it is the thinnest coat of material that will be applied in plasterwork. All common plaster thickness are multiples of ⅛″. One cubic yard of plaster (3′ × 3′ × 3′) will yield 288 slices that are ⅛″ thick by 3′ wide by 3′ long. In other words, one cubic yard of plaster will cover 2592 sq ft of surface to a depth of ⅛″ (288 slices × 9 sq ft/slice = 2592 sq ft). **See Figure 12-9.**

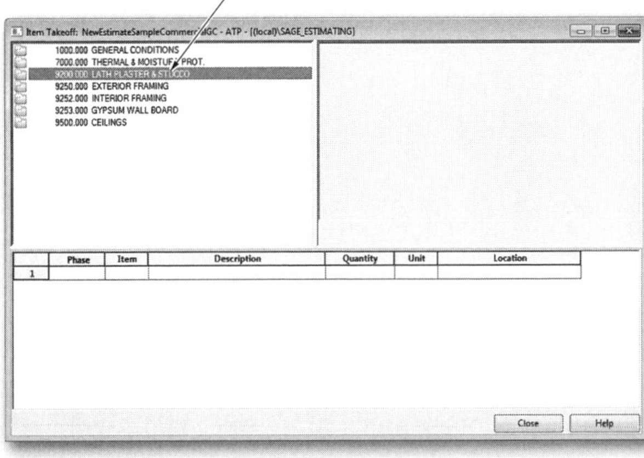

Figure 12-8. Either item or assembly takeoff may be used in an estimating program to calculate lath-and-plaster materials.

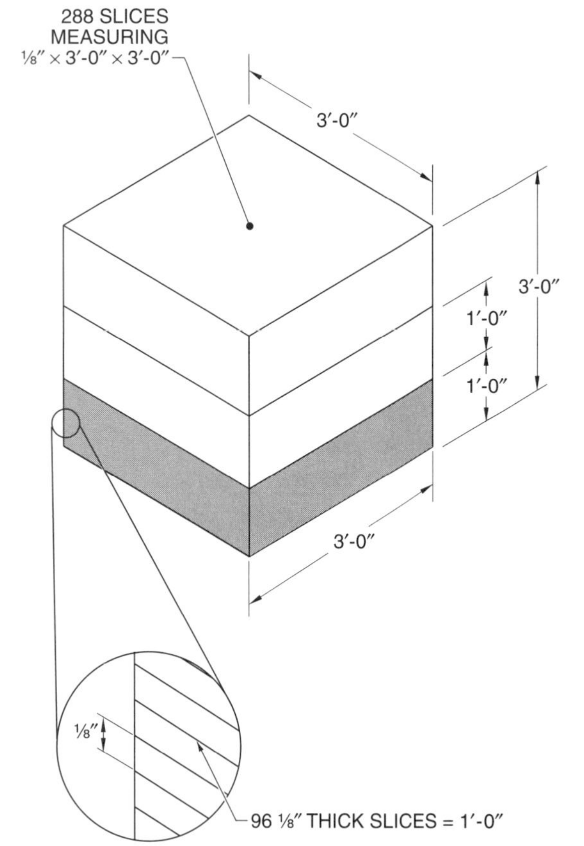

Figure 12-9. One cubic yard of plaster will cover 2592 sq ft of surface to a depth of 1/8″.

In jobs requiring sand as an aggregate, the volume (in cubic yards) of sand is determined first, and then the other material requirements are calculated. When sand and plaster are mixed in a proportion of 3 parts sand to 1 part plaster, the plaster will retain the same volume as the sand alone. Therefore, when 1 cu yd (2700 lb) of sand is mixed with 9 bags (900 lb) of plaster, the mixture will yield 1 cu yd of plaster.

To determine the amount of sand required to cover an area measuring ⅛″ × 3′ × 3′, the weight of 1 cu yd of sand is divided by 288 (2700 lb/cu yd ÷ 288 slices/cu yd = 9.375 lb). With a 3:1 ratio, the amount of plaster required to cover an area measuring ⅛″ × 3′ × 3′ is determined when the weight of the sand is divided by 3 (9.375 lb ÷ 3 = 3.125 lb). To adjust the amounts of sand and plaster for the total plaster thickness, the results are multiplied by the number of ⅛″ thick slices required.

For example, if ⅝″ thick plaster is specified, the amount of sand and plaster should be multiplied by 5 (9.375 lb × 5 = 46.875 lb sand; 3.125 lb × 5 = 15.625 lb plaster). Finally, to adjust the amounts of sand and plaster for the area to be plastered, the results are multiplied by the number of square yards of coverage required. For example, a wall measuring 8′-0″ high by 18′-0″ long (16 sq yd) requires 750 lb of sand and 250 lb of plaster (46.875 lb × 16 = 750 lb sand; 15.625 lb × 16 = 250 lb plaster). Approximately ⅓ cu yd of sand and 3 bags of plaster would be required to apply a ⅝″ thick plaster surface on a wall measuring 8′-0″ × 18′-0″.

Lath and Accessories. The amount of lath required for plaster walls is based on the area of coverage (in square feet or square yards). The type of lath is indicated on wall sections or provided in the specifications. The sheet size used for wall lath depends on the height of the ceiling and wall lengths. Plaster accessories are calculated by the linear feet of each type of material required.

Many types of fasteners are used for various types of metal lath. The type and spacing of the fasteners is indicated on detail drawings or in the specifications. Fastening requirements vary according to the type and weight of the lath. The amount of fasteners is based on the overall spacing and the square feet of lath to be installed. Fastening requirements may be specified in local building codes, manufacturer specifications, or project specifications.

TILE AND TERRAZZO MATERIALS AND METHODS

Tile is a thin building material made of cement, fired clay, glass, plastic, or stone. Tile may be applied to walls and ceilings and may be used as a ceiling-finish material in shower areas. *Terrazzo* is a mixture of colored stone, marble chips, or other decorative aggregate and a portland cement, modified portland cement, or resinous matrix. Terrazzo is used as a floor-surfacing material. It is placed wet and ground to a smooth, durable finish. Tile and terrazzo provide very durable and long-lasting finishes for interior areas. Information concerning these materials is found on floor plans, details, interior elevations, and specifications.

Tile

Tile may be ceramic tile or quarry tile. Ceramic and quarry tile is available in a variety of sizes, shapes, thicknesses, colors, and finishes. Ceramic and quarry tile may be applied with a cement base or set with mastic. **See Figure 12-10.** Where a cement base is specified, a bed of portland cement mortar is placed with reinforcement. A thin layer of mortar is applied to the top of the reinforcement. The tile is set directly into the thin layer of mortar. When mastic is specified, gypsum board, cement board, or another substrate is coated with adhesive using a grooved trowel. The tile is set into the mastic.

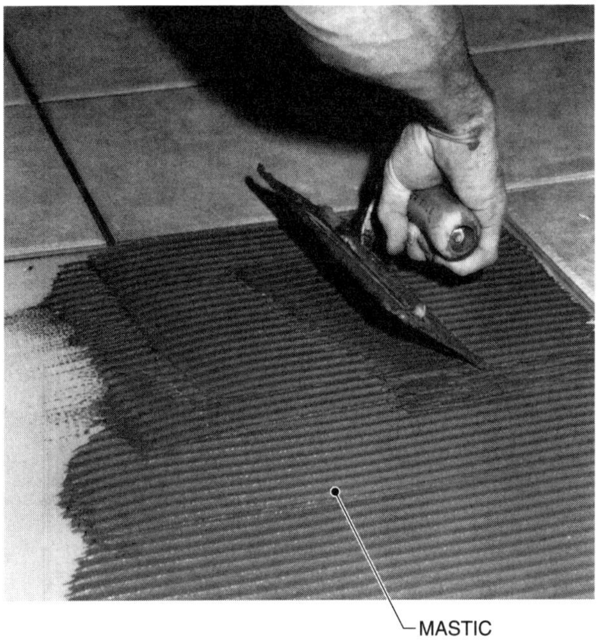

Figure 12-10. Ceramic and quarry tile may be applied with a cement base or set with mastic.

Tile installation typically starts in the middle of the work area and progresses out to the perimeter. Tile saws or tile cutters are used to cut the tile. After installation, the spaces between the tile are filled with grout. *Grout* is a thin, fluid mortar made of portland cement, fine aggregate, lime, and water. Excess grout is removed while wet, and the surface of the tile is buffed and/or polished after the grout has hardened.

Ceramic Tile. Ceramic tile is manufactured from clay powder. Ceramic tile is formed using a process that involves firing the clay while wet or using a process in which clay powder is compressed. Various manufacturers produce many different shapes and sizes of ceramic tile. The most common are square, rectangular, octagonal, hexagonal, and Spanish. Ceramic tile has a nominal thickness of 5⁄16″. Surface finishes may be bright or matte.

Many ceramic-tile trim members are available for installation at corners, caps, intersections, and bases. **See Figure 12-11.** A variety of ceramic-tile accessories are available, such as soap, towel, and toilet paper holders. Ceramic-tile accessories may be recessed or surface-mounted.

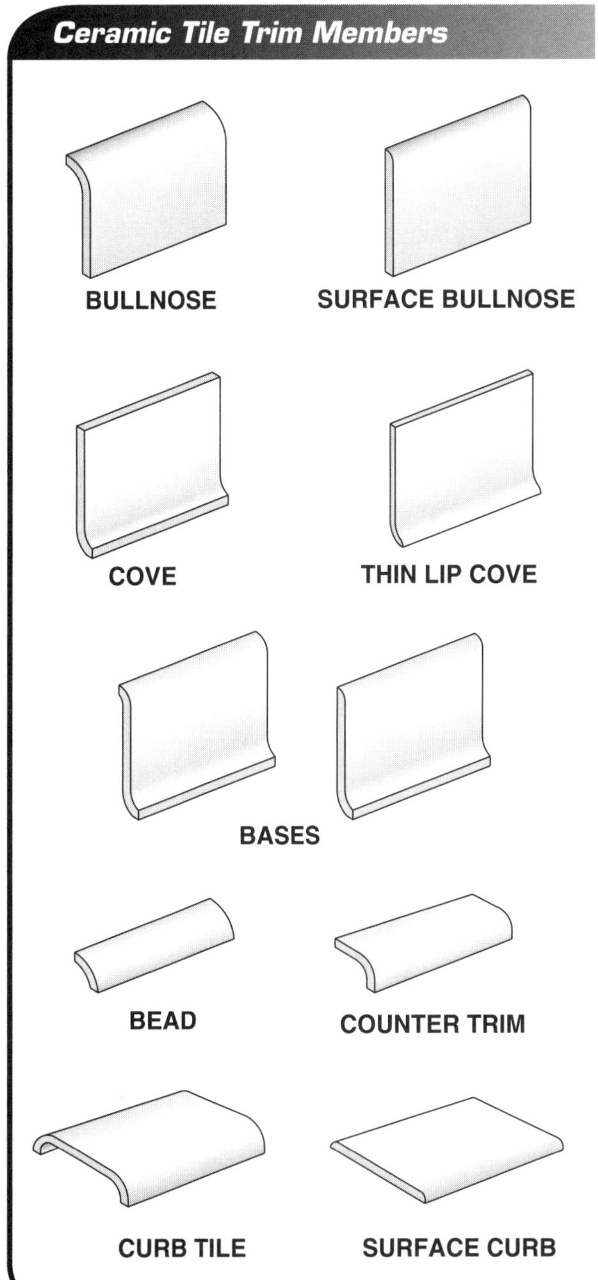

Figure 12-11. Ceramic-tile trim members are installed at corners, caps, intersections, and bases.

Quarry Tile. Quarry tile is formed of unglazed ceramic tile made of clay or shale. Quarry tile is primarily used as a floor finish material. The most common types are square, rectangular, hexagonal, and Spanish. Bullnose, cove, and windowsill trim members are also available. Nominal thicknesses for quarry tile are 3/8″, 1/2″, and 3/4″.

Adhesives and Grouts. Tile adhesives include portland cement or premixed mastics. For applications set with portland cement, materials required include sand, portland cement, and metal lath reinforcement. Premixed mastics are available in 1 gal. or 5 gal. quantities and are applied directly to wall, floor, and ceiling surfaces with a notched trowel as specified by the manufacturer.

Grout is applied in a semiliquid state to fill the joints between tiles. Grout is available in many colors and is purchased in premixed containers or powder form that is mixed with water. Grout is applied to fill all joints and is cleaned smooth after reaching a final set. The type of adhesive method and grout color are provided in the specifications or on details.

Terrazzo

Stone, marble, or glass chips are mixed with a portland cement or resinous base to form terrazzo. Terrazzo is placed in a manner similar to concrete floor slabs, formed, and screeded to the desired elevation. Terrazzo placement systems include setting the terrazzo on a sand cushion, bonding to an underbed, monolithic placement, thin set, and structural. After the terrazzo material is set, it is ground to a smooth finish, exposing the stone chips and concrete matrix. Chips used in terrazzo include marble, onyx, granite, quartz, and silica.

The four basic types of terrazzo include standard, Venetian, palladiana, and rustic. Standard terrazzo is formed with small stone chips. Venetian terrazzo is formed with large stone chips with some small chips integrated to fill voids. Palladiana terrazzo is composed of random pieces of marble, 3/8″ to 1″ thick and up to 15″ across, and set as slabs, with smaller chips filling openings. Rustic terrazzo is made by pressing the chips into the surface of the concrete base and eliminating the heavy grinding common with other applications.

Brass, copper, or aluminum dividers are installed to separate terrazzo floor areas and achieve the desired patterns on the floor surface. **See Figure 12-12.** Some precast terrazzo members such as stair treads, windowsills, and cove bases may be integrated in the installation.

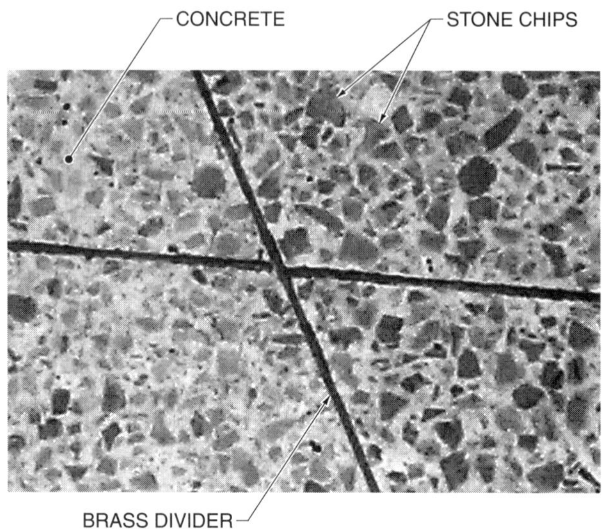

Figure 12-12. Terrazzo is formed by stone chips mixed with a concrete base. Brass, copper, or aluminum dividers may be used to provide an expansion joint and a decorative effect.

Binders. Portland cement or resinous materials are used to bind terrazzo chips together. Colors are added to the binder material to enhance the finished appearance. White or gray portland cement is used in combination with limestone or synthetic mineral pigments. Resinous binders include epoxy, polyester, polyacrylate, and latex.

TILE AND TERRAZZO QUANTITY TAKEOFF

The coverage area (in square feet) is used to determine tile and terrazzo material quantities. Estimators calculate the wall, floor, or ceiling areas to be covered and deduct areas of door, window, and floor openings. As with other materials, ledger sheets and spreadsheets contain rows and columns for various material components and labor. Estimating programs may use either the item or the assembly takeoff feature. The assembly takeoff feature includes all components required for construction. BIM use may provide square footage calculations for specific areas. Tile and terrazzo work are commonly performed by specialty contractors and are included as a bid submitted to general contractors.

Ceramic and Quarry Tile Takeoff

The quantity of ceramic and quarry tile needed is based on the coverage area (in square feet). Moldings required to finish edges are taken off in linear feet. Trim pieces such as bullnoses, caps, and base strips are also calculated in linear feet. Estimators must check each tile application to determine the type, size, tile finish, adhesion method, accessories, and grout needed. Quantities of adhesive and grout are determined according to manufacturer recommendations for the number of square feet of coverage per gallon. **See Figure 12-13.** Labor installation costs are based on company historical data or standard industry information.

PRODUCT DESCRIPTION AND USE: Color Tile 990W Ceramic Wall Tile Adhesive is designed for interior installation of ceramic and mosaic wall tile. Use over drywall, concrete, and plaster walls. Three hour open time provides easier installation. Adhesive is moisture resistant and conforms to ANSI Standard A.135.1 Type II. Do not use for setting of fixtures.
PREPARATION: No wall tile installation is better than the surface over which it is installed. Satisfactory results depend on proper preparatory work. For complete detailed instructions, refer to Color Tile Instruction Booklet. (1) Apply over clean, structurally sound surfaces. Surface must be free from all wax, dirt, and grease by cleaning with Color Tile recommended cleaner. Rinse and allow to dry. Remove poorly bonded paint by sanding or stripping. Remove old adhesive with Color Tile Old Adhesive Remover. (2) Fill all cracks with Color Tile 707 Redi-Mix or Speed Patch. Allow to harden 4–6 hours. (3) GYPSUM DRYWALL: WET AREAS AROUND TUBS. Where gypsum wallboard is used, cut 1/4″ around tub areas and fill with Color Tile Caulk or Color Tile 990W Ceramic and Mosaic Wall Tile Adhesive. (4) Apply two coats of Color Tile 880P Latex Floor and Wall Primer over all areas to be tiled. Allow final coat of 880P to dry.
ADHESIVE APPLICATION: Maintain adhesive and tile at room temperature for 24 hours before and after installation. Spread Color Tile 990W Adhesive with a Color Tile 8V notched trowel held at a 60° angle. Do not spread any more adhesive in an area than can be tiled within 3 hours. Spread adhesive with notched edge of trowel. Set tiles by using a slight twisting motion. DO NOT SLIDE TILES. If skin coat has formed on adhesive, scrape off dried adhesive and apply fresh material.
CLEANING: Remove excess adhesive by cleaning with Color Tile Cement Remover. Apply with a clean, dry cloth. Use only Color Tile Cement Remover to avoid damage to tiles. Do not attempt to grout for a period of 16–24 hours or until tiles are firmly set. Drying time is dependent on temperature and humidity.
COVERAGE: Approximately 10 square feet per quart.
CAUTION: KEEP OUT OF THE REACH OF CHILDREN. DO NOT TAKE INTERNALLY. Close cover when not in use.
NOTICE: Since the use of this product is beyond the control of seller, liability or damages, either incidental or consequential, shall be limited to replacement of the product.

Figure 12-13. Manufacturer information identifying the amount of coverage per gallon of adhesive is consulted to determine the amount of adhesive required.

Terrazzo Takeoff

Since terrazzo can be applied in many different ways, an estimator must take the type of subfloor into account when estimating terrazzo material and labor costs. If the terrazzo is to be bonded to an underbed, the volume of underbed material must be calculated. The volume of underbed material is calculated in cubic yards, similar to a cast-in-place concrete slab. The volume of terrazzo topping is then calculated separately as a cast-in-place concrete slab based on the floor area (in square feet) and the depth of the terrazzo.

When terrazzo is placed separate from the concrete slab, a thin bed of dry sand may be laid, and building paper laid on top, or a concrete primer applied to improve bonding. The underbed is placed and the terrazzo topping is applied over the underbed. The volumes (in cubic yards) of dry sand, underbed, and terrazzo topping are calculated separately, and the amount of tar paper is based on the floor area. When terrazzo is applied on wood floors, tar paper and an overlay of wire mesh are installed.

Dividers and precast terrazzo accessories are taken off in linear feet. The depth of the divider strips is considered when taking off dividers. Labor costs include preparation of the subsurface areas, layout and placement of the dividers and the terrazzo materials, and grinding and finishing. Coordination with project managers at the job site can determine if there are additional costs for cold-weather protection or protection of surrounding surfaces or areas during the placement and grinding operations.

CEILING MATERIALS AND METHODS

Ceiling applications include furred and suspended ceilings. For furred ceilings, wood or metal furring members are attached directly to structural members such as steel bar joists or concrete beams and slabs. Prefinished ceiling tile, gypsum board, or lath and plaster may be attached to the furring members. For suspended ceilings, a system of wires, tees, cross tees, tiles, and other devices are hung from structural members to create a finished ceiling.

Suspended Ceilings

A *suspended ceiling* is a ceiling hung from wires fastened to structural members. Suspended ceilings are installed a predetermined distance above the floor below and hung from structural members above to obscure the structural members and to create a finished ceiling. Hanger wires are attached to structural members above the finished area by tying them onto joists where possible or by using screws, nails, or other mechanical fasteners. The lower ends of the hanger wires are tied to a supporting metal gridwork. The gridwork is designed to support lay-in tiles, concealed grid tiles, gypsum board, or lath and plaster. The height of the finished ceiling surface above the floor is indicated on reflected ceiling plan notes, interior elevations, or room finish schedules.

Supporting Metal Gridwork. The supporting metal gridwork for suspended ceilings consists of interlocking main runners and cross tees. The height of the metal gridwork is determined after hanger wires are secured in place. L-channels are fastened to the walls around the perimeter of the area to support ceiling finish materials. **See Figure 12-14.**

Figure 12-14. Suspended ceiling systems create a finished ceiling by means of metal gridwork and decorative panels.

Luminaires (light fixtures) are set into the gridwork, and additional hanger wires are installed to support the weight as necessary. Metal runners may be designed to allow for prefinished metal channels to be clipped onto the underside of the grid. Gridwork may also be used to support gypsum board, which is attached with self-tapping screws.

Panels. Suspended ceiling panels are commonly made of fiberboard and finished with decorative surfaces. The panels are available in standard sizes of 1′ × 1′ or 2′ × 2′ squares or 2′ × 4′ rectangles. Lay-in panels are set into the gridwork, which remains exposed. Concealed-grid panels are supported at each edge by splines that tie the panels together and hide the gridwork.

In suspended gypsum board ceilings, the gypsum board sheets are applied and finished in a manner similar to other gypsum board installations. For plaster ceilings, metal or expanded wire lath is attached to the suspended gridwork. The lath is then covered with a plaster finish. Other finish panels or strips for suspended ceilings include metal pans, wood strips, and other decorative finishes.

CEILING QUANTITY TAKEOFF

A set of prints may include reflected ceiling plans. A *reflected ceiling plan* is a plan view with the viewpoint located beneath the object. Symbols are used to represent each ceiling finish and to indicate locations for air diffusers, exhaust fans and intakes, and ceiling-mounted luminaires. **See Figure 12-15.** Fire sprinkler locations may also be shown on reflected ceiling plans. Additional information concerning air diffusers and intakes is shown on mechanical prints. Ceiling-mounted luminaires are described on the electrical prints.

In order to determine the quantity of materials required to finish a ceiling, the total ceiling area (in square feet) must be determined. Areas of large openings are deducted from the ceiling area. Ledger sheet, spreadsheet, and estimating programs are commonly set up to show ceiling quantity takeoff information organized by room or area. BIM use may provide square footage amounts for various ceiling areas. Ceiling quantity takeoff is typically performed by specialty contractors and is included as a bid submitted to the general contractor.

Hanger wires are used to support the metal gridwork in suspended ceilings and must be included in an estimate.

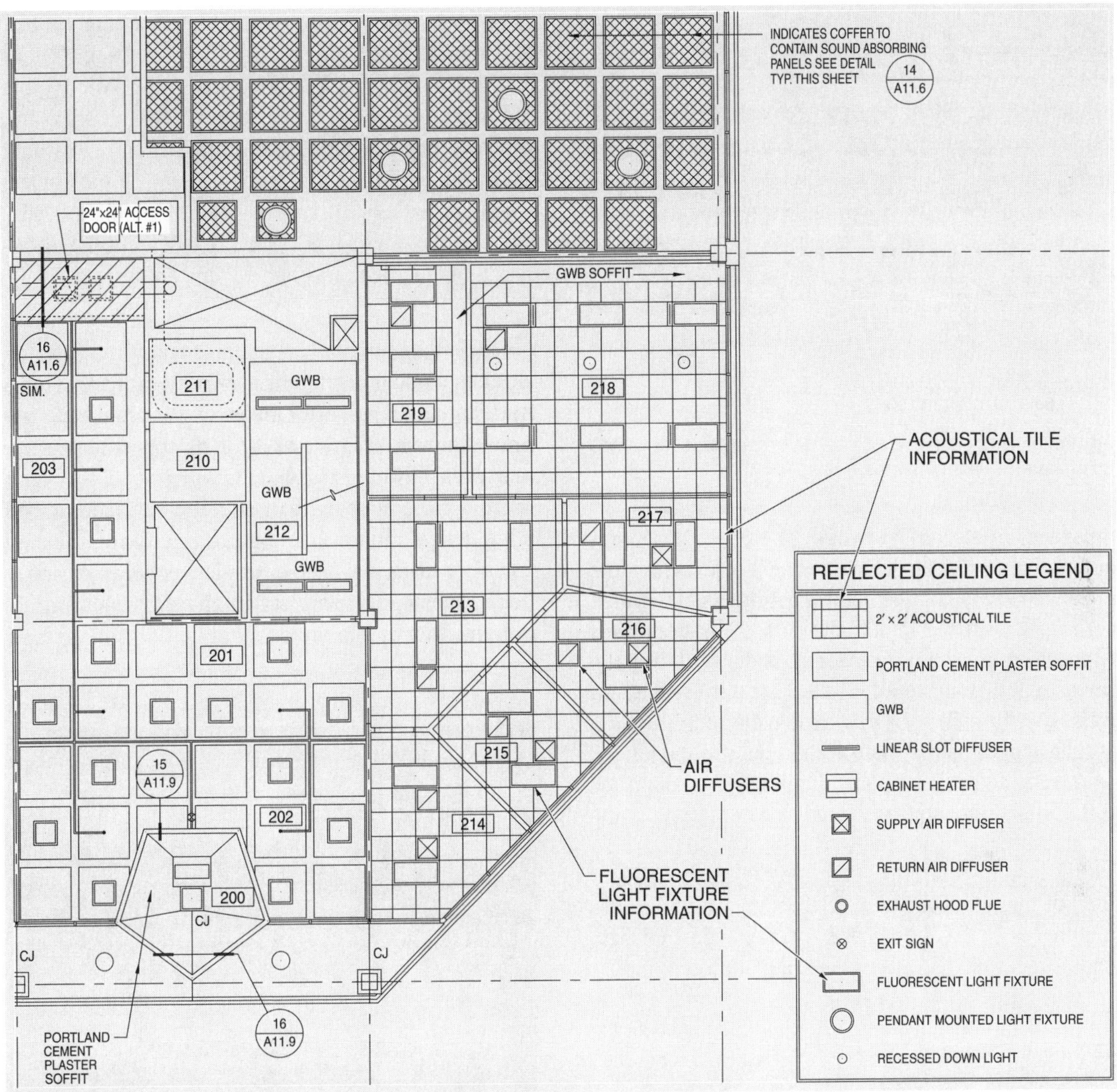

Figure 12-15. Reflected ceiling plans show ceiling finishes using symbols and a reflected ceiling legend.

Suspended Ceiling Takeoff

Material quantities to be taken off for suspended ceilings include hanger wire, clips or fasteners used to fasten the wire to the structural ceiling, main tees, cross tees, L-channel, splines, and ceiling finish materials. **See Figure 12-16.** Variables that must be considered in a suspended ceiling takeoff include the distance from the structural ceiling to the finish ceiling, the height of the finish ceiling above the floor, the grid system and spacing, and the type of finish.

Estimators must include adjustments in labor costs for suspended ceiling installation such as rolling scaffolds for standard-height ceilings, or additional scaffolds or lifting equipment for high ceilings. Total labor costs for suspended ceilings are calculated by multiplying the ceiling area (in square feet) by company historical labor cost data or standard industry costs.

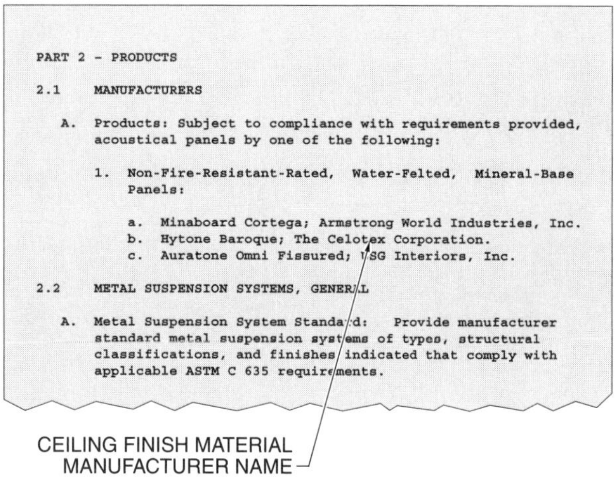

PART 2 - PRODUCTS

2.1 MANUFACTURERS

A. Products: Subject to compliance with requirements provided, acoustical panels by one of the following:

1. Non-Fire-Resistant-Rated, Water-Felted, Mineral-Base Panels:

a. Minaboard Cortega; Armstrong World Industries, Inc.
b. Hytone Baroque; The Celotex Corporation.
c. Auratone Omni Fissured; USG Interiors, Inc.

2.2 METAL SUSPENSION SYSTEMS, GENERAL

A. Metal Suspension System Standard: Provide manufacturer standard metal suspension systems of types, structural classifications, and finishes indicated that comply with applicable ASTM C 635 requirements.

Figure 12-16. Specifications may give manufacturer names for specific ceiling finish materials.

Supporting Metal Gridwork. The amount of hanger wire required is determined by calculating the distance between the structural ceiling and the finish ceiling, consulting manufacturer recommendations concerning hanger wire spacing, and calculating the linear feet of wire required. Hanger wire is ordered from suppliers in rolls of various lengths.

The length of the main tee distance is calculated from reflected ceiling plans. Each piece of main tee is 12′ long. The spacing of the tees determines the number of runs. The length of each run is multiplied by the number of runs to determine the length (in linear feet) of main tees required. The number of cross tees required is based on main tee spacing and panel size. Panels measuring 2′ × 2′ require a main tee and cross tee every 2′. Panels measuring 2′ × 4′ require a cross tee every 4′. Wall L-channel length is calculated as the perimeter of the suspended ceiling area (in linear feet).

Panels. Suspended ceiling panels and finishes are taken off by the square feet of area to be covered. Deductions are made for luminaires, skylights, and other large ceiling openings. Deductions are not made for small ceiling openings cut into panels, such as small air diffusers or small light fixtures. Estimators determine the number of square feet of ceiling finish material, the size of each panel (1 sq ft, 4 sq ft, or 8 sq ft), and the finish specified.

FLOORING MATERIALS AND METHODS

Floor finishes include masonry, hardwood, bamboo, resilient flooring, carpet, and specialized finishes such as asphalt composition floors, interlocking rubber, and various synthetic sports surfaces. Masonry floor surfaces include brick, marble, and stone. Wood floors are made of various types of hardwood that are cut, finished, and arranged in different patterns. Resilient flooring is a thin material laid onto a strong subsurface. Carpet provides a soft, sound-absorbing finish. Access flooring provides a raised surface that helps the installation of electrical, mechanical and air systems.

Masonry

Brick or stone is used for masonry-floor finish. High-traffic areas and exterior areas may use masonry as a finish floor. **See Figure 12-17.** Brick or stone flooring materials may be set in an open pattern to allow water to drain through, set with mastic, or set in a mortar bed with grout similar to ceramic or quarry tile. Sealant and filler strips may also be applied between masonry flooring materials to allow for expansion and contraction due to temperature change.

Figure 12-17. Brick is installed over structural flooring such as concrete to create durable finish flooring.

Solid brick used for flooring is commonly set on a concrete base. Patterns and joint details are shown on floor plans and details. Herringbone and basket-weave patterns are the two most common brick flooring patterns. Stone provides a durable floor surface. Marble and precast stone are the most common stone floor materials. Stone is also used for stair treads on metal or concrete supporting members.

Wood and Bamboo

Wood flooring is commonly made from hardwood materials including oak, maple, beech, and birch. Various grading systems are used with different wood species. Fastener sizes and spacing also vary among the grading systems and species. The finest grades of wood flooring minimize or exclude defects such as knots, checks, streaks, and open grain. Light-use wood flooring is 3⁄8″ to 5⁄8″ thick. Normal-use wood flooring is 25⁄32″ thick. Heavy-use wood flooring is 1 1⁄32″ to 1 21⁄32″ thick.

Bamboo is also used as a flooring material that is installed and estimated in a manner similar to wood flooring. Bamboo flooring is manufactured as tongue-and-groove, plank, or floating panels. Click-together bamboo flooring can eliminate the need for adhesives during installation. Bamboo flooring is available in a natural color, a dark color, or dyed colors. Bamboo flooring is manufactured by slicing bamboo poles into strips that are destarched, glued into boards, and milled. Various types of preservative treatments are applied during this process. Bamboo flooring is considered a more sustainable and renewable flooring product than wood because it requires a relatively small amount of urea-formaldehyde resins during the lamination process and is also a fast growing natural material. Bamboo flooring may provide additional credits for sustainable construction due to its rapid renewability. Types of bamboo flooring include solid, engineered, strand, and distressed.

Wood and bamboo flooring may be fastened to the supporting floor structure with nails, screws, or mastic depending on the floor design and manufacturer requirements. Wood and bamboo flooring may be fastened directly to the structural subfloor below or placed on furring strips. **See Figure 12-18.**

National Wood Flooring Association

Figure 12-18. Wood flooring consists of hardwood pieces fastened in place and finished with varnish or another protective finish.

Prefinished wood flooring is stained and varnished prior to installation. Unfinished wood flooring requires filling, sanding, staining, and varnishing upon completion of installation. Unfinished wood flooring requires significantly more on-site labor than prefinished wood flooring.

Parquet and Block. Wood parquet flooring is made of individual square or hexagonal panels that interlock when set in place. The most common thickness of parquet flooring panels is 5⁄16″, although panels are available in thicknesses up to 3⁄4″. Square and hexagonal dimensions vary by manufacturer, with the most common size being 12″ × 12″. These panels are available in many designs and may be unfinished or prefinished. Parquet flooring panels are most commonly set in place with mastic.

Block flooring is made of individual wood blocks set on end with the end grain of the blocks exposed. Blocks vary in size from 3″ × 6″ to 4″ × 8″ and in thickness from 2″ to 4″. Block flooring is fastened to the subfloor material with mastic. Block flooring is used in some heavy industrial applications where it is adhered to a concrete slab.

Strip. Wood-strip or bamboo-strip flooring consists of individual interlocking pieces of hardwood or bamboo fastened to concrete with mastic or nailed, screwed, or stapled to a wood deck. Wood and bamboo strips are available in various widths, wood species, lengths, and finishes. Work begins at one end of the area to be covered and proceeds across in one direction, with each strip interlocking into the previously placed piece.

Resilient Flooring

Resilient flooring includes vinyl sheets and vinyl, rubber, asphalt, or cork tile. Resilient flooring may be placed directly onto a concrete or wood floor. For wood floors, a high-grade plywood, particleboard, or hardboard underlayment is installed to provide support for the resilient surface. A moisture barrier may be set in place prior to installation of resilient flooring as required by the architect or manufacturer.

Resilient flooring is held in place with mastic that is applied with a notched trowel. Sheets or individual pieces are cut, set in place, and pressed into the mastic with weighted rollers. Fillers may be used at the seams of resilient flooring to chemically fill seams and provide a high-quality finish.

Vinyl Sheets. Vinyl sheets consist of a vinyl wear surface bonded to a backing material. The vinyl wear surface may contain PVC resins, fillers, plasticizers, pigments, or decorative vinyl chips. The vinyl resins and plasticizers together are used as a binder. Vinyl sheets are commonly manufactured in rolls 6′ wide. Widths, lengths, and patterns vary with the manufacturer. Grades of vinyl sheets vary from thin sheets used for residential applications to thick sheets used in commercial applications.

Classifications of vinyl sheets include Type I and Type II. In Type I, the binder constitutes at least 90% of the wear layer. In Type II, the binder constitutes at least 34% of the wear layer. The flooring types are divided into grades based on minimum wear-layer thickness. **See Figure 12-19.** Vinyl-sheet backings are classified as Class A (fibrous nonasbestos-formulated backing), Class B (nonfoamed plastic backing), and Class C (foamed plastic backing). The physical properties considered for vinyl-sheet flooring include overall thickness, wear-layer thickness, residual indentation, flexibility, and resistance to chemicals, heat, and light.

Tile. Resilient vinyl tile is made in 9″, 12″, 18″, or 24″ squares of various thicknesses. Similar to vinyl sheets, many different finishes of resilient vinyl tile are available. Classifications of resilient vinyl tile include Class 1 (solid-color tile), Class 2 (through-pattern tile), and Class 3 (surface-pattern tile). The physical properties considered in product selection include size, thickness, impact-deflection ability, and resistance to solvents.

VINYL SHEET GRADES		
Type	**Grade**	**Minimum Wear-Layer Thickness***
I	1	0.020
	2	0.014
	3	0.010
II	1	0.050
	2	0.030
	3	0.020

* in in.

Figure 12-19. Vinyl-sheet grades are based on minimum wear-layer thickness.

Square resilient vinyl tiles are fastened to the floor subsurface with mastic. The primary installation concern is fitting the seams between pieces and keeping the installation lines straight. Installation begins from the middle of the area to be covered and proceeds outward to the perimeter. Other types of square flooring-tile materials are installed using the same process, with the type of floor attachment based on manufacturer recommendations.

Base Moldings. A *base molding* is the molding at the bottom of a wall at the intersection of the wall and floor. Rubber and vinyl base moldings used to provide a transition between walls and floors are classified as Type I Rubber, Type II Vinyl Plastic, Class 1—Vinyl Chloride, and Class 2—Vinyl Acetate. The physical properties considered for rubber and vinyl base moldings include thickness, height, length, flexibility, and resistance to heat, light, detergents, and alkalis. Base moldings are applied using mastic after other floor finishes are in place. The mastic is applied to the back of the base molding and the base molding is pressed into place against the surface of the wall.

Carpet

Carpet is a floor covering composed of natural or synthetic fibers woven through a backing material. Variables in carpet characteristics include the traffic-load requirements, color, texture, pattern, pitch (gauge), type and height of pile and face yarns, weave (construction style), and installation method. Pitch is based on the number of yarn ends in 27″ of carpet width. A tuft is the cut or uncut loop comprising the face of a carpet.

Commercial-grade carpets are unrolled and glued to the supporting floor or stretched across padding and fastened around the perimeter of the area. Seams are sewn or connected with heat tape. The information related to carpet installation is noted in the room finish schedule. Additional carpet information of use to the estimator includes manufacturer design, weight, backing, and any special installation instructions.

Carpet sizes vary in width from 4′ to 15′. Lengths vary according to the manufacturer and type of carpet. Foam rubber or sponge rubber may be placed under a carpet as a pad to provide additional cushioning.

Weaves. A *weave* is the pattern or method used to join surface and backing carpet yarns. Carpet weaves (construction) include tufted, Axminster, fusion-bonded, knitted, velvet, and Wilton. Tufted carpet is formed by inserting face yarn through a manufactured backing, with yarns held in place by a latex coating on the backing. Tufted carpet is the most common commercial carpet. Axminster carpet has a heavy-ribbed backing that allows almost all yarn to appear at the carpet surface and allows carpet to be rolled lengthwise only.

Fusion-bonded carpet has pile yarns and backing embedded in a vinyl paste, resulting in a very dense pile cut that nearly eliminates backing deterioration. Knitted carpet is manufactured by weaving the face and back simultaneously. Velvet carpet is produced by looping yarn over wires in the looming process. Wilton carpet is similar to velvet carpet, with the difference being that additional types and colors of yarns may be incorporated into the weave.

Access Flooring

Access flooring is a raised floor system supported by short posts and runners. **See Figure 12-20.** This type of floor system allows for the placement of electrical wiring, computer cables, and air movement systems below the surface of the floor. Panels can be temporarily removed and replaced as space usages change. The area below the floor can be utilized as an additional air plenum.

Figure 12-20. Access flooring supports prefinished floor panels above the surface of the structural floor.

Installation. Pedestals are placed at regular intervals as described by the manufacturer, normally 2′ on center. Stringers are supported by these pedestals after height adjustments. Floor panels with various surface finishes are installed on this grid. The finishes on access floor panels can be made from tile, carpet, or resilient materials.

FLOORING QUANTITY TAKEOFF

Each room and area on the floor plan is numbered, coded, or named. Each room code relates to specific information given in the finish schedule that describes the finish flooring to be installed. **See Figure 12-21.** For large areas, the architect may note the floor finish on the prints. Specific details may include a note about the floor finish material at a particular location. The amount of material required to finish a floor is based on the total area to be covered (in square feet or square yards), depending on the type of flooring material. Any large openings, such as stairwells, are deducted during flooring quantity takeoff.

ROOM NO.	ROOM NAME	FLOOR	BASE	WALLS				WALLS		REMARKS
				NORTH	EAST	SOUTH	WEST	FNS'H.	HT.	
	HOUSING UNIT Ra									
H101	SALLYPORT	CS	–	PA	PA	PA	PA	PA	11′-4″	
H102	SALLYPORT	CS	–	PA	PA	PA	PA	PA	11′-4″	
H103	CONTROL	VCT	V	PA	PA	PA	PA	ACP-1	9′-0″	Paint ceiling black
H104	TOILET	CT	CT	CT	CT	PA	CT	PA	8′-0″	
H105	TOILET	CT	CT	PA	CT	CT	CT	PA	8′-0″	
H106	CONFERENCE	CS	–	PA	PA	PA	PA	ACP-1	8′-0″	
H107	COUNSELOR	VCT	V	PA	PA	PA	PA	ACP-1	8′-0″	
H108	CASE WORKER	VCT	V	PA	PA	PA	PA	ACP-1	8′-0″	
H109	ELECT SECURITY	CS	–	–	–	–	–	–	–	
H110	RECORD STORAGE	CS	–	PA	PA	PA	PA	ACP-1	8′-0″	
H111	UNIT MANAGER	CS	–	PA	PA	PA	PA	ACP-1	8′-0″	
H112	COUNSELOR	VCT	V	MET	PA	PA	SGFT	ACP-1	8′-0″	
H113	STORAGE	CS	–	–	–	–	–	–	–	
A101	HC CELL	CS	–	–	–	–	–	–	–	
A102	CELL	CS	–	PA	PA	PA	PA	EXP	8′-0″	
A102	CELL	CS	–	PA	PA	PA	PA	EXP	8′-0″	

Figure 12-21. Room finish schedules contain general information about floor finishes. Additional related information is included in Division 09 of the specifications.

The prints, schedules, and specifications for each area are studied to determine the existing or proposed subsurface material. The subsurface material affects the amount of work required to prepare the surface. Care must be taken with concrete floors to ensure that installation schedules for finish flooring allow sufficient concrete curing time. Moisture content in concrete flooring must be at an acceptable level to ensure proper finish-flooring adhesion. Moisture content is also important to consider in the application of wood and bamboo flooring. In addition to the flooring materials, the estimator must include all base moldings and adhesives in the estimate.

When using a ledger sheet or spreadsheet, each room or area is identified in the first column. Dimensions and areas (in square feet or square yards) are added in subsequent columns, depending on the floor finish material. Material quantities and costs are entered into following columns and rows including finish material, mastics, fasteners, trim, and labor installation costs. When using some estimating programs, either item takeoff or assembly takeoff may be used. The area to be covered is determined from the floor plan. BIM use may provide square footage and square yard information for individual flooring materials where intelligent modeling and product information is used. Flooring material takeoff is commonly performed by specialty contractors and is included as a bid submitted to the general contractor.

Masonry Takeoff

The area to be covered with masonry flooring is taken off in square feet. Estimators review floor plans, specifications, and details to calculate the area and determine the type of brick or stone and the fastening method needed. The subsurface material may affect labor costs. The accessibility of the area where the materials are to be delivered and the weather protection required during setting may also affect labor costs.

The size and pattern of brick placement determines the number of bricks required. A common 4″ × 8″ brick covers 32 sq in. (0.22 sq ft). The total area (in square feet) of coverage is divided by 0.22 to determine the number of bricks required. For stone, the supplying quarry is consulted to obtain information concerning the coverage per ton for the specified stone. Details or sections may also indicate the depth of the mortar bed for setting brick or stone and mortar requirements between masonry units. The volume of mortar (in cubic yards) is calculated based on mortar bed depth and the length and width of the area to be covered. Labor installation costs are based on company historical data or standard industry information.

Wood and Bamboo Takeoff

Wood and bamboo flooring is taken off by the square foot. Wood and bamboo flooring information is shown on floor plans and room finish schedules. Enlarged floor plans may be provided where specific wood or bamboo finishes and patterns are to be installed. An 8% waste factor is added for wood or bamboo flooring materials.

Parquet and Block. Parquet and block flooring is taken off by the square foot or by the number of pieces required for the area to be covered. The area of coverage for each square of parquet or piece of block is divided into the overall area to be covered. Mastic quantities are based on the area of coverage in square feet per gallon as per manufacturer specifications.

The type of subfloor used affects the wood or bamboo flooring estimate. For parquet or block flooring applied to a concrete slab, a mastic fill and damp-proofing may be required. These items are calculated in relation to the area to be covered. Labor calculations are also based on square foot of floor surface area.

Strip. The amount of material required for interlocking hardwood or bamboo plank flooring is taken off in area of floor to be covered (in square feet). Damp-proofing may be required between the rough floor and the finished floor. Fasteners or mastic quantities are based on manufacturer recommendations per square foot of floor coverage.

Resilient Flooring Takeoff

The quantity of resilient flooring required is taken off by the square foot or square yard. The amount of adhesive required is calculated by the gallon. One gallon of adhesive is usually allotted for each 100 sq ft of resilient flooring. However, the quantity may be based on the manufacturer specifications related to coverage area for various types of materials. Base and metal moldings and threshold materials are taken off by the linear feet of each wall or opening.

Vinyl Sheets. When estimating vinyl-sheet flooring, wall-to-wall dimensions are taken from the widest part of the area to be covered. Cutting is required around closets, floor drains, and other obstructions. In some instances, excess pieces of resilient sheet flooring can be used in closets and other small areas.

Roll or seamless flooring is taken off by the square yard. The width, roll length, and pattern of the material determine the waste factor. The supplier or manufacturer can provide waste information for the specific material required. Labor costs will vary based on the number of cuts required and any required seaming. Company historical data or standard labor tables are utilized for these calculations.

Tile. Vinyl, asphalt, cork, or rubber resilient floor tiles are taken off by the square foot or by the number of pieces in relation to the square feet of floor area. The area of a single tile (in square feet) is divided into the total area to be covered (in square feet) to determine the number of tiles required. The amount of mastic required is also based on the total floor area (in square feet) and on manufacturer coverage information.

Carpet Takeoff

Carpet is taken off by the square yard, with allowances made for carpet roll width, length, and pattern. Widths and lengths of carpet vary according to the style and manufacturer and may affect the final quantity required. Additional yardage may be required to keep carpet patterns in alignment. Carpet is priced by the square yard. Labor variables include additional seaming and cutting for intricate designs or unusual shaped areas. Carpet tile materials and labor are estimated with the same methods used for tile.

Determining Carpet Quantities

Determine the amount of carpet to be ordered (in square yards) for a room measuring 24′-6″ wide by 13′-0″ long.

1. Determine the floor area (in square feet).

area = width × length

A = 24.5′ × 13′

A = 318.5 sq ft

2. Convert square feet to square yards.

area (in square yards) = area (in square feet) ÷ 9

A (in square yards) = 318.5 ÷ 9

***A (in square yards)* = 35.38 sq yd; round to 36 sq yd**

Rolls. Estimators obtain information from carpet suppliers concerning roll width and length for various carpet installations. Carpet width and length affect the number of rolls required and the waste factor.

For example, in an area measuring 10′-0″ × 11′-0″, the estimator uses 12′-0″ wide carpet in the estimate because standard carpet widths are 12′-0″ and 15′-0″. The estimate would include 10 lf of 12′ wide carpet. The amount of carpet in the estimate is 120 sq ft (13.3 sq yd), whereas the actual area to be covered requires 110 sq ft (12.22 sq yd). Due to this material consideration, carpet rolls usually have a built-in waste factor.

Access Flooring Takeoff

Access flooring is taken off by the square foot of coverage area. Items to include in the quantity takeoff include the number of pedestals, normally 24″ OC in both directions throughout the area to receive the flooring; the linear feet of stringers, normally 10′ in length each; the number of flooring panels, normally 24″ square; and any additional special installations such as ramps and handrails. Special panels containing air vents or other access points may be shown on floor plans and included as individual items.

WALL-COVERING MATERIALS AND METHODS

Information concerning application of interior wall-covering materials is provided in Titles 09 91 23 and 09 93 23 of the specifications. **See Figure 12-22.** Wall-covering materials include paints, primers, stains, and sealers. Items considered include coverage properties and rates, surface preparation, application equipment and procedures, and curing requirements. Wallpaper information is specified in Title 09 72 23, and other interior wall finishes are described throughout Division 09.

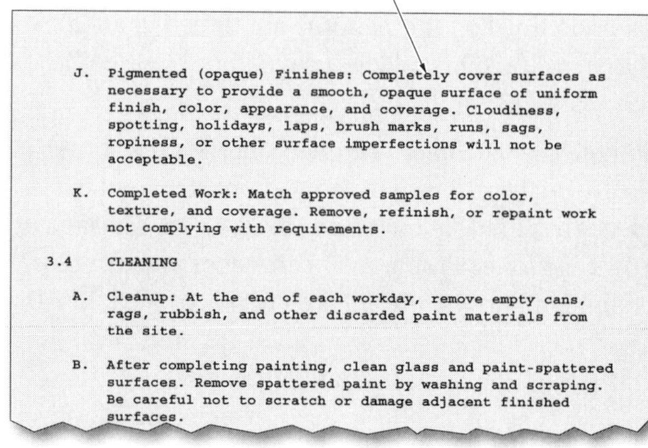

J. Pigmented (opaque) Finishes: Completely cover surfaces as necessary to provide a smooth, opaque surface of uniform finish, color, appearance, and coverage. Cloudiness, spotting, holidays, laps, brush marks, runs, sags, ropiness, or other surface imperfections will not be acceptable.

K. Completed Work: Match approved samples for color, texture, and coverage. Remove, refinish, or repaint work not complying with requirements.

3.4 CLEANING

A. Cleanup: At the end of each workday, remove empty cans, rags, rubbish, and other discarded paint materials from the site.

B. After completing painting, clean glass and paint-spattered surfaces. Remove spattered paint by washing and scraping. Be careful not to scratch or damage adjacent finished surfaces.

Figure 12-22. Specifications provide information concerning paint finish materials.

Paint Properties

Paint is a mixture of minute solid particles (pigment) suspended in a liquid medium that serves as a vehicle. The pigment provides color and hiding power. The vehicle combines a volatile solvent (thinner) and a binder that bonds the pigment particles into a cohesive paint film.

A *volatile organic compound (VOC)* is the portion of a paint mixture that evaporates after application. VOCs are associated with the paint solvent. The solvent is the volatile portion of the vehicle, such as turpentine or mineral spirits, that evaporates during the drying process. VOCs evaporate from various paints and are often limited in use due to environmental concerns.

Dry film thickness is the thickness of the resultant final mass of a paint application after the VOCs have evaporated. A paint coating must be applied to the correct dry film thickness to ensure proper performance. Paint dry film thicknesses are measured in mils. One mil is equal to 1⁄1000 of an inch (0.001″). Paints and other coatings are designed to be applied to a specific dry film thickness in order to perform in the manner required by the architect.

The proper dry film thickness of a coating depends on a number of variables, including the method of curing, type of coating, function, and substrate used. Some coatings that rely solely on evaporation of VOCs for curing produce relatively thin coatings. Examples include alkyd and waterborne acrylic primers and finishes. For the substrate, surface irregularities are a major consideration during application and for the estimator.

For example, a block filler for use over concrete masonry units must provide a sufficient film thickness to fill the pores of the substrate. The average dry film thickness of block fillers is 10 mils. However, the coating may be 30 mils thick in the pores and must cure properly at that film thickness. Dry film thickness may be critical to the function of some coatings. For example, primers with at least 75% metallic zinc in the dried film are applied at 2 mils to 3 mils minimum thickness to ensure that sufficient metallic zinc is available to protect the steel substrate.

Liquid Wall-Finish Materials

A wide range of materials and methods are used for application of the various liquid wall finish materials.

See Figure 12-23. Liquid wall finish materials include primers, paints, stains, and sealers. Each of these materials requires specific application and curing methods for proper performance. Understanding of the properties of the various types of liquid wall finish materials is required to ensure proper quantity takeoff.

Figure 12-23. Paint and other liquid finishes are rolled, brushed, or sprayed onto surfaces.

Primers. A *primer* is a coating applied to new and old surfaces to improve the adhesion and final quality of the finish coating. Primers include latex, alkyd, stain-covering, zinc, and epoxy. A *latex primer* is a water-based primer that dries when the water evaporates. Latex primers are used over gypsum board, plaster, and concrete. An *alkyd primer* is a thermoplastic or synthetic resin-based primer. Alkyd primers are used on new wood and require mineral spirits or paint thinner for cleanup.

Stain-covering primers are formulated to cover water stains or knots in bare wood that show through ordinary primers and mar the finished paint surface. Stain-covering primers are available in water-based, oil-based, and shellac-based mixtures. Zinc and epoxy primers are used in environments where harsh conditions exist or in industrial applications. Zinc and epoxy primers are also used for exterior steel applications.

Paints. Paint types include oil, latex, water-reducible, alkyd, epoxy, urethane-modified alkyd, and aluminum paint. The paint selected depends on the surface characteristics and the desired finish.

Oil paint contains pigments that are commonly suspended in linseed oil, a dryer, and mineral spirit thinner. Linseed oil serves as the binder for the pigment, the dryer controls drying time, and the thinner controls the flowability of the paint. As the thinner evaporates, the mixture of pigments and oil gradually dries to an elastic skin as the oil absorbs oxygen from the air and cures. Oil paints are used for interior and exterior applications.

Latex paint is made of a dispersion of fine particles of synthetic resin and pigment in water. Latex paint is quick-drying and thinned with water. Primer is not required for interior applications except over bare metal or wood or over highly alkaline surfaces. Exterior latex paint can be applied directly to previously painted surfaces. On new exterior wood, latex paint should be applied over a primer.

Water-reducible paints include latex paints and products based on synthetic polymers. Most latex paints dry by solvent evaporation. Synthetic polymer paints dry by a combination of solvent evaporation and chemical cross-linking. Chemical cross-linking is a curing process in which curing is dependent on the molecular action between the coating and a hardener (curing compound). Chemical cross-linking requires the blending of two materials (a two-component coating) and a digestion time before the coating is applied. The blending of specific materials results in chemical cross-linking and outstanding performance features. These products are also known as water-based or waterborne paints.

Alkyd finishes contain synthetic thermosetting resins as a binding agent. Alkyd finishes are produced in flat, semigloss (low-luster), and high-gloss sheens. Flat finishes produce a rich, softly reflective surface.

Alkyd flat finishes are often applied to painted walls and ceilings, metal, fully cured plaster, gypsum board, and woodwork without a primer. When required, the primer should be of a similar material. An alkali-resistant primer should be used for high-alkaline surfaces. Semigloss alkyd finishes contain some sheen to contrast with flat-finished wall surfaces. High-gloss alkyd enamels are often used for high service and wash applications.

Epoxy paint consists of a two-part formulation that is mixed immediately prior to application. Epoxy paint is very hard and is used for protecting materials such as steel, aluminum, and fiberglass. Epoxy paint dries to a high gloss, creating a smooth tile-like finish.

Urethane-modified alkyds are one-component coatings with high abrasion resistance for use on wood floors, furniture, paneling, and cabinets. Urethane-modified alkyds are used for interior and exterior applications requiring chemical and stain resistance and color and gloss retention. Urethane-modified alkyds are applied to primed steel, aluminum, and masonry that will be subjected to high acidity or alkalinity.

Urethane-modified alkyds have high resistance to salt, steam, grease, oils, coolants, solvents, and general-maintenance machinery fluids. A gloss or semigloss finish provides corrosion and abrasion resistance and maintains gloss and color for long periods of time.

Aluminum paint is an all-purpose coating formulated with varnish as the vehicle for aluminum-flake pigment. As the paint dries, the aluminum flakes rise to the coating surface, providing a reflective coating that is resistant to weathering. Aluminum paint is used on exterior chain-link fencing, interior wood, metal, or masonry. When formulated with an asphalt base, aluminum paint has high adhesion and water resistance when applied to asphalt materials.

Stains. A *stain* is a pigmented wood finish used to seal wood while allowing the wood grain to be exposed. Stain may be clear or contain pigments to tint the wood by penetrating into the wood grain. Stain may be oil-based, latex-based, or supplied in a gel form. Stain is commonly sealed with varnish or shellac.

Sealers. A *sealer* is a coating applied to a surface to close the pores and prevent penetration of liquids to the surface. Common wood sealers include varnish, enamel, and shellac. Varnish is a clear or tinted liquid consisting of resin dissolved in alcohol or oil that dries to a clear, protective finish. Varnish dries and hardens through evaporation of the solvent, oxidation of the oil, or both. Varnish is used for interior and exterior applications where a hard, glossy, moisture-resistant finish is required. Varnish is also commonly applied as an interior wood-floor finish.

Enamel sealers contain a large amount of varnish. Enamel has the same qualities as varnish and can be used for interior and exterior applications. A primer is required for high-quality interior work. Shellac is made from special resins dissolved in alcohol. Shellac is used as a finish for wood floors, trim, and furniture. Shellac is used as a surface-preparation coat to obtain an even tone on porous or soft wood such as pine. Instead of restaining some work, pigmented shellac (shellac enamel) is used as a sealant over stained finishes to create a uniform surface finish.

Wall-Covering Materials

Wall-covering materials include wallpaper, fabric wall covering, acoustical wall covering, and wood-veneer wall covering. Wallpaper is an interior patterned-paper or vinyl sheet wall covering. Wallpaper is manufactured in a wide variety of designs, materials, and surface finishes. Standard American wallpaper products are available in widths of 27″, 36″, and 54″. European wallpaper products are most commonly available in 20″ widths.

Wallpaper may be prepasted or unpasted. Prepasted wallpaper is supplied with the paste applied to the back. Vinyl and vinyl-coated wallpaper provide a water-resistant surface. Vinyl and vinyl-coated wallpaper is available in a variety of textures. Foil and mylar wall coverings have a highly reflective, thin metal coating. Paintable wall coverings have a neutral color and are designed to be painted after application. Grass cloths are textured wall coverings woven from natural fibers. Flocked wall coverings have raised fiber patterns. Embossed wall coverings are wallpapers stamped to create a relief pattern.

Types of acoustical materials including fabrics, cork, and sound-reflective panels are utilized as interior wall finishes. Estimators commonly consult manufacturer information to obtain specific information about applications, materials, and costs for these materials.

WALL-COVERING QUANTITY TAKEOFF

Wall-covering quantities are calculated in square feet, linear feet, or by the item. Paint coverage is based on the number of square feet covered per quart or per gallon. Estimators calculate the square feet of area to be covered and then refer to the product specifications to determine the coverage per gallon of the proper type of paint, primer, stain, or sealer. For items taken off in linear feet, such as trim members, the number of linear feet of material covered per quart or gallon is determined.

In addition to paint materials, other products included in the paint estimate include putty, fillers, scaffolds, protection of surrounding areas, application equipment such as sprayers and compressors, and personal protective equipment for tradesworkers during paint application. Estimators must also determine from the specifications whether one or more coats are required and the approximate coverage for each coat. **See Figure 12-24.**

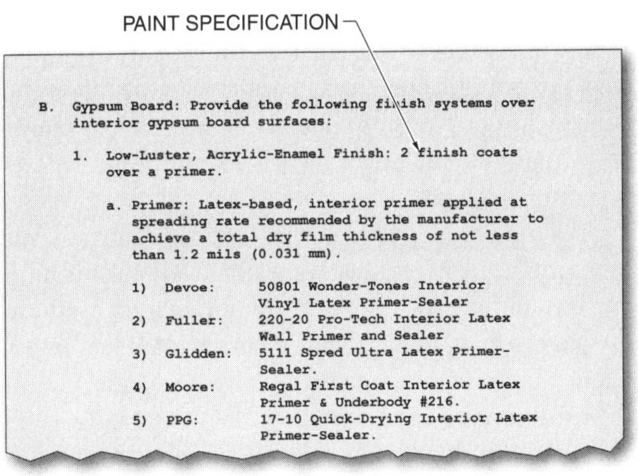


```
B. Gypsum Board: Provide the following finish systems over
   interior gypsum board surfaces:

   1. Low-Luster, Acrylic-Enamel Finish: 2 finish coats
      over a primer.

      a. Primer: Latex-based, interior primer applied at
         spreading rate recommended by the manufacturer to
         achieve a total dry film thickness of not less
         than 1.2 mils (0.031 mm).

         1) Devoe:    50801 Wonder-Tones Interior
                      Vinyl Latex Primer-Sealer
         2) Fuller:   220-20 Pro-Tech Interior Latex
                      Wall Primer and Sealer
         3) Glidden:  5111 Spred Ultra Latex Primer-
                      Sealer.
         4) Moore:    Regal First Coat Interior Latex
                      Primer & Underbody #216.
         5) PPG:      17-10 Quick-Drying Interior Latex
                      Primer-Sealer.
```

Figure 12-24. Paint specifications indicate the number of coats of finish for various materials.

Depending on the complexity of the paint finish schedule, ledger sheets and spreadsheets are set up to include the room name and the areas of the walls, floor, and ceiling. The area of each room (in square feet) and type of coating are entered into the proper row and column and totaled. BIM use may provide square footage information for wall surfaces in individual rooms or areas to assist in quantity takeoff. Labor costs and waste factors vary depending on the material, number of coats, application methods, and ease or difficulty of reaching the application location. Door and window areas are commonly included in the square footage for wall calculations for paint labor pricing.

Company historical data and standard industry information are used in estimating these items. Painting is commonly performed by specialty contractors and is included as a bid submitted to the general contractor.

Paint Coverage Takeoff

"Solids by volume" is a term used to indicate the square feet of coverage for a gallon of paint at a given dry film thickness. A 100% solids by volume coating material without solvents covers 1604 sq ft at 1 mil (0.001″) dry film thickness. A 50% solids by volume coating material covers 802 sq ft at 1 mil dry film thickness. Estimators refer to coating manufacturer specifications for each material for actual solids by volume and coverage rates. **See Figure 12-25.** With the exception of water-based coatings, the solids by volume of a coating is approximately proportional to the VOCs.

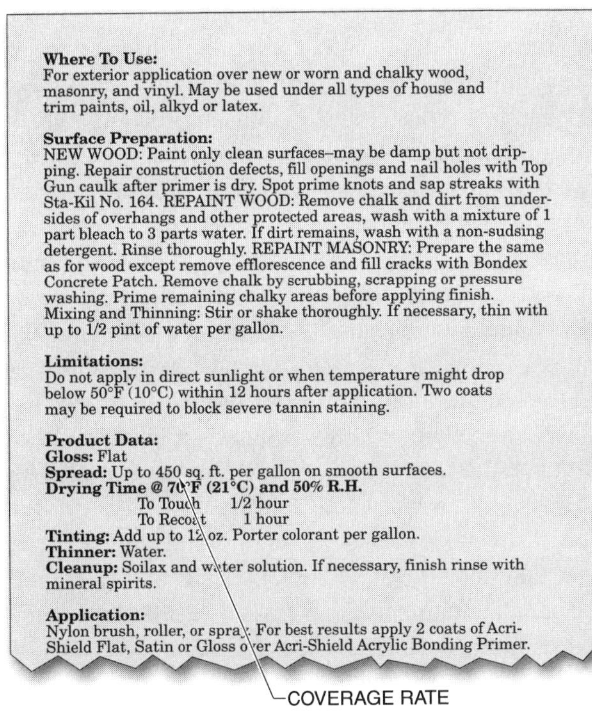

Where To Use:
For exterior application over new or worn and chalky wood, masonry, and vinyl. May be used under all types of house and trim paints, oil, alkyd or latex.

Surface Preparation:
NEW WOOD: Paint only clean surfaces–may be damp but not dripping. Repair construction defects, fill openings and nail holes with Top Gun caulk after primer is dry. Spot prime knots and sap streaks with Sta-Kil No. 164. REPAINT WOOD: Remove chalk and dirt from undersides of overhangs and other protected areas, wash with a mixture of 1 part bleach to 3 parts water. If dirt remains, wash with a non-sudsing detergent. Rinse thoroughly. REPAINT MASONRY: Prepare the same as for wood except remove efflorescence and fill cracks with Bondex Concrete Patch. Remove chalk by scrubbing, scrapping or pressure washing. Prime remaining chalky areas before applying finish.
Mixing and Thinning: Stir or shake thoroughly. If necessary, thin with up to 1/2 pint of water per gallon.

Limitations:
Do not apply in direct sunlight or when temperature might drop below 50°F (10°C) within 12 hours after application. Two coats may be required to block severe tannin staining.

Product Data:
Gloss: Flat
Spread: Up to 450 sq. ft. per gallon on smooth surfaces.
Drying Time @ 70°F (21°C) and 50% R.H.
To Touch 1/2 hour
To Recoat 1 hour
Tinting: Add up to 12 oz. Porter colorant per gallon.
Thinner: Water.
Cleanup: Soilax and water solution. If necessary, finish rinse with mineral spirits.

Application:
Nylon brush, roller, or spray. For best results apply 2 coats of Acri-Shield Flat, Satin or Gloss over Acri-Shield Acrylic Bonding Primer.

Figure 12-25. Paint manufacturers provide information concerning paint coverage rates.

The actual coverage rate of any coating is less than the theoretical coverage rate calculated from the solids by volume. Waste may occur from improper mixing, coating residue left in containers and applicators, variations in the methods of application (spray, brush, or roller), environmental conditions such as wind, inconsistent film thickness, and surface porosity and texture. Material spread rates vary depending on the surface being coated. Estimators must include a waste factor to compensate for these losses based on industry standards or company historical data.

Primers, Paints, Stains, and Sealers. The quantity of primer, paint, stain, or sealer is determined by calculating the area to be covered (in square feet) and dividing by the coverage per quart or gallon of material. For walls, the areas of windows and doors should be deducted if the windows and doors are not to be painted. Labor rates for various applications vary for wall, ceiling, door, window, or other trim work. Labor rates for walls are commonly based on overall wall area without door or window deductions. Additional costs may be incurred where scaffolds, lifts, or other equipment is required to safely reach the areas to be painted.

Coordination with the project manager and other contractors and subcontractors on the job is important in minimizing labor costs for painting. The ability to apply some types of surface finishes in coordination with other construction processes, such as fixture installation, can reduce the amount of time required for finishing around other objects or protecting other finishes. Labor costs are commonly based on company historical data or standard industry information.

Standard latex primer for gypsum board typically covers 270 sq ft/gal. to 300 sq ft/gal. Latex finish coatings cover approximately 300 sq ft/gal. For interior door frames and window openings, approximately 1 qt of paint is required per coat for 4 openings. The areas of doors are calculated by multiplying door length by width and then multiplying by two (the two door sides). An additional 2 sq ft to 3 sq ft may be added to compensate for panel bevels and irregular surfaces. Stain and sealer for doors covers about 500 sq ft/gal. Estimators generally add 1 qt of stain and sealer each for every 5 stained doors. Doors may also be estimated individually.

Approximately 3 qt of paint per coat is required for 20 windows or 20 openings. Ten openings require approximately ½ gal. of paint per coat. Approximately ⅓ the amount of paint required for wood window frames is taken off for metal window frames. For painting shutters, the area of both sides of the shutters is determined to calculate the total shutter area. When enamel paint is applied to window frames and other openings, 1 qt of primer is allowed for every 4 openings. Enamel provides approximately the same or slightly less coverage than ordinary paint. For painted stairs, the areas of treads, risers, and stringers are calculated by multiplying length and width.

Wall-Covering Material Takeoff

The amount of wall-covering materials required may be taken off as the number of rolls, linear yards, or square yards needed to cover an area. The number of rolls is determined by dividing the total wall area (in square feet) by the area of coverage for each roll (in square feet). Deductions may be made for large door and window openings. Small openings are cut around and are not deducted from overall wallpaper coverage due to the need to keep patterns aligned. Wood-veneer materials for wall covering are taken off and bid according to the area (in square feet).

A common waste factor for wall coverings is approximately 15%. A small repeating wallpaper pattern results in less waste than a large repeating pattern. A pattern with a horizontal match wastes less paper than a pattern with a drop or alternate match.

Wallpaper is commonly available in single, double, and triple rolls. The coverage area is determined from individual product information. The per roll coverage value is divided into the total square feet of area to be covered. The answer is rounded up to the nearest whole number to determine the required number of rolls.

Borders are ordered by the linear yard. Paste is calculated by the pound. The amount of paste required depends on the weight of the wallpaper. Labor rates are determined based on the type of the surface to be covered and the accessibility of the surfaces by scaffold or aerial lifts if required. Labor costs are based on company historical data or standard industry information.

SAGE TIMBERLINE ESTIMATING

ASSEMBLY TAKEOFF ITEM SUBSTITUTION

Sage Timberline Estimating allows items to be substituted in an assembly. The quantity, unit, and calculation method in the takeoff grid remain the same after a substitution. If required, enter a quantity or invoke a command from the right-click shortcut menu to update the calculation method. If the unit or formula of the substitute item is different from the original item, a message is displayed. The unit or formula can then be replaced.

1. Start a new estimate named **Modesto - Ceiling** and link it to the **Sample Ext Commercial Drywall** database.
2. Double-click on assembly **0947 - DRYWALL – FURDOWNS & FRAMING** and **0952 - Ceiling - Layin 2×4 or 2×2**. Any changes made to the assembly for this estimate will not affect the original assembly.
3. Take off the wall by entering **100** for **Length LF**, **45** for **Width LF**, **20** for **# of Lights 2×4**, **10** for **Ht. Framing Above LF**, and **8** for **Ht. of Ceiling AFF**.
4. Pick the **Add Pass** button to generate a takeoff item list.
5. In the item grid pane, pick any cell in the row containing item **9511.010 120 Main Tee Intermediate White**.
6. Right-click in the highlighted cell.
7. Select **Substitute Item...** from the shortcut menu to display the **Item List** dialog box.
8. Double-click item **9511.010 110 Main Tee Intermediate Color** to place it in the item grid pane and substitute it for the white tee.

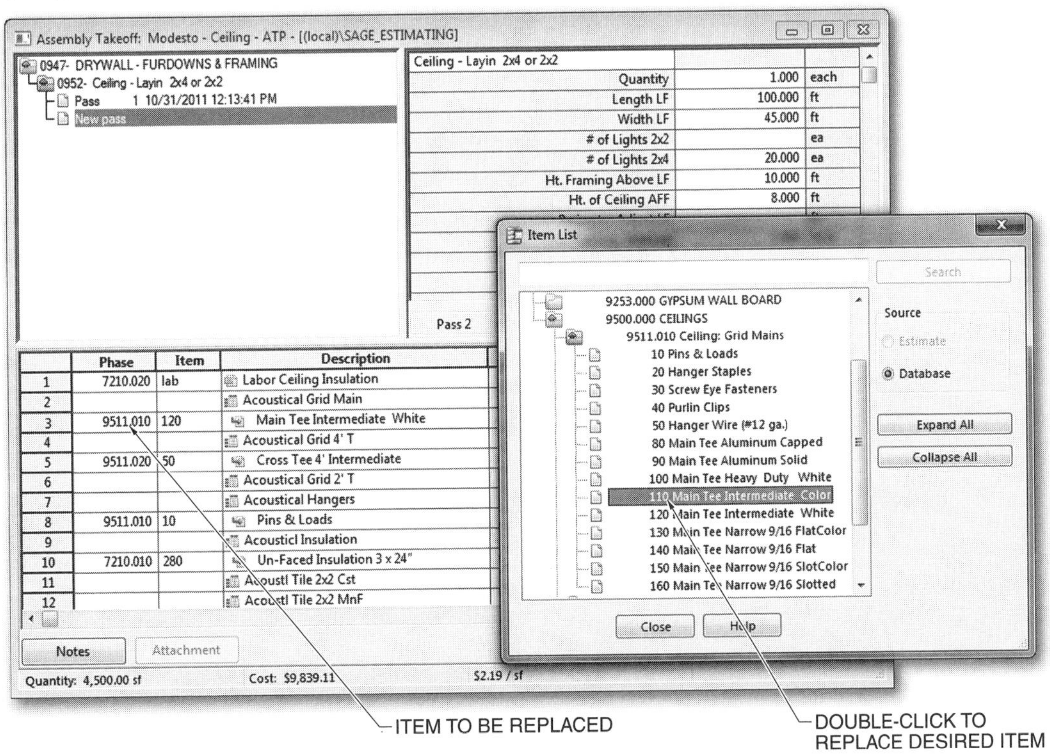

9. Pick OK and Close to send these items and their quantities to the spreadsheet.

Quick Quiz®

Quick Quiz®

Refer to the CD-ROM for the Quick Quiz® questions related to chapter content.

Key Terms

Illustrated Glossary

- *access flooring*
- *base molding*
- *carpet*
- *control joint*
- *corner bead*
- *dry film thickness*
- *grout*
- *lath*
- *paint*
- *plaster*
- *primer*
- *reflected celing plan*
- *sealer*
- *stain*
- *suspended ceiling*
- *terrazzo*
- *tile*
- *volatile organic compounds (VOCs)*
- *weave*

Web Links

Web Links

Association of the Wall and Ceiling Industries International
www.awci.org

Building Stone Institute
www.buildingstoneinstitute.org

Carpet and Rug Institute
www.carpet-rug.org

Ceramic Tile Institute of America
www.ctioa.org

Floor Covering Installation Contractors Association
www.fcica.com

Gypsum Association
www.gypsum.org

Marble Institute of America
www.marble-institute.com

Mason Contractors Association of America
www.masoncontractors.org

National Hardwood Lumber Association
www.nhla.com

National Paint and Coatings Association
www.paint.org

National Wood Flooring Association
www.nwfa.org

Operative Plasterers' and Cement Masons' International Association
www.opcmia.org

Painting and Decorating Contractors of America
www.pdca.org

Resilient Floor Covering Institute
www.rfci.com

Tile Council of North America
www.tileusa.com

Wallcoverings Association
www.wallcoverings.org

Finishes 12

Review Questions

Estimating

______________ 1. ___ is the backing fastened to structural members onto which plaster is applied.

______________ 2. ___ is a mixture of portland cement, water, and sand used as an interior and exterior wall finish.

T F 3. Tile is a thick building material made of cement, fired clay, glass, plastic, or stone.

______________ 4. A(n) ___ ceiling is a ceiling hung on wires from structural members.

______________ 5. ___ is a floor covering composed of natural or synthetic fibers woven through a backing material.

______________ 6. ___ thickness is the thickness of the resultant mass after the VOCs have evaporated.

______________ 7. A(n) ___ primer is a water-based primer that dries by the evaporation of water.

______________ 8. A(n) ___ is a wood or metal shape used as an edge for plaster material and a gauge for plaster thickness.

______________ 9. ___ is a mixture of cement and water with colored stone, marble chips, or other decorative aggregate embedded in the surface.

______________ 10. ___ is wood or metal strips fastened to a structural surface to provide a base for fastening finish material.

______________ 11. A(n) ___ molding is the molding at the bottom of a wall at the intersection with the floor.

T F 12. A weave is the pattern or method used to join surface and backing carpet yarns.

T F 13. A stain is a coating applied to new and old surfaces to improve the adhesion and final quality of the finish coating.

______________ 14. A(n) ___ is a coating applied to a surface to close off the pores and prevent penetration of liquids to the surface.

______________ 15. A(n) ___ is a thin strip of perforated metal applied on lath and plaster surfaces to relieve stress resulting from expansion and contraction in large ceiling and wall surfaces.

______________ 16. A(n) ___ primer is a thermoplastic or synthetic resin-based primer.

_______________ **17.** Chemical ___ is a curing process in which the curing is dependent on molecular action between the coating and a hardener.

T F **18.** Shellac is a coating made from special resins dissolved in alcohol.

_______________ **19.** A(n) ___ is a pigmented wood finish used to seal wood while allowing the wood grain to be exposed.

Short Answer

1. When estimating a gypsum board finish for a structure, what other items must be estimated in addition to the gypsum board?

2. Describe the purpose of a reflected ceiling plan.

3. Identify materials that should be included in an estimate for a suspended ceiling.

Finishes 12 Activities

Activity 12-1—Ledger Sheet Activity

Refer to Print 12-1 and Quantity Sheet No. 12-1. Take off the square feet of field tile VCT-1, field tile VCT-2, and border tile VCT-3 for corridor 314.

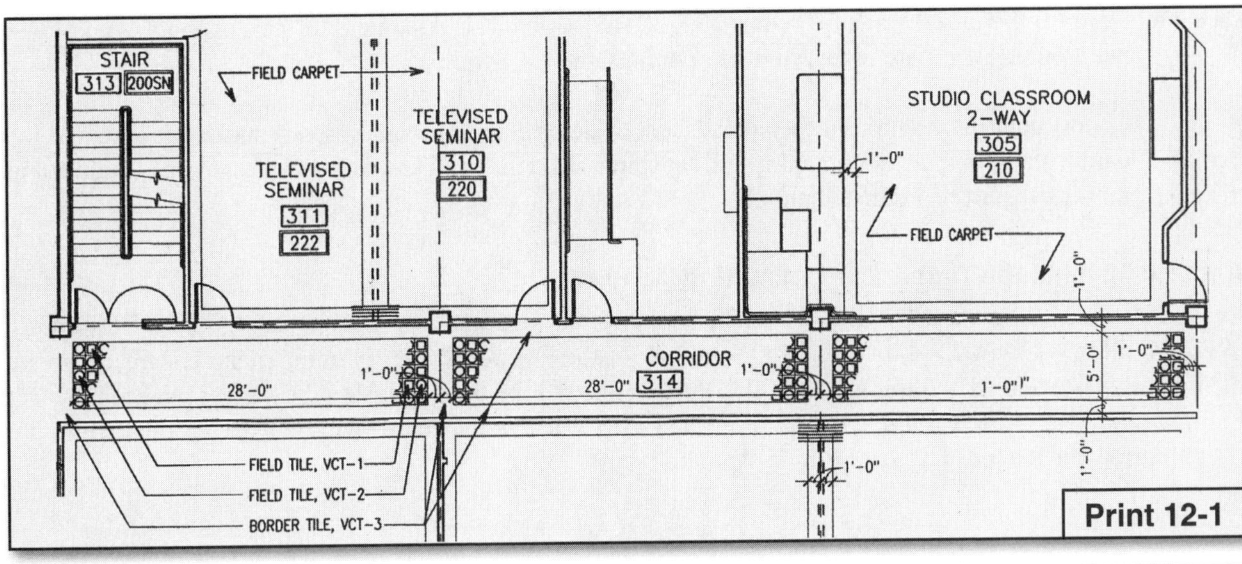

Print 12-1

QUANTITY SHEET

Project: ______________________
Estimator: ______________________

Sheet No. 12-1
Date: ______________________
Checked: ______________________

No.	Description	Dimensions						Unit		Unit		Unit		Unit
		L	W											
	Field tile VCT-1													
	Field tile VCT-2													
	Border tile VCT-3													

Activity 12-2—Spreadsheet Activity

Refer to Print 12-2 and Quantity Spreadsheet No. 12-2 on the CD-ROM. Rooms 131, 132, and 133 have a suspended ceiling 8′-0″ above the floor consisting of 2′ × 2′ acoustical tile. Take off the number of 2′ × 2′ acoustical tiles needed for all three rooms.

Room 132 is finished with 6″ × 6″ ceramic tile from floor to ceiling on all four walls. Take off the number of tiles required. Do not make deductions for the door opening.

Rooms 131 and 133 are finished with two coats of paint. Take off the gallons of paint for the walls in rooms 131 and 133. Assume 400 sq ft of coverage per gallon.

Activity 12-3—Estimating Gypsum Board Finishes

Refer to Print 12-3 and Quantity Spreadsheet No. 12-3 on the CD-ROM. Take off the number of gypsum-board sheets required for the type 4 walls for rooms 305, 308, 310, and 311. Assume 8′-0″ ceiling heights and ½″ × 4′ × 8′ gypsum-board sheets are used. All door openings are 3′-0″ wide.

Refer to the cost data information in the spreadsheet to determine the total material and labor costs for the gypsum-board installation, including taping and finishing. (Material and labor cost for fasteners is included in the cost data for gypsum-board installation.)

Activity 12-4—Sage Timberline Estimating Activity

Refer to Print 12-4 on the CD-ROM. Create a new estimate and name it Activity 12-4. Take off assembly **0952 – Ceiling – Layin 2 × 4 or 2 × 2** for the ceiling of the Control Room. Framing members are 12′ above the finished floor and the ceiling height is 10′. Purlin clips will be used to fasten the hanger wire to the steel framing members of the building. Tegular fire-rated tile (2′ × 2′ × ¾″ thick) will be used for the ceiling tile with a narrow, float acoustical grid. Print a standard report.

Mechanical Systems 13

Key Concepts

- Estimates for mechanical and plumbing systems include all pipes, fittings, valves, supports, and fixtures.
- Unit estimating—estimating of material and labor costs in a single step—may be used on small jobs. Additional costs are added for complicated installations or where high-quality fixtures are specified.
- Estimating labor for mechanical and plumbing work can be performed by calculating labor costs based on a percentage of material costs, plus adding detailed calculations for trenching and other items with little material cost, or can be based on a percentage cost of an average job.
- The estimator must include piping for water in a hydronic system, piping for any additional solar or geothermal installations, piping for fuel to the boiler or furnace, and piping for expansion tanks and flues in the estimate.
- Heating equipment considered by the estimator includes solar units, boilers, expansion tanks, furnaces, and terminal units.
- Air-handling equipment considered by the estimator includes fans, motors, drive units, and the necessary ductwork to facilitate airflow.

Introduction

Mechanical systems include piping, fixtures, and equipment for plumbing, fire protection, heat generation, refrigeration, and air distribution systems. Mechanical system materials include many types of pipe, accessories, fittings, valves, and fixtures. Mechanical prints show piping, connection, valve, and fixture information. Mechanical specifications or prints may also contain a schedule detailing the equipment to be installed in each occupied space.

Heating, ventilating, and air conditioning systems control the temperature and humidity of various building areas. Mechanical specifications and prints provide information for the various components of heating, cooling, and ventilating systems. Plan views, elevations, and details provide installation and fabrication information.

MECHANICAL SYSTEM MATERIALS AND METHODS

The three primary mechanical systems are fire suppression, plumbing, and heating, ventilating, and air conditioning (HVAC). Information regarding mechanical system components is included in CSI MasterFormat™ Division 21 (Fire Suppression), Division 22 (Plumbing), Division 23 (Heating, Ventilating, and Air Conditioning), Division 40 (Process Integration) and Division 42 (Process Heating, Cooling, and Drying Equipment).

These divisions include information on fixtures, piping, instrumentation and controls, and operation and maintenance. Using a variety of symbols and abbreviations, mechanical prints and specifications provide information about the piping, connection, valve, and fixture information required for the structure. (See the Appendix for common symbol and abbreviation tables.)

When plumbing and mechanical systems are very complex and require detailed descriptions, portions of a print may be drawn to a large scale and provided on separate details or shop drawings. Mechanical specifications and prints may also contain a schedule that lists information about the equipment in various areas. **See Figure 13-1.**

General contractors often obtain bids for fire suppression, plumbing, and HVAC systems from subcontractors who specialize in these trades. These subcontractors are often required to be licensed in the construction and installation of these systems. For sustainable construction, these subcontractors may also need specially trained employees who are knowledgeable and certified in the design and installation of sustainable plumbing and HVAC systems. Estimators for mechanical systems must have specialized knowledge of piping systems, installation and labor requirements, equipment requirements, available fixtures and fittings, and other technical aspects of these systems.

PLUMBING FIXTURE SCHEDULE

ITEM	FIXTURE	MANUFACTURER	MODEL #	MOUNTING	TYPE	MATERIAL
P1	WATER CLOSET	AMERICAN STANDARD	2174.138	FLOOR	SIPHON-ACTION	VITREOUS CHINA
P2	"HANDICAPPED" WATER CLOSET	AMERICAN STANDARD	2218.143	FLOOR	SIPHON-ACTION	VITREOUS CHINA
P3	LAVATORY	AMERICAN STANDARD	0476.028	COUNTER	SELF-PRIMING	VITREOUS CHINA
P4	"HANDICAPPED" LAVATORY	AMERICAN STANDARD	0476.029	COUNTER	SELF-PRIMING	VITREOUS CHINA
P5	SHOWER	LASCO	1866-C	WALL/FLOOR	MODULE	FIBERGLASS
P6	"HANDICAPPED" SHOWER	LASCO	3636-BPS	WALL/FLOOR	MODULE	FIBERGLASS
P7	SINK	ELKAY	LRAD 1722	COUNTER	SELF-PRIMING	STAINLESS STEEL
P8	KITCHEN SINK	DAYTON	GE-23321	COUNTER	SELF-PRIMING	STAINLESS STEEL
P9	LAUNDRY SINK	MUSTEE	MODEL NO. 10	COUNTER	SELF-PRIMING	FIBERGLASS
P10	SERVICE SINK	STERN-WILLIAMS	HL-1300	FLOOR	SQUARE	TERRAZO
P11	"HANDICAPPED" LAVATORY	AMERICAN STANDARD	0373.080	WALL	BACKSPALSH	VITREOUS CHINA
P12	BATH TUB	LASCO	2008-80	FLOOR	–	FIBERGLASS
P13	WASHING MACHINE SUPPLY BOX	GUY GRAY	B-200	WALL	CORNER	STEEL WITH ENAMEL FINISH

NOTES:
① OFFSET TRAP BACK AGAINST WALL ② MOUNT PER ADA & WAC REQUIREMENTS ③ PROVI
⑤ PROVIDE GRAB BARS & FOLD UP SEATS ⑥ PROVIDE T-35 HOSE &WALL HOOK, T-40 STAINLESS STEEL

Figure 13-1. Schedules on mechanical prints indicate specific equipment for each building space.

Mechanical System Materials

Mechanical systems include piping, accessories, fittings, valves, and fixtures as part of a complete system. For drainage and waste piping, details of the system, including roof and floor drains, are identified with architectural notes and identifying letters and/or numbers that refer to a schedule contained in the general notes on the prints or in the specifications. Information such as diameters for roof and floor drains is noted on floor plans and isometric drawings. For water supply systems and compressed air and gas systems, similar schedules are included along with isometric drawings showing piping, valves, and fixtures. **See Figure 13-2.**

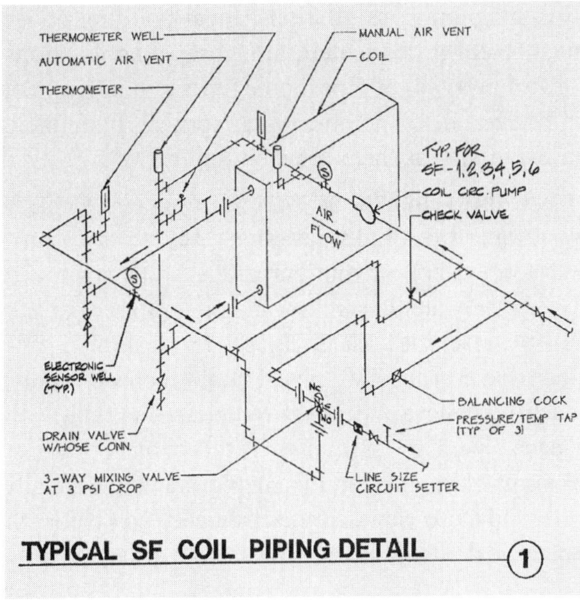

Figure 13-2. An isometric piping drawing shows piping configurations, valves, accessories, and connections.

Piping Materials. Piping for mechanical systems is made from a variety of materials including copper, brass, aluminum, steel, stainless steel, cast iron, ductile iron, polyvinyl chloride (PVC), and cross-linked polyethylene (PEX). Piping is available in various diameters and grades for different applications. Estimators consult the specifications to ensure that the correct types and grades of pipe are used in developing the estimate.

Copper pipe is available in Types K, L, M, G, DWV, ACR, and medical gas. Type K is used for underground water supply, fire protection, and HVAC applications. Type L is used for interior water supply, fire protection, liquefied petroleum (LP) gas, and HVAC applications. Type M is used in domestic water, service and distribution, fire protection, solar, fuel oil, HVAC, and snow-melting applications.

Type G copper pipe is used for natural gas and LP gas. Type DWV is used for drain, waste, and vent applications. Type ACR is used for air conditioning and refrigeration applications. Medical gas copper pipe is used for medical gas applications, such as in hospital installations. Nominal diameters of copper pipe vary from ¼″ to 6″ with a variety of wall thicknesses based on type and diameter.

Brass pipe is available in standard and extra-strong weights. Nominal diameters for brass pipe vary from ⅛″ to 6″ with a variety of wall thicknesses based on grade and diameter. Aluminum pipe is a lightweight pipe. Nominal diameters vary from ¼″ to 3″. As a comparison, ½″ nominal brass pipe weighs 0.934 lb/ft and ½″ nominal aluminum pipe weighs 0.294 lb/ft.

Steel pipe may be seamless or welded. Steel pipe is available in a variety of types defined by ASTM specifications. For example, Type A53 is a seamless or welded steel pipe, black or hot-dip galvanized, and used for general service. Type A106 is a seamless carbon steel pipe used for high-temperature service. Seamless steel pipe is available in nominal diameters from ⅛″ to 48″.

Welded steel pipe is available in nominal diameters from ⅛″ to 24″ and is used for standard water, gas, air, or steam applications. Weights for seamless and welded steel pipe include standard, extra-strong (XS), and double extra-strong (XXS). Stainless steel pipe is installed where corrosion resistance and freedom from contamination are required. Stainless steel pipe is available in nominal diameters from ⅛″ to 12″.

Fabrication costs vary depending on the complexity of the pipe installation and should be reflected in the estimate.

Cast iron pipe is produced by pouring molten iron into a mold, which defines the final shape. Centrifugal force may be used to ensure a high-quality pipe with greater consistency. Iron is identified by color or physical properties. The most common iron from which cast iron soil pipe and fittings are produced is gray iron. Different grades of cast iron pipe are available based on pipe diameter and the design at each end of the pipe. Cast iron pipe is available in hubless, single-hub, and double-hub designs. **See Figure 13-3.**

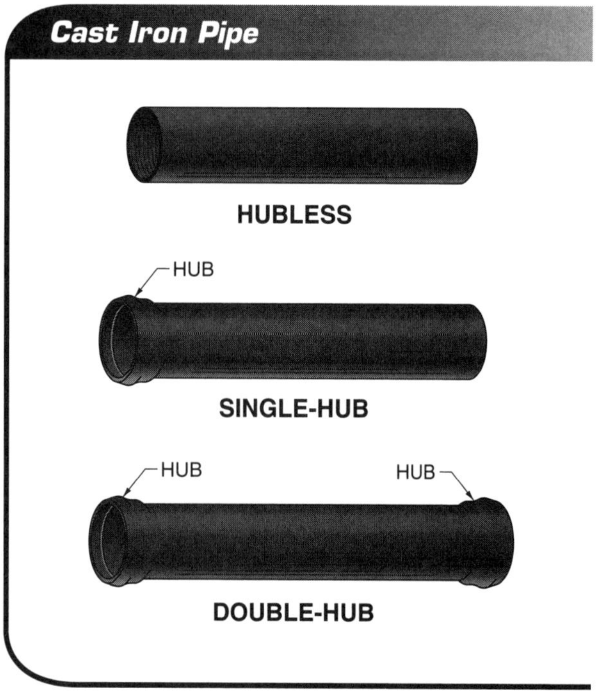

Figure 13-3. Cast iron pipe is available in hubless, single-hub, and double-hub designs.

Ductile iron is a form of cast iron that has magnesium added to the molten gray iron mixture. This causes the graphite flakes in gray iron to reform into spherical particles. Standard wall thickness of ductile iron pipe is based on the pressure class of the pipe. Pressure classes include 150 psi, 200 psi, 250 psi, 300 psi, and 350 psi. Ductile iron pipe is available in nominal diameters from 4″ to 64″.

PVC pipe is available in cementing-only grades or in cementing or threading grades. Wall thickness is greater for cementing or threading grades. PVC pipe is available in nominal diameters from ¼″ to 6″. PVC pipe is a general-purpose pipe used for a variety of waste and water supply applications.

PEX, or cross-linked polyethylene, is a material manufactured from high-density polyethylene in accordance with ASTM specifications F876 and F877. PEX tubing is utilized in hot and cold water systems and radiant heat applications. PEX tubing is manufactured through the extrusion process in diameters ranging from ¼″ to 2″ and is available in coils or straight lengths. Fittings and manifolds for distributing the water may be required to complete a PEX assembly.

In all piping installations, local building codes must be consulted to determine proper applications for various types of pipe. Some types of pipe may not be permitted for potable water service, flammable gases, or various heat transfer applications.

Waste and Vent Piping. Waste pipes collect sanitary wastewater from sinks, lavatories, and water closets, and stormwater from roof, surface, and floor drains and convey the water to sanitary and/or storm sewers and treatment facilities. Materials used for waste piping include cast iron and PVC pipe. Plumbing specifications and prints indicate the size, purpose, type, and elevation of each pipe. This information is shown on floor plans, sections, elevations, and isometric drawings. Details related to these views provide reference points for the isometric drawings and sections. **See Figure 13-4.**

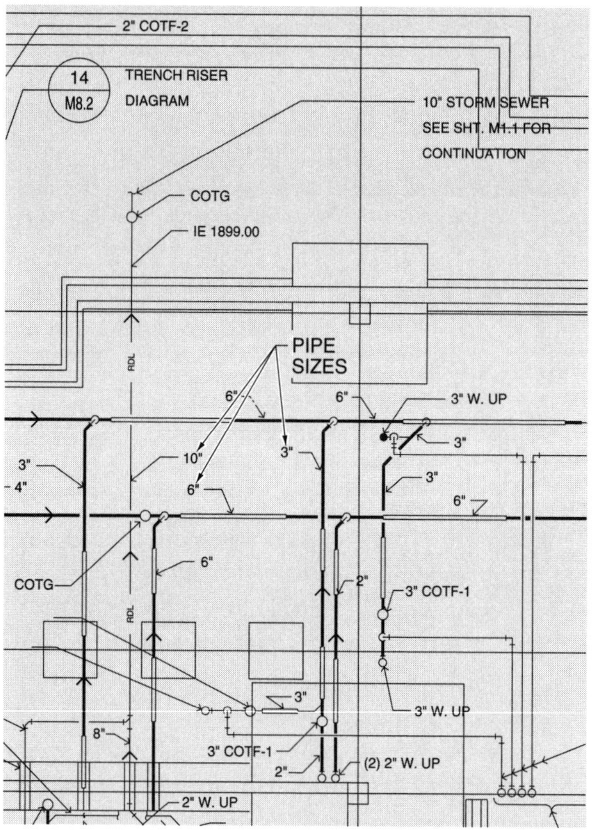

Figure 13-4. Estimators review details to ensure all piping information is determined for each location.

Waste and vent pipe sizes are expressed in inches and indicate the inside diameter of the pipe to be installed. The outside diameters of pipes will vary depending on the material used for the pipe and the pipe schedule.

Abbreviations and symbols indicate the purpose of the pipe. For example, a dashed line with an abbreviation of RDL on a plan view indicates a roof drain line. Specifications contain information concerning the pipe types used for various applications and the means for connecting the pipes. General notes on the plumbing portion of the mechanical prints may provide information about the material used for waste pipes.

A vent pipe removes gases and odors from inhabited areas. Vent pipes branch off from the liquid-carrying waste pipes and commonly terminate above the roofline. Vent pipes are typically shown on prints using dashed lines. **See Figure 13-5.** Plumbing prints also indicate locations for placement of cleanouts in the waste piping.

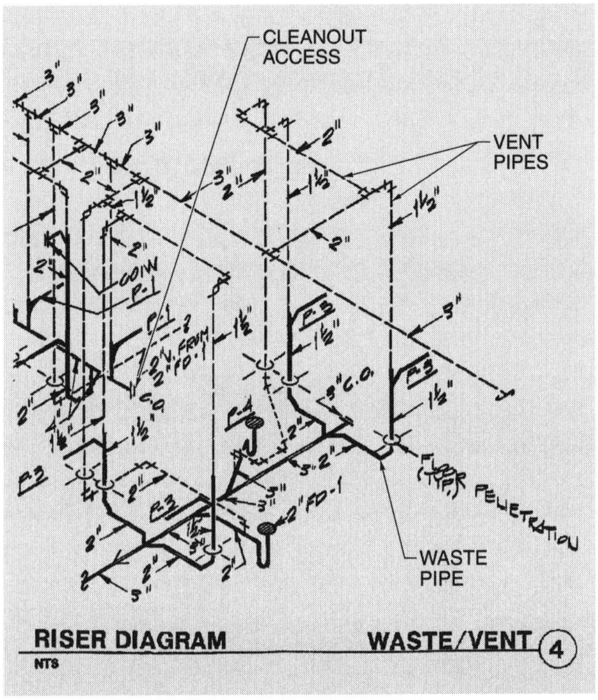

Figure 13-5. Vent pipes are shown on isometric drawings using dashed lines. Waste pipes are shown using solid lines.

Supply Piping. *Supply piping* is piping that delivers water from a source to the point of use. Materials for supply piping include copper, PVC, and PEX. As with waste piping, many different designs and types of fittings and valves are used to join the pipe and control fluid flow. The material and diameter of the supply piping is indicated on plumbing and mechanical prints and in the specifications.

Process Piping. *Process piping* is piping installed in commercial and industrial facilities for transporting compressed air, vacuum, gas, or fuel. Pipes for process piping are described in the same manner as waste and water supply piping. Special abbreviations and symbols are used on plumbing prints to describe process piping. Details are provided where process piping systems are connected to tanks for fuel supply or vacuums. Plan views and isometric drawings of the piping, fitting, and valve configurations are also provided.

Fittings, Accessories, and Valves. Various fittings, accessories, and valves are available to create plumbing piping systems according to mechanical prints. A *pipe fitting* is a device fastened to the end of a pipe to terminate it or connect it to another pipe. **See Figure 13-6.**

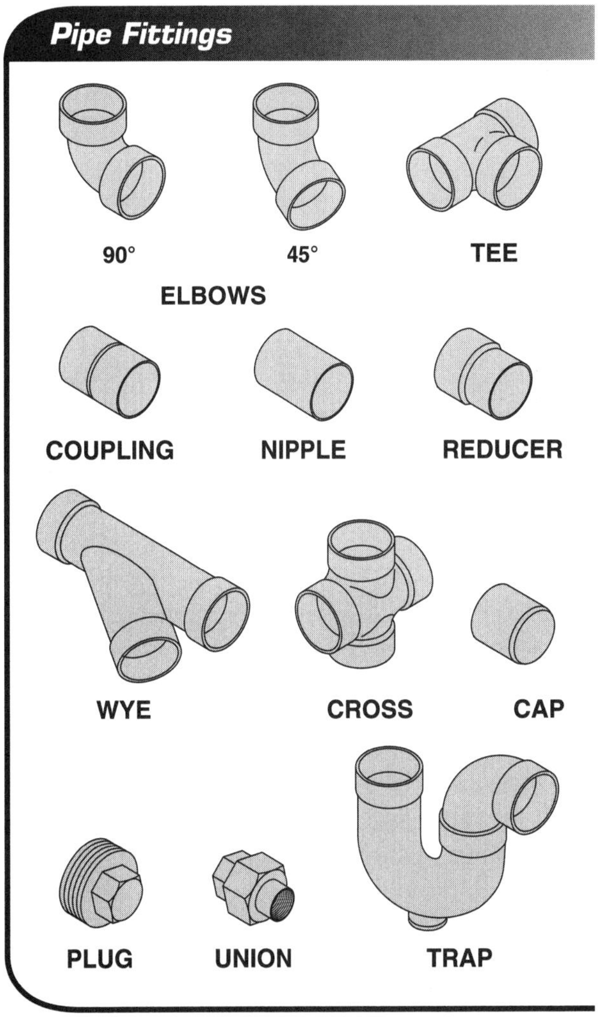

Figure 13-6. A pipe fitting is a device fastened to the end of a pipe to terminate it or connect it to another pipe.

Waste and vent pipes and fittings are connected by various methods depending on the pipe used. Cast iron waste pipes and fittings are connected with couplings or molten lead. PVC waste pipes are fitted and solvent-welded together with a variety of fittings. Steel pipes may be joined with threaded connections or by welding.

Waste pipes are supported by structural members, with hangers and fasteners holding them at the proper elevations. Details of these support connections are not included in the plumbing portion of the mechanical prints. A general note concerning connections to structural members may be included in the mechanical notes or the specifications. Minimum pipe support requirements are commonly included in the local building code.

A *valve* is a device that controls the pressure, direction, or rate of fluid flow. Different valves are used depending on the flow regulation required and the material being regulated. Valves used in mechanical systems include gate, butterfly, globe, pressure-reducing, check, and pressure-relief valves. **See Figure 13-7.**

A gate valve has an internal gate that slides over the opening that the fluid flows through. A butterfly valve controls fluid flow by a square, rectangular, or round disc mounted on a shaft that seats against a resilient housing. A 90° rotation of the shaft moves the valve from fully opened to fully closed. A globe valve controls the flow of liquid by means of a circular disc. Globe valves are used where throttling of flow is required. A pressure-reducing valve limits the maximum pressure at its outlet, regardless of the inlet pressure. Globe valves and pressure-reducing valves are used to reduce the pressure of liquid, air, or steam in a piping system.

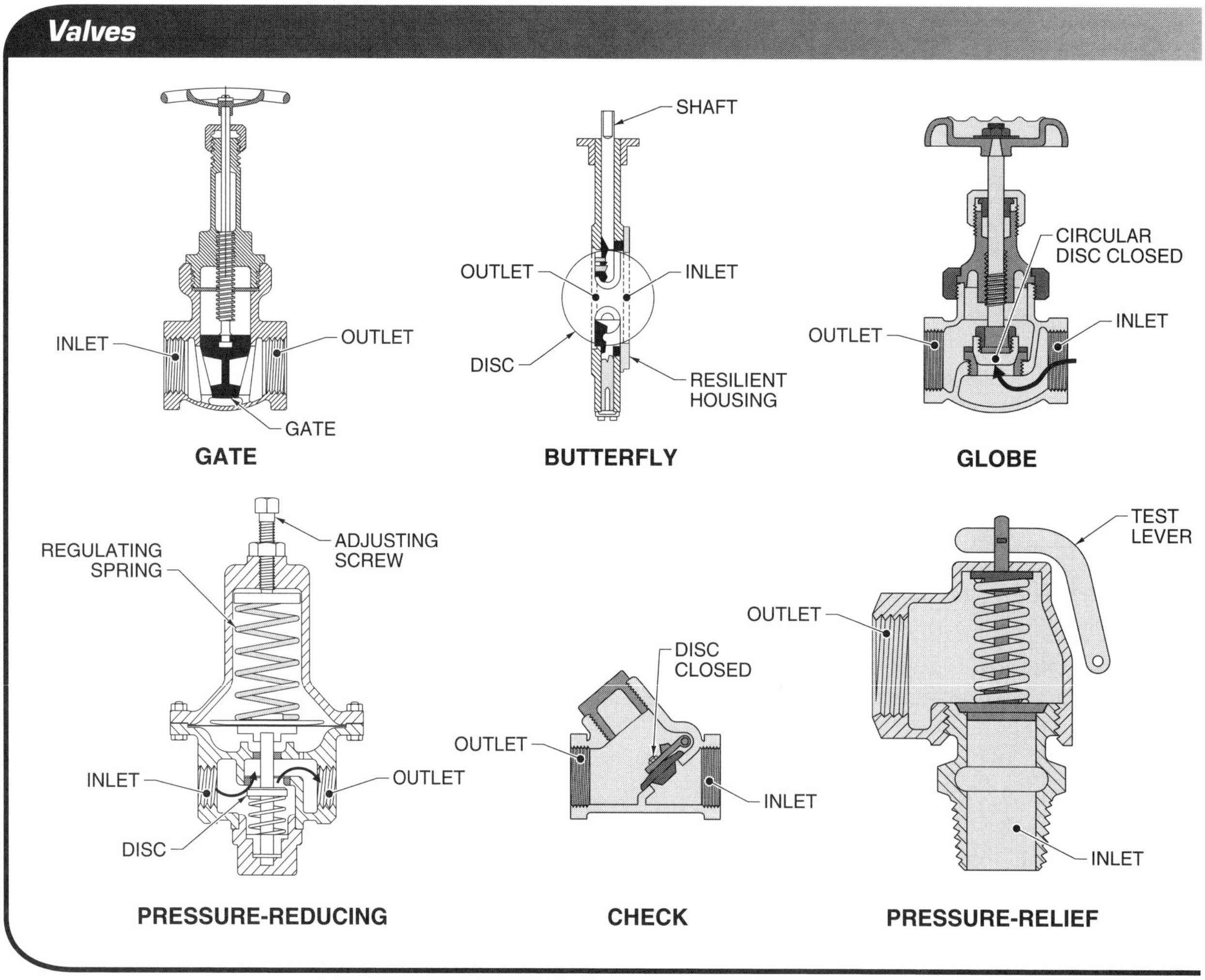

Figure 13-7. Valves used in mechanical systems include gate, butterfly, globe, pressure-reducing, check, and pressure-relief valves.

A check valve allows flow in only one direction and is designed to prevent the reversal of flow. A *pressure-relief valve* is a device that sets a maximum operating pressure level for a piping circuit to protect the circuit from overpressure. Pressure-relief valves open automatically when internal pipe pressure exceeds a preset limit.

Fixtures. Plumbing and mechanical system fixtures include items such as drinking fountains, sinks, water closets, water heaters, and bathtubs. Bathtubs and shower enclosures are commonly made of molded fiberglass. Other fixtures can be made from porcelain or porcelain-coated iron. Suppliers offer a wide variety of styles, grades, and models of plumbing fixtures.

A plumbing fixture schedule is included in the plumbing and mechanical specifications or prints. Notes on plan views and isometric plumbing drawings relate to the plumbing fixture schedule. **See Figure 13-8.**

Schedule information may include the manufacturer names and product numbers, sizes, fittings, methods for attachment to the structure, sizes of vent, waste, and supply piping, and general notes. Details for connections to fixtures or equipment, such as hot water tanks, are provided where necessary.

Mechanical System Installation Methods

Piping for mechanical systems is installed after structural members are in place to support piping systems. Integration of the piping system into the structure must continue during structural construction. For cast-in-place concrete floors, walls, and roofs, piping may be set in place prior to placement of the concrete, or chases may be provided for piping to pass through the floor, wall, or roof. Pipe chases may also be created using a core drill after the concrete has hardened. Fixtures are installed during structural and finish construction.

Features:

- 20″ × 17″ Counter-mounted lavatory
- 4″ Faucet centers
- Front overflow
- Self rimming
- Supplied with template and color-matched sealant

Nominal Dimensions:

- 20⅜″ × 17⅜″

Bowl Sizes:

- Height .... 5⅝″
- Width .... 16″
- Depth .... 10″

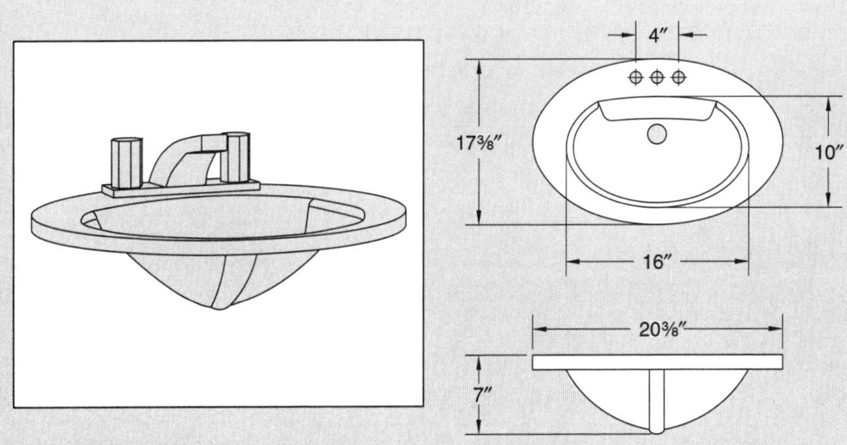

PLUMBING FIXTURE SCHEDULE													
Symbol	**Fixture**	**Mfr**	**Model No.**	**Mounting**	**Type**	**Material**	**Size**	**Drain**	**Trap**	**W**	**V**	**HW**	**CW**
P1	Water closet	American Standard "Afwall"	2477.016	Wall	Siphon jet	White vitreous china	Elongated bowl	—	—	4″	2″	—	1″
P2	Urinal	American Standard "Jetbrook"	6570.022	Wall	Blowout	White vitreous china	21″ × 14½″ × 15⅛″	—	—	2″	2″	—	1″
P3	Lavatory	American Standard "Aqualyn"	0476.028	Counter	Self-rimming	White vitreous china	Oval 20″ × 17″	American Standard #7723.018	1½ × 17 GA	1½″	1½″	⅜″	⅜″

Figure 13-8. Estimators rely on manufacturer information for plumbing fixture options. A plumbing fixture schedule identifies the fixtures installed at each location noted on floor plans.

Piping Installation. Piping is installed prior to the application of wall, ceiling, and floor finish materials. Water supply pipes are commonly stubbed and capped as necessary where they project out of the structural materials and through the finish surface area where plumbing fixtures are to be attached. Copper pipes are connected by solder and fittings. PVC pipes and fittings are solvent-cemented. PEX tubing and pipes are connected with fittings. Piping for high-pressure applications may be joined by welding. All of these connections are designed to be leakproof for the amount of pressure placed on them.

The ends of waste and vent pipes may be stubbed and capped where fixtures are to be connected after finishes are applied. A piping system may be subjected to a pressure test in which a certain amount of pressure is applied inside the piping system for a certain period of time. Leaks are detected and fixed as required. These tests are required prior to the application of finish materials so that a full inspection of the installation can be made.

Care should be taken to consult local building codes and specifications concerning minimum support requirements for various piping installations. For example, for cast iron soil pipe with multiple joints within a 4′ length, hangers are required at alternate couplings or hubs. Vertical stacks require support at each floor.

Building codes establish the minimum standards required to protect the health, safety, and welfare of persons who will be living or working in the building. Often, a separate code book exists for plumbing requirements.

Fixture Installation. Plumbing and mechanical fixtures such as pumps, storage tanks, and filtration equipment are attached to the piping and supporting structure as part of the piping system installation. Manufacturer specifications concerning types of support and lubrication are used during installation. Finish fixtures such as water closets, drinking fountains, and lavatories are commonly installed after floor, ceiling, and wall finishes have been applied. **See Figure 13-9.** Fixture installation for commercial applications must conform to Americans with Disability Act (ADA) requirements.

Figure 13-9. Plumbers install fixtures after floor, ceiling, and wall finishes have been applied.

MECHANICAL SYSTEM QUANTITY TAKEOFF

Estimates for mechanical and plumbing systems include all pipes, fittings, valves, supports, and fixtures. After the list is created, the most current and applicable prices are obtained from a supplier. Overhead expenses such as transportation, storage, waste, and any other company-determined items are then added to material costs.

Unit estimating may be used on small jobs. Unit estimating is estimating material and labor costs in a single step based on unit prices. Unit prices are based on company historical data using average labor costs and average-priced fixtures. For example, unit pricing for plumbing includes the cost of the pipes, fixtures, fittings, valves, supports, and labor for an average installation based on historical data.

Additional costs are added for complicated installations or when high-quality fixtures are specified. Estimators should consult with project managers and field personnel to ensure that piping installations are coordinated with structural construction to minimize additional drilling, aerial lifting, and inspections.

Labor and Material Takeoff

Labor costs are calculated based on company historical records or standard industry information. Variables include the local labor-market supply and the installation conditions. Additional labor costs may be required based on the need to dig trenching for pipes, hang pipes at high heights, or perform any other atypical installations. Material takeoff includes all piping, accessories, and fixtures. Complete and accurate plan markup and calculation is essential. The type, material, diameter, size, and grade of each pipe and fixture must be determined for the final bid.

Labor. Estimating labor for mechanical and plumbing work can be performed by calculating labor costs based on a percentage of material costs, plus adding detailed calculations for trenching and other items with little material costs, or can be based on a percentage cost of an average job. Labor costs may be based on the number of pipe joints plus a certain percentage of fixtures and detailed calculations for trenching. For long pipe installations, labor costs may also be based on the number of linear feet of pipe to be installed. Company historical data or standard labor rate information is utilized for linear foot labor rates.

As a general rule, the installation costs for a basic plumbing fixture may be taken as 50% of the total fixture cost. For high-quality fixtures and materials, a rate of approximately 40% of the total fixture cost is used for determining labor costs. For water heaters and pumps, labor is calculated at approximately 30% of material cost.

For example, a basic water closet and tank with fittings and hardware may cost $800. A labor cost of $400 (50% × $800 = $400) is added to the estimate for a total finish installation cost of $1200. Where high-quality fittings or expensive gold or high-quality metals are used, material costs for the same job may be $1000. In this instance, a labor cost of $400 (40% × $1000 = $400) is added for a total finish installation cost of $1400.

Waste Piping. The specifications and mechanical prints are reviewed to determine the type, diameter, and grade of pipe required at each waste pipe location. Elevations for waste pipes are given to the inside elevation at the bottom of the flow line of the pipe. Elevations are specific to a location and provide reference points for pipe placement used in calculating pipe length. Elevations for waste piping between these elevation reference points are indicated as a percentage of slope. A slope of 1% is equal to a change in elevation of 1′ over a distance of 100′. For example, the change in elevation on a 100′ run of pipe with a 6% slope is 6′ (100′ × 0.06 = 6′). The length of pipe required for the 100′ run is calculated using the Pythagorean theorem.

Determining Pipe Length

Determine the length of pipe required for a 100′ pipe run with a 6% slope.

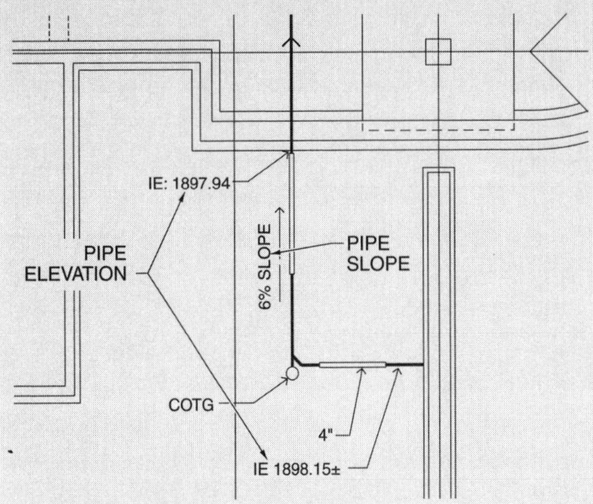

1. Determine change in elevation.

change in elevation = pipe run (in feet) × slope

change in elevation = 100′ × 0.06

change in elevation = 6′

2. Determine pipe length.

$$PL = \sqrt{(100')^2 + (6')^2}$$

$$PL = \sqrt{10{,}036}$$

$$\mathbf{PL = 100.18' = 100'\text{-}2\tfrac{1}{8}''}$$

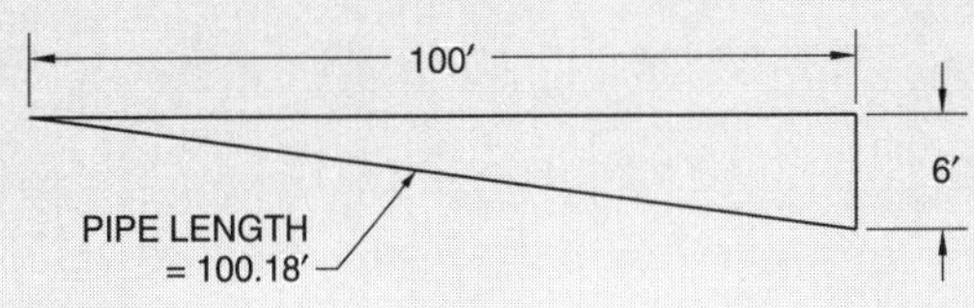

When using a spreadsheet or various estimating programs, these formulas may be set up so that the program will automatically determine the length of pipe required after the length and slope are entered.

On plan views, vertical waste pipes are indicated with open circles and architectural notes. Exact dimensions for the placement of waste pipes are not given on the mechanical prints. Gridlines and some major structural members are shown on the mechanical prints. Exact locations for piping installations are determined by obtaining dimensions from architectural and structural prints, including the foundation and floor plans. For example, where a waste pipe for a sink is to be placed through an interior wall, the location of the wall and the pipe installation are determined from dimensions provided on the architectural prints.

Supply Piping. Mechanical prints indicate the sizes and locations of pipes for carrying hot and cold water. Pipe diameters are expressed as the inside diameter of the supply piping. Exact pipe locations are not indicated. Dimensions for supply piping locations are obtained from architectural prints. Pipe information is given on floor plans and isometric drawings in a manner similar to waste piping diagrams. Information about the type and grade of pipe used is provided in the specifications.

An additional set of drawings may be included for fire suppression piping indicating the location and type of piping and the number and types of sprinkler heads and valves required. The number of feet of length of each type of pipe is entered into the appropriate cell on a ledger sheet, spreadsheet, or estimating program. **See Figure 13-10.**

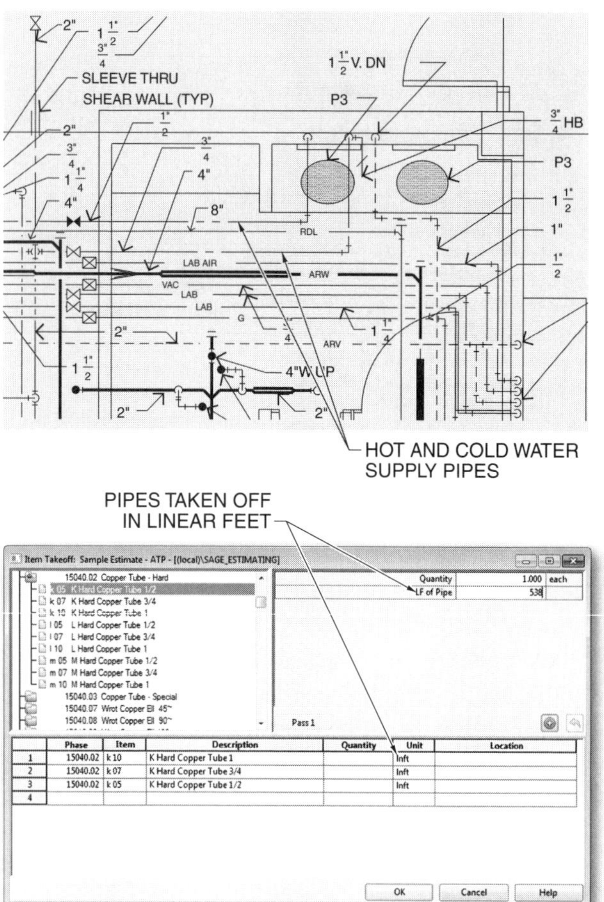

Figure 13-10. The length of hot and cold water pipes shown on mechanical prints is estimated based on linear feet.

Fittings, Accessories, and Valves. Types and locations of fittings, accessories, and valves are described in the specifications and shown on floor plans and isometric drawings. Standard symbols indicate the types of fittings and valves to be installed. Estimators may need to determine some of the types of pipe fittings and accessories required based on the length of pipe, number of joints, number of turns, pipe support system, and number of changes of elevation of the piping.

Each type of fitting and accessory is entered into a ledger sheet, spreadsheet, or estimating program. Valves, fittings, and pipe supports are counted as individual unit items based on the detail drawings and information provided in the specifications. **See Figure 13-11.**

QUANTITY SHEET

Project: Office/Warehouse

Estimator: KMD

No.	Description	Material			
		Quantity	Unit	Price/Unit	Total
1	4" butterfly valve	1	ea	$ 117.00	$ 117.00
2	4" gate valve	23	ea	$ 223.00	$ 5,129.00
3	4" check valve	2	ea	$ 149.00	$ 298.00
5	4" tee	20	ea	$ 74.50	$ 1,490.00
6	4" 90 degree elbow	18	ea	$ 52.80	$ 950.40
7	4" union	6	ea	$ 43.80	$ 262.80

VALVES TAKEN OFF
AS INDIVIDUAL ITEMS

Figure 13-11. Valves in piping systems are taken off as individual items.

Fixtures. Each plumbing fixture is counted and entered in the estimate as an individual unit item. A plumbing fixture schedule is included in the specifications or plumbing prints. BIM projects may provide the capacity to provide quantity amounts for various types of fixtures required. Notes on plan views and isometric drawings relate to the fixture schedule. Schedule information may include specific manufacturer names and product numbers, sizes, fittings, methods for attachment to the structure, vent sizes, waste and supply piping, and general notes.

HVAC SYSTEM MATERIALS AND METHODS

HVAC systems control the temperature and humidity of the area to be conditioned. HVAC systems include different components based on the system design, fuel, and air circulation requirements. HVAC systems include hydronic and forced-air systems. Estimators determine the system used and interpret specifications and drawings concerning the materials and installation requirements for each system.

A *hydronic system* is a system that uses water, steam, or another liquid to condition the building spaces. A *forced-air system* is a system that uses the circulation of warm or cool air to condition building spaces. Hydronic and forced-air systems require various types of automatic controls to provide efficient operation. The controls regulate fuel consumption and temperature levels and humidity, and provide safety in case of system emergencies. HVAC system estimators include all controls, piping, and wiring requirements in the takeoff and estimate in addition to installation labor costs.

General contractors commonly obtain subcontractor bids for HVAC systems. HVAC system estimators are specialized in their knowledge of heating and cooling equipment, piping, ductwork, filtration systems, installation, labor, equipment requirements, available fixtures and fittings, and other environmental conditions, sustainability requirements, and technical aspects of HVAC systems.

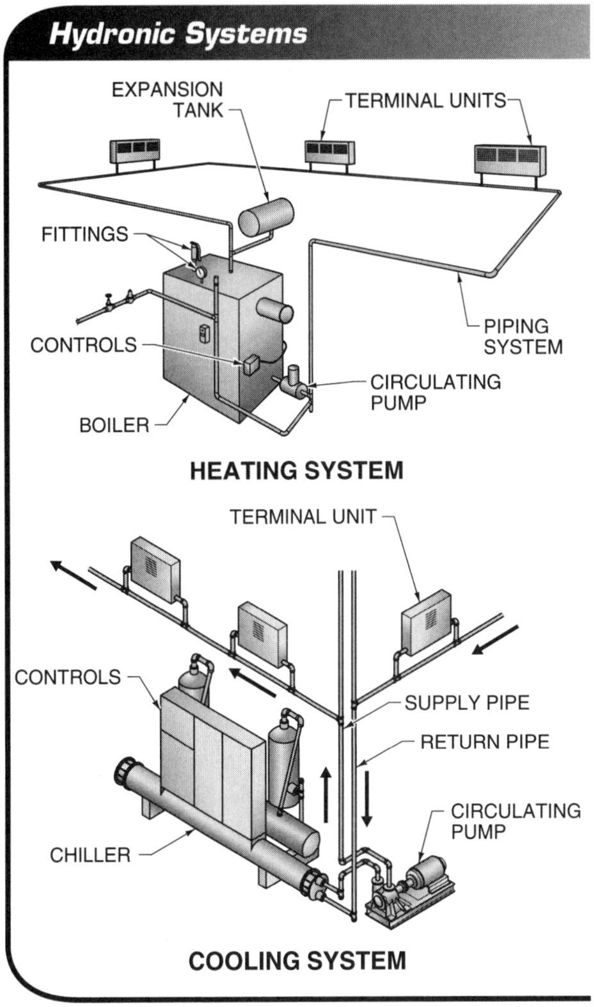

Figure 13-12. Hydronic systems consist of a boiler, chiller, circulating pump, piping system, terminal units, and controls.

Hydronic Systems

Hydronic systems consist of a heat source such as a boiler or solar array, a cooling source such as a chiller, circulating pump, piping system, terminal units, and controls. **See Figure 13-12.** A boiler is used to produce heated water using natural gas, heating oil, or electric heating elements. A chiller is used to cool water that is then used to cool the air. Geothermal or solar system components may be included in hydronic systems to preheat the water or liquid in the system to reduce energy consumption. The heated or cooled water is pumped through a piping system to the terminal units in the areas to be conditioned. A terminal unit transfers heat or cold from the water in a piping system to the air in building spaces. One or more circulating pumps move the conditioned water through the heating or cooling devices, piping system, and terminal units. Water is pumped back and recirculated through the heating and cooling equipment.

The entire process is controlled by thermostats and other control devices. A thermostat is a temperature-actuated electric switch that controls heating and/or cooling equipment. Thermostats at each terminal unit indicate the need for heating or cooling and activate valves or fans to bring the temperature to the level indicated by the thermostat setting. Each portion of the hydronic system is shown on plan views and elevations.

Boilers. Boilers may be low-pressure or high-pressure boilers. Low-pressure boilers have a maximum allowable working pressure (MAWP) of up to 15 psi and are commonly used in warehouses, factories, schools, and residential applications. High-pressure boilers have an MAWP above 15 psi and over 6 BHP (boiler horsepower). High-pressure boilers are commonly used for industrial operations and electricity generation.

Cleaver-Brooks
Boilers and their related equipment provide hot water and/or steam for heat and industrial processes.

Solar and Geothermal Systems. Utilizing naturally occurring systems to preheat water to a higher temperature can reduce the amount of temperature increase required in a hydronic heating system. Solar systems circulate water through panels that are exposed to sunlight, thereby raising the temperature of the water prior to it entering a boiler system. Geothermal systems circulate water through a series of underground pipes, again raising water temperature. The reduction in the differential between the water temperature entering the boiler and the required exit water temperature reduces boiler energy consumption.

> Solar and geothermal systems can require special training for equipment installation and setup. Check with the manufacturer for specific requirements.

Expansion Tanks. An expansion tank allows the water in a hydronic heating system to expand without raising the water pressure to dangerous levels. Expansion tanks are installed in hydronic heating systems. The size of the expansion tank is determined by the total heating capacity of the system.

Chillers. A chiller uses mechanical compression refrigeration or absorption refrigeration to cool water. Mechanical compression refrigeration uses mechanical equipment to produce a refrigeration effect. Absorption refrigeration uses the absorption of one chemical by another chemical and heat transfer to produce a refrigeration effect. Chillers are used for air conditioning systems that produce large amounts of cooling. Geothermal systems may also be used in conjunction with a chiller to reduce water temperature entering the chiller. Water is piped through a chiller, pumped to building areas to be cooled, passed through terminal units through which air is circulated to distribute the cooling, and returned to the chiller to be cooled again.

Piping Systems. In a hydronic system, piping systems distribute the water from the boiler or chiller, through the circulating pumps, to the terminal units, and back to the boiler or chiller. The piping system is full of water at all times. Piping systems may be one-, two-, three-, and four-pipe systems.

In a one-pipe system, water passes through each terminal unit in a continuous flow back to the boiler or chiller. A one-pipe system is economical to install and provides good temperature control when used in small buildings. In a two-pipe system, separate piping systems are installed to supply water to the terminal units and return water from the terminal units to the boiler or chiller. Two-pipe systems are used in medium- to large-size residential and commercial buildings. In a three-pipe system, two supply pipes and one return pipe are used. Three-pipe systems are used when different parts of a system require heating and cooling simultaneously.

In a four-pipe system, separate piping systems are installed for supply and return for heating and cooling. Four-pipe systems are expensive to install but provide excellent control of air temperature. Pipes are clearly identified as heating hot-water supply, heating hot-water return, chilled-water supply, and chilled-water return. Chilled-water supply and return pipes are used for cooling operations. The piping, valves, and fittings used are similar to those used in plumbing and process piping systems. **See Figure 13-13.**

A flue is a heat-resistant passage in a chimney that conveys smoke or other gases of combustion to a safe dispersion area. Piping is normally installed at the bottom of the flue or chimney to remove condensation. Flues may be made of masonry materials, stainless steel, or sheet metal. Design engineers determine the stack diameter based on the boiler output.

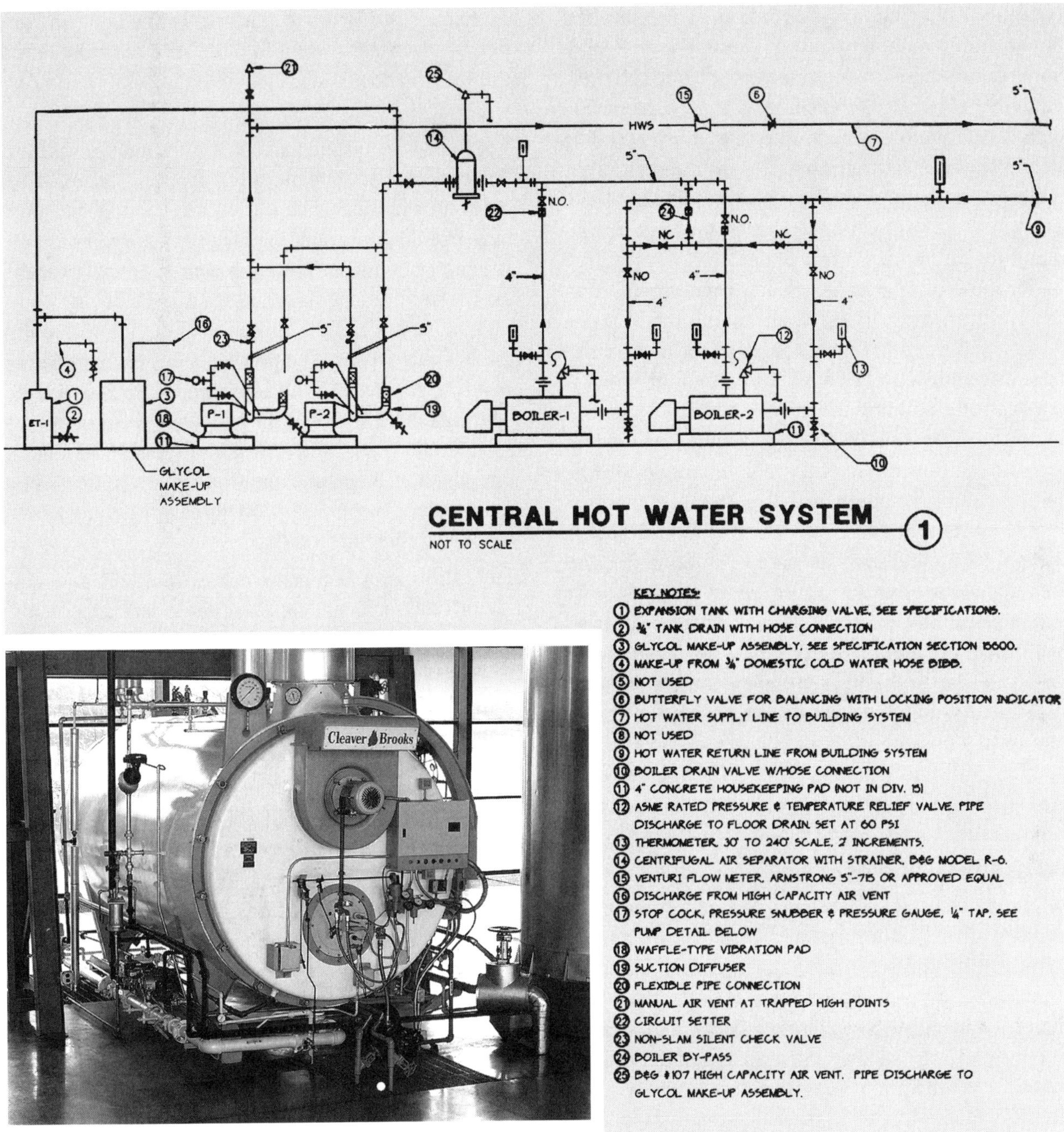

Figure 13-13. Mechanical prints include identification for each pipe run and piece of equipment in an HVAC system.

Controls. Hydronic system control devices include solar, geothermal, boiler operating, safety, and terminal unit controls. Sensors in solar and geothermal systems measure water temperature and regulate valves to increase or decrease water flow into the boiler or chiller. Boiler operating controls include an aquastat and a pressure control. An aquastat senses temperature of boiler water and controls the burner firing to maintain the correct boiler water temperature. A pressure control is a pressure-actuated mercury switch that controls the boiler burner or burners by starting and stopping the burners based on the pressure inside the boiler.

Boiler safety controls include high-temperature limit controls and safety valves. The high-temperature limit control senses boiler water temperature and shuts off the boiler burners if the temperature gets excessively high. A safety valve prevents the boiler from exceeding its MAWP by exhausting hot water and steam from the boiler when the pressure becomes excessively high. The safety valve functions as the last pressure regulation safeguard device on the boiler.

Terminal unit controls include zone valves, low-temperature limit controls, thermostats, and blowers. A zone valve regulates the flow of water in a control zone or terminal unit of a building. A *control zone* is any part of a building in which the hydronic system is controlled by one controlling device. Zone valves regulate the flow of water through the piping system and terminal units based on thermostat readings.

A low-temperature limit control energizes the damper motor and shuts the damper if the ventilation air temperature drops below a preset point. A low-temperature limit control allows outside air to be used in a system but shuts off the airflow when the temperature drops below freezing to prevent system damage. A thermostat controls valves, pumps, and blower motors to regulate the temperature in each area of the building. Blowers in terminal units diffuse air into each area of a building and are controlled by thermostats.

Forced-Air Systems

A forced-air system uses warm or cool air to condition building spaces. In forced-air systems, air is warmed by a heat source such as natural gas, hot water, or electric coils, or cooled by refrigeration. The air is then circulated and distributed through ductwork to condition the building spaces. Forced-air systems consist of furnaces, air conditioners, air handlers, ductwork, and controls. Humidifiers may be installed to add moisture to the air. Air filters are installed in forced-air systems to remove dust and other airborne particles.

Furnaces. A *furnace* is a self-contained heating unit that includes a blower, burner or burners, a heat exchanger or electric heating elements, and controls. A furnace safely and efficiently transfers heat from the point of combustion or the electrical coils to the surrounding air. In combustion furnaces, fuel is burned in a firebox. In electric furnaces, electric current flows through a high-resistance wire to create heat. The heat exchanger is utilized to transfer the heat of combustion or electrical current into the air that is circulated through the duct system and into the building spaces.

Air Conditioners. An *air conditioner* is the component in a forced-air system that cools and conditions the air. Air conditioners contain an evaporator, compressor, condenser, and expansion valve. An evaporator absorbs heat from the surrounding air to evaporate liquid refrigerant into a gas. The cooled air is passed over the evaporator and distributed to areas to be cooled.

A *compressor* is a mechanical device that compresses refrigerant. The pressurized refrigerant vapor from the compressor is discharged into the condenser. The condenser removes the heat from the pressurized refrigerant vapor. The refrigerant condenses and is circulated through an expansion valve. An expansion valve reduces the pressure on liquid refrigerant, allowing the refrigerant to expand.

Heat Pumps. A heat pump moves heat from unconditioned outdoor air and transfers it to indoor air systems. Refrigerant systems are utilized to separate warmer air from cooler air. A reversing valve changes the direction of flow of refrigerant, allowing the heat pump to operate in two directions. This allows for movement of heated and cooled air in two directions. In the heating mode, refrigerant moves cooled air out of the living spaces and warmer air in. When the refrigerant flow direction is reversed, it absorbs heat on the indoor side of the unit and moves it to the outside. The heat is released outside, which allows the refrigerant to cool down and flow back inside to pick up more heat.

Air Handlers. An *air handler* is a device used to distribute conditioned air to building spaces. Air handlers include fans, filters, movable dampers, and heating and cooling coils. Air handler drive system requirements vary depending on the horsepower required and the speed of each fan. Ceiling and exhaust fans are also included in this portion of the estimating process. Ceiling and exhaust fan information is based on the horsepower, speed, and required diameter of the fan blades.

Mechanical specifications and prints provide information about air handlers and fans on schedules that are cross-referenced to mechanical plans and elevations. Each air handler and fan is shown by overall diameter of the blades, the size and speed of the motor, and the volume of air the fan moves in a given time. Air handlers are integrated into the ductwork system. **See Figure 13-14.**

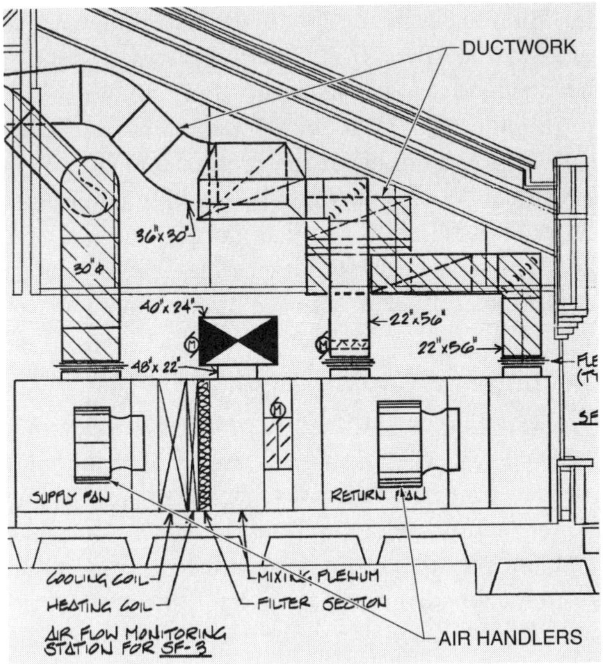

Figure 13-14. Air handlers include fans to move large volumes of air through HVAC systems.

Ductwork. *Ductwork* is the distribution system for forced-air heating and cooling systems. Ductwork is formed from galvanized sheet metal or plastic and distributes air throughout an HVAC system. Ductwork for heating and cooling may be internally or externally insulated to minimize heat transfer to unconditioned spaces. Sheet metal ductwork is most commonly prefabricated at a shop and transported to the job site in sections for installation. Standard sizes and types of plastic ductwork are also available. Plastic ductwork is generally used for applications where flexible ductwork is necessary.

In forced-air systems, return-air ductwork is installed to remove air from conditioned areas and return it for reheating or recooling. In ventilation systems, ductwork is installed to exhaust air from the building to the outside or to areas where it can be conditioned and recycled. For food preparation areas and some types of manufacturing operations, large quantities of air may need to be removed quickly and replaced by makeup air.

A *louver* is a cover for the end of a portion of ductwork and consists of horizontal slats that allow for the passage of air and prevent rain or other objects from entering the ductwork. Louvers are installed where ventilation ductwork passes through exterior walls or roofs.

A register covers the opening of supply-air ductwork and may contain a damper for controlling airflow. A damper is a movable plate or series of fins that control and balance airflow in a forced-air system. A grille covers the opening of return-air ductwork. In some applications, return air may utilize unducted building spaces as return-air channels. In these instances, return-air grilles are sized and located at various locations in these channels to return the proper amount of supply air to the system.

Filtration of air within a structure is a function of an HVAC system. An *air filter* is a porous device that removes particles from air. Mechanical or electrostatic air filters may be installed in ductwork to remove dust and fumes from the air. Electronic filter systems use electrostatic energy that attracts particles and removes them as air flows across the energized filters.

> HVAC systems condition air by maintaining the proper temperature, humidity, and air quality. HVAC systems should operate at optimum energy efficiency while maintaining the desired environmental conditions.

Controls. Forced-air system controls include power controls, operating controls, and safety controls. Power controls include disconnects, fuses, and circuit breakers. A disconnect is a switch that deenergizes electrical circuits for motors and machines. Disconnects stop the flow of electricity when manually activated.

Fuses are overcurrent protection devices with fusible links that melt and open the electrical circuit when an overload condition or short circuit occurs. A circuit breaker is an overcurrent protection device with a mechanism that automatically opens the electrical circuit when an overload condition or short circuit occurs. Fuses and circuit breakers stop the flow of electricity when activated by an overcurrent condition.

Forced-air system operating controls include transformers, thermostats, and blower controls. Transformers change the voltage in an electrical circuit to allow for operation of low-voltage control devices. Thermostats turn heating or cooling equipment on or off based on the thermostat setting in relation to the temperature in the area to be heated or cooled. A blower control is a temperature-actuated switch that controls the blower motor in a furnace.

Forced-air system safety controls include limit switches, pilot safety controls, pressure switches, and stack switches. A limit switch contains a bimetallic element that senses the temperature of the surrounding air. Limit switches turn a furnace off in case of overheating. A pilot safety control determines if the pilot light is burning and does not allow fuel to flow if the pilot light is extinguished.

A pressure switch is operated by the amount of pressure acting on a diaphragm or bellows element. Pressure switches are used to turn off air conditioning systems if refrigerant pressure becomes excessively high. A stack switch is a mechanical combustion safety control containing a bimetallic element that senses flue gas temperature and turns off the fuel flow to a furnace when the flue gas temperature drops below a predetermined level.

HVAC SYSTEM QUANTITY TAKEOFF

HVAC plants are complex systems that provide the temperature control and air circulation required for large public buildings, offices, meeting rooms, classrooms, and work areas. Mechanical specifications and prints describe HVAC systems. Floor plan, elevations, and details provide material, installation and fabrication information.

Specifications, general notes, and schedules contain information concerning the manufacturer designs and equipment codes. Shop drawings may also be produced by the subcontractor or the supplier to guide in the proper fabrication and installation of equipment, piping, and ductwork. For temperature control systems, mechanical prints relate to other components of the construction documents, such as architectural, structural, and electrical prints. BIM systems may be integrated with manufacturing equipment and utilize computer-controlled manufacturing equipment to cut, shape, and mark HVAC ductwork. BIM systems may also provide specific information concerning quantity take-off for louvers, grills, and other individual items.

When calculating HVAC system labor costs, estimators use company historical data or standard industry labor data. BIM systems can help detect potential HVAC equipment, piping, and ductwork installation clashes at the job site. The effective use of BIM clash detection in an integrated model can help reduce labor costs by maximizing the ability to prefabricate HVAC components and minimize installation conflicts at the project site. Company historical data is generated by tracking labor units on each installation and maintaining company records concerning installation times under various conditions for various pieces of equipment.

Heating and Cooling Piping Takeoff

The estimator must include piping for water in a hydronic system, piping for any additional solar or geothermal installations, piping for fuel to the boiler or furnace, and piping for expansion tanks and flues in the estimate. Estimators determine from mechanical prints if the hydronic piping system is a one-, two-, three-, or four-pipe system. The fuel used to fire the boiler or furnace is determined, along with the location of the utility service meter. The distance between a boiler and expansion tank determines the amount of piping required for the expansion tank. Flue requirements are determined by the architect and HVAC engineer based on a variety of building and fire protection codes.

Hydronic Systems. Locations for pumps for the hot-water supply and hot-water return are shown on mechanical prints. Expansion tanks connected to boilers are also shown. Schedules on mechanical prints or specifications provide pump information, including the manufacturer name and model number, flow rate in gallons per minute, and motor size and speed. These items are taken off for the estimate as individual units.

Piping connections between the boiler, any auxiliary heat supply sources, the pump or pumps, and the expansion tank are also shown on mechanical prints. Piping information between the boiler, solar or geothermal systems, pumps, and expansion tank includes the inside diameter of the pipes and valve and meter types and locations. Pipe notations indicate the purpose of the pipe and the direction of water flow. **See Figure 13-15.** The estimator makes calculations concerning the amount of pipe and elbows, valves, tees, and other fittings based on this information. Flue details are provided by the architect, HVAC engineer, or boiler manufacturer.

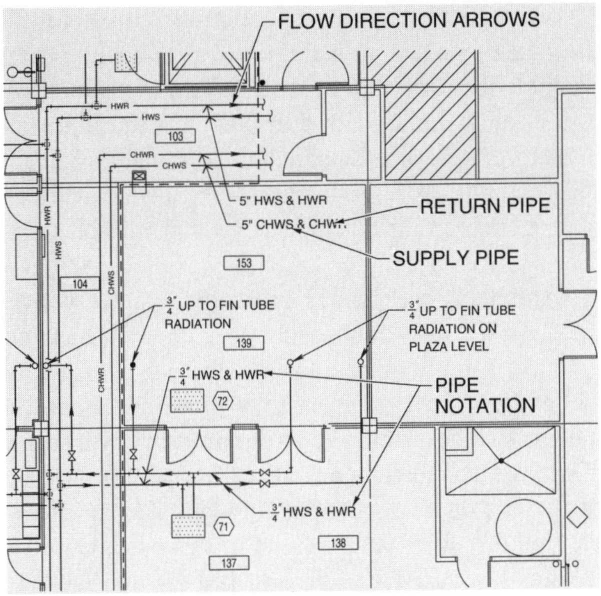

Figure 13-15. The water flow direction noted on mechanical prints helps to determine pump and fitting requirements.

Pipe is taken off and entered into a ledger sheet, spreadsheet, or estimating program based on linear feet of each pipe for each diameter and pipe material. Valves and fittings are taken off as individual items. Flue members are based on the vertical linear feet of flue required. BIM systems may provide linear feet of pipe and fitting information. Care must be taken to account for any degree of fall angle required in piping installation.

When calculating labor installation costs for piping, the estimator should consult with a fabrication shop or the project manager. In some instances, portions of hydronic piping systems can be prefabricated in a shop in a more efficient and productive manner, reducing the amount of labor required at the job site. Additional transportation costs and reduced labor costs may be incurred when prefabricated assemblies are used.

Heating Equipment Takeoff

Heating equipment considered by the estimator includes solar units, boilers, expansion tanks, pumps, furnaces, and terminal units. Specifications in CSI MasterFormat™ Title 23 52 00 are often very detailed concerning gross output capacity, MAWP, and approved manufacturers of boilers. Boiler and furnace information, including information on auxiliary devices, is located throughout Division 23. Boiler and furnace information is typically provided on mechanical, plumbing, and HVAC prints.

Solar Units. A variety of solar units are available to collect and direct solar energy in order to heat liquid. These solar units are included in a schedule or on mechanical prints. Solar panels or collectors are taken off as individual units, ensuring that all mounting devices, piping and pumping systems are included.

Boilers. Items taken into account by design engineers in the selection of a boiler include the number of passes that hot gases make through the system, the input rating, gross output rating, net unit output, and efficiency. The input rating of a boiler is equal to the input in British thermal units per hour (Btu/hr) per unit of fuel. The gross output rating of a boiler is equal to the heat output of a boiler when fired continuously. Net unit output is equal to the gross output rating multiplied by the percentage of heat loss due to initial heat-up. Efficiency factors are based on the fuel used. Coal has an efficiency rating of 65% to 75%. Natural gas and oil have efficiency ratings of 70% to 80%. Electricity has an efficiency rating of 95% to 100%.

Information about the boiler type, size, capacity, and heating load may be included on a schedule in the mechanical prints or in the specifications. Boiler locations and types may be shown on plumbing, mechanical, or HVAC plan views and elevations. Specific dimensions for boiler location are typically not shown on plumbing, mechanical, or HVAC prints. These dimensions are obtained by cross-referencing the plumbing, mechanical, or HVAC prints with the architectural and structural prints. In addition to boiler location, mechanical and electrical prints show flue piping, fuel and water piping, and power and control wiring connections to boilers.

Boilers, pumps, and expansion tanks are taken off as individual unit items and entered into the estimate. **See Figure 13-16.** Boiler manufacturers provide information concerning the necessary installation and safety equipment requirements for various sizes and types of boiler systems. Estimators should consult suppliers to obtain current information and pricing on these items. Company historical records can provide information concerning labor costs for various boiler equipment and pump installations. Coordinating with project managers can help to schedule placement of boilers and pumps when access to boiler rooms is convenient. This can reduce labor and transportation costs for setting the boilers and pumps in place.

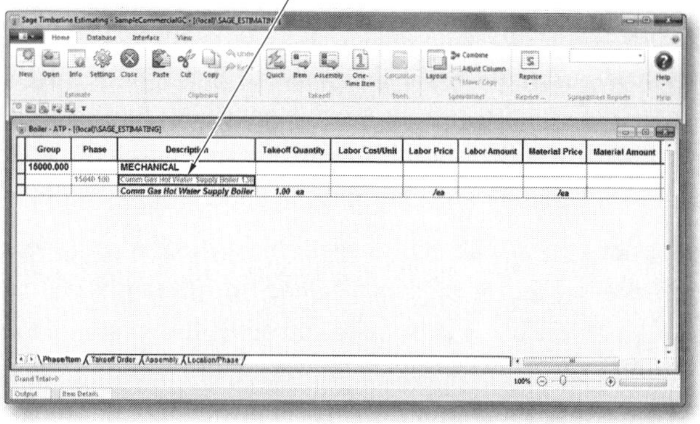

Figure 13-16. Estimating programs allow entry of boiler information into an estimate as an individual item.

Furnaces. Mechanical prints indicate the general location of unit heaters and the required fuel piping and flue information. **See Figure 13-17.** Information concerning electric furnaces and unit heaters is contained in the electrical prints. Estimators should determine the types of furnaces required from the specifications and mechanical prints.

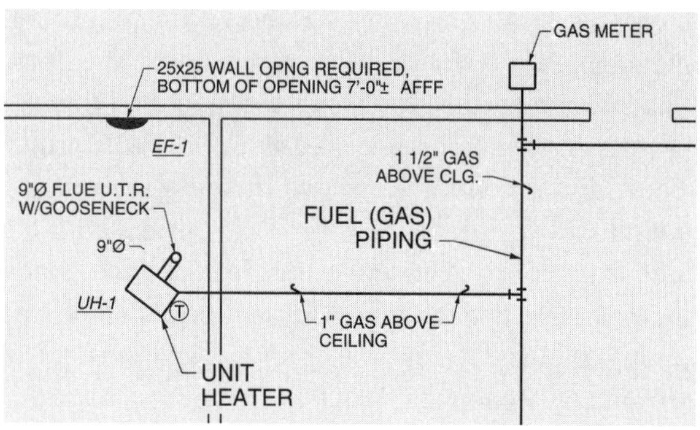

Figure 13-17. Fuel piping and unit heater location are shown on mechanical prints.

As with boilers, each furnace is taken off as an individual unit item and entered into a ledger sheet, spreadsheet, or estimating program. Supplier information is required for selecting the make and model of a furnace that will meet the heating requirements. Labor installation costs are based on company historical data or standard industry information. Coordination of installation with project managers can minimize labor installation and transportation costs.

Terminal Units. Pipe connections are made to terminal units, which transfer heat and/or cooling from water to the surrounding air. Mechanical prints show the location and type of terminal units throughout a structure. Specifications include terminal unit information that describes the water- and airflow requirements and may also include a terminal unit schedule. **See Figure 13-18.**

Architects use various geometric shapes such as hexagons, ovals, and circles for easy identification and cross-referencing of symbols and numbers to schedules. BIM projects may provide a quantity take-off of the terminal units required and their types. Terminal unit details are also provided to indicate water and duct connections. Individual item takeoff is used for all terminal units. Proper coordination of terminal unit installation with project managers as related to structural and finish material installation can minimize labor installation costs.

When taking off HVAC systems, the cost of the standard temperature control system is typically included in the cost of the equipment. An upgraded temperature control system would require additional costs.

Air Conditioning Equipment Takeoff

Mechanical specifications and prints provide plan views, details, and schedules of all components of cooling systems, including chillers, air conditioners, piping, and pumps. Cooling tower locations are shown on plan views and site plans. Air conditioners are shown on electrical prints. In commercial applications, air conditioners may be placed on rooftops.

Chillers. In a manner similar to boilers, specifications note detailed chiller information. Chiller information includes approved compressors, pumps, required circuit breakers, insulation requirements, refrigerant circuit details, accessories, and approved manufacturers. Estimators use information in the specifications and on plumbing, mechanical, and HVAC prints to take off each chiller and pump as an individual item for entry into the estimate. Access to installation areas for installation is coordinated with structural and finish construction to minimize transportation and installation labor costs.

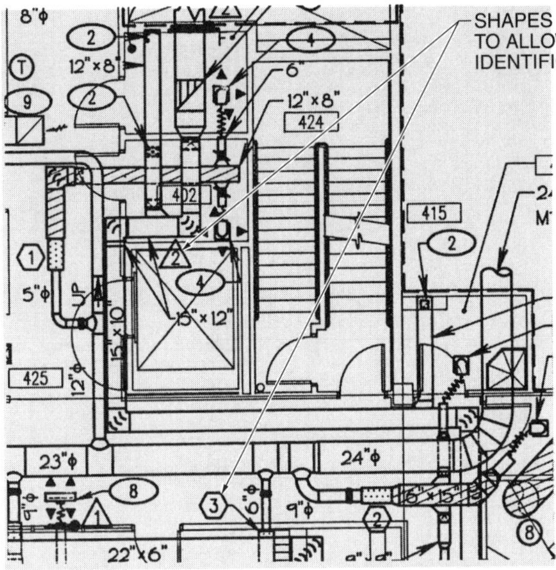

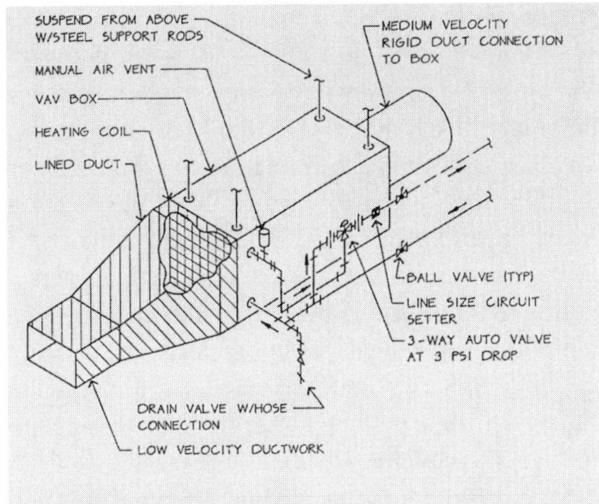

TERMINAL UNIT COIL PIPING DETAIL (4)

TERMINAL UNIT SCHEDULE

MFR: TITUS

#	INLET DIA. (IN)	MAX CFM	MIN CFM	MIN SP (IN)	NOISE CRI AT ROOM(2)	NOISE(1) CRITERIA RADIATED	EAT = 55°	HOT WATER COIL		EWT = 180°			MODEL NO.	NOTES
							LAT	LWT	GPM	WATER PD (IN)	AIR PD (IN)	# OF ROWS	MBH		
TU-1	5	180	180	0.13	37	27	99	160	0.87	0.63	-	1	-	DESV-3000	SF-4 (6)
2	9	760	760	0.19	39	27	105	158	3.8	1.05	-	2		"	"
3	6	290	110	0.17	33	25	102	159	0.54	0.20		1		"	" (6)
4	9	700	400	0.16	39	26	115	153	1.9	0.31		2		"	"
5	5	215	80	0.11	36	27	109	156	0.39	0.11		1		"	" (6)
6	6	300	100	0.18	40	27	102	153	0.38	0.11		1		"	" (6)
7	7	400	200	0.09	35	26	94	157	0.79	0.53		1		"	" (6)
8	8	750	375	0.23	41	27	117	153	1.7	0.25		2		"	"

Figure 13-18. Estimators rely on mechanical plans views, details, and schedules to take off the proper equipment for mechanical systems.

Cooling Towers. Cooling tower takeoff requires consideration of the airflow design, water basin requirements, fan and motor types, water distribution system, control system, and location of the cooling towers in relation to chillers and other mechanical equipment. As with boilers and chillers, the architect may provide a list of approved manufacturers. Item takeoff is required for cooling towers with assistance from manufacturers or individuals specializing in cooling tower construction. Labor costs are based on company historical data or standard industry information.

Air-Handling Equipment Takeoff

Air-handling equipment considered by the estimator includes fans, motors, drive units, and the necessary ductwork to facilitate airflow. Additional items include accessories such as dampers, filters, louvers, and grilles. Specialized air-handling requirements that may be included in this portion of the estimate include dust collection systems, paint booths, smoke and fire dampers, fume hoods, and special ductwork linings. BIM systems may provide quantity take-off information for many of these individual items.

Air Handlers. Air handler considerations for estimators include fan dimensions and capacities, power requirements, drive systems required for various horsepower levels, housings, accessories, and permissible manufacturers. Weather protection considerations are required for exterior applications. Fans and air handler accessories are entered into the ledger sheet, spreadsheet, or estimating program as individual unit items. Information is obtained from Title 23 06 00 of the specifications and the mechanical or HVAC prints. Labor rates are based on company historical records or industry standards for installation of the various types of equipment.

Ductwork. Ductwork is commonly prefabricated using standard-size pieces and fittings. Ductwork designs are shown with plan views and elevations on mechanical prints. CAD and BIM systems can integrate with computer-aided manufacturing systems that cut, shape, and fabricate ductwork directly from the electronic model. **See Figure 13-19.** Shop drawings may be developed for use in the fabrication shop. Diameters are indicated for round ductwork. Width and height are indicated for rectangular or square ductwork. Ductwork components are commonly marked with a code to identify their job-site placement. Methods for attachment of ductwork to structural members are not shown. Hanging information is included in the specifications or the general plan notes.

Figure 13-19. Ductwork bends and supports can be estimated and fabricated based on the electronic model information.

Diffusers are used at the ends of ductwork. A diffuser is an air outlet in an HVAC system that directs air in a wide pattern. Standard sizes and types are available from various manufacturers. Each diffuser is identified by size and type using symbols on mechanical plan views or on a diffuser schedule. **See Figure 13-20.** Diffusers are taken off as individual unit items or may appear in a quantity take-off schedule in BIM systems.

Ductwork drawings may be provided for large commercial buildings or generated electronically from the HVAC drawings. All ductwork layout drawings should be drawn to scale, and this scale should be used to determine the lengths of the various pieces of ductwork required. Stack lengths of all vertical ductwork are determined from vertical dimensions shown on the drawings plus the distance above each floor level. The number and types of various fittings required for fastening ductwork are taken off by each unit item throughout various ductwork lengths.

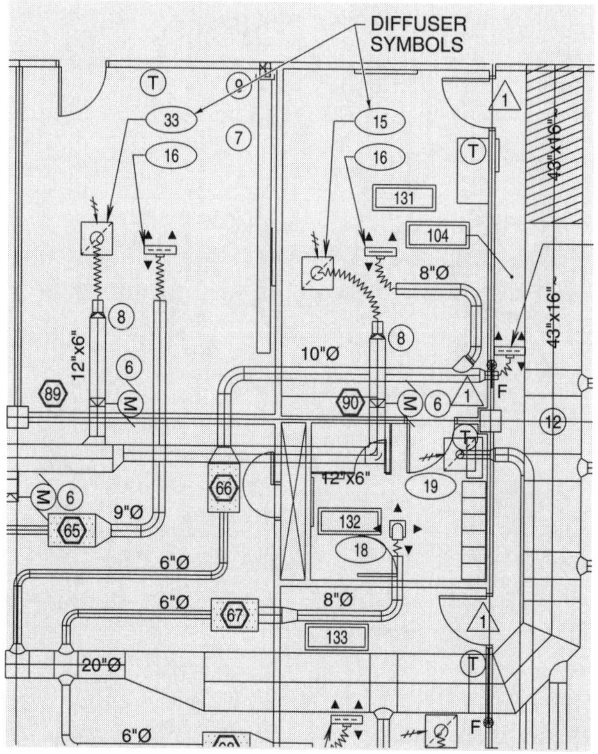

DIFFUSER SCHEDULE

#	DIFFUSER			#	DIFFUSER		
1	24"	1	210	18	8"Ø	6	200
2	10"Ø	7	260	19	8"Ø	16	225
3	48"	9	550	20	24"	1	260
4	10"Ø	11		21	16"X12"	14	600
5	9"X6"	12	200	22	8"X6"	13	140
6	14"X8"	13	200	23	12"X10"	2	140
7	8"Ø	6	80	24	12"X8"	2	100
8	6"X6"	2	90	25	6"Ø	3	100
9	18"X18"	2	330	26	10"X4"	2	100
10	18"X18"	17	300	27	6"Ø	6	100
11	24"	4	115	28	48"	4	140
12	6"Ø	15	115	29	24"	4	60
13	24"	4	115	30	21"X10"	2	750
14	24"	1	200	31	10"Ø	16	
15	10"Ø	7	260	32	24"X8"	2	550
16	24"	1	240	33	10"Ø	7	300
17	24"	4	100	34	8"Ø	15	200

Figure 13-20. Air diffuser symbols on plan views are referenced in a diffuser schedule that indicates diffuser size and type.

Manufacturers of prefabricated ductwork provide listings of all parts and fittings together with figure or part numbers. As each size or part is taken off of the drawings, it is noted in the quantity takeoff column on a ledger sheet, spreadsheet, or estimating software.

HVAC contractors normally keep historical records for labor costs of installing ductwork on previous jobs to help determine the cost of installing ductwork per linear foot. This allows for determination of an average labor cost for ductwork based on the number of linear feet of ductwork or the number of registers. Estimators review prints to note any possible irregularities in ductwork runs. Ductwork labor costs are impacted by job-site accessibility, aerial lift requirements, coordination with structural and finish construction, custom ductwork fabrication at the job site, or other labor-intensive operations.

Each type of ductwork run and fitting should be noted by the architect on an individual line, column, cell, or row in the estimate to indicate the number of feet of each type of ductwork and quantity of each type of fitting to be installed. Shop drawings where standard-size duct components are used allow for some consolidation of material costs.

Registers, grilles, and louvers are taken off as individual unit items according to their size, type, and material, such as sheet metal, brass, or other decorative metal. Sizes and types are indicated on schedules or shown on elevations and details. Labor cost rates are based on the standard labor cost per unit installation time. Ventilation hoods for the collection of fumes and smoke inside buildings are shown in details. The details indicate dimensions of the hood and materials for construction. Hoods are taken off as individual unit items for both material costs and labor installation. Locations and types of materials for filters within the air circulation system are shown on ductwork details.

Registers, grilles, louvers, and filters are taken off as individual items. Each type is separated into an individual line, column, cell, or row and added for a total for the entire job. Labor installation costs for each of these items are based on company historical data or standard labor-rate information.

Controls. Mechanical system control considerations include the type and amount of equipment being controlled and manufacturer and designer specifications. Many control devices are bundled into the cost of the equipment at the time of purchase. Other controls may require additional material. Estimators should check the specifications for controls for each piece of equipment to ensure all control devices are included in the estimate. Other items such as thermostats are taken off as individual items. In any case, labor costs for installation of control devices are based on the installation time per unit and labor-rate cost as determined by company historical data or standard labor-rate tables.

SAGE TIMBERLINE ESTIMATING . . .

TAKEOFF USING ONE-TIME ITEMS

The One-time Item feature in Sage Timberline Estimating allows an estimator to take off items that are not currently in the database. Details can be added through the Detail dialog box and, if desired, one-time items can be saved to the current database.

1. Start a new estimate called **Sprecher Apartments** and link it to the **Sample Ext Commercial GC** database.
2. Open the **One-time Item** dialog box by picking the **One-time Item** button in the toolbar.

. . . SAGE TIMBERLINE ESTIMATING . . .

3. Right-click in the **Phase** field and select **Edit Phase...** from the shortcut menu to display the **Estimate Phase** dialog box. Pick the **Add** button to create the phase record. Enter **15100** for **Phase** and **Copper Pipe** for **Description**. Pick **OK** and then pick **Close**.

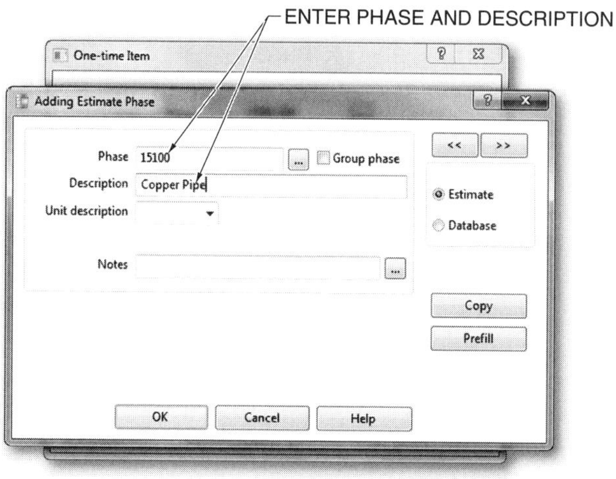

4. Create the item record by completing the fields in the **One-time Item** dialog box. Enter **Copper Pipe 1″** for **Description**, **125** for **Quantity**, and **lf** for **Takeoff Unit**.
5. Select the **Labor** and **Material** category checkboxes.

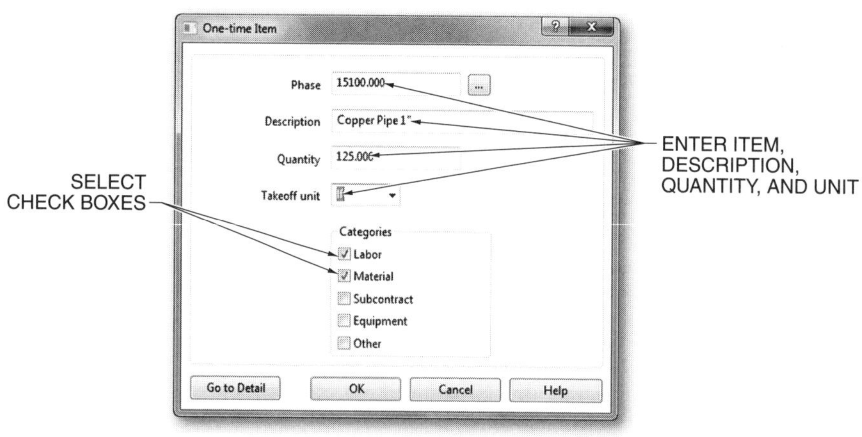

. . . SAGE TIMBERLINE ESTIMATING

6. Pick the **Go to Detail** button and enter **0.64** for **Labor Unit Price** and **5.67** for the **Material Unit Price** for the item.

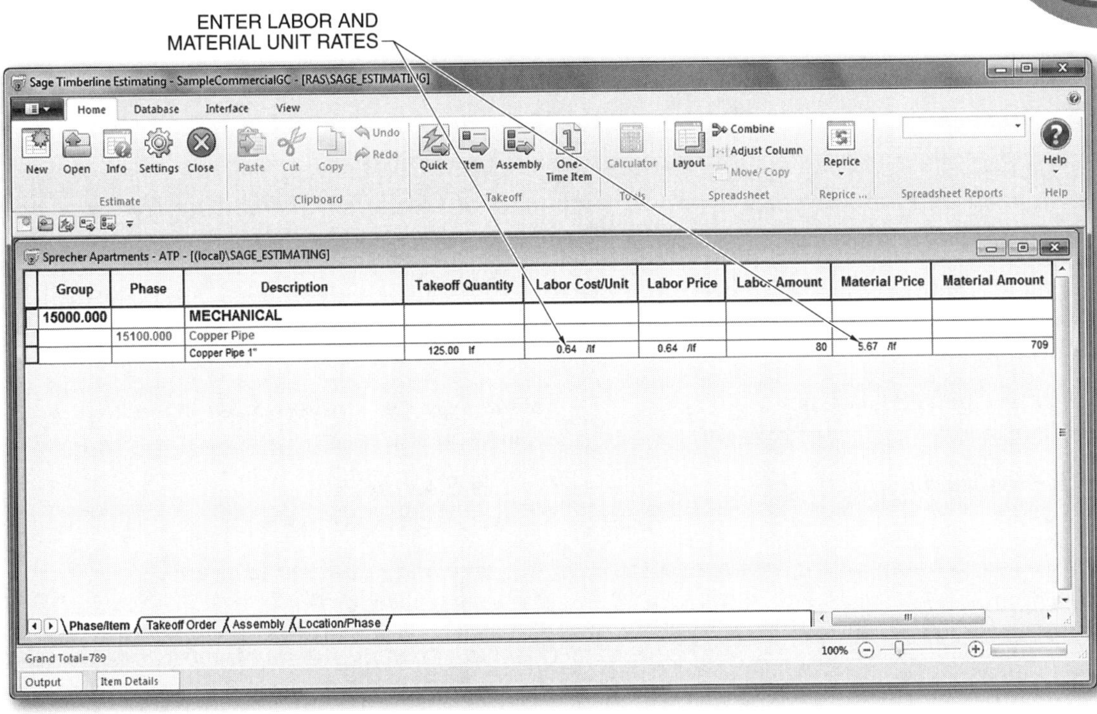

7. Exit the **Detail** dialog box by picking **OK** and then **Close**. The one-time item displays in the spreadsheet.

Quick Quiz®

Quick Quiz®

Refer to the CD-ROM for the Quick Quiz® questions related to chapter content.

Key Terms

Illustrated Glossary

- *air conditioner*
- *air filter*
- *air handler*
- *compressor*
- *control zone*
- *ductwork*
- *forced-air system*
- *furnace*
- *hydronic system*
- *louver*
- *pipe fitting*
- *pressure relief valve*
- *process piping*
- *supply piping*
- *valve*

Web Links

Air Conditioning Contractors of America
www.acca.org

Air-Conditioning, Heating, and Refrigeration Institute
www.ahrinet.org

Air Diffusion Council
www.flexibleduct.org

Air Movement and Control Association International
www.amca.org

American Boiler Manufacturers Association
www.abma.com

American Fire Sprinkler Association
www.firesprinkler.org

American Gas Association
www.aga.org

American Society of Heating, Refrigerating and Air Conditioning Engineers
www.ashrae.org

American Society of Mechanical Engineers
www.asme.org

American Society of Plumbing Engineers
www.aspe.org

American Water Works Association
www.awwa.org

Canadian Institute of Plumbing and Heating
www.ciph.com

Cast Iron Soil Pipe Institute
www.cispi.org

Cooling Tower Institute
www.cti.org

Fluid Controls Institute
www.fluidcontrolsinstitute.org

Industrial Heating Equipment Association
www.ihea.org

International Association of Plumbing and Mechanical Officials
www.iapmo.org

Mechanical Contractors Association of America
www.mcaa.org

National Association of Plumbing-Heating-Cooling Contractors
www.phccweb.org

National Fire Sprinkler Association
www.nfsa.org

Plastic Pipe and Fittings Association
www.ppfahome.org

Plastics Pipe Institute
www.plasticpipe.org

Plumbing and Drainage Institute
www.pdionline.org

Plumbing-Heating-Cooling Contractors Association
www.phccweb.org

Plumbing Manufacturers Institute
www.pmihome.org

Sheet Metal and Air Conditioning Contractors National Association
www.smacna.org

Sump and Sewage Pump Manufacturers
http://sspma.org

Uni-Bell PVC Pipe Association
www.uni-bell.org

Water Systems Council
www.watersystemscoucil.org

CONSTRUCTION ESTIMATING

Mechanical Systems 13

Review Questions

Estimating

T F 1. Brass pipe is available in standard and extra-strong weights.

T F 2. Steel pipe may be seamless or welded.

__________ 3. ___ piping is piping that delivers water from the source to the point of use.

__________ 4. ___ piping is piping installed in industrial facilities for compressed air, vacuum, gas, or fuel.

__________ 5. A(n) ___ system is a system that uses water, steam, or another liquid to condition building spaces.

__________ 6. A(n) ___ pipe is a pipe that removes gases and odors from the waste piping and exhausts them away from inhabited areas.

__________ 7. A(n) ___ system is a system that uses warm or cool air to condition building spaces. forced-air.

__________ 8. A(n) ___ is used to produce heated water or steam.

__________ 9. A(n) ___ is the component in a hydronic air conditioning system that cools water, which in turn cools the air.

__________ 10. A(n) ___ is a temperature-actuated electric switch that controls heating and/or cooling equipment.

T F 11. A low-pressure boiler is a boiler that has a maximum allowable working pressure of up to 15 psi.

T F 12. A high-pressure boiler is a boiler that has a maximum allowable working pressure above 15 psi and over 4 BHP (boiler horsepower).

__________ 13. A(n) ___ is a heat-resistant passage in a chimney that conveys smoke or other gases of combustion.

__________ 14. A(n) ___ is a self-contained heating unit that includes a blower, burner(s), heat exchanger or electric heating elements, and controls.

__________ 15. A(n) ___ is the component in a forced-air system that cools the air.

__________ 16. A(n) ___ is an electrical device that uses electromagnetism to change voltage from one level to another.

_______________ **17.** A(n) ___ is an overcurrent protection device with a mechanism that automatically opens the circuit when an overload condition or short circuit occurs.

_______________ **18.** A(n) ___ is a switch that disconnects electrical circuits from motors and machines.

_______________ **19.** A(n) ___ is an overcurrent protection device with a fusible link that melts and opens the circuit when an overload condition or short circuit occurs.

T F **20.** A chiller is an air outlet in an HVAC system that directs air in a wide pattern.

Short Answer

1. Define "unit estimating" and explain how it is used for estimating mechanical and plumbing systems.

2. Discuss the difference between one-, two-, three-, and four-pipe hydronic systems.

Mechanical Systems 13

Activities

Activity 13-1—Ledger Sheet Activity

Refer to the cost data, Print 13-1, and Estimate Summary Sheet No. 13-1. Take off the total item count for each fixture. Determine the total material and labor costs.

COST DATA			
Material	**Unit**	**Material Unit Cost***	**Labor Unit Cost***
1. Water closet, one piece, wall-hung	ea	478.50	97.35
2. Water closet, bowl only, with flush valve, wall-hung	ea	379.50	89.10
3. Lavatory, vitreous china, 20″ × 16″, single bowl	ea	179.30	95.70
4. Shower, built-in, 4 gpm valve	ea	74.25	71.50

* in $

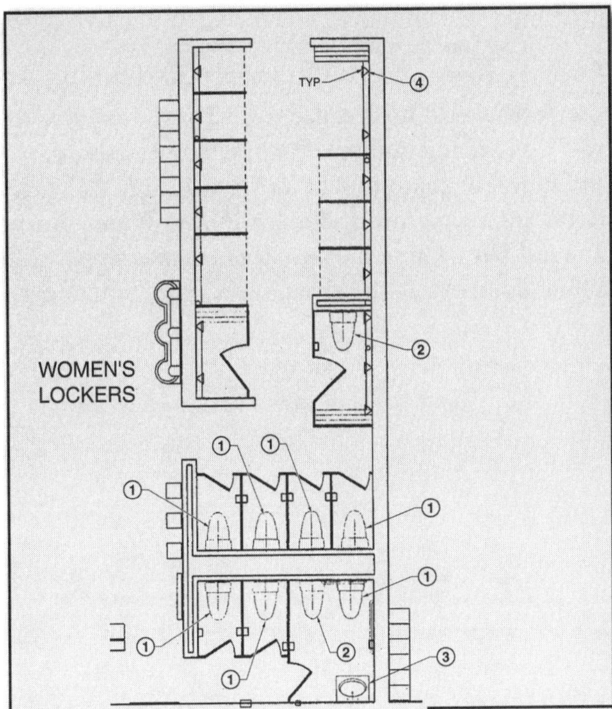

Print 13-1

ESTIMATE SUMMARY SHEET

Sheet No. 13-1
Date: ______
Checked: ______

Project: ______
Estimator: ______

No.	Description	Dimensions			Quantity		Material		Labor		Total	
						Unit	Unit Cost	Total	Unit Cost	Total	Unit Cost	Total
	Water closet, one-piece											
	Water closet, bowl only											
	Lavatory											
	Shower											
	Total											

Activity 13-2—Spreadsheet Activity

Refer to Print 13-2 and Quantity Spreadsheet No. 13-2 on the CD-ROM. Take off the length of ductwork (in linear feet) and number of supply and return air grilles. Round all values up to the next whole unit. Do not make allowances for angular runs.

Activity 13-3—Plumbing Takeoff

Refer to Prints 13-3A and 13-3B and Estimate Summary Spreadsheet No. 13-3 on the CD-ROM. Take off the number of plumbing fixtures (including floor drains) identified in the Plumbing Fixture Schedule. Then take off the length (in linear feet) of the horizontal 2″ and ¾″ gas lines shown on the Plumbing Plan to the nearest 2′ increment.

Activity 13-4—Sage Timberline Estimating Activity

Create a new estimate called Activity 13-4 and link it to the Sample Ext. Commercial GC database. Add phase 15400 Plumbing Equipment by picking **Phases** from the **Database** menu. Pick **Add** and enter **15400** for **Phase** and **Plumbing Equipment** for **Description**. Pick **OK** and **Close**. Add a one-time item by picking the **One-time Item** button entering **15400** for **Phase**, **Water Storage Tank** for **Description**, **1** for **Quantity**, and **each** for **Takeoff Unit**. Close the **One-time Item** dialog box. Input a **labor cost unit** of **$7882** and a **material price** of **$1500** directly into the spreadsheet. Print a standard estimate report.

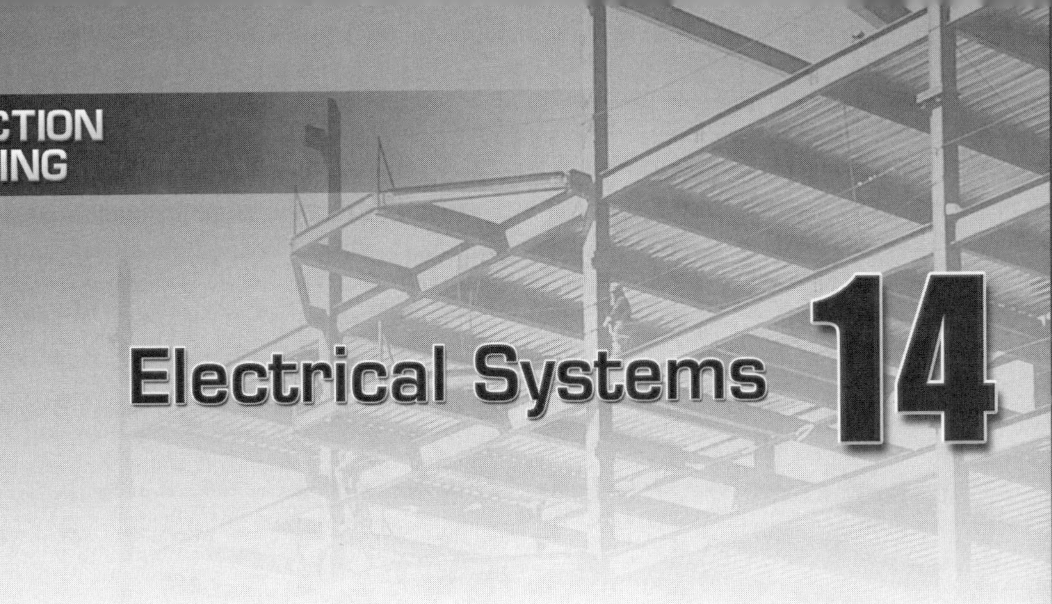

Electrical Systems 14

Key Concepts

- Specific dimensions for electrical wiring, equipment, and fixture installation are not provided on electrical plans. Rather, dimensions are obtained from architectural and structural prints and elevations, and the applicable electrical or fire protection codes.
- Electrical estimators must be familiar with all applicable building and fire protection codes related to electrical installation.
- The two methods of estimating used for electrical takeoff are detailed takeoff and averaging takeoff.
- Detailed takeoff is a precise method used for industrial and commercial work and wherever unique job conditions exist.
- Averaging takeoff is sufficiently accurate for small takeoff jobs on basic electrical wiring systems.

Introduction

Many divisions and numbered titles within the CSI MasterFormat™ address topics related to electrical systems. Divisions 25 (Integrated Automation), 26 (Electrical), 27 (Communications), 28 (Electronic Safety and Security), and 48 (Electrical Power Generation), as well as numbered titles 33 70 00 (Electrical Utilities), 33 80 00 (Communications Utilities), 34 20 00 (Traction Power), and 34 40 00 (Transportation Signaling and Control Equipment) include electrical information.

Electrical specifications and prints contain information about the wiring to be installed, electrical equipment, and electrical finish materials. Electrical prints may be divided into

separate sections for lighting, power supply, and signals such as fire alarms and smoke detectors. Connections to electrical finish fixtures are made after panelboards, conduit, and wiring are installed. Electrical finish fixtures include luminaires (light fixtures), communication systems, and alarm systems.

POWER DISTRIBUTION SYSTEM MATERIALS AND METHODS

Electrical power distribution systems include service provisions from the power source into a structure, wiring and distribution throughout the structure, controls and switches throughout the power system, fixtures, and other equipment that transfers the power to the intended application. As both high- and low-voltage electrical systems have become more complex, a variety of electrical specifications and prints may be required to describe an entire electrical installation.

Power sources are noted on site plans and electrical specifications and prints. Locations for transformers and panelboards are shown on site plans and electrical prints along with circuit configurations. Conduit, cable, and conductor installations and types are indicated on the prints and in the specifications.

Conduit is hollow pipe or tubing that supports and protects electrical conductors. Conduit is installed prior to the placement of cast-in-place concrete, masonry, or finish wall materials and after the placement of structural members such as steel columns or metal or wood framing. Exact conduit locations are not indicated on electrical prints.

Conduit is placed according to architectural and structural print dimensions, local building codes, and the necessary connections that are designated for the end runs of the conduit. During concrete placement, the ends of preplaced conduit are capped and sealed to prevent the conduit from being filled with concrete or mortar. Conduit is fastened together with a variety of fittings.

After structural members are in place and concrete and masonry have adequately set, conductors and cables are pulled through the conduit. The types of conductors and cables to be installed are noted in the specifications and on the electrical prints. Switches, receptacles, fixtures, and other electrical equipment are installed and tested to ensure safe and proper operation according to all applicable building codes and electrical specifications and prints.

A review of all portions of electrical work must be completed by the estimator to develop an accurate takeoff and final estimate. Electrical specifications and prints may be divided into separate sections for high-voltage systems such as lighting and power supply, and low-voltage systems such as signals and controls for communications equipment, fire alarms, and smoke detectors.

A general contractor commonly obtains bids for electrical systems from electrical estimators. Electrical estimators are specialized in their knowledge of electrical power and wiring systems, electrical codes, installation methods, labor rates, equipment requirements, available materials, types of fixtures and fittings, and the other technical aspects of electrical work.

Wiring

Electrical system wiring can be located underground, fastened in place to structural members and hidden behind finish materials, placed in conduit, or left as exposed cable. Electrical system wiring may be installed in walls, above ceilings, or under floors. Electrical loads vary from low-voltage loads for items such as computer equipment and programmable thermostats to high-voltage loads for welding equipment or heavy manufacturing machinery. Each wiring application is shown on the electrical specifications and prints. **See Figure 14-1.**

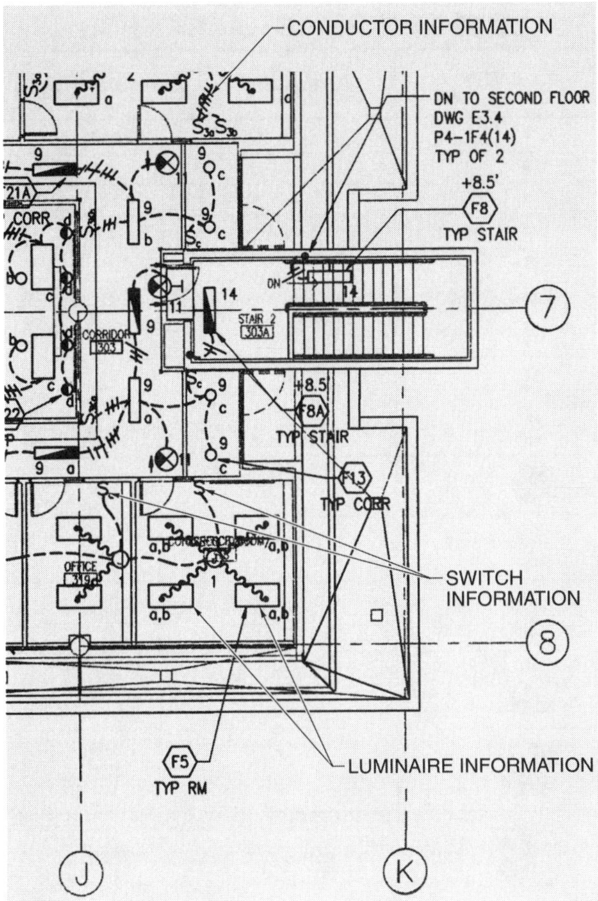

Figure 14-1. Electrical prints show conductor, luminaire, and switch information.

Raceways and Conduit. A *raceway* is an enclosed channel for conductors. **See Figure 14-2.** A raceway includes all of the enclosures used for running conductors for different components of an electrical system. Applicable building and fire codes should be consulted for the proper raceway application and installation. Raceways include intermediate metal conduit, rigid metal conduit, rigid nonmetallic conduit, electrical metallic tubing, and flexible metal conduit.

Intermediate metal conduit (IMC) is a medium-weight galvanized metal conduit that is available in diameters ranging from ½″ to 4″ in 10′ lengths. IMC is available with a 10-, 20-, or 40-mil-thick PVC coating to resist corrosion (1 mil = 0.001″). This coating is partially removed at the ends of conduit to be joined to fittings and couplings. Rigid metal conduit (RMC) is heavy conduit made of metal and is regarded as the universal raceway material. RMC is permitted for use in all atmospheric conditions and types of occupancies. RMC is available in diameters ranging from ½″ to 6″ in 10′ lengths with a coupling on one end.

RACEWAYS			
	Abbr	**NEC® Article**	**Diameter/ Length**
Intermediate Metal Conduit	IMC	342	½″ to 4″ in 10′ lengths
Rigid Metal Conduit	RMC	344	½″ to 6″ in 10′ lengths
Rigid Nonmetallic Conduit	RNC	352	½″ to 6″ in 25′ lengths
Electrical Metallic Tubing	EMT	358	½″ to 4″ in 10′ lengths
Flexible Metal Conduit	FMC	348	⅜″ to 4″ in 25′, 50′, or 100′ coils

Figure 14-2. Raceways include intermediate metal conduit, rigid metal conduit, rigid nonmetallic conduit, electrical metallic tubing, and flexible metal conduit.

Rigid nonmetallic conduit (RNC) is a conduit made of materials other than metal, such as polyvinyl chloride. Rigid polyvinyl chloride (PVC) conduit is a nonconductive conduit that can replace metallic conduit in some applications. PVC conduit cannot be installed in hazardous locations or air plenums. PVC conduit is available in diameters ranging from ½″ to 6″ in 10′ lengths. PVC conduit is available in thin wall (Schedule 40) and thick wall (Schedule 80). Couplings and other fittings are fastened to PVC conduit with an adhesive that is formulated for this purpose. PVC conduit is also available in P&C grade for use in power and communication applications. P&C grade PVC conduit is available in diameters ranging from ½″ to 6″ in 25′ lengths.

Electrical metallic tubing (EMT), also known as thin wall or steel tube conduit, is lightweight tubular steel raceway without threads on the ends. EMT is available in diameters ranging from ½″ to 4″ in 10′ lengths. EMT is galvanized or electroplated to resist corrosion. A variety of EMT fittings are available to facilitate connections, including compression and set screw couplings and connectors. A wide variety of hangers and supports are available to hold EMT in place when attaching it to structural members. EMT may also be encased in concrete or installed in masonry.

Flexible metal conduit (FMC), also known as Greenfield, is a raceway composed of circular metal strips. The metal strips are helically wound, formed, and interlocked. FMC is available in diameters ranging from ⅜″ to 4″ in 25′, 50′, or 100′, depending on its diameter. A variety of connectors are available.

Special-purpose conduit includes fiber duct, soapstone duct, wrought-iron pipe, clay conduit, liquid-tight flexible metal conduit (LFMC), liquid-tight flexible nonmetallic conduit (LFNC), and flexible metallic tubing (FMT). Estimators should take note of any special-purpose conduit shown in the specifications and on electrical prints. Manufacturers and suppliers can provide information about these and other conduit types.

Wires, Conductors, and Cables. A *conductor* is a slender rod or wire that is used to control the flow of electrons in an electrical circuit. Conductors are composed of a circular, single-strand (solid) or multistrand (stranded) conductive material (copper or aluminum). Conductors may be bare or insulated.

Conductors are commonly called wires. A gauge number is used to designate the diameter of the wire (without insulation). The higher the gauge number, the smaller the wire diameter. **See Figure 14-3.** A cable is two or more conductors grouped together within a common protective cover and used to connect individual components. A metallic shield may also be formed around conductors within the outer jacket.

COPPER CONDUCTOR RATINGS*		
AWG	Ampacity	Diameter (Mils)
18	—	40
17	—	45
16	—	51
15	—	57
14	20	64
12	25	81
10	30	102
8	40	128
6	55	162
4	70	204
3	85	229
2	95	258
1	110	289
0	125	325
00	145	365
000	165	410

*at 60°C

Figure 14-3. Conductors are sized by gauge number.

Wire used in construction applications is identified by the type of insulation protecting the conductors. Wire designations include the following:

- moisture-resistant thermoplastic
- heat-resistant thermoplastic
- moisture- and heat-resistant thermoplastic
- moisture- and heat-resistant thermoset
- moisture-, heat-, and oil-resistant thermoplastic
- thermoplastic-covered fixture wire
- thermoplastic-covered fixture wire—flexible stranded
- heat-resistant thermoplastic-covered fixture wire
- underground service entrance cable—single-conductor

Cables used in construction include hard service cord, junior hard service cord, nonmetallic sheathed cable, underground feeder and branch-circuit cable, service-entrance cable, aluminum wire, bare copper wire, armored cable, and high-voltage wire and cable.

Moisture-resistant thermoplastic wire (TW) is used in wet and dry applications. TW wire has no outer covering and is made with No. 14 gauge solid or stranded conductors. TW has a maximum operating temperature of 140°F.

Heat-resistant thermoplastic wire (THHN) has a nylon or equivalent coating and is used in dry locations. THHN is made with No. 14 gauge solid or stranded conductors and has a maximum operating temperature of 194°F.

Moisture- and heat-resistant thermoplastic wire (THW) is used in wet and dry locations. THW has no outer covering and is made with No. 14 gauge solid or stranded conductors. THW has a maximum operating temperature of 194°F when used with lighting equipment of 1000 V or less.

Moisture- and heat-resistant thermoset wire (XHHW) has a fire-resistant synthetic polymer insulation and no outer covering. XHHW is used in wet and dry locations, is made with No. 14 gauge solid or stranded conductors, and has a maximum operating temperature of 167°F in wet locations and 194°F in dry locations.

Moisture-, heat-, and oil-resistant thermoplastic wire (MTW) has thick (Type A) or thin (Type B) insulation. Type A insulation has no outer covering. Type B insulation has a nylon covering. MTW wire is commonly used for machine tool wiring in wet locations and has an operating temperature limit of 140°F. MTW wire is made with No. 14 gauge stranded conductors.

Thermoplastic-covered fixture wire (TF) has thermoplastic insulation with no other covering. TF wire is made with No. 16 or No. 18 gauge solid or stranded conductors and has a maximum operating temperature of 140°F. TF wire is commonly used for wiring electrical fixtures. Thermoplastic-covered fixture wire—flexible stranded (TFF) has similar characteristics to TF wire with the exception that there is no solid conductor. Heat-resistant thermoplastic-covered fixture wire (TFFN) is also used for fixture wiring and has a heat resistance of 194°F.

Hard service cord may be Type S, Type SO, or Type STO. Type S has two or more conductors ranging in gauge from No. 2 to No. 18. The insulation and covering are the same as SJ cable. Type SO has the same number and type of conductors and insulation as Type S but has an outer coating of oil-resistant plastic. Type STO is similar to Type SO with the exception that the insulation and covering may be thermoplastic.

Junior hard service cord (SJ) is cable made with 2, 3, or 4 conductors of No. 10 gauge or No. 18 gauge wire. SJ cable has thermoset plastic insulation and covering. SJ cable is used for hanging fixtures in damp locations.

Nonmetallic sheathed cable (NM) is made with 2, 3, or 4 conductors in gauges ranging from No. 2 to No. 14. NM cable is made with or without a ground wire and may be insulated or bare. The outer covering is fire- and moisture-resistant plastic.

Underground feeder and branch circuit cable (UF) is made with 2, 3, or 4 conductors with gauges ranging from No. 14 to No. 4/0. UF cable is made with or without a ground wire. Insulation thickness varies according to the gauge of the conductor wire. The outer covering is moisture- and heat-resistant.

Underground service entrance cable—single conductor wire (USE) is made with a covering that is moisture-resistant and sized from No. 12 gauge solid or stranded conductors. USE wire has a maximum operating temperature of 167°F per the NEC®.

Service-entrance cable (SE) is made with one or more conductors with gauges ranging from No. 12 to No. 4/0. The outer covering is fire- and moisture-resistant.

Aluminum wire is a lightweight wire used on feeder circuits, such as electrical runs from distribution switchboards to power panels. Aluminum wire is susceptible to oxidation when exposed to the atmosphere. Oxidation results in resistance to current flow that may lead to overheating at connections. Antioxidation materials may be applied prior to completion of the installation to guard against overheating.

Bare copper wire is available in gauges ranging from No. 14 to No. 4/0. In its uninsulated form, copper wire is most commonly used for electrical system grounding. Three grades of copper wire are soft drawn, medium-hard drawn, and hard drawn temper.

Armored cable, commonly referred to as Type AC or BX, is made of insulated wires wrapped in a metallic flexible covering. Armored cable is available with single or multiple conductors ranging from No. 4 to No. 14 gauge.

High-voltage wire and cable are specified by electrical engineers for specialized applications. In some situations, detailed specifications from the electrical engineer concerning installation requirements, safe applications, and testing are required. Cable manufacturers can provide technical data and pricing for high-voltage wire and all types of cable. Independent testing agencies provide quotes for specialized high-voltage cable testing.

A circuit run consists of a cable or conductors either in conduit or without conduit. A circuit run is often indicated on electrical prints by a solid or dashed line running between points of connection. The number of slash marks across the line indicates the number of conductors in the circuit. For example, an electrical circuit line crossed by three slash marks indicates a three-conductor cable.

Normally, a circuit line on the electrical prints without any slash marks indicates a two-conductor circuit run. **See Figure 14-4.** This method of identifying conductors is usually applied to circuits that require conductors no larger than No. 10 gauge. A symbol chart included with the electrical prints should be consulted for information on the method used for designating conductor sizes and quantities in a circuit run.

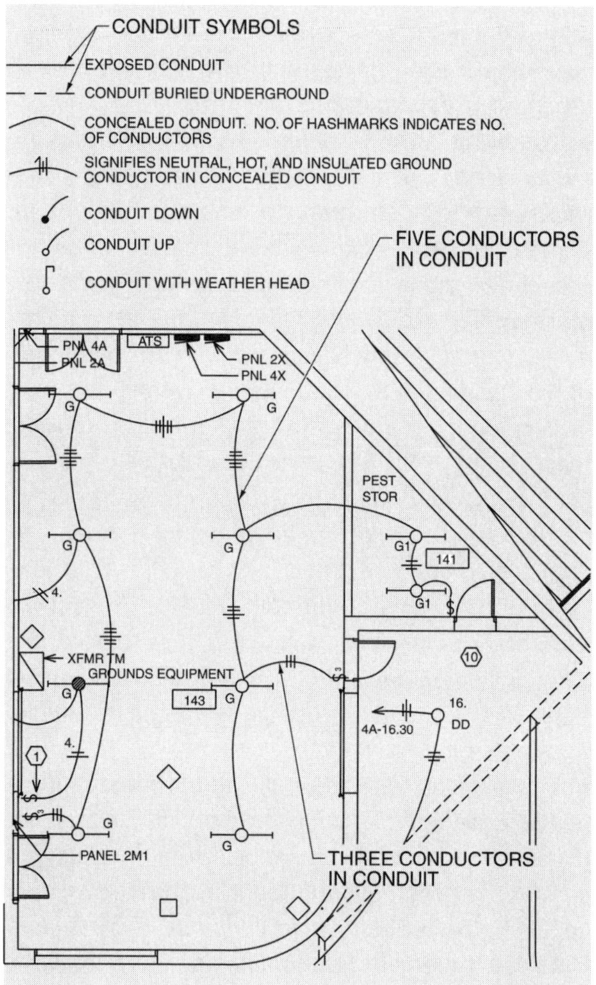

Figure 14-4. The number of slash marks on an electrical circuit run denotes the number of conductors.

Large conductors that carry electrical service from the main power source to panelboards are specified by number of conductors, wire gauge, number and gauge of grounding conductor, and overall cable diameter. For example, a notation of "4#4, 1#10 GRD, 1¼" indicates 4 conductors of No. 4 gauge, one No. 10 gauge grounding conductor, and an overall cable diameter of 1¼″. **See Figure 14-5.**

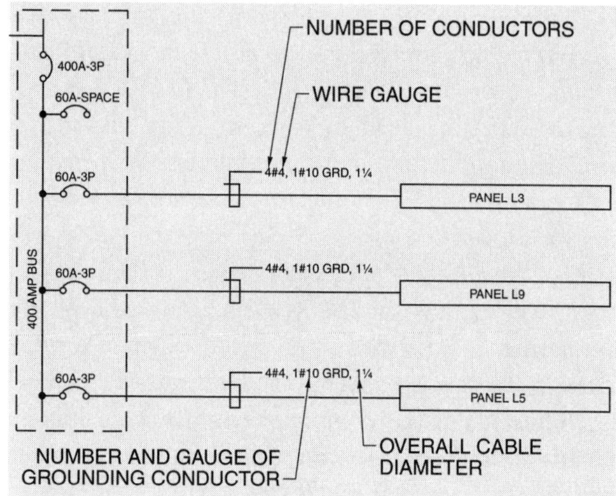

Figure 14-5. Heavy cables are specified according to the number of conductors, wire gauge, number and gauge of grounding conductors, and the overall cable diameter.

Cable Supports. Electrical cables are fastened to structural members, suspended behind walls or above ceilings, placed in cable trays, buried underground, or run within conduit. Clamps and various types of cable supports are used to fasten conduit and cables to structural members. Specific support and fastening information for cable installation is not commonly provided on electrical prints, but it is described in the specifications by reference to the applicable electrical code and fire code requirements.

The NEC® has established standards for cable and conduit fastening requirements. According to the NEC®, cables and conduit must be secured to wall studs or structural members within a certain distance of switches, receptacles, or other electrical fixtures. Estimators should be familiar with these fastening requirements so they can include the correct number of fasteners in the quantity takeoff.

Sections on electrical prints provide information concerning cable trays and sleeves for electrical cable and conduit. A *cable tray* is an open grid rack suspended from structural members to support a series of cables. Cable trays include ladders, troughs, and channel trays. Cable trays are commonly manufactured in 12′ sections with a variety of fittings and connectors. Cable tray locations are shown on floor plans or electrical prints. **See Figure 14-6.**

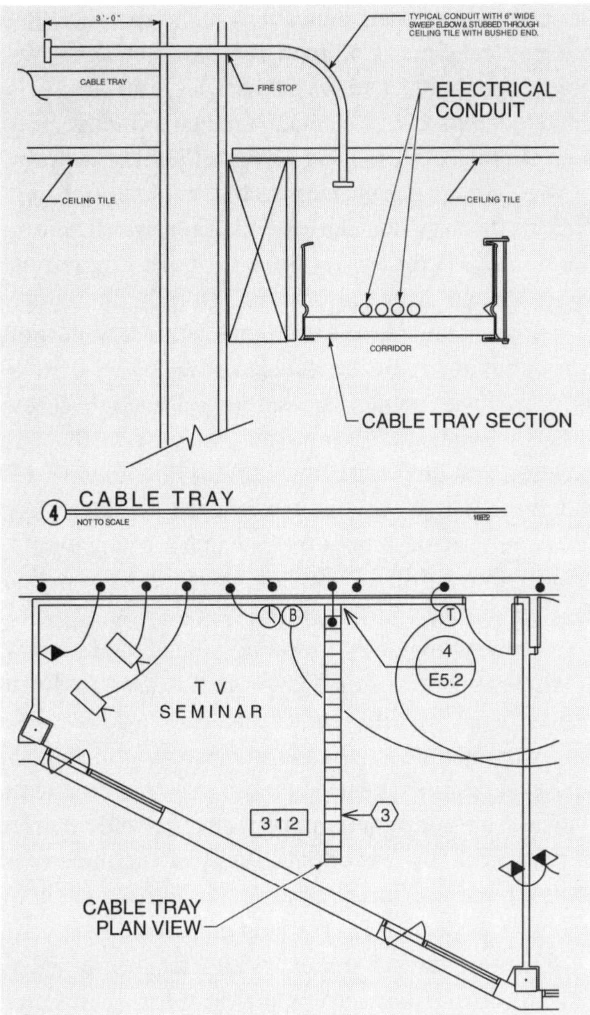

Figure 14-6. Cable tray information is provided on cable tray sections and electrical plan views.

Support and hanging requirements vary based on the class of material used and the code requirements. There are four classes of cable trays, with Class 1 being the lightest grade and Class 4 being the heaviest. Cable trays are commonly supported by threaded rods hanging from overhead structural supports.

Boxes. A *box* is a metallic or nonmetallic electrical enclosure installed in an electrical system to protect or encase equipment, devices, connections, and pulling and terminating conductors. Boxes protect connections, give access to wiring, and provide a method for mounting switches, receptacles, and some fixtures. Boxes are commonly made of stamped galvanized steel or PVC with punched holes for mounting and knockouts to allow for insertion of wiring and conduit. Common electrical boxes include utility (handy), square, switch, octagonal, and masonry boxes. **See Figure 14-7.**

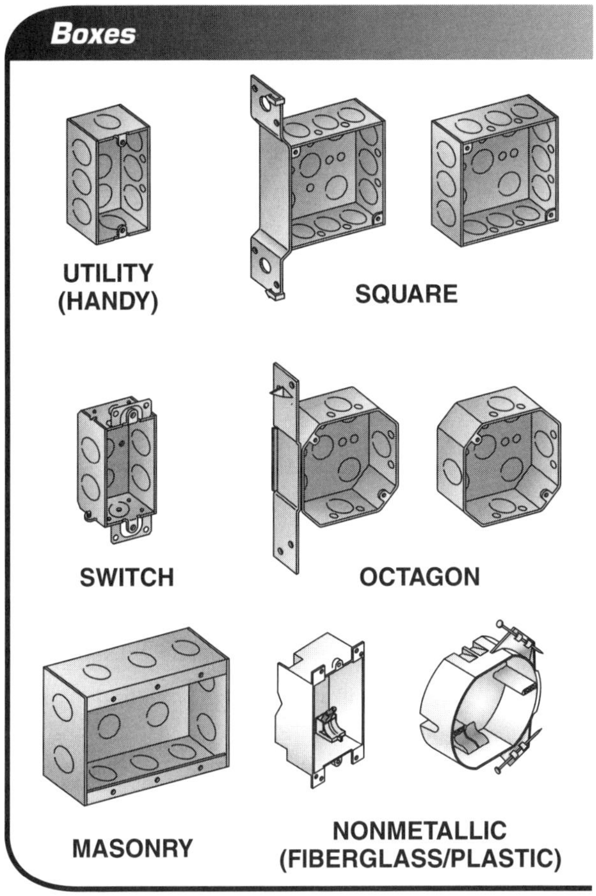

Figure 14-7. A box is a metallic or nonmetallic electrical enclosure used for equipment, devices, and pulling or terminating conductors.

Utility (handy) boxes are designed for surface mounting, measure 2″ wide by 4″ high, and are available in depths ranging from 1½″ to 2³⁄₁₆″. Square boxes range in size from 4″ square to 4¹¹⁄₁₆″ square and in depths ranging from 1¼″ to 2⅛″.

Switch boxes are sectional boxes or welded boxes. Sectional switch boxes are installed flush with the wall surface finish and can be joined to form large boxes by removing the box sides. Welded switch boxes are similar to sectional switch boxes except that the sides cannot be removed. Octagonal boxes are most commonly installed flush with the wall or ceiling surface and may be mounted onto a bar hanger that allows for placement between structural members.

Items taken off during electrical estimating include cable, wire nuts, boxes, box covers, fittings, and mounting hardware.

A masonry box is a box that has punched mounting holes, is stamped with knockouts, and has the top and bottom edges of the box turned inward. Masonry boxes are deep enough so they can be flush mounted and still contact conduit that is run through the cavity of a masonry unit.

A wide variety of covers and plates are designed to be installed on the faces of the different boxes. Covers and plates provide protection for the conductors.

Switches and Receptacles. A *switch* is a device that is used to start, stop, or redirect the flow of current in an electrical circuit. A *receptacle* is a device used to connect equipment with a cord and plug into an electrical system. The types and locations of switches and receptacles are shown on electrical prints using various symbols and abbreviations. (See the Appendix for symbol and abbreviation tables.) Electrical prints show general receptacle locations and the type of receptacle at each location. Specific dimensions are typically not provided, as minimum outlet spacing requirements are provided in the NEC® and other applicable electrical codes.

Switches may be classified according to their number of poles, number of closed positions, method of operation, or other indications in the specifications or electrical plans. Common switches include single-pole, double-pole, two-way, three-way, four-way, key-operated, momentary-contact, maintain-contact, dimmer, photoelectric, motion sensor, and safety switches.

A pole indicates the number of completely isolated circuits that a switch can control. A throw is the number of closed contact positions per pole. Single-pole switches can carry current through only one circuit at a time. Double-pole switches can carry current through two circuits simultaneously.

A two-way switch is a single-pole, single-throw (SPST) switch that allows current flow in the ON position and does not allow current flow in the OFF position. A two-way switch is used to connect or break one circuit. A three-way switch is a single-pole, double-throw (SPDT) switch. A three-way switch is used to divert power to one of two circuit paths. A four-way switch is a double-pole, double-throw (DPDT) switch. A four-way switch is used to divert power from two different circuits.

A key-operated switch is an electrical control device that requires the use of a key to activate a circuit and provide additional safety protection and security at the switch in areas of public access. A contactor is a control device that uses a small control current to energize or deenergize the load connected to it.

A momentary-contact switch is an electrical control device that activates current for a brief moment. Momentary-contact switches control items such as lighting contactors for large banks of lamps. A maintain-contact switch is an electrical control device that allows current flow in the up and down positions and must be manually returned to a center position to stop the flow of current.

A dimmer switch is an electrical control device that changes the brightness of a lamp by regulating the amount of voltage applied to the lamp. A photoelectric (daylight) switch is an electrical control device that is activated or deactivated by ambient light levels. A photoelectric switch is used to control dusk to dawn lighting.

Motion sensors energize or deenergize a circuit based on movement in an area near the switch. Motion sensors are common for lighting controls in sustainable construction where circuits can be automatically deenergized after a certain amount of time has passed and no movement in an area has occurred.

A *safety switch (disconnect)* is an electrical control device used periodically to remove electrical circuits from their supply source. Safety switches are used to stop current flow in an emergency situation and may be equipped with a locking mechanism so the power cannot be activated or deactivated by accident.

A variety of receptacles are available to meet requirements of design, wiring methods, grounding, use, and other factors. The most common receptacle is the duplex receptacle. Other receptacle types include multioutlet, clock hanger, locking, and ground-fault circuit interrupter receptacles.

A duplex receptacle is an electrical device that has two separate spaces for connecting two individual plugs. A multioutlet assembly is a metal raceway with factory-installed conductors and attachment plug receptacles. Multioutlet assemblies provide a series of fixed placement receptacles and power sources along a continuous strip. A clock hanger receptacle is an electrical device that has a slight recess in the face to allow for the electrical plug to be installed and recessed into the face of the finish material. A locking receptacle is an electrical device designed for a plug to be twisted and locked into the face to complete the connection.

A *ground-fault circuit interrupter (GFCI)* receptacle is a device with a disconnect that interrupts the flow of current when a ground fault exceeding a predetermined value of current occurs. GFCIs are designed for installation in areas where exposure to water or moisture is possible. GFCIs automatically stop current flow to prevent electrical shock.

Service

A *service* is the electrical supply in the form of conductors and equipment that provides electrical power to a building or structure. The service components to be estimated include transformers, panelboards, fuses and circuit breakers, and busways and are described in specifications and shown on electrical prints. Electrical schedules and schematics indicate the various electrical loads, circuits, and demands on each leg of the electrical system. **See Figure 14-8.**

Transformers. A *transformer* is an electric device that uses electromagnetism to change voltage from one level to another. Transformers are sized by the number of kilovolt-amperes (kVA) they can safely handle. The size of a transformer for an application is based on the amount of electrical power required. The most common transformers are air-cooled. Oil-cooled transformers are generally used by utilities.

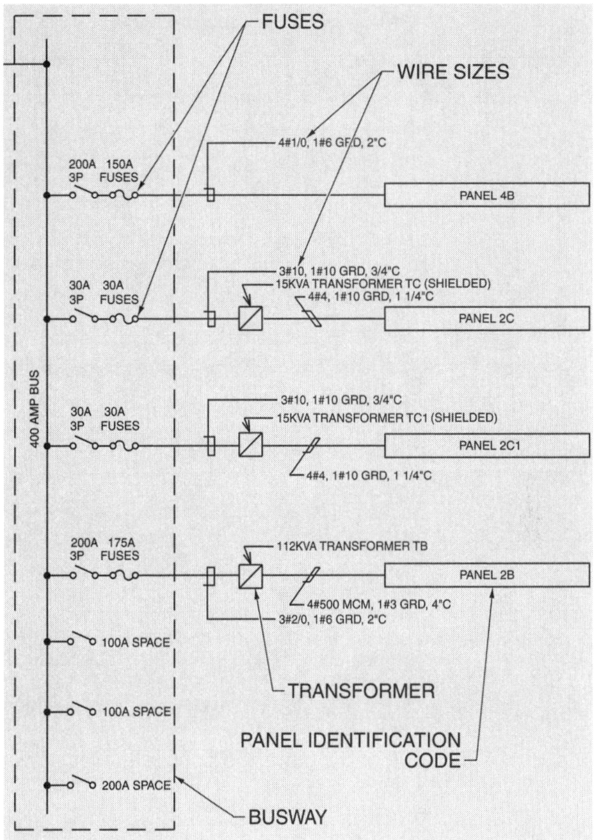

Figure 14-8. Estimators use schematics to determine wire gauge, transformer, and fuse requirements.

Panelboards. A *panelboard* is a wall-mounted distribution cabinet containing overcurrent and short-circuit protection devices for lighting, heating, or power circuits. Panelboards consist of an enclosure (tub) and an interior on which circuit breakers or fuses are mounted and connected to busbars.

Panelboard interiors are commonly assembled at a factory in accordance with specific requirements as designed by an electrical engineer. The enclosures and interiors may be shipped from the electrical equipment supplier as separate items. This allows the enclosures to be installed on the project while the interiors are being assembled. The interiors are then shipped at a later time in the project construction and installed into the enclosures at the appropriate time. **See Figure 14-9.** Panelboards are rated by their total safe load capacity (in amperes).

Figure 14-9. Panelboards provide a connection for power distribution and circuit protection.

Fuses and Circuit Breakers. A *fuse* is an overcurrent protection device with a fusible link that melts and opens a circuit when an overload condition or short circuit occurs. A *circuit breaker* is an overcurrent protection device with a mechanism that automatically opens a circuit when an overload condition or short circuit occurs. Fuses and circuit breakers are rated by the current that can flow through the device without interrupting the circuit.

The interrupting (trip) mechanism within a circuit breaker can be operated thermally, magnetically, or by a thermal/magnetic combination. A thermally sensitive component operates the trip mechanism when an electrical overload causes a gradual increase in circuit current that exceeds the current rating of the circuit breaker. A magnetically sensitive component operates the trip mechanism when an instantaneous increase in circuit current caused by a short circuit occurs. Most circuit breakers use a combination thermal/magnetic trip mechanism.

Busways. A *busway* is a metal-enclosed distribution system of busbars available in prefabricated sections. **See Figure 14-10.** Busways are used where conventional wiring methods are not practical or cost-effective. Busways include feeder and plug-in busways. Feeder busways are rated from 800 A to 5000 A and are used to distribute power from a large central power distribution point, such as the main switchgear, to several smaller distribution points, such as panelboards. A feeder busway is tapped using special fittings to provide electrical power at many locations throughout the system. Feeder busway tap ratings range from 200 A to 1600 A.

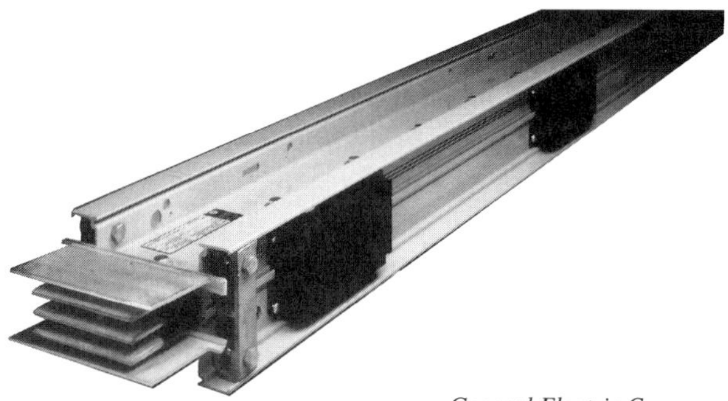

General Electric Company

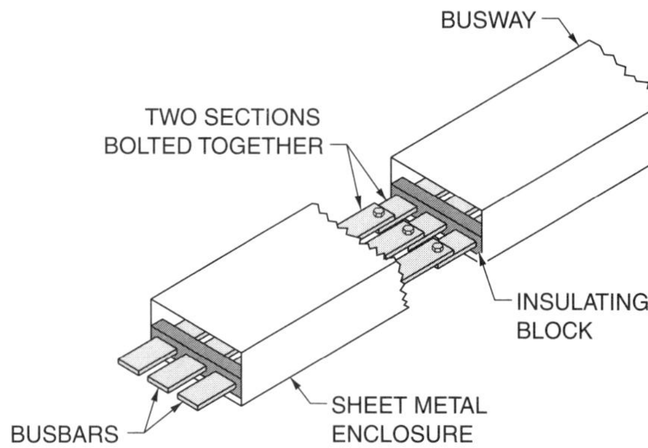

Figure 14-10. Busways allow flexibility in power distribution systems where locations of electrical equipment change often.

Plug-in busways are rated from 225 A to 600 A and are frequently installed overhead where heavy power loads are necessary and outlet locations must be flexible and movable. Plug-in busway tap ratings range from 30 A to 100 A. Many different fittings and delivery devices are used with busways. Electrical prints and specifications provide detailed information about busway equipment.

Low-Voltage Systems

Low-voltage electrical installations for control and signal systems include motor control circuits, temperature control circuits, communication circuits, and various alarm and signaling systems. Low-voltage electrical control and signaling systems require transformers and wiring that are different from high-voltage systems. Manufacturers and suppliers provide specific information about control and signal system requirements.

Temperature Control. Temperature control systems are commonly installed in conjunction with the HVAC system and fall under Title 23 09 33 of the specifications as organized by the CSI MasterFormat™. General locations for thermostats that control HVAC systems are shown on electrical HVAC prints. **See Figure 14-11.** Connections from the thermostat to the heating or air conditioning equipment or to computerized control systems may be shown on plan views.

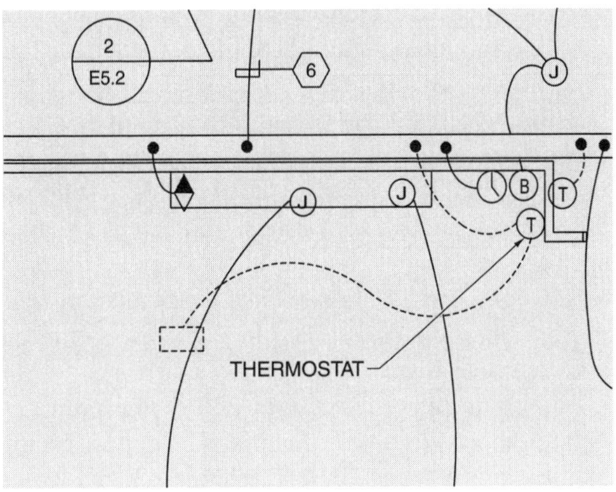

Figure 14-11. Locations for thermostats that control HVAC systems are shown on electrical HVAC prints.

Communication Systems. Communication systems include antennas, closed-circuit televisions, and computer, communications, and telephone equipment. Symbols and abbreviations on electrical prints indicate connection locations for all the various communications devices that are installed in conjunction with these systems. Each of these systems is highly specialized, with changes in technology requiring the estimator to work closely with manufacturers or suppliers to ensure that all components are powered, installed, and priced according to the most current information related to the specifications and prints.

> It is better to oversize a conductor than to undersize it. A larger conductor has less resistance than a smaller conductor.

Alarms. Configurations of electrical systems for smoke detection, heat detection, and security systems are also shown on electrical prints. Fire alarm and security systems are often required to be tested prior to the local government agency granting an occupancy permit. Nonstaining smoke may be released in the interior of a building to ensure that all fire warning systems are operable.

Security system equipment includes card readers; scanners; patrol tour systems; door and window alarms; photoelectric or infrared detection systems; ultrasonic systems; microwave systems; vibration detection systems; closed circuit and internet-based video systems; and audio communication systems. The security system installed depends on the time of response required, facility location, and the degree of security needed.

FIXTURE MATERIALS AND METHODS

After panelboards, conduit, and wiring are installed, connections to electrical finish fixtures are made. Electrical finish fixtures include items such as appliances and luminaires (light fixtures). Low-voltage systems such as communication systems and alarm systems include transformers that are powered by a high-voltage circuit. When estimating a large panelboard with many circuits and finish fixtures, the estimator should ensure that proper circuit wiring and connections are provided for the finish fixtures. Symbols on the prints indicate electrical devices and general switch, receptacle, and fixture locations.

Lighting and Motor Controls

Common types of luminaires include surface-mounted, ceiling-mounted, recessed, track lighting, bracket-mounted, pole-mounted, and exit lights. Common lamps include incandescent, fluorescent, and high-intensity discharge lamps. Separate prints and schedules may be provided for light fixture installation, depending on the complexity of the project. Symbols and abbreviations indicate the luminaire to be installed at each location. **See Figure 14-12.** Specifications, schedules, and electrical prints indicate the type of lamp for each luminaire.

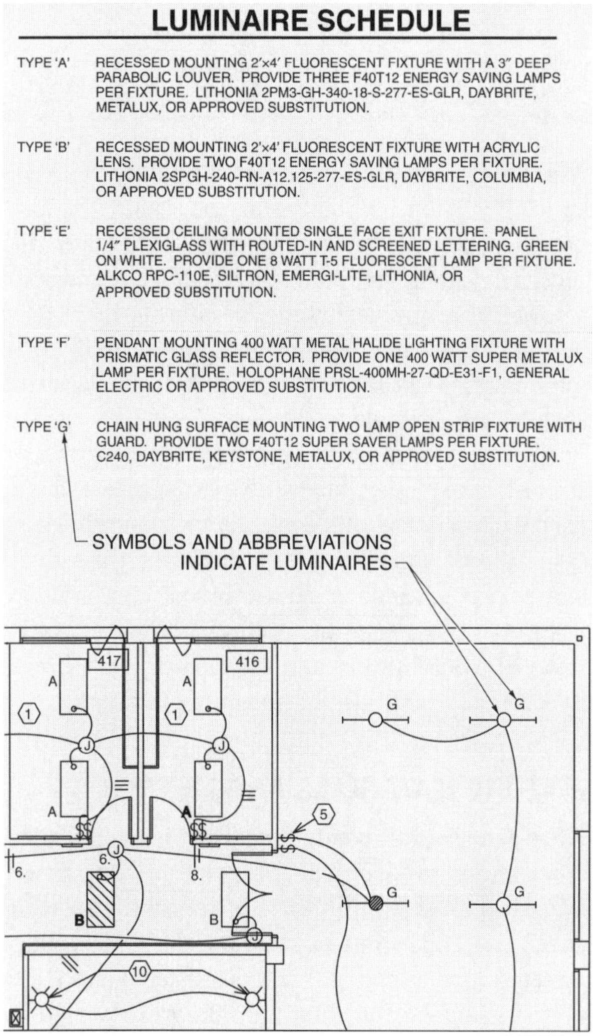

Figure 14-12. Luminaire (light fixture) schedules and electrical print symbols indicate the general location and fixture required.

Luminaires and Poles. A wide variety of designs are available for luminaires and poles. Manufacturers and suppliers provide luminaires and poles in a broad range of designs, materials, colors, shapes, and application functions. Estimators should compare information provided by manufacturers and suppliers to the specifications and electrical schedules and prints to ensure that the size, style, and material of the fixture are per the architectural design.

Surface-mounted luminaires are fastened to walls, ceilings, or other structural members. Ceiling-mounted luminaires may be surface-mounted or pendant light fixtures. Recessed luminaires are designed so that the housing is behind the surface of the finish material, leaving the face of the fixture flush with the finish surface. Track lighting is composed of a frame along which multiple lamp receptacles are mounted.

Bracket-mounted fixtures allow for directional adjustment of lights and are commonly used for floodlights. Pole-mounted luminaires are mounted on steel, precast concrete, aluminum, or fiberglass poles and are commonly used for street lighting and exterior applications. Exit lights are normally surface-mounted fixtures used to indicate the direction to a safe exit route in case of an emergency.

Lamps. An *incandescent lamp* is a lamp that produces light by the flow of current through a tungsten filament inside a gas-filled, sealed glass bulb. Incandescent lamps include general service, parabolic reflector, and tungsten-halogen lamps. Incandescent lamps are identified by shape. Shape designations include A for a standard lamp, G for globe-shaped, PAR for parabolic-shaped, and PS for pear-shaped. The designation ES indicates extended service lamps that are used in hard-to-reach locations. Tungsten-halogen lamps maintain a constant light output during the life of the lamp.

A *fluorescent lamp* is a low-pressure discharge lamp in which ionization of mercury vapor transforms ultraviolet energy generated by the discharge into light. Fluorescent lamps are available in standard tube design, high-output design, and special shapes such as U or circle. Fluorescent lamps are available in tube lengths from 18″ to 96″ and diameters from ⅝″ to 2⅛″. Energy-efficient fluorescent lamps operate on less electrical power than standard lamps.

A high-intensity discharge (HID) lamp produces light from an arc tube, which is the light-producing element of the lamp. High-intensity discharge lamps include mercury-vapor lamps, metal-halide lamps, and high-pressure sodium lamps. A mercury-vapor lamp is an HID lamp that produces light by an electric discharge through mercury vapor. Mercury-vapor lamps are installed where long life is required and replacement of the lamp is difficult.

A *metal-halide lamp* is an HID lamp that produces light by an electric discharge through mercury vapor and metal halides in the arc tube. Metal-halide lamps provide clear white light and are used in interior and exterior applications. A high-pressure sodium lamp is an HID lamp that produces light when current flows through sodium vapor under high pressure and high temperature. High-pressure sodium lamps are commonly used for street lamps and other exterior lighting applications.

Motors and Starters. A portion of the electrical work for a construction project includes equipment connections to motors and starters. Motors include two-speed, variable-speed, wound-rotor, synchronous, and direct current motors. The type of motor used is based on the requirements of the application.

Motors are normally controlled by a motor starter. Motor starters may be manual or magnetic. A manual motor starter has overload protection and uses pushbuttons to energize or deenergize the load (motor) connected to it. Manual motor starters are rated in horsepower according to the motor to be controlled. A magnetic motor starter has overload protection and uses a small control current to energize or deenergize the load (motor) connected to it. AC magnetic motor starters are the most common motor starters used for single-phase and three-phase motors.

Many different magnetic motor starters are available. Estimators should consult the project specifications and equipment schedule for the starter requirements for each motor or device to be controlled.

ELECTRICAL SYSTEM QUANTITY TAKEOFF

Electrical specifications, schedules, and prints contain information about the types of wiring to be installed, types and locations of electrical equipment, and electrical finish materials. **See Figure 14-13.** BIM usage may generate quantity takeoff information for various types of electrical equipment such as switches, receptacles, and luminaires. In most instances, specific dimensions for electrical wiring, equipment, and fixture installation are not provided on electrical plans. Rather, dimensions are obtained from architectural and structural prints and elevations and the applicable electrical or fire protection codes. Electrical information that may be shown on site plans includes power sources and exterior lighting.

Electrical estimators must be familiar with all applicable building and fire protection codes related to electrical installation. Electrical estimators should check the specifications carefully. There may be occasions when information on architectural and electrical prints may not be in accordance with these codes. Estimates taken only from specifications and prints without reference to the applicable electrical codes could result in an inaccurate bid. Any discrepancies between the prints and codes should be brought to the attention of the architect, engineer, and contractor prior to bid completion.

An electrical equipment circuit schedule may be included in the electrical prints. This schedule names each piece of equipment and shows the amount of electric power necessary, as well as the circuit assignments. **See Figure 14-14.**

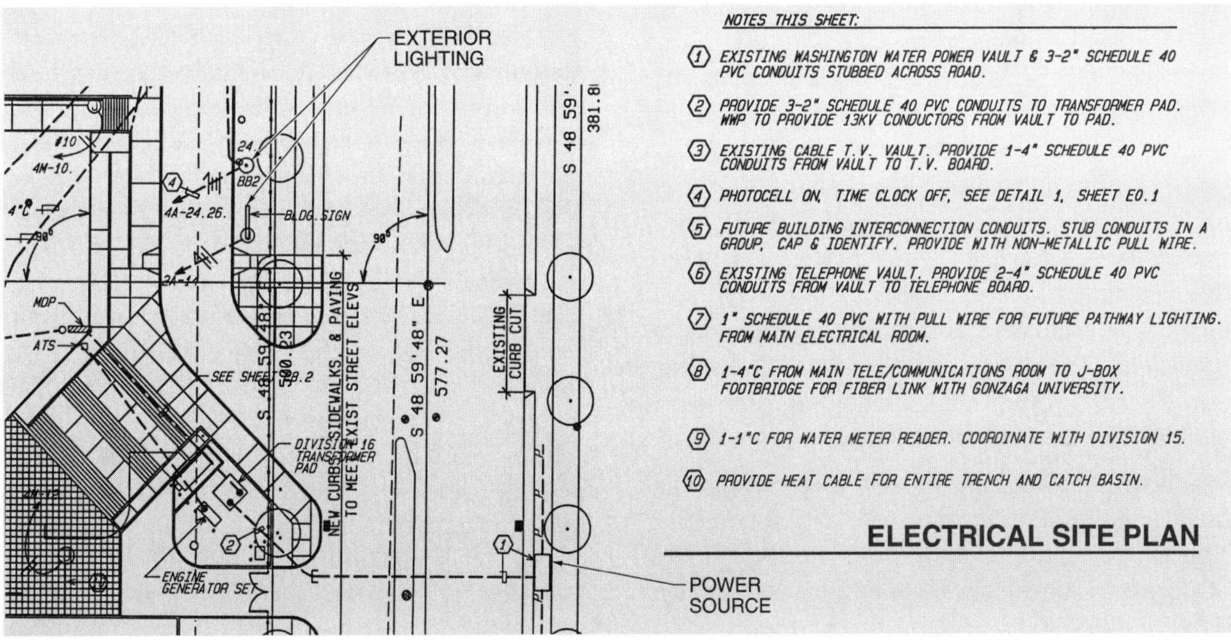

Figure 14-13. Site plans show power source and exterior lighting information.

ELECTRICAL EQUIPMENT CIRCUIT SCHEDULE								
ITEM	NAME	VOLT/PHASE	HP/AMP	SAFETY SWITCH	FUSE	CIRCUIT	CIRCUIT No.	MAG. STARTER
AC-1	AIR COMP.	480-3Ø	40 HP	3P-100 A	80 A	3#4, 1 1/4"C	4M-38.40.42.	DIV 15
	AIR DRYER	480-3Ø	40 HP	3P-30 A	50 A	3#6, 1"C	4M-43.45.47.	SIZE 1, FVN
AS-1	AIR SHOWER	460-1Ø	2 HP	3P-30 A	5.6 A	3#12, 3/4"C	4M-32.34.36.	SIZE 0
B 1	BOILER	480-3Ø	3 HP	3P-30 A	8 A	3#12, 3/4"C	4M-1.3.5	DIV 15
B 2	BOILER	480-3Ø	3 HP	3P-30 A	8 A	3#12, 3/4"C	4M-7.9.11	DIV 15
BCP-1	CIRC PUMP	480-3Ø	3 HP	3P-30 A *	8 A	3#12, 3/4"C	MCC-1A	SIZE 1, FVNR
BCP-2	CIRC PUMP	480-3Ø	3 HP	3P-30 A *	8 A	3#12, 3/4"C	MCC-1F	SIZE 1, FVNR
CAB-1	CABINET HEATER	120-1Ø	1/60 HP	DIV 15	N/A	2#12, 3/4"C	2M-9.	N/A
CAB-2	CABINET HEATER	120-1Ø	1/60 HP	DIV 15	N/A	2#12, 3/4"C	2M-9.	N/A
CAB-3	CABINET HEATER	480-3Ø	1/60 HP	3P-30 A	N/A	2#12, 3/4"C	4B-22.24.26.	N/A
CAB-4	CABINET HEATER	480-3Ø	1/60 HP	3P-30 A	N/A	2#12, 3/4"C	4B-28.30.32.	N/A
CCHWP-1	PUMP	480-3Ø	3 HP	3P-30 A *	8 A	3#12, 3/4"C	MCC-1B	SIZE 1, FVNR
CCHWP-2	PUMP	480-3Ø	1 HP	3P-30 A	2.8 A	3#12, 3/4"C	MCC-1B	SIZE 1, FVNR
CCHWP-3	PUMP	480-3Ø	1.5 HP	3P-30 A *	4 A	3#12, 3/4"C	MCC-2A	SIZE 1, FVNR
CCHWP-4	PUMP	480-3Ø	1.5 HP	3P-30 A *	4 A	3#12, 3/4"C	MCC-2A	SIZE 1, FVNR
CCHWP-5	PUMP	480-3Ø	3/4 HP	3P-30 A *	2.25 A	3#12, 3/4"C	MCC-1A	SIZE 1, FVNR
CCHWP-6	PUMP	120-1Ø	1/3 HP	1P-30 A	12 A	2#12, 3/4"C	2M-2	SIZE 00, FVNR
CH-1	CHILLER	480-3Ø	309 MCA	3P-400 A	400 A	SEE RISER	MPD	DIV 15
CH-2	FUTURE CHILLER	480-3Ø	309 MCA			SEE RISER	MPD	
CHP-1	PUMP	480-3Ø	10 HP	3P-30 A *	20 A	3#12, 3/4"C	MCC-1F	SIZE 1, FVNR
CHP-2	PUMP	480-3Ø	10 HP	-	-	3/4"C ONLY	MCC-1B	
CHWP-1	PUMP	480-3Ø	3 HP	3P-30 A *	8 A	3#12, 3/4"C	MCC-1B	DIV 16
CHWP-2	PUMP	480-3Ø	3/4 HP	3P-30 A *	2.25 A	3#12, 3/4"C	MCC-2A	DIV 16
CHWP-3	PUMP	480-3Ø	3/4 HP	3P-30 A *	2.25 A	3#12, 3/4"C	MCC-1A	DIV 16
CHWP-4	PUMP	120-1Ø	1/2 HP	3P-30 A *	15 A	2#12, 3/4"C	MCC-1E	SIZE 1, FVNR
CT-1	COOLING TOWER	480-3Ø	15 HP	3P-60 A	30 A	3#10, 3/4"C	4M-13,15,17	DIV 15
		480-3Ø	10 KW	3P-30 A	N/A	3#12, 3/4"C	4M-20,22,24	N/A
CT-2	COOLING TOWER	480-3Ø	15 HP			2)3/4"CO.		
CU-1	CONDENSER	208-1Ø	14 FLA	3P-30 A	20 A	3#10, 3/4"C	2X1-2.4.	SIZE 1, FVNR
	PAINT BOOTH	120-1Ø	15 A	N/A	N/A	2#12, 3/4"C	2A-30.	
	PAINT BOOTH	480-3Ø	3/4 HP	3P-30 A	225 A	2#12, 3/4"C	4A-23.25.27.	DIV 15
CWP-1	PUMP	480-3Ø	20 HP	3P-60 A	40 A	3#8, 1"C	4M-2,4,6	DIV 15
CWP-2	PUMP	480-1Ø	20 HP			1"C ONLY	4M	N/A

Figure 14-14. Electrical equipment circuit schedules provide information for each piece of electrical equipment required for a construction project.

Takeoff Methods

To ensure estimating accuracy, electrical estimators create separate worksheets for each type of electrical item needed, such as cables, luminaires, lamps, and switches. A room number or area designation is noted on a worksheet as the location of each item. Electrical estimators may also mark up electrical plans to ensure that all items are included in the quantity takeoff. Two methods of estimating used for electrical takeoff are detailed takeoff and averaging takeoff. BIM usage may provide much of this information when the model is properly designed and implemented.

Detailed Takeoff. Detailed takeoff is a precise method used for industrial and commercial work and wherever unique job conditions exist. It is a very accurate method of electrical estimating work. A detailed item-by-item quantity takeoff is derived from electrical prints, schedules, and specifications.

If detailed electrical prints and specifications are not provided as part of the contract documents, the electrical estimator can prepare electrical layouts, with or without an electrical engineer, and have them approved by the architect before beginning the estimate. These electrical layouts should show all necessary wiring and equipment and give the sizes and types of electrical service and the sizes and locations of all feeders and distribution panelboards. Electrical layouts should show the sizes and locations of all motors and controllers and the types, number, and locations of all fixtures, switches, and receptacles.

In detailed takeoff, a list showing each type of material required is created. The list includes cables; boxes; receptacles; conduit and fittings; labor required for installation and equipment connection; motor controls; service and distribution equipment; switches; and trenching and excavation, and may include low-voltage components such as alarms and communication systems. Each of these items is thoroughly checked by the estimator to ensure the items being taken off are included in the scope of work required in the electrical bid to create a complete and comprehensive takeoff. A ledger sheet, spreadsheet, or estimating program may be used for a detailed takeoff of an electrical system. **See Figure 14-15.**

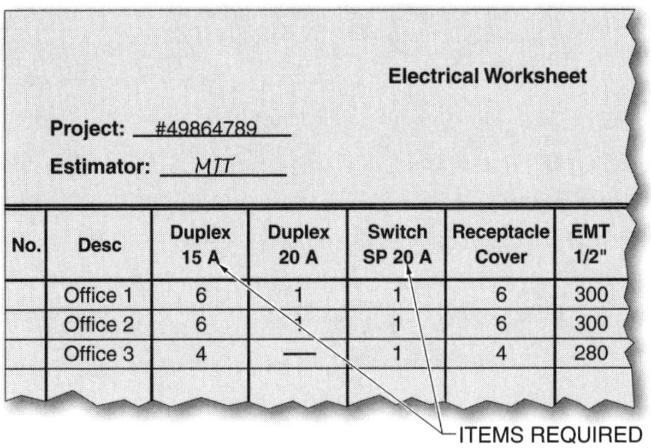

Electrical Worksheet

Project: #49864789

Estimator: MTT

No.	Desc	Duplex 15 A	Duplex 20 A	Switch SP 20 A	Receptacle Cover	EMT 1/2"
	Office 1	6	1	1	6	300
	Office 2	6	1	1	6	300
	Office 3	4	—	1	4	280

Figure 14-15. A detailed takeoff includes a list of each item required for the electrical system.

The total cost of materials is obtained by consulting information from various manufacturers and suppliers. The amount of electrical labor required to install wiring and equipment is determined by applying labor units to each type of material.

Electrical labor includes the testing required to confirm proper equipment installation.

A *labor unit* is the average time it takes an average tradesperson to install a specific type of material. Labor units are obtained from standard estimating references or company historical data. **See Figure 14-16.** Labor units are commonly related to each item or material unit. After the total amounts of electrical material and equipment are known, the estimator applies the appropriate labor units and calculates the total cost to complete the installation.

Variables considered during preparation of the electrical estimate include items such as the potential need for temporary service or temporary wiring for power tools during construction, job-site conditions where equipment access may be restricted or work at heights is required, schedules for material and equipment delivery and installation, and climatic conditions that affect labor production such as rain or snow. After all variables have been evaluated in conjunction with the project manager and job-site personnel, an electrical estimator prepares an accurate bid for the complete electrical installation.

CABLE INSTALLATION LABOR

5 kV Unshielded	Unit*	Labor Hours per Unit
No. 8	100	2.3
No. 6	100	2.5
No. 4	100	2.7
No. 2	100	3.4
No. 1/0	100	3.6
No. 2/0	100	4.0
No. 3/0	100	4.2

* in lf

BUSWAY INSTALLATION LABOR

Copper Busway in Steel Case*	Unit†	Labor Hours per Unit
225	10	4.7
400	10	5.7
600	10	6.8
800	10	8.1
1000	10	9.0
1350	10	11.5
1600	10	16.9
2000	10	24.0

* rating in A
† in lf

Figure 14-16. The amount of electrical labor required to install wiring and equipment is determined by applying labor units to each type of material.

Averaging Takeoff. *Averaging takeoff* is a simplified electrical takeoff method that is useful for a nonspecialist in electrical estimating. Averaging takeoff is sufficiently accurate for small takeoff jobs on basic electrical wiring systems. In averaging takeoff, material and labor costs are calculated as an average cost per outlet.

Estimators determine the cost of the materials and labor required to install a typical electrical outlet. This is done for each different type of outlet in addition to service entrance conduit, meters, and panelboards. The total cost of the job is determined from these figures. Variables considered include the wiring methods used, local codes and ordinances, material costs, and labor costs.

Due to local variations, it is not possible to quote exact cost figures for these items. Outlet, receptacle, and switch costs are commonly figured on the basis of a 15′ run of cable or conduit. This provides a general cost estimate for electrical work, but it is not applicable to large commercial installations.

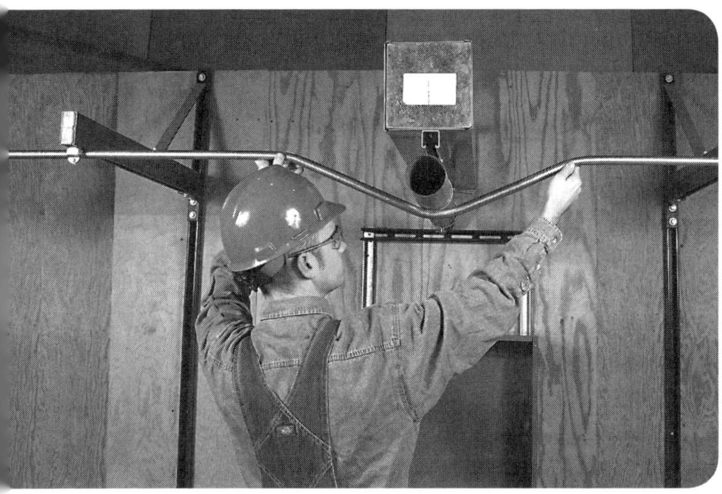

Three-bend saddles are shaped to fit around an obstruction and require additional time to fabricate.

Wiring Takeoff

Wiring, cable, and conduit takeoff begins by listing each type of conductor, cable, and conduit noted on the electrical prints. A paper or electronic template already containing this listing may be used to start the process. After all types are listed, quantity takeoff is made in the quantity of linear feet for each type of conductor, cable, or conduit. From scaled prints, the length of each circuit is measured to determine the total length for conduit and wiring. **See Figure 14-17.**

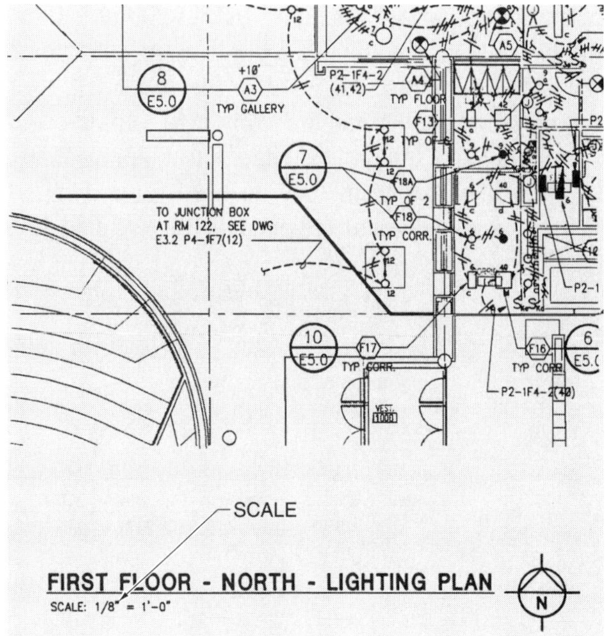

Figure 14-17. For conductor, cable, and conduit takeoff, scaled plans are used to determine the length of each circuit run.

Additional lengths of conductors and cable are added to the total length to allow for placement through and around structural and finish members and for connections at fixtures, junction boxes, devices, and panelboards. The conductors, cable, and conduit are listed in separate portions of the takeoff. Material and labor costs for wiring and conduit are commonly calculated based on the cost per 100′ or 1000′ of conduit or conductor. Detailed labor installation costs are available from industry standard information or company historical data.

Estimators must also include some allowance for connectors and clamps for electrical conductors, cable, and conduit. Electrical code requirements for fastener spacing will help with quantity takeoff. An electrical estimate should include clips, wire nuts, bushings, nipples, gaskets, washers, special pulling equipment required for cables, and connecting lugs. Knowledge about these specialized fittings and connectors and the labor required for installation is required to complete the wiring and conduit estimate.

Conduit. Conduit size is determined based on the number of conductors or cables allowed by the NEC®. NEC® tables provide information concerning the allowable percentage of conduit fill. **See Figure 14-18.** Estimators may allow additional space in conduit beyond the values given in the NEC® tables to minimize labor costs. Conduit with a higher percentage fill makes pulling wire more difficult and may increase the amount of labor required for wire or cable installation. It is common to increase the conduit size after reaching approximately 80% of the allowable conduit fill.

RIGID METAL CONDUIT ALLOWABLE FILL					
Size*	Internal Diameter*	Total Area† 100%	2 Wires 31%†	Over 2 Wires 40%†	1 Wire 53%†
½	0.632	0.314	0.097	0.125	0.166
¾	0.836	0.549	0.170	0.220	0.291
1	1.063	0.888	0.275	0.355	0.470
1¼	1.394	1.526	0.473	0.610	0.809
1½	1.624	2.071	0.642	0.829	1.098
2	2.083	3.408	1.056	1.363	1.806
2½	2.489	4.866	1.508	1.946	2.579
3	3.090	7.499	2.325	3.000	3.975
3½	3.570	10.010	3.103	4.004	5.305
4	4.050	12.283	3.994	5.153	6.828

* in in.
† in sq in.

Figure 14-18. To stay within NEC® requirements, electrical conduit can only be filled to a predetermined capacity.

Conduit fittings are counted for each connection and termination of conduit. One coupling is allowed for each 10′ length of conduit. Two additional couplings are added for each 90° elbow. Special attention is required to ensure that the proper fittings are provided in special conduit situations.

Estimators must also take wall height and fixture height into account during conductor, cable, and conduit takeoff. Additional length is added to the totals obtained from prints for switches, receptacles, and fixtures based on heights above floors or below ceilings. Conductor, cable, and conduit installation labor may also be impacted by job-site schedules. Proper coordination of installation with project managers and field personnel can favorably impact the amount of labor needed on a particular job.

Boxes and Cable Supports. Electrical boxes are taken off as individual units. Estimators create a list or use a template to count each box of each type that is shown on the electrical prints. Each type of box is listed and quantities are noted during takeoff. Estimators refer to plan symbols and specification notes to ensure proper takeoff of each type of box. **See Figure 14-19.** Estimators must also include the proper type of cover for each box in the takeoff according to the specifications and electrical prints.

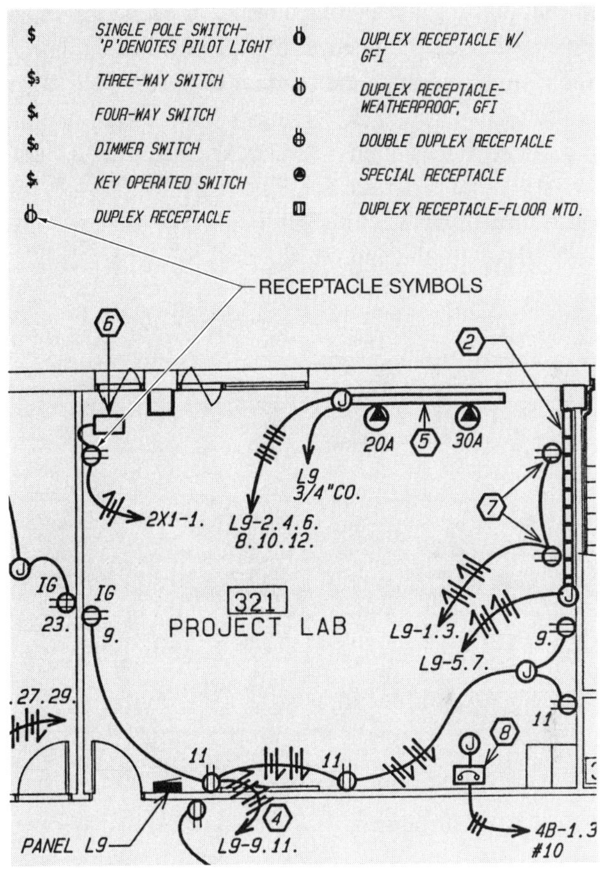

Figure 14-19. Electrical receptacle boxes are noted on electrical prints using standard symbols.

Cable tray system takeoff is based on the linear feet of tray to be installed and the number of each type of fitting required to complete the system. An important element to consider during cable tray system takeoff is the accessibility of the area into which the tray system is being installed. There may be potential conflicts with HVAC piping, sprinkler systems, and other overhead devices. BIM usage that integrates the cable tray placement into the model will identify any potential clashes with these systems and minimize installation costs. There may also be height requirements that make hanging the cable tray difficult. Coordination of installation with project managers and field personnel, as well as other contractors, is important for accurate determination of labor costs.

Wiring Devices. Wiring devices included in the quantity takeoff include switches, receptacles, and other indicated devices. As with luminaires, switches, receptacles, and other common devices are taken off as individual units, with the total number of each individual type of device included in the takeoff. For example, the estimator counts the number of single-pole 20 A switches and duplex 20 A receptacles in each location. The location is entered in the ledger sheet, spreadsheet, or estimating program. **See Figure 14-20.** The total of each type of device for each location of the structure is entered into the appropriate row, column, or cell.

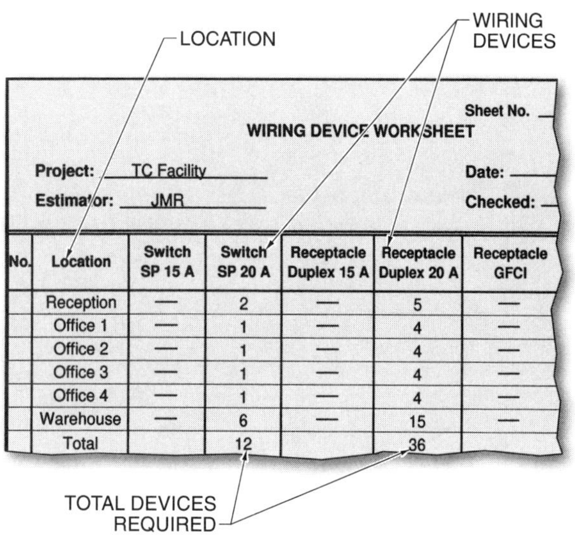

WIRING DEVICE WORKSHEET

Sheet No.

Project: TC Facility — Date:

Estimator: JMR — Checked:

No.	Location	Switch SP 15 A	Switch SP 20 A	Receptacle Duplex 15 A	Receptacle Duplex 20 A	Receptacle GFCI
	Reception	—	2	—	5	—
	Office 1	—	1	—	4	—
	Office 2	—	1	—	4	—
	Office 3	—	1	—	4	—
	Office 4	—	1	—	4	—
	Warehouse	—	6	—	15	—
	Total		12		36	

Figure 14-20. Estimators list the type of each electrical device followed by the quantity takeoff for each device.

The estimator must be familiar with the various electrical symbols used by architects to depict electrical devices. Cover plates for each type of switch and receptacle are also included in this portion of the takeoff. Material and labor rates for supplying and installing each device are determined from manufacturer and supplier information, standard industry references, or company historical data.

Lighting Takeoff

Takeoff for luminaires may be the first step taken by the electrical estimator in the bid process. This portion of the estimate may also be included in the BIM systems or completed by electrical suppliers under some conditions where there is a large fixture package on a large job. Estimators should check the scope of work in the contract to include all applicable portions of the lighting estimate, including luminaires, poles, lamps, fixture guards, safety wires, cable supports, suspended fixture safety clips, and other special support equipment that may not be readily apparent from the electrical prints.

Luminaires, Poles, and Lamps. Each type of luminaire should be indicated on the estimate ledger sheet or spreadsheet or entered as an item into the estimating program. Each fixture is counted as an individual item. The number of each type of fixture is totaled and entered into the appropriate location on the estimate. **See Figure 14-21.**

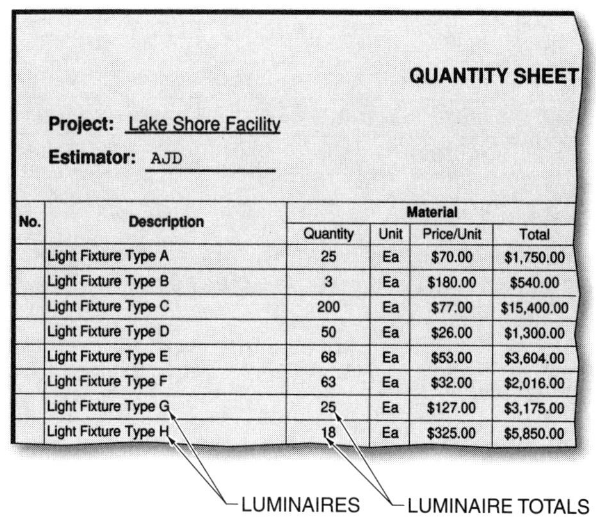

QUANTITY SHEET

Project: Lake Shore Facility

Estimator: AJD

No.	Description	Material			
		Quantity	Unit	Price/Unit	Total
	Light Fixture Type A	25	Ea	$70.00	$1,750.00
	Light Fixture Type B	3	Ea	$180.00	$540.00
	Light Fixture Type C	200	Ea	$77.00	$15,400.00
	Light Fixture Type D	50	Ea	$26.00	$1,300.00
	Light Fixture Type E	68	Ea	$53.00	$3,604.00
	Light Fixture Type F	63	Ea	$32.00	$2,016.00
	Light Fixture Type G	25	Ea	$127.00	$3,175.00
	Light Fixture Type H	18	Ea	$325.00	$5,850.00

Figure 14-21. Luminaires are counted as individual items during electrical takeoff.

Luminaires are priced using a number of different methods. *Unit cost pricing* is a method where each type of fixture is priced individually. Lump sum pricing is a method where a lot price is determined for all fixtures on a project. Pricing may be performed with a combination of unit cost and lump sum pricing for various types of fixtures. Some fixtures may be supplied by the owner of the project and the material costs excluded from the bid. Knowledge of the scope of work is important for accurate luminaire and labor estimates.

JLG Industries, Inc.

Job-site conditions, such as high ceilings, may increase luminaire labor and installation costs.

Some job-site conditions that may increase luminaire labor costs include a job site with excessive debris, high ceilings, difficult-to-access work areas due to small rooms or other obstructions, and a final installation area into which it is difficult to move fixtures. Scheduling of fixture installation in conjunction with other finish materials can also impact labor costs.

Material and labor costs for support poles are taken off as individual units. Labor costs for poles may include footings, excavation, and equipment costs for installation and aerial lifts to reach the top of the poles for fixture installation.

Lamps are taken off and totaled on a separate quantity takeoff. The number and types of lamps for each fixture are determined and multiplied by the number of fixtures to obtain a total lamp quantity.

Power, Low-Voltage Control, and Signal System Takeoff

All power, control, and signal systems (such as alarms, communications, and video equipment) are taken off as individual units. Factors considered include the service voltage and size; single-phase, three-phase, or low-voltage installation; interior or exterior installation; grounding requirements; overhead or underground service; corrosion-resistance requirements; fuses and circuit breakers; transformers; metering equipment; utility company connection fees; and other equipment support provisions.

Power. Panelboards, transformers, and other main service equipment units are based on the total safe electrical load with some added capacity for potential service expansion. This portion of the takeoff includes panelboards, fuses and circuit breakers, and any metering equipment. This bid segment may be prepared by a manufacturer or supplier who specializes in service equipment.

Labor unit costs are based on industry standard information or company historical data for installation of each unit of material. Labor costs for these items include the connections of all circuits to the main service equipment. The number of circuits to be connected to a panelboard impacts the labor time required to complete the connections.

Locations of panelboards are shown on electrical prints. Plan views relate to the overall electrical service schematic, schedules, and detail drawings. **See Figure 14-22.** Panelboards are noted by an architectural symbol and may have a letter or number code relating to a schedule. A schedule of panelboards indicates the size of the panelboard, mounting directions, and equipment to be serviced by each circuit. Panelboards are taken off as individual units and listed in the estimate individually.

Connection. Electrical connections for mechanical, electrical, and kitchen equipment and low-voltage systems can be complex. Design and electrical engineers should be consulted concerning special connection requirements for these specialized systems and machinery. Items considered include size; voltage; transformers; horsepower of equipment; motor protection requirements; disconnect and lockout switches; grounding provisions; fan requirements; fire protection connections; and any potential hazardous installation conditions.

Each of these items could add to the labor hours required for proper and safe installation and connection. Labor units for connection and installation of various low-voltage systems and other pieces of electrical equipment are based on standard industry information or company historical data for each item. As with other electrical installations, coordination with project management for scheduling of connections can impact final labor costs.

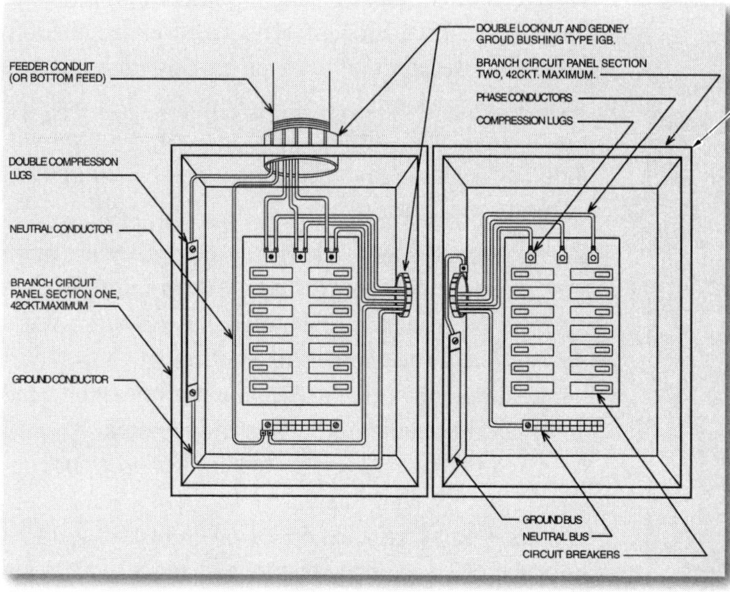

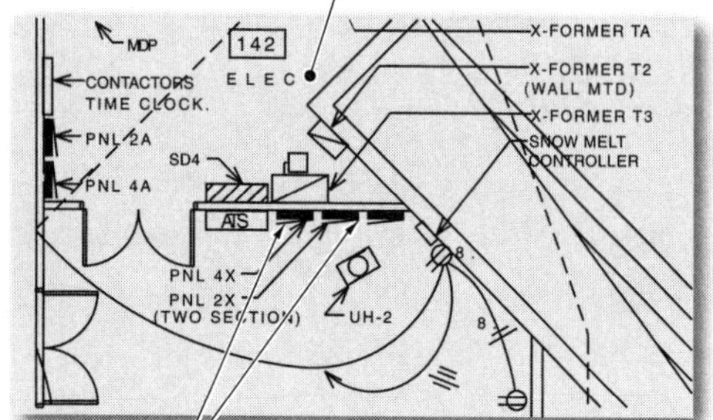

GENERAL PANELBOARD INFORMATION
CIRCUIT BREAKER IDENTIFICATION CODE
PANELBOARD SCHEDULE

NUMBER 2X (SECTION ONE)
LOCATION 1ST FLOOR
FED FROM 4X

PANEL SCHEDULE

Riley Engineering, Inc.

CCT #	BKR A/P	CCT DESCRIPTION	LTG	RECEP	MOTOR	APPLI	CCT #	BKR A/P	CCT DESCRIPTION	LTG	RECEP	MOTOR	APPLI	PHASE LOADS A	B	C
1	20/1	TREATMENT PANEL				0.2	2	15/1	EF-8			0.2		0.4		
3	20/1	TEMP.CONTROL				0.2	4	20/1	DT-2				0.4		0.6	
5	20/1	SPARE					6	20/1	TV-2				0.2			0.2
7	20/1	SPARE					8	20/1	MASTER CLK.				0.2	0.2		
9	20/1	UPS				0.8	10	20/1	FA PANEL				0.2		1.0	
11	20/1	UPS				0.8	12	20/1	SECURITY				0.2			1.0
13	20/1 *	UPS				0.8	14	20/1	INTRUSION				0.2	1.0		
15	20/1 *	UPS				0.8	16	20/1 *	DT-2				0.4		1.2	
17	20/1 *	UPS				0.8	18	20/1 *	DT-2				0.4			1.2
19	20/1 *	UPS				0.8	20	20/1 *	DT-2				0.4	1.2		
21	20/1 *	TEL.RM.				0.4	22	20/1 *	DT-2				0.4		0.8	
23	20/1 *	TEL.RM.				0.4	24	20/1 *	DT-1				0.4			0.8
25	20/1 *	TEL.RM.				0.4	26	20/1 *	DT-1				0.4	0.8		
27	20/1 *	TEL.RM.				0.4	28	20/1 *	DT-1				0.4		0.8	
29	20/1 *	TEL.RM.				0.4	30	20/1 *	DT-1				0.4			0.8
31							32							0.0		
33							34								0.0	
35							36									0.0
37							38							0.0		
39							40								0.0	
41							42									0.0
														4	4	4

LOAD DESCRIPTION

	LTG	RECEP	MOTOR	APPLI	TTL KVA	TTL AMPS
TOTAL CONNECTED	0	0	0	12	12	33
LOAD FACTOR (NEC)	1.25			1		
CALCULATED LOAD	0	0	0.2	11.8	12	33

PROVIDE ISOLATED
GROUND BAR.

VOLTS: 208 Y/120 3 PHASE 4 WIRE
SIZE: 100 AMPS
MAIN: 60 AMPS
AIC: 10,000
MOUNTG SURFACE
30 POLE PANE
*ISOLATED GROUND CIRCUIT.
SECTION ONE

Figure 14-22. Panelboard information is provided on details, electrical plan views, and electrical equipment schedules.

SAGE TIMBERLINE ESTIMATING . . .

BUILDING DATABASES USING ONE-TIME ITEMS

One-time items may be taken off and automatically added to the database in Sage Timberline Estimating.

1. Pick the **One-time Item** button on the toolbar to open the **One-time Item** dialog box.
2. Create the phase record by right-clicking in the **Phase** field and selecting **Edit Phase...** from the shortcut menu.
3. Enter **16100** in the **Phase** field and **Subcontractors** in the **Description** field. Pick **OK** and close the window.

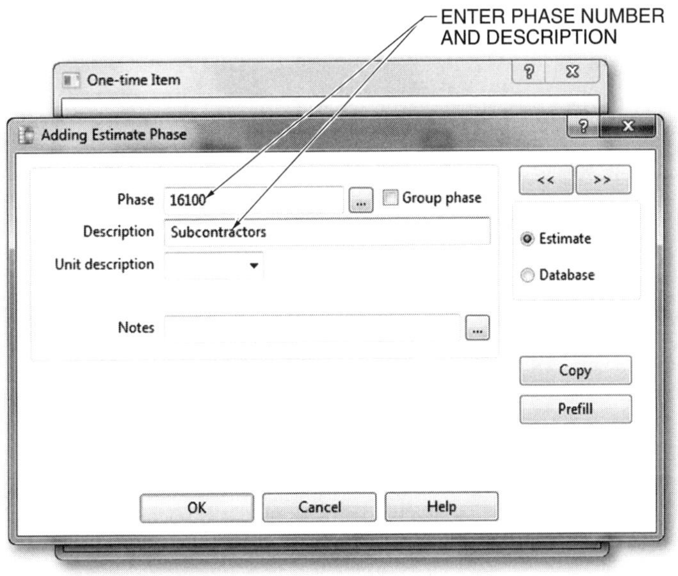

4. Complete the fields in the **One-time Item** dialog box. Enter **16100** for **Phase**, **Electrical Rough-In** for **Description**, **1** for **Quantity**, and **ls** for **Takeoff unit**.
5. Deselect the **Labor** and **Material** categories and select the **Subcontract** category checkbox. Pick **OK** to dismiss the dialog box and display the item on the spreadsheet.

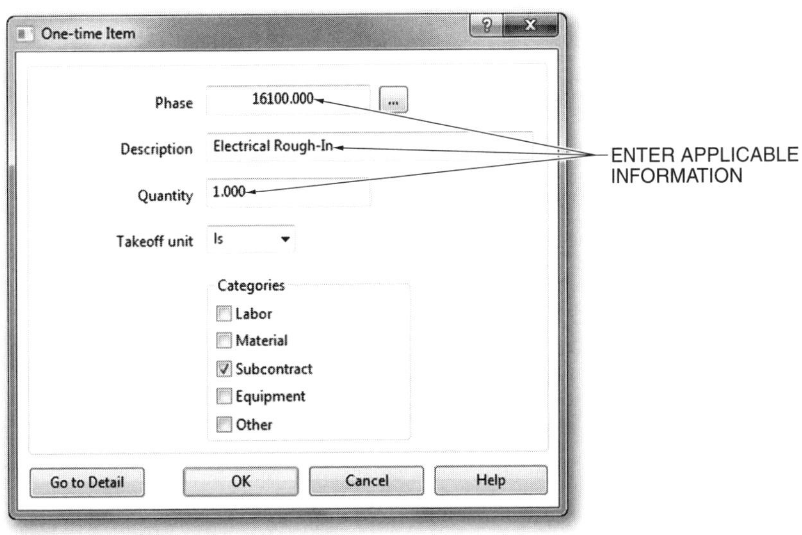

. . . SAGE TIMBERLINE ESTIMATING

6. Pick the **Electrical Rough-In** cell and press **Ctrl+S** to automatically move to the **Sub Amount** field. Enter **15000** and press **Enter**.
7. Hold down the **Ctrl** key and press the **Left Arrow** key to automatically jump to the far left column on the spreadsheet.
8. Save the one-time item associated with this estimate to the database. **Select Database → Update Database... → Save One-time Items to Database** from the pull-down menus to display the dialog box. Each one-time item included in the estimate is identified in the list pane. Default item numbers have been generated for the one-time items in the list. Review and revise the item number and description as desired. To add the item to the database, ensure the checkbox in the **Add** column is selected and pick the **Update** button. Multiple items can be saved to the database by selecting the desired items in the list pane.

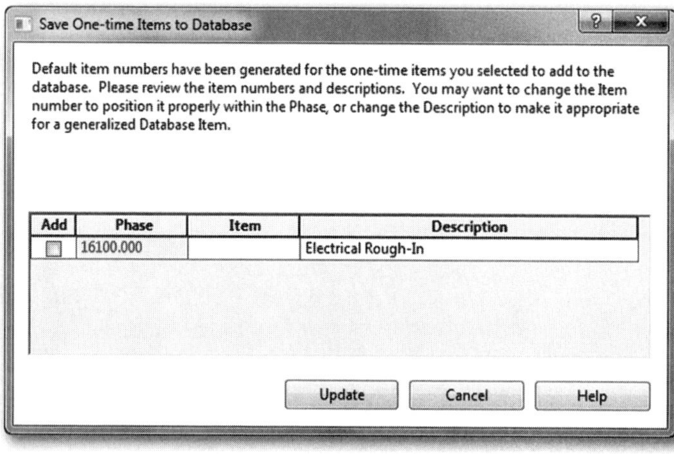

Quick Quiz®

Quick Quiz®

Refer to the CD-ROM for the Quick Quiz® questions related to chapter content.

Key Terms

Illustrated Glossary

- *averaging takeoff*
- *box*
- *busway*
- *cable tray*
- *circuit breaker*
- *conductor*
- *conduit*
- *fluorescent lamp*
- *fuse*
- *ground-fault circuit interrupter (GFCI)*
- *incandescent lamp*
- *labor unit*
- *metal-halide lamp*
- *panelboard*
- *raceway*
- *receptacle*
- *safety switch (disconnect)*
- *service*
- *switch*
- *transformer*
- *unit cost pricing*

Web Links

Web Links

Alliance for Telecommunications Industry Solutions
www.atis.org

American Lighting Association
www.americanlightingassoc.com

American Public Power Association
www.publicpower.org

Architectural Lighting
www.archlighting.com

Audio Engineering Society
www.aes.org

Cable Tray Institute
www.cabletrays.com

Canadian Electricity Association
www.canelect.ca

Electrical Generating Systems Association
www.egsa.org

Illuminating Engineering Society of North America
www.iesna.org

Independent Electrical Contractors, Inc.
www.ieci.org

Institute of Electrical and Electronics Engineers
www.ieee.org

Insulated Cable Engineers Association
www.icea.net

International Association of Lighting Designers
www.iald.org

International Brotherhood of Electrical Workers
www.ibew.org

Lightning Protection Institute
www.lightning.org

National Electrical Contractors Association
www.necanet.org

National Electrical Manufacturers Association
www.nema.org

Telecommunications Industry Association
www.tiaonline.org

Electrical Systems 14

Review Questions

Estimating

______________ 1. ___ is hollow pipe that supports and protects electrical conductors.

______________ 2. A(n) ___ is a slender rod or wire that is used to control the flow of electrons in an electrical circuit.

______________ 3. Conductors may be bare or ___.

T F 4. A cable is two or more conductors grouped together within a common protective cover and used to connect individual components.

T F 5. A cable tray is a closed grid rack suspended from structural members to support a series of cables.

______________ 6. A(n) ___ is a metallic or nonmetallic electrical enclosure used for equipment, devices, and pulling or terminating conductors.

______________ 7. A(n) ___ is a device that is used to start, stop, or redirect the flow of current in an electrical circuit.

______________ 8. A(n) ___ is a device used to connect equipment with a cord and plug to an electrical system.

______________ 9. A(n) ___ is the number of completely isolated circuits that a switch can control.

______________ 10. A(n) ___ receptacle is a receptacle that has two spaces for connecting two different plugs.

T F 11. A multioutlet assembly is a metal raceway with on-the-job installed conductors and attachment plug receptacles.

T F 12. A service is the electrical supply in the form of conductors and equipment that provides electrical power to a building or structure.

______________ 13. A(n) ___ is a wall-mounted distribution cabinet containing overcurrent and short-circuit protection devices for lighting, heating, or power circuits.

______________ 14. A(n) ___ is an overcurrent protection device with a fusible link that melts and opens a circuit when an overload condition or short circuit occurs.

______________ 15. A(n) ___ is an overcurrent protection device with a mechanism that automatically opens a circuit when an overload condition or short circuit occurs.

______________ 16. A(n) ___ is a metal-enclosed distribution system of busbars available in prefabricated sections.

_______________ **17.** A(n) ___ lamp is an electric lamp that produces light by the flow of current through a tungsten filament inside a gas-filled, sealed glass bulb.

_______________ **18.** A(n) ___ lamp is a low-pressure discharge lamp in which ionization of mercury vapor transforms ultraviolet energy generated by the discharge into light.

_______________ **19.** A(n) ___ lamp is a lamp that produces light from an arc tube.

_______________ **20.** A(n) ___ motor starter is a control device that has overload protection and uses pushbuttons to energize or deenergize the load (motor) connected to it.

Raceways

_______________ **1.** Rigid metal conduit (RMC)

_______________ **2.** Rigid nonmetallic conduit (RNMC)

_______________ **3.** Electrical metallic tubing (EMT)

_______________ **4.** Flexible metal conduit (FMC)

_______________ **5.** Intermediate metal conduit (IMC)

A. Metal conduit with a thicker wall than EMT, but a thinner wall than RMC

B. Lightweight tubular steel raceway without threads on the ends

C. Heavy conduit made of metal

D. Conduit made of materials other than metal

E. Raceway of metal strips that are formed into a circular cross-sectional raceway

Short Answer

1. Explain the importance of being familiar with electrical codes when preparing an electrical estimate.

____

____

____

____

____

____

____

2. Compare the two methods of electrical takeoffs: detailed takeoffs and averaging takeoffs.

3. Explain why a scaled set of electrical prints may be required when preparing an electrical estimate.

Electrical Systems 14

Activities

Activity 14-1—Ledger Sheet Activity

Refer to the cost data, Print 14-1, and Estimate Summary Sheet No. 14-1. Take off the number of fixtures and lamps. Determine the total material and labor costs.

COST DATA			
Material	**Unit**	**Material Unit Cost***	**Labor Unit Cost***
A. Interior lighting fixture, 2′ W × 4′ L two 40 W	ea	84.70	45.10
B. Interior lighting fixture, 2′ W × 4′ L four 40 W	ea	100.10	52.80
C. Interior lighting fixture, acrylic lens 1′ W × 4′ L, two 40 W	ea	78.10	40.15
P. Pendant mounted	ea	47.30	49.50
Lamp, fluorescent, 4′ L, 40 W	100	247.50	312.40
Lamp, twin tube, compact	100	445.50	312.40

* in $

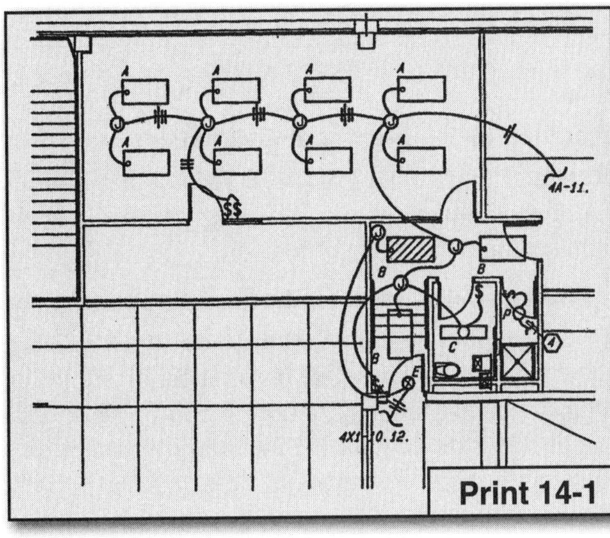

Print 14-1

ESTIMATE SUMMARY SHEET

Sheet No. 14-1
Date: ____
Checked: ____

Project: ____
Estimator: ____

No.	Description	Dimensions			Quantity		Material		Labor		Total	
						Unit	Unit Cost	Total	Unit Cost	Total	Unit Cost	Total
	Lighting fixtures											
	Type A											
	Type B											
	Type C											
	Type P											
	Lamps											
	Fluorescent											
	Twin tube compact											
	Total											

Activity 14-2—Spreadsheet Activity

Refer to the cost data below, and Print 14-2 and Estimate Summary Spreadsheet No. 14-2 on the CD-ROM. Take off the number of duplex receptacles, number of smoke detectors, number of GFCI receptacles, number of fire alarms, and number of junction boxes for Suite 103. Determine the total material and labor costs.

COST DATA			
Material	**Unit**	**Material Unit Cost***	**Labor Unit Cost***
Receptacle, duplex Type NM cable	ea	5.94	19.31
Smoke detector, ceiling type	ea	82.50	45.10
GFCI receptacle, Type NM cable	ea	37.95	26.40
Fire alarm	ea	134.20	35.20
Junction box, 4″	ea	8.64	11.22

* in $

Activity 14-3—Electrical Estimating

Refer to Prints 14-3A and 14-3B and Estimate Summary Spreadsheet No. 14-3 on the CD-ROM. Take off the number of duplex receptacles in Circuit 31A. In addition, determine the length of run for Circuit 31A, including the home run to the panelboard.

Activity 14-4—Sage Timberline Estimating Activity

Create a new estimate and name it Activity 14-4. Create a new phase record and save a new one-time item to the database. Pick the **One-time Item** button. Right-click in the **Phase** field and select **Edit Phase…**. Enter **16500** for **Phase** and **Lighting** for **Description**. Pick **OK** and **Close**. In the **One-time Item** dialog box, enter **Emergency Lighting** for **Description**, **23** for **Quantity**, and **ea** for **Takeoff unit**. Pick **OK** to send the item to the spreadsheet. Enter a labor unit cost of $60 each and a material price of $30 each. Print a standard estimate report.

CONSTRUCTION ESTIMATING

Specialty Items and Final Bid Preparation 15

Key Concepts

- Most specialty items are taken off as individual units and individual components.
- Manufacturer names and model numbers are commonly specified for appliances.
- Estimators review the prints and specifications for manufacturer references concerning the required conveying systems.
- The final bid is a total sum of the costs of all items included in the construction project; it is the price that the contractor submits to the owner to cover all costs for materials, labor, overhead, and profit.
- After an owner accepts a bid from a contractor to perform the construction work, a contract is signed.

Introduction

Specialized applications require unique construction materials and equipment. Many divisions of the CSI MasterFormat™, such as Divisions 10 through 14, are dedicated to specialized appliances, equipment, furnishings, various types of unique construction, and conveying systems. A variety of specialized items are also distributed throughout other divisions of the MasterFormat. Specialty items require highly specific estimating skills.

The final bid is the price the contractor or construction team submits to the owner to cover all expenses for materials, labor, overhead, and profit that are necessary to complete the job.

Accurate recordkeeping by the estimator and construction company during the construction process is a good practice to ensure accurate historical data for future projects.

SPECIALTY ITEMS AND QUANTITY TAKEOFF

Unique construction materials and equipment are required for specialized applications. Each of these specialty items requires highly specific estimating skills based on up-to-date equipment and manufacturer recommendations and information. Bids for many of these items are commonly developed by subcontractors and specialty contractors who specialize in a particular type of work. As with all other portions of the bid, estimators should be familiar with all of the items required in each of these categories to ensure an accurate bid.

Division 10—Specialties

Division 10 of the CSI MasterFormat™ includes specialty items such as signage, partitions, chalkboards and tackboards, toilet compartments, fireplaces, flagpoles, and lockers and shelving. Other specialty items in this portion of the bidding documents include service walls, wall and corner guards, pest control, postal and telephone specialties, and exterior protection. Each item requires specific takeoff quantities and labor costs based on the specialty item required. Specifications for these materials commonly use manufacturer names and model numbers. **See Figure 15-1.**

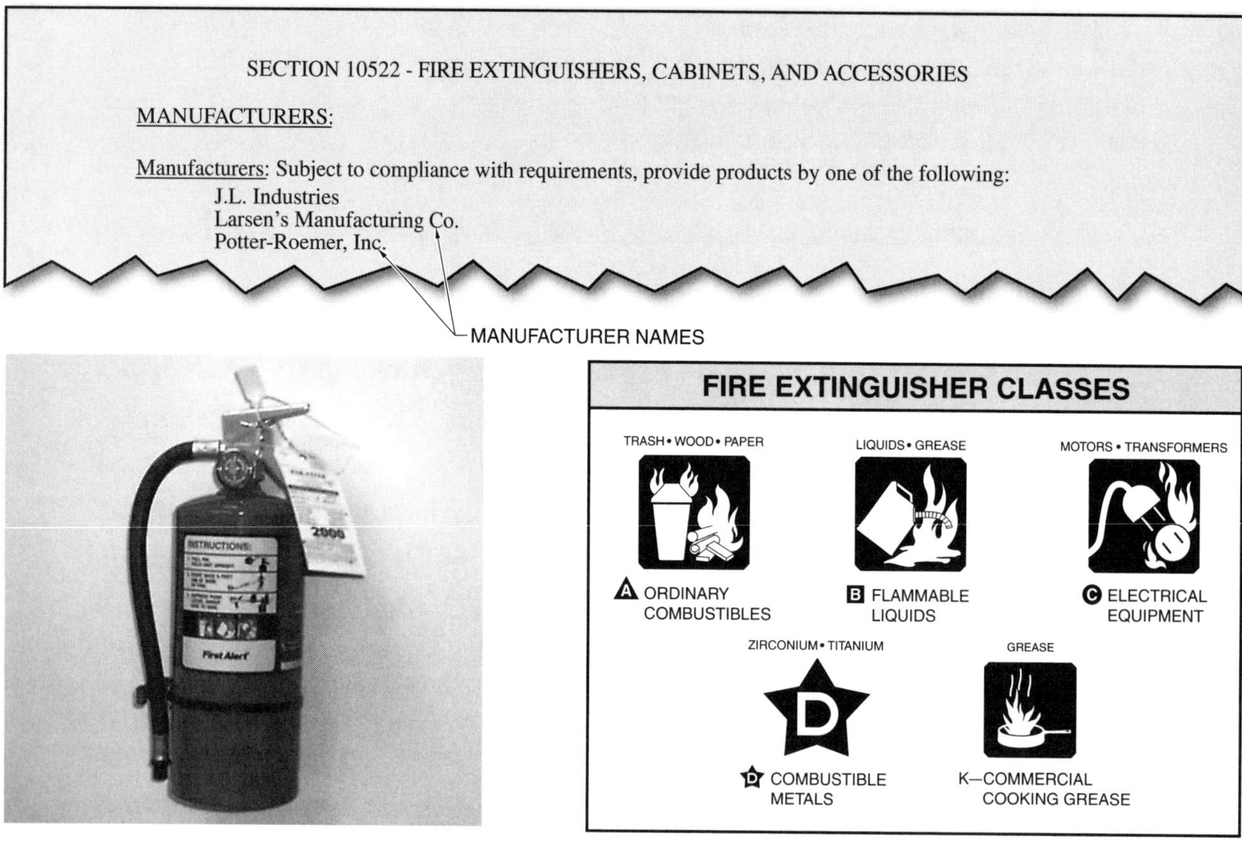

Figure 15-1. Manufacturer names are commonly used to describe specialty items such as fire extinguishers.

Most specialty items are taken off as individual units and individual components. BIM usage may be helpful in generating accurate quantity takeoffs for these individual specialty items. Individual item takeoff is used for storage units, flagpoles, signage, and lockers. Measurements in linear feet are used for takeoff of folding or sliding partitions. Interior elevations are provided to show chalkboards, projection screens, lockers, marker boards, and miscellaneous fixtures. The sizes and manufacturer codes for many of these fixtures are noted in the specifications and room finish schedule. Common specialty items include partitions (Division 10), access flooring (Division 9), and louvers (Division 8).

Partitions. A *partition* is a vertical divider that separates one area from another. Partitions detailed in Division 10 include operable, demountable, and wire mesh partitions and portable partitions, screens, and panels.

An *operable partition* is a panel or partition that is moved by hand or motor along an overhead track. Operable partitions include folding (accordion), coiling, and sliding partitions. Folding, coiling, and sliding partitions are commonly suspended from an overhead track attached to structural members. They may also be floor-supported. Folding, coiling, and sliding partitions are installed in large rooms to allow for open spaces to be divided into smaller individual areas for use by multiple groups. **See Figure 15-2.** Panel thickness, width, height, finish, and acoustic properties vary depending on architectural and owner requirements and manufacturer designs.

Figure 15-2. Partitions allow for temporary division of large spaces into smaller areas.

A sign such as a menu board is taken off as an individual unit in an estimate.

A *demountable partition* is a type of room divider that is designed to be disassembled and reassembled as needed to form new room layouts. Demountable partitions are fastened to adjoining permanent construction and can be moved with a minimal amount of damage to the supporting structure. Tracks are mounted to the finished floor and ceiling. Demountable wall panels, doors, lights, and frames are set in the tracks. Demountable partition panels may be made of gypsum board or other composite materials. Takeoff for demountable partitions includes the number of panels, door units, and window units required, the linear feet of top and bottom track, and other hardware such as stiffeners installed between panels.

Wire mesh partition is a partition fabricated of wire mesh. Portable partitions, screens, and panels include partitions, wall screens, and operable panels that are not connected to permanent construction.

Partition installation items considered by the estimator include the quality and type of partition installed, delivery into the installation area, and any unique installation requirements. For operable and demountable partitions, the setting and fastening of the track is a key item considered in labor costs.

Demountable partitions are commonly installed in existing finished spaces, so some additional labor costs may need to be taken into consideration for working around existing office operations. Installation during evening or nonworking hours may be required by the owner to minimize work disruption. Labor rates are commonly calculated based on the linear feet of wall to be installed.

Access Flooring. *Access flooring* is a raised floor system supported by short posts and runners. Access flooring posts and runners support prefinished floor panels above the surface of the structural floor. **See Figure 15-3.** The open area below the prefinished floor panels is used for routing electrical or computer wiring and ventilation ducts, or it may serve as a plenum for air circulation in a structure. Access flooring systems may use tracks between the pedestals to support the prefinished floor panels. Prefinished floor panels are commonly manufactured in 24″ squares.

Figure 15-3. Access flooring supports prefinished floor panels above the surface of the structural floor.

Access flooring takeoff is commonly calculated in square feet or square yards of surface area. Items considered by the estimator include the method of fastening the pedestals to the structural floor, track requirements, the type of finished surface on the floor panels, and the requirements for any floor panel vents and grilles, ramps, or edge treatments around the perimeter of the raised floor.

Louvers. A louver is a ductwork cover containing horizontal slats that allow the passage of air but prevent rain from entering. Louvers allow airflow from one area to another while providing a level of weather protection, visual protection, or security. Louver styles include standard, storm-proof, operating, acoustic, and sight-proof. Louvers are commonly made of galvanized or cold-rolled steel or extruded aluminum alloy. Louver finishes may be baked enamel, clear lacquer, or anodic.

Louver takeoff is performed by the item based on height, width, depth, type of louver, and finish characteristics. The number of each type of louver is entered into the proper ledger sheet, spreadsheet, or estimating program cell. Labor costs are calculated on a unit basis, with consideration given to any aerial lift equipment that may be necessary to reach the louver installation.

Division 11—Equipment

Division 11 of the MasterFormat includes food service, athletic, banking, photographic, office, audiovisual, laboratory, laundry, library, and health-care equipment. The most common equipment encountered is food service equipment. Food service equipment includes a wide variety of beverage and food handling equipment. Food service equipment in Division 11 includes food storage, preparation, and cooking equipment. **See Figure 15-4.**

Appliances include coolers and freezers, ranges and ovens, bar equipment, and ice machines. Manufacturer names and model numbers are commonly specified. The unit takeoff method is used to enter the quantity of each type of food service equipment into the proper ledger sheet, spreadsheet, or estimating program. Labor costs include delivery, any fastening requirements, and hookups to power sources or plumbing. Labor costs are estimated on a per unit basis, with the estimator verifying that the proper power sources and plumbing are provided at a location close to the required installation.

QUANTITY SHE

Project: *Restaurant*

Estimator: *RDJ*

No.	Description	Mat Quantity	Unit
1	Ice bin	2	ea
2	Baked potato oven w/ stand and timer	1	ea
3	Ice tea dispenser	2	ea
4	Hot chocolate machine	1	ea
5	Chicken pressure fryer assembly	1	ea
6	Walk-in cooler/freezer	1	ea
7	Salad bar base unit	1	ea
8	Salad bar sneeze guard	1	ea
9	Exhaust hood	2	ea
10	Electric modular range top w/ stand	1	ea

Carlisle

Figure 15-4. Food service equipment is taken off according to the number and type of each piece of equipment required.

Athletic equipment may include basketball hoops, boxing and weightlifting equipment, nets and protective devices, and seating for grandstands and stadiums. Banking equipment includes depositories, teller windows, vaults, and automatic banking apparatus. Photographic equipment includes processing equipment and transfer cabinets. Laboratory equipment includes any specialized cabinetry or venting required for laboratory work, sterilizers, and emergency safety appliances. Laundry room equipment includes washers, dryers, exhaust equipment, and dry-cleaning and ironing equipment. Library equipment includes shelving, automated book storage and retrieval systems, and book depositories. Healthcare equipment includes patient equipment, dental equipment, optical equipment, and radiology equipment.

Division 12—Furnishings

Division 12 includes furnishings such as open office systems furniture, manufactured wood casework, storage units, file cabinets and filing systems, tables and chairs, and window blinds and curtains. Information determined by the estimator includes unit item counts based on variations of materials, finishes, hardware, fabrication, and installation. Additional information concerning wood casework is found in Division 6.

Open Office Systems Furniture. *Open office systems furniture* is furniture that consists of panels that are connected in various configurations to provide working office space such as desktops, storage units, and lighting in a compact, flexible configuration. Open office systems furniture is designed for use in an open office plan that uses few fixed floor-to-ceiling partitions. **See Figure 15-5.** Estimators must be familiar with all the available components of each type of open office systems furniture. Panel finishes include plastic laminate, fabric, wood veneer, and plastic. Common panel heights vary from 50″ to 80″, and common panel widths vary from 12″ to 48″.

Figure 15-5. Open office systems furniture allows the integration of wall units, working surfaces, storage, and lighting.

> Manufacturers or suppliers of open office system furniture often provide assistance when estimating large office installations. However, it is the responsibility of the estimator to verify the quantities of office furniture components and ensure that the correct items and quantities are included in the estimate.

Items considered during open office systems furniture takeoff include panel finish, height, and width; attachments such as shelving and storage units; lighting requirements; and provisions for integrated electrical or computer wiring. Labor considerations are similar to those of demountable walls in that installation may be required during evening hours or at times that will minimize disruption to office operations. Delivery of equipment and accessibility to the installation area may add some labor costs. Installation labor rates are commonly calculated based on company historical data per unit.

Special Construction

Special construction included in other various divisions of the CSI MasterFormat™ includes air-supported structures (Division 13), radiation, sound, and seismic control systems (Division 13), solid waste control equipment (Division 44), swimming pools (Division 13), and storage tanks (Divisions 21, 22, 23, and 33).

Hazardous material remediation inside buildings, such as lead and asbestos abatement, is included in Division 2. Measurement and control instrumentation is included in Division 40. Solar energy equipment, such as solar panels, is included in Divisions 23 and 48. Wind energy equipment is described in Division 48. Various automated control systems such as communications equipment, electronic safety and security systems, and conveying systems are included in Division 25. Fire suppression system information is included in Division 21.

Seismic Construction Systems. Seismic construction refers to new construction built to specific seismic requirements or to retrofitting existing structures to improve their ability to withstand earthquake damage. Earthquake damage results from random vertical and horizontal vibrations. Engineering research is ongoing to determine new and improved methods of designing and retrofitting various structures such as bridges and buildings to resist earthquake damage. Seismic construction systems require close adherence to design tolerances for items such as reinforcing steel, concrete, weld connections, bracing, and flexible pipe couplings. Estimators should check all portions of the drawings for seismic control devices and safeguards.

Fire Suppression Systems. Fire suppression systems include the pipe materials that join water supplies, sprinkler systems, and attachments to safety valves and alarms. Mechanical prints indicate locations of pipes, sprinkler head requirements, valve types, alarm box placement, and areas to be protected. **See Figure 15-6.**

Fire suppression systems include wet-pipe, dry-pipe, and gaseous. In wet-pipe systems, fire suppression pipes are connected to a main water supply source. Sprinkler system piping carries water throughout the structure to various locations where it is necessary to provide fire suppression protection. Sprinkler pipes are shown on plan views with a pipe symbol and abbreviation, such as WP for wet pipe. Isometric drawings of fire suppression piping are also provided.

Sprinkler pipes are made of cast iron or any pipe material that does not fail due to heat from a fire. Valves are placed throughout the piping system to allow periodic testing and provide maximum safety. Sprinkler heads are installed at various locations throughout the fire suppression piping system as required by the applicable fire protection code. Sprinkler heads in a wet-pipe system are designed to open and release water when activated by heat. Sprinkler head trim members are installed after ceiling tiles are in place in suspended ceiling systems.

Dry-pipe system pipes are noted on plan views with the abbreviation DP for dry pipe. Dry-pipe systems are connected to the same fire suppression water supply as wet-pipe systems. The portion of the system that is dry consists of pipes filled with air. When a fire occurs in an area protected with a dry-pipe system, the sprinkler heads are designed to release the air from the system piping and then distribute water from the fire service main. Additional information concerning pipes, sprinkler heads, and valves is found in the specifications.

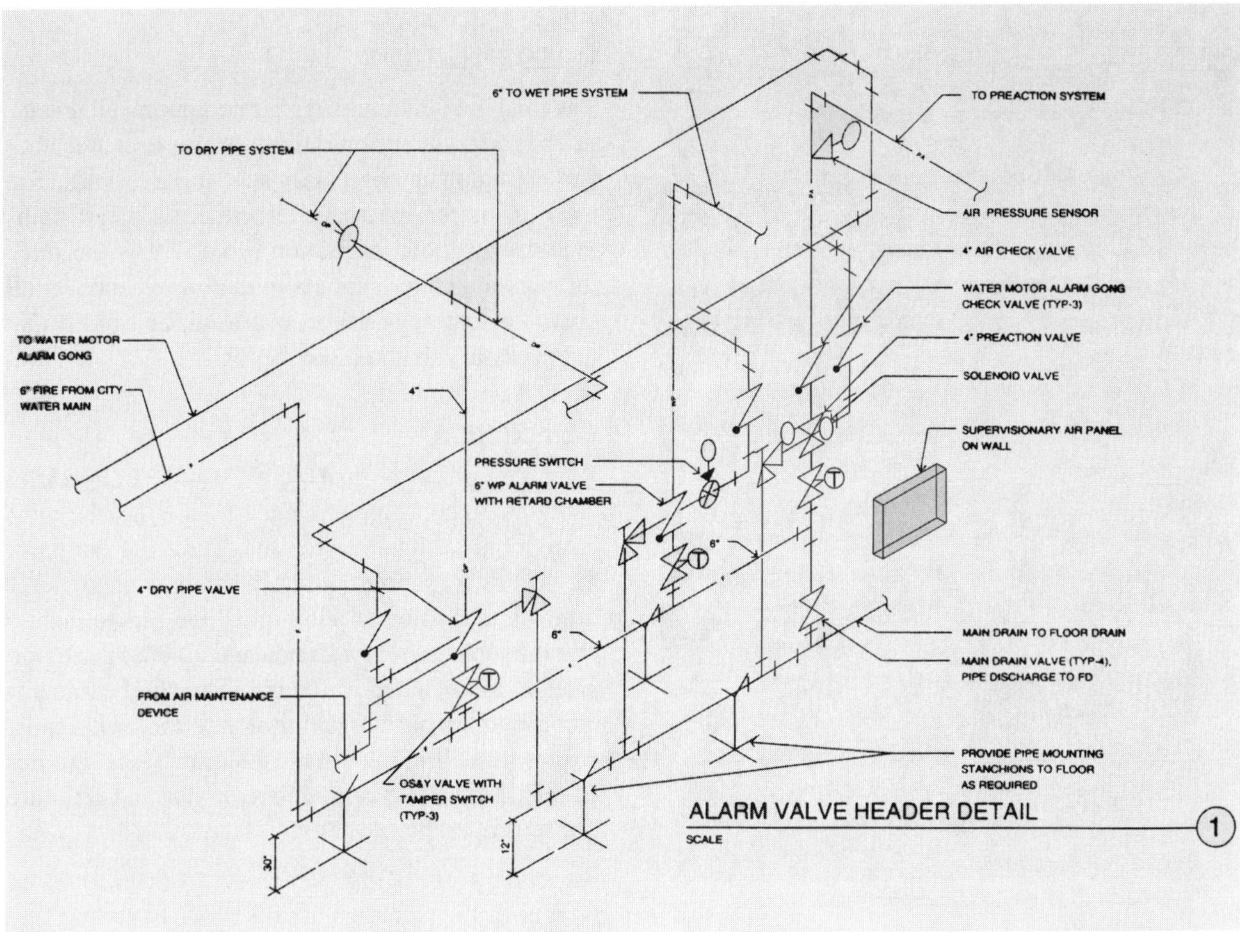

Figure 15-6. Fire suppression piping is shown on mechanical prints and included in Division 21 of the CSI MasterFormat™.

Gaseous systems consist of a fire-suppressing gas such as halon piped into an area where electronic equipment, such as computer equipment, is installed and cannot be exposed to water. Sprinkler pipes are connected to a tank filled with halon. Gaseous system piping is not connected to fire protection water mains.

Estimates for these systems are similar to other plumbing estimates, with the linear feet of each type of pipe based on floor plans and isometric fire suppression system drawings. Estimators perform unit takeoff for elbows, valves, sprinkler heads, and any necessary storage tanks. Labor costs are determined from company historical data for the size of the installation or from standard labor cost tables.

Conveying Systems

Conveying systems include elevators, material-handling equipment such as cranes (Division 41), hoists (Division 41), escalators (Division 14), conveyors (Division 41), and monorail systems (Division 34). The MasterFormat also includes information concerning conveying systems in Division 14. Conveying systems are specialized according to manufacturer design and load-carrying capacity.

Estimators review the prints and specifications for manufacturer references concerning the required conveying system. Cross-referencing with electrical, mechanical, and structural drawings may provide additional information concerning power, piping, and support requirements and tolerances. In addition to material costs, accessibility of the installation area for conveyor installation is coordinated with the project manager. The need for aerial lifts or attachments to structural supports can be coordinated with the construction team to minimize additional labor costs.

Southern Forest Products Association

Conveyors, such as those used to transport lumber through a lumber mill, are included in Division 41 of the CSI MasterFormat™.

Elevators. Elevators may be powered electrically or hydraulically. The elevator car is guided by vertical rails fastened to the walls of the elevator shaft. The car is hoisted in the shaft by wire ropes and counterweights for an electric elevator or a post for a hydraulic elevator. A machine room is commonly located at the top of the elevator shaft for electric elevators and at the bottom of the shaft in an elevator pit for hydraulic elevators. Items considered in the selection of an elevator system include the building height, use, population, and elevator entrance requirements.

Elevator prints and specifications required by an estimator include the net capacity, speed, travel distance, car size, and interior finish. Labor rates are determined based on the height of the elevator shaft, the type of elevator being installed, and the coordination with other structural construction on the project. Company historical data is commonly used for this type of specialized construction.

FINAL BID ASSEMBLY AND PRESENTATION

The final bid is assembled after computing all quantities of materials and amounts of equipment and labor and obtaining any necessary subcontractor bids. The final bid is the total sum of the costs of all items included in the construction project. It is the price that the contractor submits to the owner to cover all costs for materials, labor, overhead, and profit that are necessary to complete the job.

Compilation

When compiling the final bid, the contractor, specialty contractors, and suppliers must ensure that the most current bid documents are used. The chief estimator for the project should contact the architect and owner immediately prior to submitting the bid to ensure that the most current addenda and all changes to the project are included in the bid. The chief estimator compiling the bid for submission to the owner must ensure that all items are included, no items are duplicated, and all construction materials and labor are priced properly.

Subcontractor Bids. Subcontractors must often wait until the final bid time is close to ensure that they are working with current bid documents and to ensure confidentiality of their bid. Throughout the bidding process, estimators may include plug numbers for subcontractor bids. A *plug number* is a monetary amount used as a temporary placeholder in an estimate during bid development. Plug numbers are based on historical estimate information. Plug numbers allow a general overall project cost value to be developed until final subcontractor bids are received. If subcontractor bids are not received before the bid submission deadline, a plug number can be used in a bid until final costs for various items can be determined.

The chief estimator assembling the final bid, including all subcontractor bids, must carefully and completely check subcontractor bids even though the time requirements may be short. The lowest-price subcontractor bid for any particular portion of the project may exclude items that are required in the subcontractor scope of work.

The chief estimator must also ensure that two subcontractors with similar-type portions of the project, for example, plumbing and fire suppression systems, do not have the same installation included in both bids, which would double the material and labor costs. The estimator assembling the final bid must ensure that all items are properly covered and not duplicated prior to developing a final cost. A system based on the CSI MasterFormat™ and specification sections aids in tracking subcontractor and supplier bids during the final compilation on bid day. Many bids are received at the last minute by email or fax, which requires planning for proper communication-system access on bid day.

Takeoff Quantities. All takeoff quantities for work being performed by the contractor assembling the bid should be included. A common estimating error in bidding is unintentionally omitting one or more items in the bid. It is very important to double-check all bidding documents to ensure that all items have been included in the final bid price and that none are duplicated.

Low Bid. In a competitive, low-bid situation, contractors work hard to determine the lowest responsible bid that allows for a fair profit while ensuring all material, equipment, labor, and overhead costs are covered. Estimators should quickly analyze subcontractor and supplier prices to detect pricing trends and should not accept a low bid that may be inaccurate. An inaccurate bid that omits important work processes or overlaps the work included in another bid can put both the contractor and subcontractor in a difficult situation later when the construction work cannot be completed as required by the owner or additional costs are incurred.

To ensure that the bid is complete and as low as necessary to win the work, a checklist may be used. A subcontractor and supplier quotation recap may also be used to allow the chief estimator to quickly recognize incomplete bids or pricing trends. Some estimating programs include a scan function that shows areas where blank cells or zeros appear in the final bid documents. **See Figure 15-7.**

A spreadsheet estimate recap sheet can also be formatted according to the MasterFormat with some automatic tabulation formulas added in the proper cells. This information can include descriptions for general contractor, subcontractor, and supplier items. This document may be maintained in a separate file from the final estimate to prevent inclusion of an incomplete price.

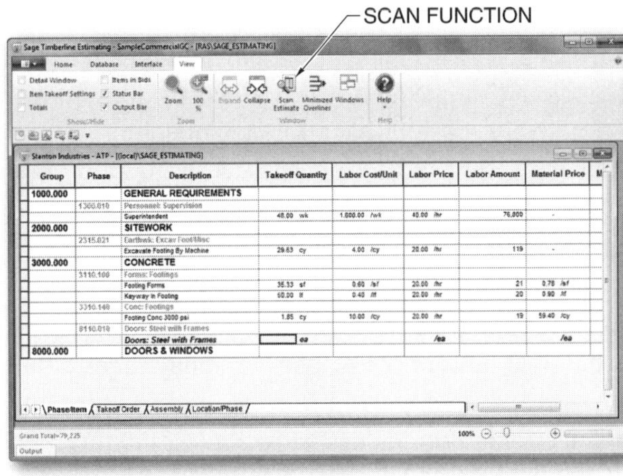

Figure 15-7. The scan function in an estimating program alerts the estimator to blank cells in the bid documents.

Some estimators use a cut-and-add system to help track final changes on bid day. When subcontractors or suppliers make final changes to their pricing, cut-and-add systems show the original pricing and the revised pricing for each change, including the name of the subcontractor or supplier. These cost changes are then made in the final estimate calculations.

A complicated project with many bid items and sources commonly requires more than one estimator to check the overall compilation and ensure that all costs are included. All of the costs for materials, labor, overhead, and profit are carefully analyzed to ensure that no items are missed and no items are duplicated or overpriced. Estimators should allow for the possible final input of senior management in the construction firm. This can affect the final bid price and the timing of the final compilation of the bid.

Negotiated Construction Work. *Negotiated construction work* is a project delivery system in which a qualified contractor submits a bid to an owner based on the plans and specifications without a competitive bid process. In negotiated construction work, the owner and architect may review the final bid documents with the contractor to determine if there are areas for cost savings. When there are significant differences between the conceptual bid and the actual final bid, discussions between the owner and the contractor can be used to resolve these differences prior to award of the contract.

Value engineering is a process whereby the owner and contractor agree on changes to the original design documents that will result in a comparable finished construction project with some substitutions and cost savings. Negotiated construction work still requires estimators to ensure that all items, materials, labor, overhead, and profit are calculated accurately and completely.

Bid Presentation. The presentation of the final bid to the owner is an item that should be included in the overhead cost that may make a significant difference in the final bid price. Some owners may require the use of a standardized bid form for bid submission. Other owners may allow contractors to submit their bids in the format determined by the contractor.

One of the primary advantages of computerized bidding systems is the ability to continue to work with the bidding data and costs close to the bid date and time, and still be able to produce a high-quality, accurate, comprehensive bid presentation. **See Figure 15-8.** Information can be condensed and simplified as needed prior to the final presentation to the owner.

Contract Award. After an owner reviews and accepts a bid from a contractor to perform the construction work, a contract is signed. Once the contract is signed, the contractor is obligated to perform the construction work as specified for the sum of money shown in the contract. The construction contract binds the contractor to furnish all the labor, materials, and subcontracted work to completely finish the project as detailed in the specifications, prints, and all related construction documents.

The terms of the contract concerning work to be performed and the allocation of risk between the owner, architect, engineers, contractor, subcontractors, and suppliers should be carefully reviewed by all parties both during the estimating and bidding process and prior to the signing of the contract to avoid potential problems during construction. Risk allocation may have a significant impact on overhead costs for various types of insurance and bonding coverage.

The process for making any changes to the project during the course of construction should also be clearly understood during the estimating and final contracting processes. *Change orders* are any changes to the construction project requested by the owner, contractor, or subcontractor. These are commonly subject to review and approval or rejection by the owner.

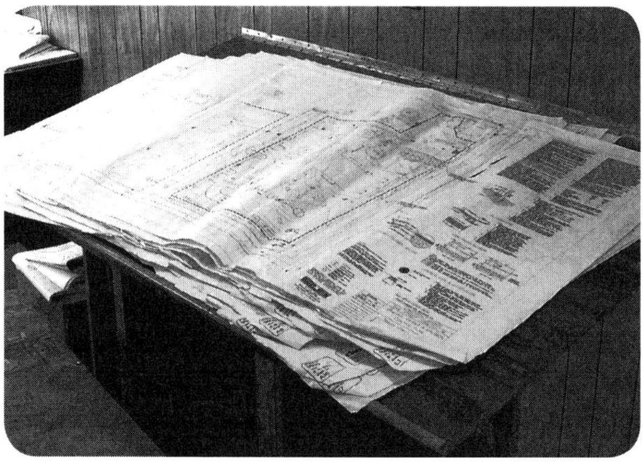

The contractor needs to track the change orders on the construction set of prints.

Simpson Industries

Spreadsheet Report

SI98-Refrigeration

Page 1

8:29 AM

Group	Description	Labor Amount	Material Amount	Sub Amount	Equip Amount	Other Amount	Total Amount
1100.00	GENERAL REQUIREMENTS	53,260	5,317	29,485	2,530	2,940	93,532
2100.00	SITEWORK	24,468	1,448	88,700	3,149		117,765
3000.00	CONCRETE	101,503	142,999	4,500			249,003
4000.00	MASONRY	79,290	32,545				111,834
5000.00	METALS	5,428	931	265,000			271,359
6000.00	WOOD & PLASTICS	3,812	853	4,400			9,066
7000.00	THERM \ MOISTURE PROTECT.	37,168	111,659	195,000			343,827
8000.00	DOORS & WINDOWS	2,834	7,029	10,000			19,863
9000.00	FINISHES	29,728	47,297	20,000			97,026

Figure 15-8. Successful estimating requires that the final presentation of the estimate be in a clear, readable, and understandable format.

The contract also binds the owner to make payments based on satisfactory completion according to the agreed-upon contract terms and prices, and may describe any payment retention that the owner may hold pending satisfactory completion of portions of the project or the entire construction project.

POST-BID TRACKING

Accurate recordkeeping by the project manager in conjunction with the estimator and construction company staff during the construction process helps to ensure accurate bidding in the future. A successful construction estimating department works closely with the construction project management team to monitor the various costs that are incurred during a project in an attempt to keep company historical cost and labor data accurate and find efficiencies in the bidding and costing processes. Estimators track quantities and pricing for materials, labor, overhead, profit, and other items that affect the final construction cost. Change order quantities and types may also help indicate areas where improvements in the estimating and bidding processes can be realized. Variances from estimates provide important information for future bidding activities.

Application of overhead costs to a construction project can have a significant effect on the success of a construction project and an accurate bid. Variances in overhead costs in relation to the original estimate should be carefully tracked. Where costs for items such as insurance, bonding, security, taxes, or other overhead items are not consistent with the estimate, determinations should be made concerning whether these are unusual variations or items that should be calculated differently on future estimates.

Materials

All materials used on the construction project should be allocated to a particular account and job-site operation, commonly referred to as cost codes, work codes, or the work breakdown structure (WBS). Tracking of material costs compared to estimated costs shows where material prices have changed over the course of the construction process. New materials or new pricing may cause variances in material costs that impact the final cost of a construction project. Coordination with the project management team related to scheduling and projected material-cost changes can impact the accuracy of the bid and profitability of the completed project.

Quantities. Comparison of estimated material quantities with actual material quantities used indicates areas where material takeoffs are excessively high or low. Cases where material takeoff quantities vary greatly from actual materials required for construction may indicate excessive waste, theft, items missed on the takeoff, inaccurate quantities being delivered, or improper waste percentages applied in the estimating formulas. Adjustments should be made to company historical data where applicable to ensure greater accuracy in material quantity takeoff on future projects.

Labor

One of the highest risk variables in the construction process is labor cost. Estimators use a variety of standard industry tables and company historical records for pricing labor. Labor production rates and costs must be tracked in conjunction with the construction project manager throughout the construction process to ensure minimal variance from the original labor estimate. Labor costs should be allocated according to cost codes in a manner similar to materials. Adjustments in scheduling or job-site management should be coordinated with the project management team to allow for ongoing comparison of the accuracy of labor-rate and production estimates.

Crew Performance. Work at a construction site is based on unit costs tied to the amount of production for various material units. Where estimated labor costs and crew performances vary from the allocation based on the number of material units, items to be considered include climatic conditions, crew training, provision of proper equipment and tools, and job-site safety, management, and scheduling. Accurate tracking of labor production rates for various material units in conjunction with onsite project management can help estimators keep their labor-rate and production information more accurate for future bids based on the construction crews that are performing the field work.

SAGE TIMBERLINE ESTIMATING . . .

FINALIZING AN ESTIMATE

In Sage Timberline Estimating, an estimate is finalized after the takeoff is complete. Here, the Modesto - Ceiling estimate created in Chapter 12 will be finalized and prepared for printing.

1. Ensure the estimate is maximized by pressing the maximize button in the upper-right corner of the estimate.
2. Hide columns by right-clicking on the column heading and selecting **Hide Column** from the shortcut menu. Hide all columns on the spreadsheet except **Group, Phase**, **Description**, **Labor Amount**, **Material Price**, **Material Amount**, **Sub Amount**, and **Total Amount**. Several columns can be hidden simultaneously by holding down the left mouse button, dragging the highlight across the column headings, right-clicking, and selecting **Hide Column** from the shortcut menu.

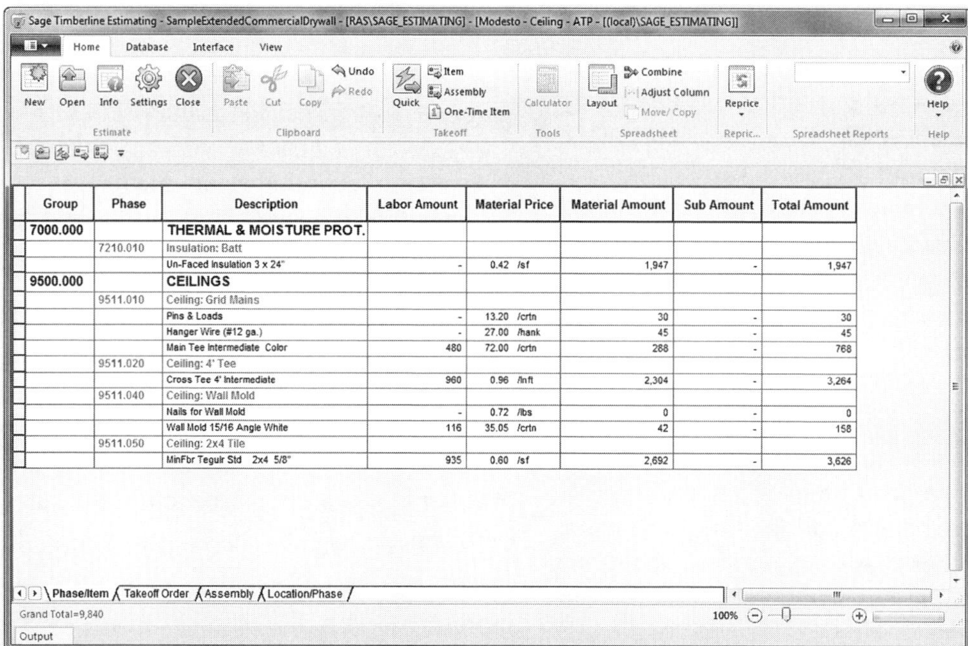

Group	Phase	Description	Labor Amount	Material Price	Material Amount	Sub Amount	Total Amount
7000.000		THERMAL & MOISTURE PROT.					
	7210.010	Insulation: Batt					
		Un-Faced Insulation 3 x 24"	-	0.42 /sf	1,947	-	1,947
9500.000		CEILINGS					
	9511.010	Ceiling: Grid Mains					
		Pins & Loads	-	13.20 /crtn	30	-	30
		Hanger Wire (#12 ga.)	-	27.00 /hank	45	-	45
		Main Tee Intermediate Color	480	72.00 /crtn	288	-	768
	9511.020	Ceiling: 4' Tee					
		Cross Tee 4' Intermediate	960	0.96 /lnft	2,304	-	3,264
	9511.040	Ceiling: Wall Mold					
		Nails for Wall Mold	-	0.72 /lbs	0	-	0
		Wall Mold 15/16 Angle White	116	35.05 /crtn	42	-	158
	9511.050	Ceiling: 2x4 Tile					
		MinFbr Tegulr Std 2x4 5/8"	935	0.60 /sf	2,692	-	3,626

3. Make price changes by placing the cursor in the relevant material price cell, inputting the change, and pressing **Enter**. For example, an increase in the price of the ceiling tile is reflected by placing the cursor in the **Material Price** cell for tile, changing the price from **$0.60/sf** to **$1.10/sf**, and pressing **Enter**. The **Material Amount** and the **Total Amount** columns are adjusted automatically.

. . . SAGE TIMBERLINE ESTIMATING . . .

4. Reduce the level of detail shown on the spreadsheet by picking the **View** tab, then picking the **Collapse** button on the toolbar. The **Collapse** button may be picked again for a more summarized view. Expand the estimate by picking the **Expand** button twice to show full detail.

TOTALS CHECKBOX

EXPAND AND COLLAPSE BUTTONS

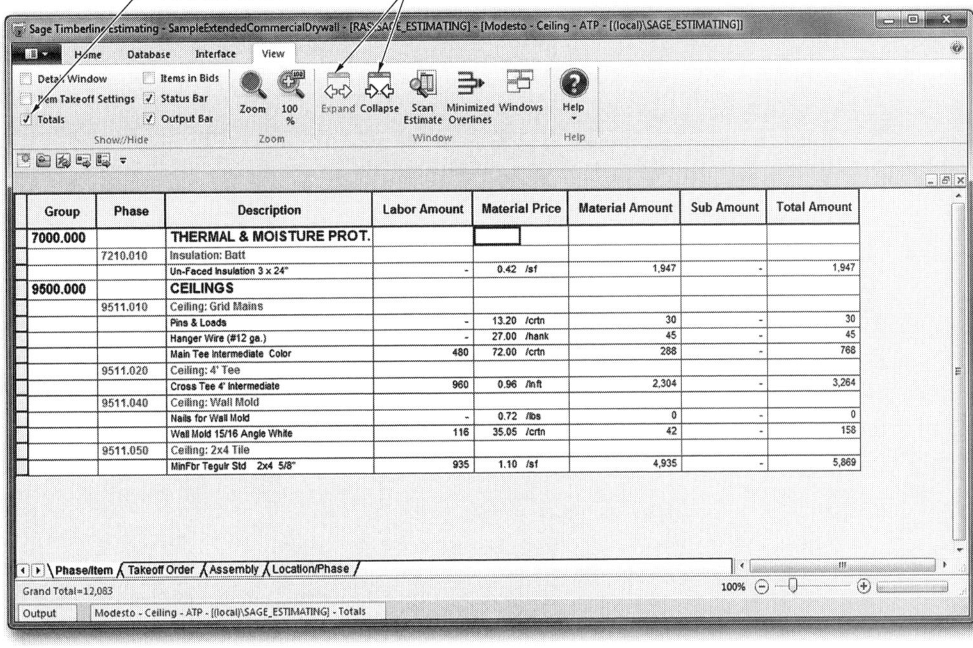

5. Display estimate totals by category (labor, material, etc.) by checking the **Totals** checkbox on the toolbar. The **Totals** window appears below the **Spreadsheet** window.

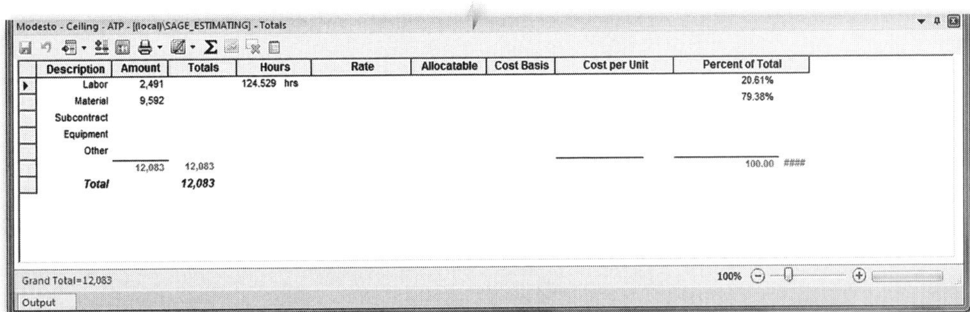

. . . SAGE TIMBERLINE ESTIMATING . . .

6. Pick the **Database** tab then pick the **Addons** button. The **Adding Database Addon** dialog box is then displayed. Enter **18** for **Addon** and **Profit & Overhead** for **Description**. Select the **Allocatable** checkbox and select **Total** from the **Cost basis** dropdown list. Select the **Estimate total** radio button and enter a rate of **19%**.

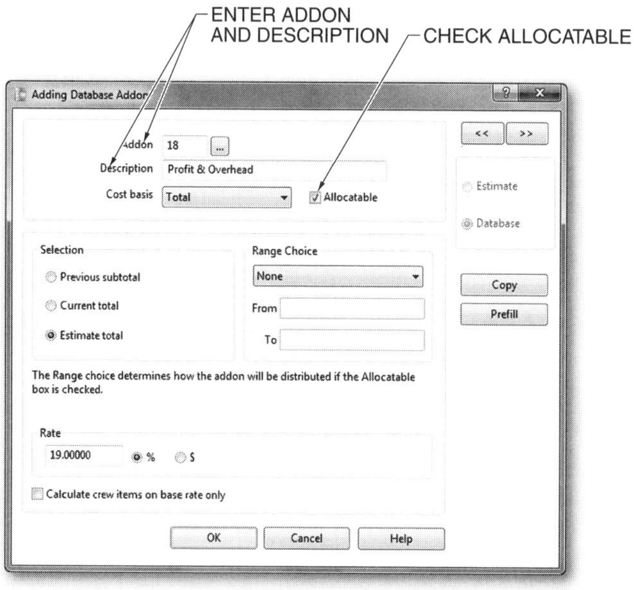

7. Pick **OK** and **Close** to exit the dialog box.
8. Pick the **Insert addon** button to display the **Addon List** dialog box. Select addon **18** by double-clicking on it and close the dialog box. The addon amounts are calculated automatically and displayed in the **Totals** dialog box.
9. Close the **Totals** dialog box by picking **OK** and **Close**. Note that the addon has been proportionally spread across all estimated items. The **Addon Amounts** and **Grand Total** columns can be displayed by right-clicking the **Total Amount** column heading and selecting **Show Hidden Columns** from the shortcut menu.
10. Generate a user-defined spreadsheet report by selecting **Spreadsheet...** from the **Reports** pulldown menu.
11. Pick the **Report Options...** button to display the **Spreadsheet Report Options** dialog box. Pick the **Spreadsheet** button. From the **Spreadsheet Report** window, deselect the **Print cover page** checkbox. Select the **Shaded headings** and **Minimize overline columns** box. Select the **Preview** button to preview the spreadsheet report.

. . . SAGE TIMBERLINE ESTIMATING

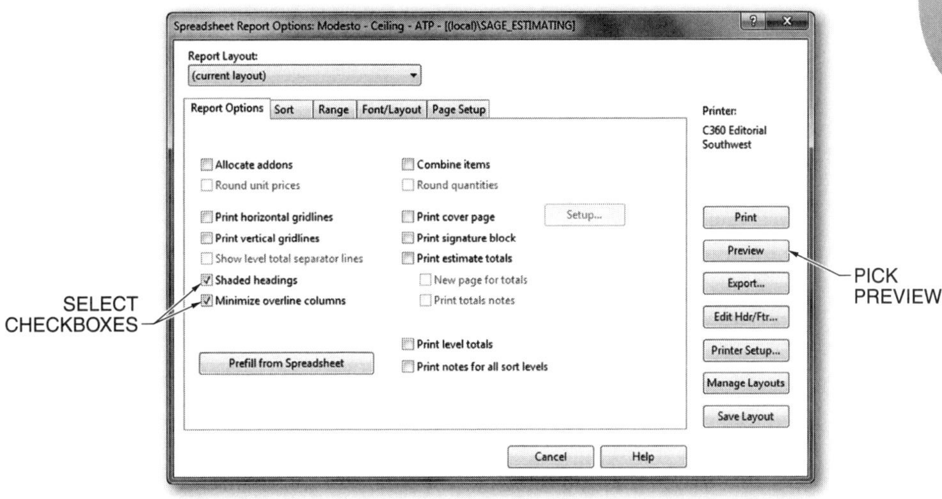

12. Pick anywhere on the report to zoom in. Pick the **Close** button to return to the **Spreadsheet Report Options** dialog box. Pick **Cancel** and **Close** to return to the spreadsheet.

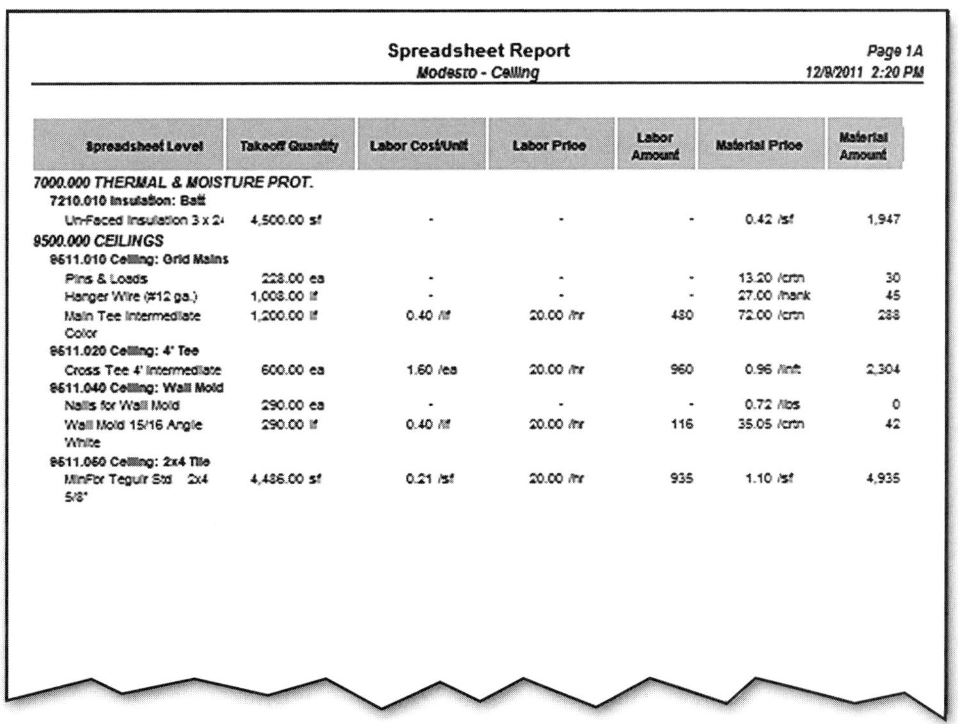

Spreadsheet Report
Modesto - Ceiling

Page 1A
12/8/2011 2:20 PM

Spreadsheet Level	Takeoff Quantity	Labor Cost/Unit	Labor Price	Labor Amount	Material Price	Material Amount
7000.000 THERMAL & MOISTURE PROT.						
7210.010 Insulation: Batt						
Un-Faced Insulation 3 x 24	4,500.00 sf	-	-	-	0.42 /sf	1,947
9500.000 CEILINGS						
9511.010 Ceiling: Grid Mains						
Pins & Loads	228.00 ea	-	-	-	13.20 /crtn	30
Hanger Wire (#12 ga.)	1,008.00 lf	-	-	-	27.00 /hank	45
Main Tee Intermediate Color	1,200.00 lf	0.40 /lf	20.00 /hr	480	72.00 /crtn	288
9511.020 Ceiling: 4' Tee						
Cross Tee 4' Intermediate	600.00 ea	1.60 /ea	20.00 /hr	960	0.96 /lnft	2,304
9511.040 Ceiling: Wall Mold						
Nails for Wall Mold	290.00 ea	-	-	-	0.72 /lbs	0
Wall Mold 15/16 Angle White	290.00 lf	0.40 /lf	20.00 /hr	116	35.05 /crtn	42
9511.060 Ceiling: 2x4 Tile						
MinFbr Tegulr Std 2x4 5/8"	4,486.00 sf	0.21 /sf	20.00 /hr	935	1.10 /sf	4,935

Quick Quiz®

Quick Quiz®

Refer to the CD-ROM for the Quick Quiz® questions related to chapter content.

Key Terms

Illustrated Glossary

- *access flooring*
- *change orders*
- *demountable partition*
- *negotiated construction work*
- *open office systems furniture*
- *operable partition*
- *plug number*
- *partition*
- *value engineering*
- *wire mesh partition*

Web Links

Web Links

American Solar Energy Society
www.ases.org

American Wind Energy Association
www.awea.org

Association of Pool and Spa Professionals
www.apsp.org

Business and Institutional Furniture Manufacturers Association
www.bifma.org

Cleaning Equipment Trade Association
www.ceta.org

Fire Apparatus Manufacturers Association
www.fama.org

Gas Appliance Manufacturers Association
www.gamanet.org

International Safety Equipment Association
www.safetyequipment.org

International Sign Association
www.signs.org

Material Handling Industry of America
www.mhia.org

Metal Building Manufacturers Association
www.mbma.com

Modular Building Institute
www.modular.org

National Alarm Association of America
www.naaa.org

National Association of Elevator Contractors
www.naec.org

National Association of Food Equipment Manufacturers
www.nafem.org

National Association of Store Fixture Manufacturers
www.nasfm.org

National Burglar and Fire Alarm Association
www.alarm.org

National Elevator Industry, Inc.
www.neii.org

National Fire Protection Association
www.nfpa.org

National Kitchen and Bath Association
www.nkba.org

National Restaurant Association
www.restaurant.org

Safety Equipment Distributors Association
www.safetycentral.org

Seismological Society of America
www.seismosoc.org

Steel Tank Institute
www.steeltank.com

Specialty Items and Final Bid Preparation 15

Review Questions

Estimating

______ **1.** A(n) ___ is a vertical divider that separates one area from another.

______ **2.** ___ flooring is a raised floor system supported by short posts and runners.

______ **3.** Prefinished floor panels are commonly manufactured in ___″ squares.

______ **4.** A(n) ___ partition is a type of room divider that is designed to be disassembled and reassembled as needed to form new room layouts.

______ **5.** A(n) ___ partition is a panel or partition that is moved by hand or motor along an overhead track.

______ **6.** A(n) ___ partition is a partition fabricated of wire mesh.

______ **7.** A(n) ___ is a ductwork cover containing horizontal slats that allow the passage of air but prevent rain from entering.

______ **8.** ___ is furniture that consists of panels that are connected in various configurations to provide working office space such as desktops, storage units, and lighting in a compact, flexible configuration.

T F **9.** The highest risk variable in construction is labor cost.

______ **10.** ___ construction work is a project delivery system in which a qualified contractor submits a bid to an owner based on the plans and specifications without a competitive bid process.

T F **11.** The preliminary bid is the price the contractor submits to the owner to cover all expenses for materials, labor, services, overhead, and profit that are necessary to complete the job.

T F **12.** Most specialty items are taken off as individual units.

Specialty Items and Final Bid Preparation 15

Activities

Activity 15-1—Ledger Sheet Activity

Refer to the cost data, equipment schedule, Print 15-1, and Estimate Summary Sheet No. 15-1. Take off the number of wire shelf units. Determine the total material and labor costs, including 8% for overhead and profit.

COST DATA			
Material	Unit	Material Unit Cost*	Labor Unit Cost*
Wire shelf units, 4-tier			
18″ × 48″	ea	1210.00	31.35
18″ × 60″	ea	1375.00	31.35
24″ × 36″	ea	423.50	31.35
24″ × 42″	ea	440.00	31.35
24″ × 48″	ea	484.00	31.35
24″ × 60″	ea	654.50	31.35

* in $

EQUIPMENT SCHEDULE	
ID Number	Equipment Description
63	Wire shelf — cooler
64	Wire shelf units — freezer
65	Wire shelf — dry storage

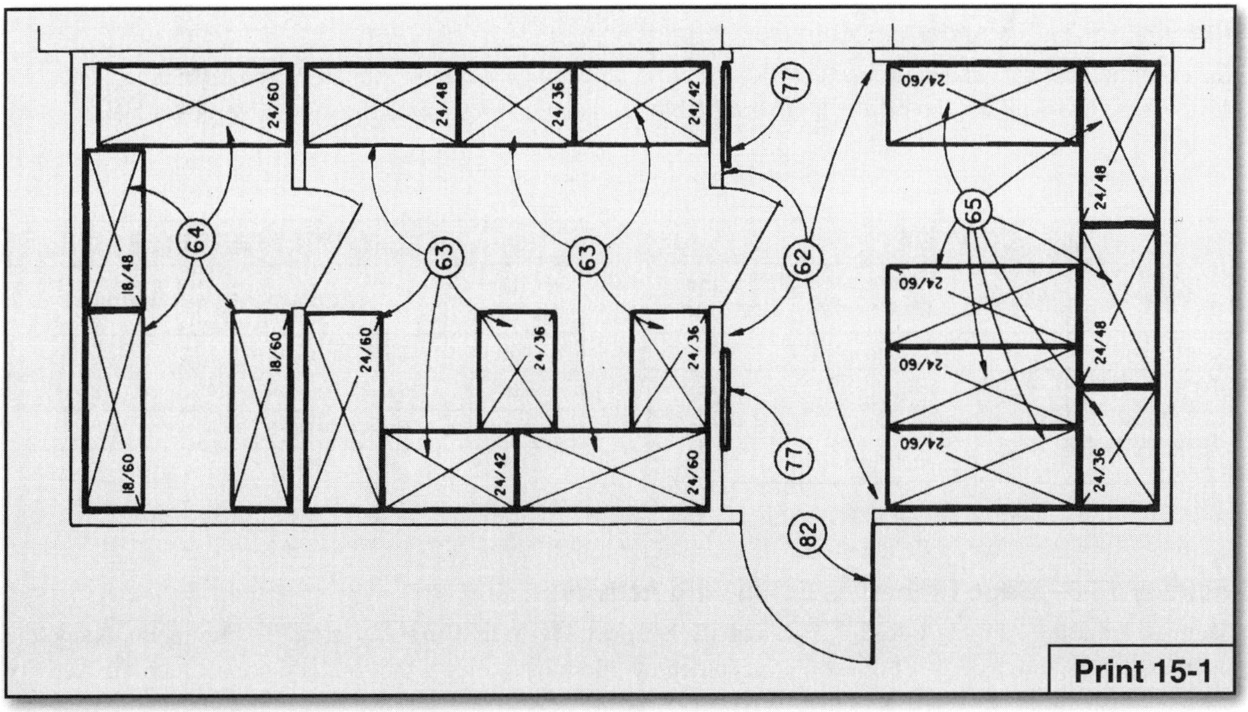

Print 15-1

ESTIMATE SUMMARY SHEET

Project: ____________________ Sheet No. 15-1

Estimator: ____________________ Date: ____________________

Checked: ____________________

No.	Description	Dimensions			Quantity		Material		Labor		Total	
						Unit	Unit Cost	Total	Unit Cost	Total	Unit Cost	Total
	Wire shelf units											
	18″ × 48″											
	18″ × 60″											
	24″ × 36″											
	24″ × 42″											
	24″ × 48″											
	24″ × 60″											
	Total											
	Total including 8% overhead and profit											

Activity 15-2—Spreadsheet Activity

Refer to the cost data and equipment schedule below, and Print 15-2 and Estimate Summary Spreadsheet No. 15-2 on the CD-ROM. Take off the number of tables and chairs. Determine the total material and labor costs, including 12% for overhead and profit.

COST DATA

Material	Unit	Material Unit Cost*	Labor Unit Cost*
Chair	ea	72.05	—
Table, two seat, 20″× 24″	ea	456.50	28.60
Table, four seat, 30″× 30″	ea	456.50	30.80

* in $

EQUIPMENT SCHEDULE

ID Number	Equipment Description
105	Chairs
108	Pedestal table, 20″× 24″
109	Pedestal table, 30″× 30″

Activity 15-3—Sage Timberline Estimating Activity

Merge estimating Activity 6-4, 7-4, 9-4, and 12-4. Select **Merge Estimate...** from the **File** menu. Name the newly merged Activity 15-3. Follow the on-screen instructions to merge the four activities. Open the Activity 15-3 estimate and print a bill of materials report.

Appendix

PLOT PLAN SYMBOLS

NORTH	FIRE HYDRANT	WALK	ELECTRIC SERVICE
POINT OF BEGINNING (POB)	MAILBOX	IMPROVED ROAD	NATURAL GAS LINE
UTILITY METER OR VALVE	MANHOLE	UNIMPROVED ROAD	WATER LINE
POWER POLE AND GUY	TREE	BUILDING LINE	TELEPHONE LINE
LIGHT STANDARD	BUSH	PROPERTY LINE	NATURAL GRADE
TRAFFIC SIGNAL	HEDGE ROW	PROPERTY LINE	FINISH GRADE
STREET SIGN	FENCE	TOWNSHIP LINE	EXISTING ELEVATION

REFRIGERATION SYMBOLS

GAUGE	PRESSURE SWITCH	DRYER
SIGHT GLASS	HAND EXPANSION VALVE	FILTER AND STRAINER
HIGH SIDE FLOAT VALVE	AUTOMATIC EXPANSION VALVE	COMBINATION STRAINER AND DRYER
LOW SIDE FLOAT VALVE	THERMOSTATIC EXPANSION VALVE	EVAPORATIVE CONDENSOR
IMMERSION COOLING UNIT	CONSTANT PRESSURE VALVE, SUCTION	HEAT EXCHANGER
COOLING TOWER	THERMAL BULB	AIR-COOLED CONDENSING UNIT
NATURAL CONVECTION, FINNED TYPE EVAPORATOR	SCALE TRAP	WATER-COOLED CONDENSING UNIT
FORCED CONVECTION EVAPORATOR	SELF-CONTAINED THERMOSTAT	

ARCHITECTURAL SYMBOLS . . .

Material	Elevation	Plan View	Section
EARTH			
BRICK	WITH NOTE INDICATING TYPE OF BRICK (COMMON, FACE, ETC.)	COMMON OR FACE FIREBRICK	SAME AS PLAN VIEWS
CONCRETE		LIGHTWEIGHT STRUCTURAL	SAME AS PLAN VIEWS
CONCRETE MASONRY UNITS		OR	OR
STONE	CUT STONE RUBBLE	CUT STONE RUBBLE CAST STONE (CONCRETE)	CUT STONE CAST STONE (CONCRETE) RUBBLE OR CUT STONE
WOOD	SIDING PANEL	WOOD STUD REMODELING DISPLAY	ROUGH MEMBERS FINISHED MEMBERS PLYWOOD
PLASTER		WOOD STUD, LATH, AND PLASTER METAL LATH AND PLASTER SOLID PLASTER	LATH AND PLASTER
ROOFING	SHINGLES	SAME AS ELEVATION	
GLASS	OR GLASS BLOCK	GLASS GLASS BLOCK	SMALL SCALE LARGE SCALE

. . . ARCHITECTURAL SYMBOLS			
Material	**Elevation**	**Plan View**	**Section**
FACING TILE	CERAMIC TILE	FLOOR TILE	CERAMIC TILE LARGE SCALE CERAMIC TILE SMALL SCALE
STRUCTURAL CLAY TILE			SAME AS PLAN VIEW
INSULATION		LOOSE FILL OR BATTS RIGID SPRAY FOAM	SAME AS PLAN VIEWS
SHEET METAL FLASHING		OCCASIONALLY INDICATED BY NOTE	
METALS OTHER THAN FLASHING	INDICATED BY NOTE OR DRAWN TO SCALE	SAME AS ELEVATION	SMALL SCALE STEEL CAST IRON ALUMINUM BRONZE OR BRASS
STRUCTURAL STEEL	INDICATED BY NOTE OR DRAWN TO SCALE	OR	REBARS SMALL SCALE LARGE SCALE L-ANGLES, S-BEAMS, ETC.

ELECTRICAL SYMBOLS . . .

LIGHTING OUTLETS

OUTLET BOX AND INCANDESCENT LUMINAIRE	CEILING WALL
INCANDESCENT TRACK LIGHTING	
BLANKED OUTLET	B B
DROP CORD	D
EXIT LIGHT AND OUTLET BOX; SHADED AREAS DENOTE FACES	
OUTDOOR POLE-MOUNTED FIXTURES	
JUNCTION BOX	J J
LAMPHOLDER WITH PULL SWITCH	L_{PS} L_{PS}
MULTIPLE FLOODLIGHT ASSEMBLY	
EMERGENCY BATTERY PACK WITH CHARGER	B
INDIVIDUAL FLUORESCENT FIXTURE	
OUTLET BOX AND FLUORESCENT LIGHTING TRACK FIXTURE	
CONTINUOUS FLUORESCENT FIXTURE	
SURFACE-MOUNTED FLUORESCENT FIXTURE	

PANELBOARDS

FLUSH-MOUNTED PANELBOARD AND CABINET	
SURFACE-MOUNTED PANELBOARD AND CABINET	

CONVENIENCE OUTLETS

SINGLE RECEPTACLE OUTLET	
DUPLEX RECEPTACLE OUTLET	
TRIPLEX RECEPTACLE OUTLET	
SPLIT-WIRED DUPLEX RECEPTACLE OUTLET	
SPLIT-WIRED TRIPLEX RECEPTACLE OUTLET	
SINGLE SPECIAL-PURPOSE RECEPTACLE OUTLET	
DUPLEX SPECIAL-PURPOSE RECEPTACLE OUTLET	
RANGE OUTLET	R
SPECIAL-PURPOSE CONNECTION	DW
CLOSED-CIRCUIT TELEVISION CAMERA	
CLOCK HANGER RECEPTACLE	C
FAN HANGER RECEPTACLE	F
FLOOR SINGLE RECEPTACLE OUTLET	
FLOOR DUPLEX RECEPTACLE OUTLET	
FLOOR SPECIAL-PURPOSE OUTLET	
UNDERFLOOR DUCT AND JUNCTION BOX FOR TRIPLE, DOUBLE, OR SINGLE DUCT SYSTEM AS INDICATED BY NUMBER OF PARALLEL LINES	

BUSDUCTS AND WIREWAYS

SERVICE, FEEDER, OR PLUG-IN BUSWAY	B B B
CABLE THROUGH LADDER OR CHANNEL	C C C
WIREWAY	W W W

SWITCH OUTLETS

SINGLE-POLE SWITCH	S
DOUBLE-POLE SWITCH	S_2
THREE-WAY SWITCH	S_3
FOUR-WAY SWITCH	S_4
AUTOMATIC DOOR SWITCH	S_D
KEY-OPERATED SWITCH	S_K
CIRCUIT BREAKER	S_{CB}
WEATHERPROOF CIRCUIT BREAKER	S_{WCB}
DIMMER	S_{DM}
REMOTE CONTROL SWITCH	S_{RC}
WEATHERPROOF SWITCH	S_{WP}
FUSED SWITCH	S_F
WEATHERPROOF FUSED SWITCH	S_{WF}
TIME SWITCH	S_T
CEILING PULL SWITCH	S
SWITCH AND SINGLE RECEPTACLE	S
SWITCH AND DOUBLE RECEPTACLE	S
A STANDARD SYMBOL WITH AN ADDED LOWERCASE SUBSCRIPT LETTER IS USED TO DESIGNATE A VARIATION IN STANDARD EQUIPMENT	a.b / a.b / $S_{a.b}$

. . . ELECTRICAL SYMBOLS

COMMERCIAL AND INDUSTRIAL SYSTEMS

PAGING SYSTEM DEVICE	
FIRE ALARM SYSTEM DEVICE	
COMPUTER DATA SYSTEM DEVICE	
PRIVATE TELEPHONE SYSTEM DEVICE	
SOUND SYSTEM	
FIRE ALARM CONTROL PANEL	FACP

SIGNALING SYSTEM OUTLETS FOR RESIDENTIAL SYSTEMS

PUSHBUTTON	
BUZZER	
BELL	
BELL AND BUZZER COMBINATION	
COMPUTER DATA OUTLET	
BELL RINGING TRANSFORMER	BT
ELECTRIC DOOR OPENER	D
CHIME	CH
TELEVISION OUTLET	TV
THERMOSTAT	T

UNDERGROUND ELECTRICAL DISTRIBUTION OR ELECTRICAL LIGHTING SYSTEMS

MANHOLE	M
HANDHOLE	H
TRANSFORMER-MANHOLE OR VAULT	TM
TRANSFORMER PAD	TP
UNDERGROUND DIRECT BURIAL CABLE	
UNDERGROUND DUCT LINE	
STREET LIGHT STANDARD FED FROM UNDERGROUND CIRCUIT	

ABOVE-GROUND ELECTRICAL DISTRIBUTION OR LIGHTING SYSTEMS

POLE	
STREET LIGHT AND BRACKET	
PRIMARY CIRCUIT	
SECONDARY CIRCUIT	
DOWN GUY	
HEAD GUY	
SIDEWALK GUY	
SERVICE WEATHERHEAD	

PANEL CIRCUITS AND MISCELLANEOUS

LIGHTING PANEL	
POWER PANEL	
WIRING—CONCEALED IN CEILING OR WALL	
WIRING—CONCEALED IN FLOOR	
WIRING EXPOSED	
HOMERUN TO PANELBOARD Indicate number of circuits by number of arrows. Any circuit without such designation indicates a two-wire circuit. For a greater number of wires indicate as follows: (3 wires) (4 wires), etc.	
FEEDERS Use heavy lines and designate by number corresponding to listing in feeder schedule	
WIRING TURNED UP	
WIRING TURNED DOWN	
GENERATOR	G
MOTOR	M
INSTRUMENT (SPECIFY)	I
TRANSFORMER	T
CONTROLLER	
EXTERNALLY-OPERATED DISCONNECT SWITCH	
PULL BOX	

PLUMBING SYMBOLS . . .

FIXTURES . . .	
STANDARD BATHTUB	
OVAL BATHTUB	
WHIRLPOOL BATH	
SHOWER STALL	
SHOWER HEAD	
TANK-TYPE WATER CLOSET	
WALL-MOUNTED WATER CLOSET	
FLOOR-MOUNTED WATER CLOSET	
LOW-PROFILE WATER CLOSET	
BIDET	
WALL-MOUNTED URINAL	
FLOOR-MOUNTED URINAL	
TROUGH-TYPE URINAL	
WALL-MOUNTED LAVATORY	
PEDESTAL LAVATORY	
BUILT-IN LAVATORY	
WHEELCHAIR LAVATORY	
CORNER LAVATORY	
FLOOR DRAIN	
FLOOR SINK	

. . . FIXTURES	
LAUNDRY TRAY	
BUILT-IN SINK	
DOUBLE OR TRIPLE BUILT-IN SINK	
COMMERCIAL KITCHEN SINK	
SERVICE SINK	SS
CLINIC SERVICE SINK	
FLOOR-MOUNTED SERVICE SINK	
DRINKING FOUNTAIN	DF
WATER COOLER	
HOT WATER TANK	HW T
WATER HEATER	WH
METER	M
HOSE BIBB	HB
GAS OUTLET	G
GREASE SEPARATOR	G
GARAGE DRAIN	
FLOOR DRAIN WITH BACKWATER VALVE	

PIPING . . .	
SOIL, WASTE, OR LEADER—ABOVE-GRADE	————
SOIL, WASTE, OR LEADER—BELOW-GRADE	— — — —
VENT	- - - - - - - - -
COMBINATION WASTE AND VENT	—— SV ——
STORM DRAIN	—— SD ——
COLD WATER	—·——·——

. . . PIPING	
CHILLED DRINKING WATER SUPPLY	—— DWS ——
CHILLED DRINKING WATER RETURN	—— DWR ——
HOT WATER	—— - - —— - - ——
HOT WATER RETURN	—— - - - —— - - - ——
SANITIZING HOT WATER SUPPLY (180°F)	
SANITIZING HOT WATER RETURN (180°F)	
DRY STANDPIPE	—— DSP ——
COMBINATION STANDPIPE	—— CSP ——
MAIN SUPPLIES SPRINKLER	—— S ——
BRANCH AND HEAD SPRINKLER	——o——o——
GAS—LOW PRESSURE	—— G —— G ——
GAS—MEDIUM PRESSURE	—— MG ——
GAS—HIGH PRESSURE	—— HG ——
COMPRESSED AIR	—— A ——
OXYGEN	—— O ——
NITROGEN	—— N ——
HYDROGEN	—— H ——
HELIUM	—— HE ——
ARGON	—— AR ——
LIQUID PETROLEUM GAS	—— LPG ——
INDUSTRIAL WASTE	—— INW ——
CAST IRON	—— CI ——
CULVERT PIPE	—— CP ——
CLAY TILE	—— CT ——
DUCTILE IRON	—— DI ——
REINFORCED CONCRETE	—— RCP ——
DRAIN—OPEN TILE OR AGRICULTURAL TILE	═ ═ ═ ═

. . . PLUMBING SYMBOLS

PIPE FITTING AND VALVE SYMBOLS

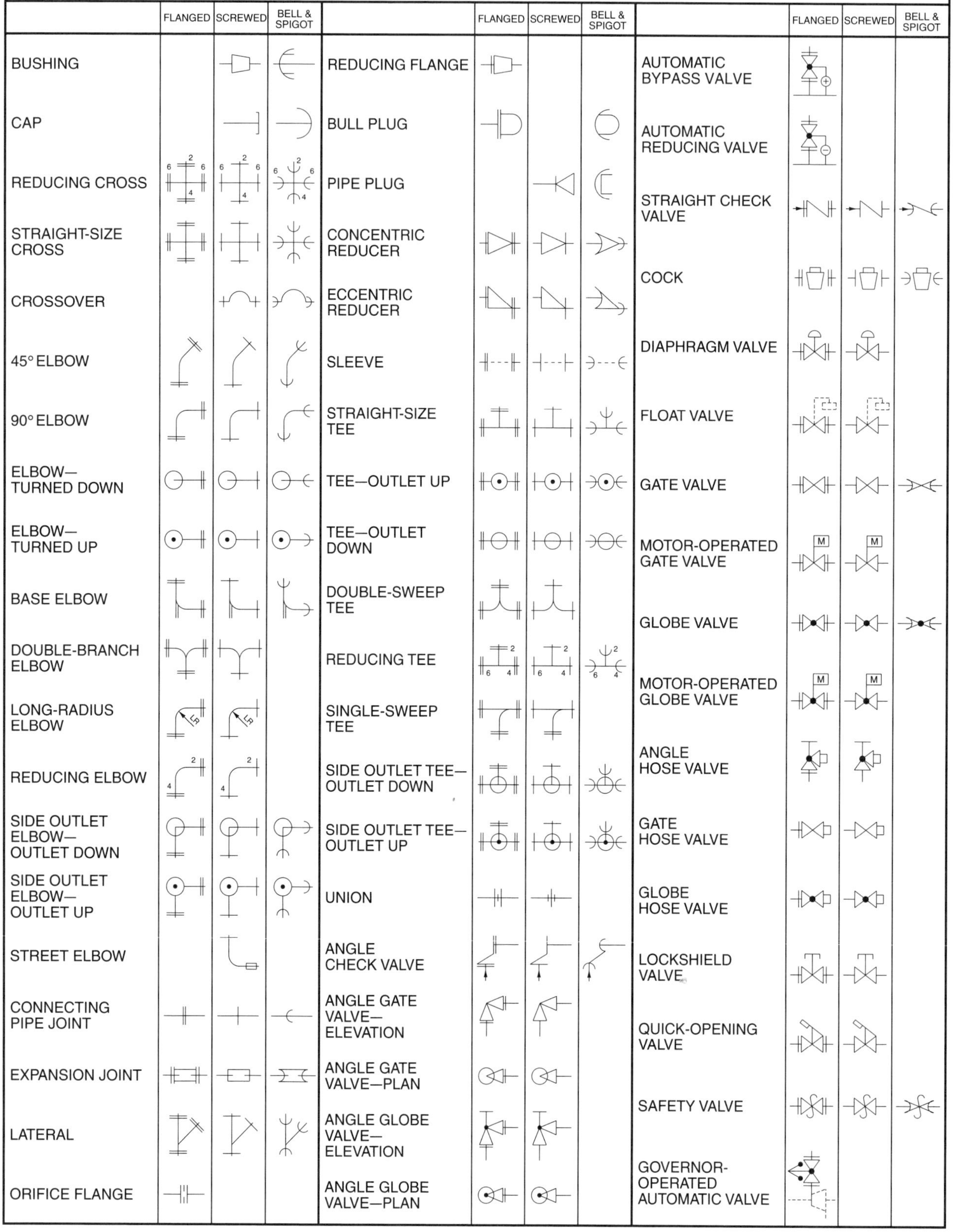

	FLANGED	SCREWED	BELL & SPIGOT
BUSHING			
CAP			
REDUCING CROSS			
STRAIGHT-SIZE CROSS			
CROSSOVER			
45° ELBOW			
90° ELBOW			
ELBOW—TURNED DOWN			
ELBOW—TURNED UP			
BASE ELBOW			
DOUBLE-BRANCH ELBOW			
LONG-RADIUS ELBOW			
REDUCING ELBOW			
SIDE OUTLET ELBOW—OUTLET DOWN			
SIDE OUTLET ELBOW—OUTLET UP			
STREET ELBOW			
CONNECTING PIPE JOINT			
EXPANSION JOINT			
LATERAL			
ORIFICE FLANGE			
REDUCING FLANGE			
BULL PLUG			
PIPE PLUG			
CONCENTRIC REDUCER			
ECCENTRIC REDUCER			
SLEEVE			
STRAIGHT-SIZE TEE			
TEE—OUTLET UP			
TEE—OUTLET DOWN			
DOUBLE-SWEEP TEE			
REDUCING TEE			
SINGLE-SWEEP TEE			
SIDE OUTLET TEE—OUTLET DOWN			
SIDE OUTLET TEE—OUTLET UP			
UNION			
ANGLE CHECK VALVE			
ANGLE GATE VALVE—ELEVATION			
ANGLE GATE VALVE—PLAN			
ANGLE GLOBE VALVE—ELEVATION			
ANGLE GLOBE VALVE—PLAN			
AUTOMATIC BYPASS VALVE			
AUTOMATIC REDUCING VALVE			
STRAIGHT CHECK VALVE			
COCK			
DIAPHRAGM VALVE			
FLOAT VALVE			
GATE VALVE			
MOTOR-OPERATED GATE VALVE			
GLOBE VALVE			
MOTOR-OPERATED GLOBE VALVE			
ANGLE HOSE VALVE			
GATE HOSE VALVE			
GLOBE HOSE VALVE			
LOCKSHIELD VALVE			
QUICK-OPENING VALVE			
SAFETY VALVE			
GOVERNOR-OPERATED AUTOMATIC VALVE			

HVAC SYMBOLS

EQUIPMENT SYMBOLS

EXPOSED RADIATOR	
RECESSED RADIATOR	
FLUSH ENCLOSED RADIATOR	
PROJECTING ENCLOSED RADIATOR	
UNIT HEATER (PROPELLER)—PLAN	
UNIT HEATER (CENTRIFUGAL)—PLAN	
UNIT VENTILATOR—PLAN	
STEAM	
DUPLEX STRAINER	
PRESSURE-REDUCING VALVE	
AIR LINE VALVE	
STRAINER	
THERMOMETER	
PRESSURE GAUGE AND COCK	
RELIEF VALVE	
AUTOMATIC 3-WAY VALVE	
AUTOMATIC 2-WAY VALVE	
SOLENOID VALVE	S

DUCTWORK

DUCT (1ST FIGURE, WIDTH; 2ND FIGURE, DEPTH)	12 X 20
DIRECTION OF FLOW	→
FLEXIBLE CONNECTION	
DUCTWORK WITH ACOUSTICAL LINING	
FIRE DAMPER WITH ACCESS DOOR	FD, AD
MANUAL VOLUME DAMPER	VD
AUTOMATIC VOLUME DAMPER	
EXHAUST, RETURN OR OUTSIDE AIR DUCT—SECTION	20 X 12
SUPPLY DUCT—SECTION	20 X 12
CEILING DIFFUSER SUPPLY OUTLET	20" DIA CD 1000 CFM
CEILING DIFFUSER SUPPLY OUTLET	20 X 12 CD 700 CFM
LINEAR DIFFUSER	96 X 6-LD 400 CFM
FLOOR REGISTER	20 X 12 FR 700 CFM
TURNING VANES	
FAN AND MOTOR WITH BELT GUARD	
LOUVER OPENING	20 X 12-L 700 CFM

HEATING PIPING

HIGH-PRESSURE STEAM	——HPS——
MEDIUM-PRESSURE STEAM	——MPS——
LOW-PRESSURE STEAM	——LPS——
HIGH-PRESSURE RETURN	——HPR——
MEDIUM-PRESSURE RETURN	——MPR——
LOW-PRESSURE RETURN	——LPR——
BOILER BLOW OFF	——BD——
CONDENSATE OR VACUUM PUMP DISCHARGE	——VPD——
FEEDWATER PUMP DISCHARGE	——PPD——
MAKEUP WATER	——MU——
AIR RELIEF LINE	——V——
FUEL OIL SUCTION	——FOS——
FUEL OIL RETURN	——FOR——
FUEL OIL VENT	——FOV——
COMPRESSED AIR	——A——
HOT WATER HEATING SUPPLY	——HW——
HOT WATER HEATING RETURN	——HWR——

AIR CONDITIONING PIPING

REFRIGERANT LIQUID	——RL——
REFRIGERANT DISCHARGE	——RD——
REFRIGERANT SUCTION	——RS——
CONDENSER WATER SUPPLY	——CWS——
CONDENSER WATER RETURN	——CWR——
CHILLED WATER SUPPLY	——CHWS——
CHILLED WATER RETURN	——CHWR——
MAKEUP WATER	——MU——
HUMIDIFICATION LINE	——H——
DRAIN	——D——

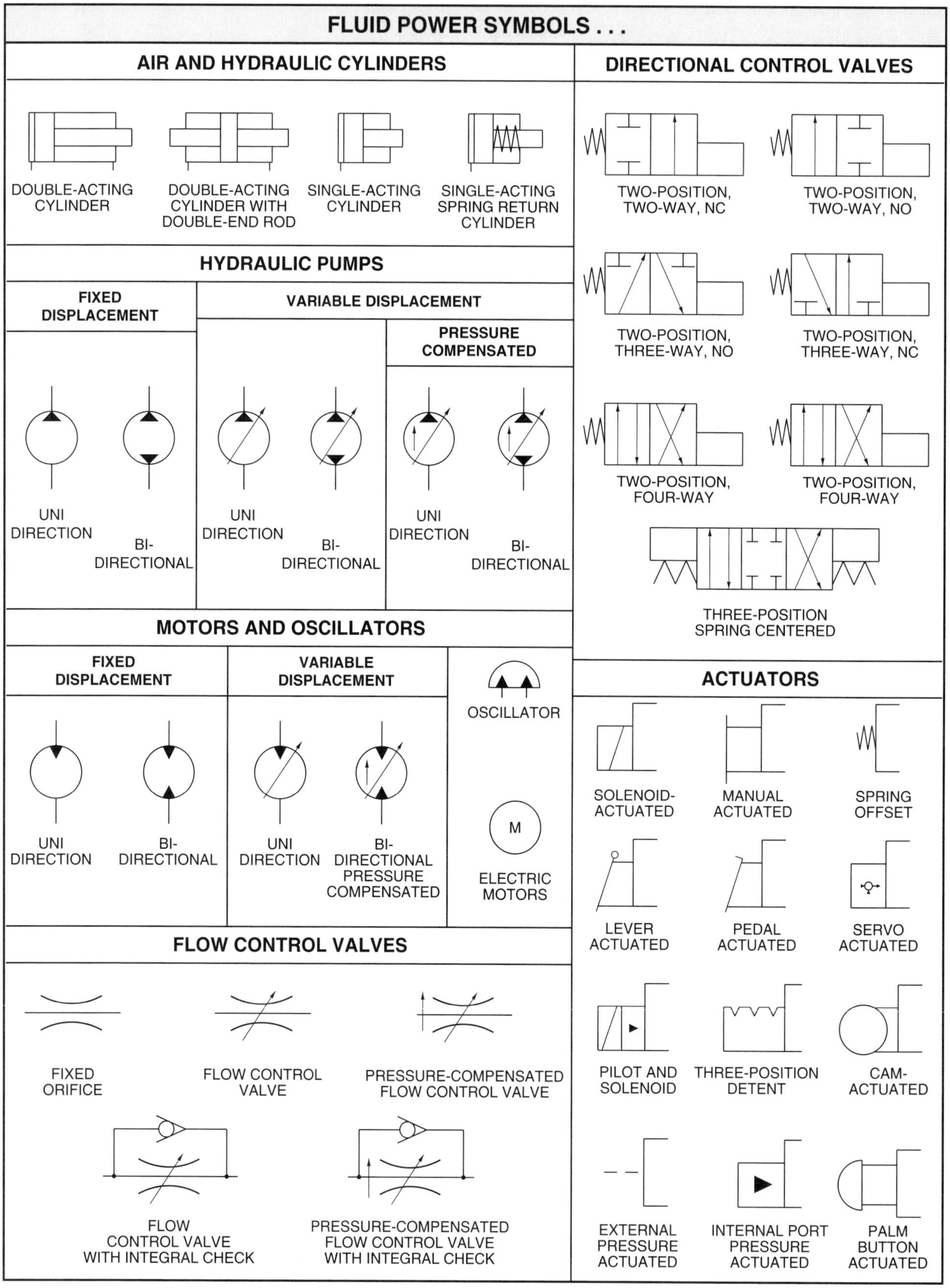
FLUID POWER SYMBOLS . . .
AIR AND HYDRAULIC CYLINDERS
DOUBLE-ACTING CYLINDER
DOUBLE-ACTING CYLINDER WITH DOUBLE-END ROD
SINGLE-ACTING CYLINDER
SINGLE-ACTING SPRING RETURN CYLINDER
HYDRAULIC PUMPS
FIXED DISPLACEMENT
VARIABLE DISPLACEMENT
PRESSURE COMPENSATED
UNI DIRECTION
BI-DIRECTIONAL
UNI DIRECTION
BI-DIRECTIONAL
UNI DIRECTION
BI-DIRECTIONAL
MOTORS AND OSCILLATORS
FIXED DISPLACEMENT
VARIABLE DISPLACEMENT
OSCILLATOR
UNI DIRECTION
BI-DIRECTIONAL
UNI DIRECTION
BI-DIRECTIONAL PRESSURE COMPENSATED
M
ELECTRIC MOTORS
FLOW CONTROL VALVES
FIXED ORIFICE
FLOW CONTROL VALVE
PRESSURE-COMPENSATED FLOW CONTROL VALVE
FLOW CONTROL VALVE WITH INTEGRAL CHECK
PRESSURE-COMPENSATED FLOW CONTROL VALVE WITH INTEGRAL CHECK
DIRECTIONAL CONTROL VALVES
TWO-POSITION, TWO-WAY, NC
TWO-POSITION, TWO-WAY, NO
TWO-POSITION, THREE-WAY, NO
TWO-POSITION, THREE-WAY, NC
TWO-POSITION, FOUR-WAY
TWO-POSITION, FOUR-WAY
THREE-POSITION SPRING CENTERED
ACTUATORS
SOLENOID-ACTUATED
MANUAL ACTUATED
SPRING OFFSET
LEVER ACTUATED
PEDAL ACTUATED
SERVO ACTUATED
PILOT AND SOLENOID
THREE-POSITION DETENT
CAM-ACTUATED
EXTERNAL PRESSURE ACTUATED
INTERNAL PORT PRESSURE ACTUATED
PALM BUTTON ACTUATED

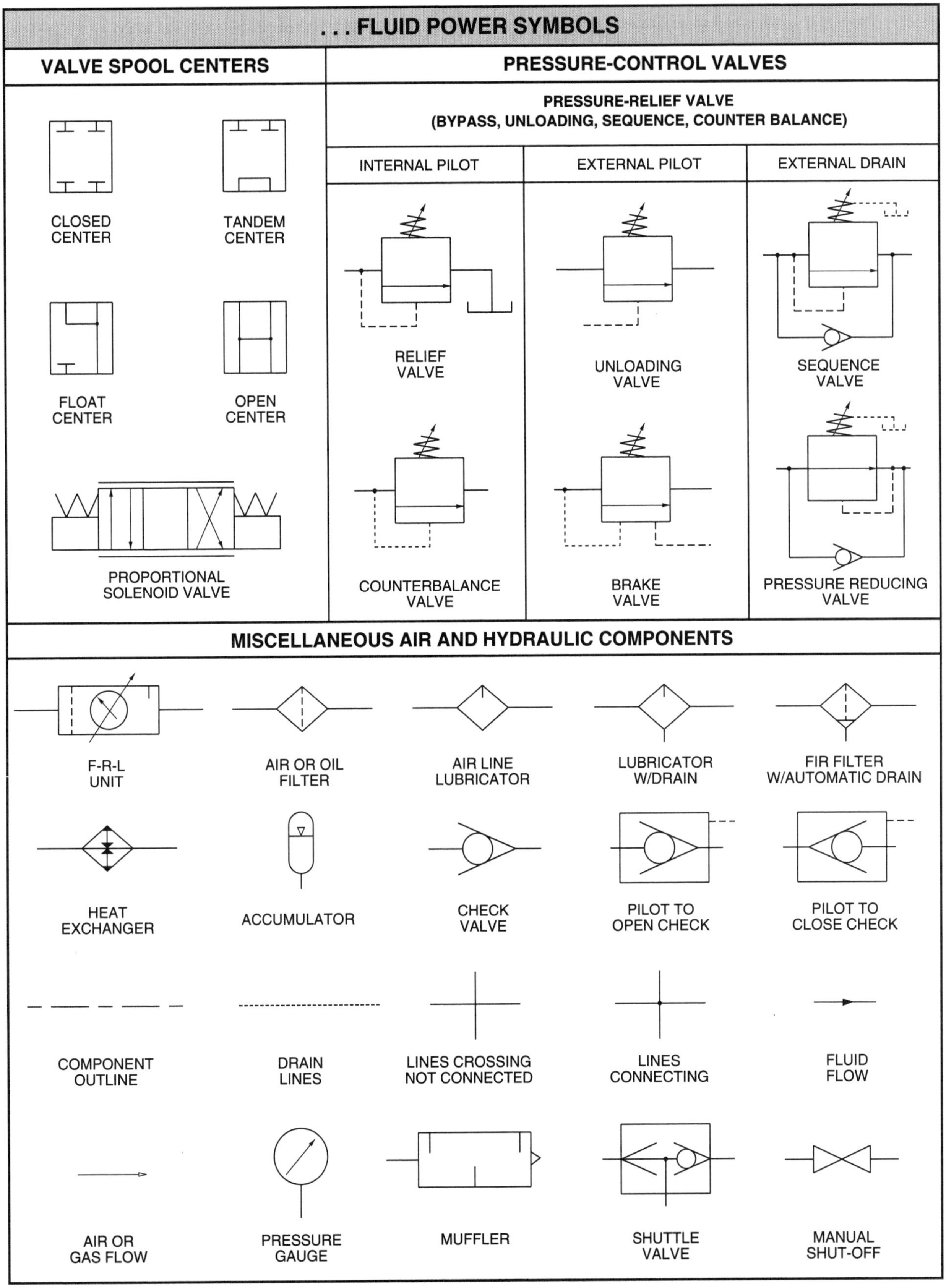
. . . FLUID POWER SYMBOLS
VALVE SPOOL CENTERS
CLOSED CENTER
TANDEM CENTER
FLOAT CENTER
OPEN CENTER
PROPORTIONAL SOLENOID VALVE
PRESSURE-CONTROL VALVES
PRESSURE-RELIEF VALVE
(BYPASS, UNLOADING, SEQUENCE, COUNTER BALANCE)
INTERNAL PILOT
EXTERNAL PILOT
EXTERNAL DRAIN
RELIEF VALVE
UNLOADING VALVE
SEQUENCE VALVE
COUNTERBALANCE VALVE
BRAKE VALVE
PRESSURE REDUCING VALVE
MISCELLANEOUS AIR AND HYDRAULIC COMPONENTS
F-R-L UNIT
AIR OR OIL FILTER
AIR LINE LUBRICATOR
LUBRICATOR W/DRAIN
FIR FILTER W/AUTOMATIC DRAIN
HEAT EXCHANGER
ACCUMULATOR
CHECK VALVE
PILOT TO OPEN CHECK
PILOT TO CLOSE CHECK
COMPONENT OUTLINE
DRAIN LINES
LINES CROSSING NOT CONNECTED
LINES CONNECTING
FLUID FLOW
AIR OR GAS FLOW
PRESSURE GAUGE
MUFFLER
SHUTTLE VALVE
MANUAL SHUT-OFF

AMERICAN WELDING SOCIETY
Welding Symbol Chart

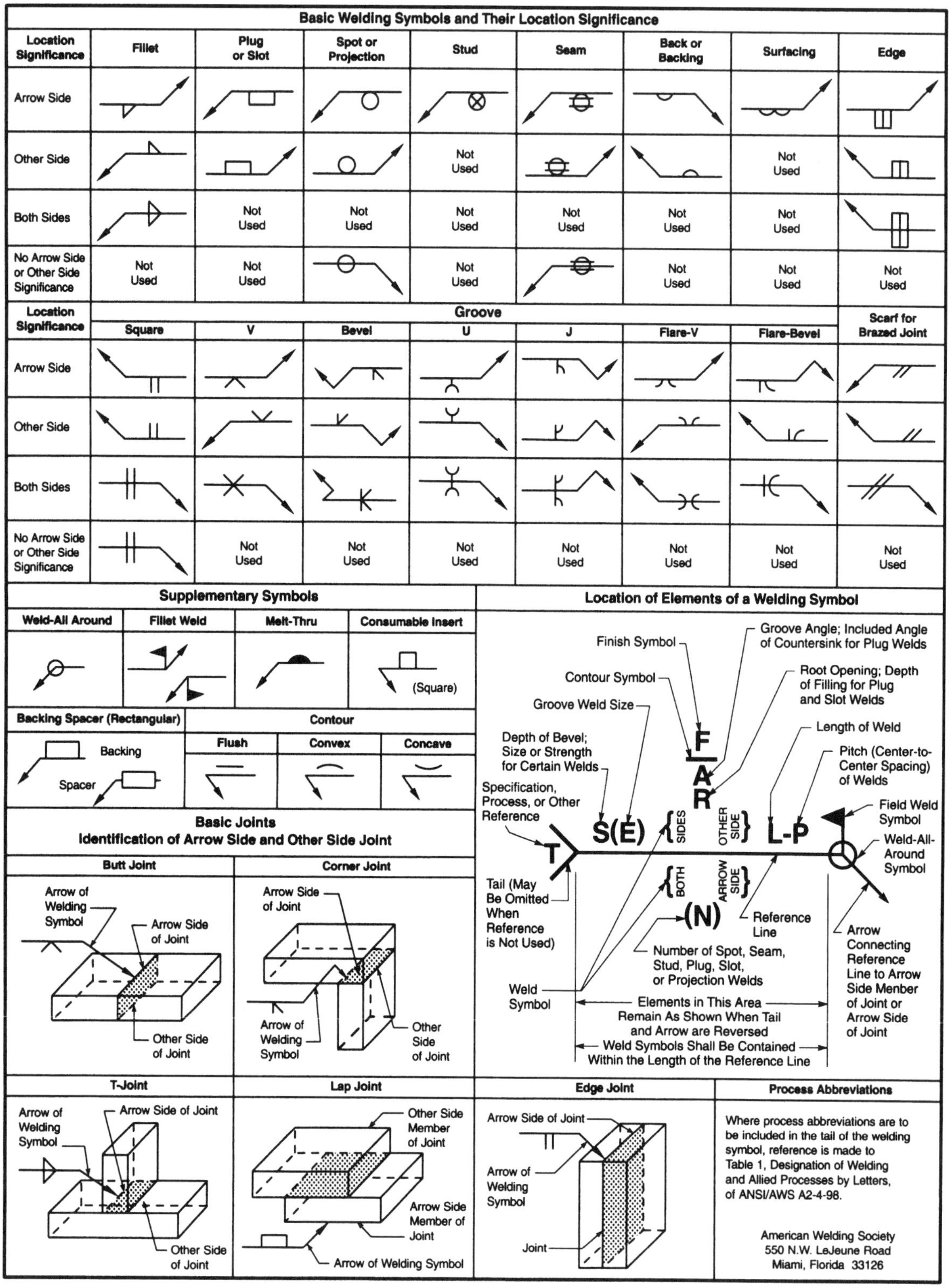

ABBREVIATIONS . . .

Term	Abbreviation
A	
above	ABV
access	ACS
access panel	AP
acoustic	AC or ACST
acoustical plaster ceiling	APC
acoustical tile	AT. or ACT.
adjacent	ADJ
adjustable	ADJT or ADJ
aggregate	AGG or AGGR
air circulating	ACIRC
air conditioner	AIR COND
air conditioning	A/C or AIR COND
alloy	ALY
alloy steel	ALY STL
alternate	ALTN
alternating current	AC
aluminum	AL
ambient	AMB
American National Standard	AMER NATL STD
American National Standards Institute	ANSI
American Steel Wire Gauge	ASWG
American Wire Gauge	AWG
ampere	A or AMP
anchor	AHR
anchor bolt	AB
appearance	APP
apartment	APT.
approximate	APX or APPROX
architectural	ARCH.
architecture	ARCH.
area	A
area drain	AD
as drawn	AD
asphalt	ASPH
asphalt roof shingles	ASPHRS
asphalt tile	AT.
automatic	AUTO.
auxiliary	AUX
avenue	AVE
azimuth	AZ
B	
barrier	BARR
barrier, moisture vapor-proof	BMVP
barrier, waterproof	BWP
basement	BSMT
bathroom	B
bathtub	BT
batten	BATT
beam	BM
bearing	BRG
bearing plate	BPL or BRG PL
bedroom	BR
below	BLW
bench mark	BM
beveled wood siding	BWS
bituminous	BIT
blocking	BLKG
blueprint	BP
board	BD
board foot	BF or BD FT
boiler	BLR
bookcase	BC
book shelves	BK SH
boulevard	BLVD
boundary	BDRY
brass	BRS
breaker	BRKR
brick	BRK
British thermal unit	BTU
bronze	BRZ
broom closet	BC
building	BLDG or BL
building line	BL
built-in	BLTIN
built up roofing	BUR
C	
cabinet	CAB.
cable	CA
canopy	CAN.
caulking	CK or CLKG
cantilever	CANV
cased opening	CO
casing	CSG
cast iron	CI
cast-iron pipe	CIP
cast steel	CS
cast stone	CST or CS
catch basin	CB
caulked joint	CLKJ
cavity	CAV
ceiling	CLG
cellar	CEL
cement	CEM
cement floor	CF
cement mortar	CEM MORT
center	CTR
centerline	CL
center matched	CM
center-to-center	C TO C
central	CTL
ceramic	CER
ceramic tile	CT
ceramic-tile base	CTB
ceramic-to-metal (seal)	CERMET
chamfer	CHAM or CHMFR
channel	CHAN
check valve	CV
chimney	CHM
chord	CHD
cinder block	CINBL
circle	CIR
circuit	CKT
circuit breaker	CB or CIR BKR
circuit interrupter	CI
circumference	CRCMF
cleanout	CO
clear glass	CL GL
closet	C, CL, CLO, or CLOS
coaxial	COAX.
cold air	CA
cold-rolled	CR
cold-rolled steel	CRS
cold water	CW
collar beam	COL B
color code	CC
combination	COMB.
combustible	COMBL
combustion	COMB.
common	COM
composition	COMP
concrete	CONC
concrete block	CCB or CONC BLK
concrete floor	CCF, CONC FLR, or CONC FL
concrete masonry unit	CMU
concrete pipe	CP
concrete splash block	CSB
condenser	COND
conductor	CNDCT
conduit	CND
construction	CONSTR
construction joint	CJ or CONSTR JT
continuous	CONT
contour	CTR
contract	CONTR or CONT
contractor	CONTR
control joint	CJ or CLJ
conventional	CVNTL
copper	CU
corner	COR
cornice	COR
corrugate	CORR
counter	CNTR
county	CO
cubic	CU
cubic feet	CFT or CU FT
cubic foot per minute	CFM
cubic foot per second	CFS
cubic inch	CU iN.
cubic yard	CU YD
current	CUR
cutoff	CO
cutoff valve	COV
cut out	CO
D	
damper	DMPR
datum	DAT
decibel	DB

. . . ABBREVIATIONS . . .

Term	Abbreviation
degree	DEG
depth	DP
design	DSGN
detail	DTL or DET
diagonal	DIAG
diagram	DIAG
dimension	DIM.
dimmer	DIM. or DMR
dining room	DR or DNG RM
direct current	DC
direction	DIR
disconnect	DISC.
disconnect switch	DS
dishwasher	DW
distribution panel	DPNL
ditto	DO.
door	DR
door stop	DST
door switch	DSW
dormer	DRM
double-acting	DA or DBL ACT
double-hung window	DHW
double-pole double-throw	DPDT
double-pole double-throw switch	DPDT SW
double-pole single-throw	DPST
double-pole single-throw switch	DPST SW
double-pole switch	DP SW
double-strength glass	DSG
down	DN or D
downspout	DS
dozen	DOZ
drain	D or DR
drain tile	DT
drawer	DWR
drawing	DWG
dryer	D
drywall	DW
dwelling	DWEL
E	
each	EA
east	E
elbow	ELB
electric or electrical	ELEC
electrical metallic tubing	EMT
electric operator	ELECT. OPR
electric panel	EP
electromechanical	ELMCH
elevation	ELEV
enamel	ENAM
end-to-end	E to E
entrance	ENTR
equipment	EQPT
equivalent	EQUIV
estimate	EST
example	EX
excavate	EXCA or EXC
exchange	EXCH
exhaust	EXH

Term	Abbreviation
exhaust vent	EXHV
existing	EXST
expanded metal	EM
expansion joint	EXP JT
exterior	EXT
exterior grade	EXT GR
F	
face brick	FB
faceplate	FP
Fahrenheit	°F
fiberboard, solid	FBDS
finish	FIN. or FNSH
finish all over	FAO
finish grade	FG
finish one side	FIS
finish two sides	F2S
finished floor	FIN. FLR, FIN. FL, or FNSH FL
firebrick	FBRK or FBCK
fireplace	FPL or FP
fireproof	FP or FPRF
fire resistant	FRES
fixed transom	FTR
fixed window	FX WDW
fixture	FIX. or FXTR
flashing	FIG or FL
floor	FLR or FL
floor drain	FD
flooring	FLR or FLG
fluorescent	FLUR or FLUOR
flush	FL
footing	FTG
foundation	FND or FDN
frame	FR
frostproof hose bibb	FPHB
full scale	FSC
full size	FS
furnace	FURN
furred ceiling	FC
furring	FUR
fuse	FU
fuse block	FB
fusebox	FUBX
fuseholder	FUHLR
fusible	FSBL
G	
gallon	GAL.
gallon per hour	GPH
gallon per minute	GPM
galvanized iron	GI or GALVI
galvanized steel	GS or GALVS
garage	GAR.
gas	G
gate valve	GTV
gauge	GA
general contractor	GEN CONT
girder	G
glass	GL
glass block	GLB or GL BL

Term	Abbreviation
glaze	GLZ
glued laminated	GLULAM
grade	GR
grade line	GL
gravel	GVL
grill	G
gross weight	GRWT
ground	GRD
ground (outlet)	G
ground-fault circuit interrupter	GFCI
ground-fault interrupter	GFI
gypsum	GYP
gypsum board	GYP BD
gypsum-plaster ceiling	GPC
gypsum-plaster wall	GPW
gypsum sheathing board	GSB
gypsum wallboard	GWB
H	
hardboard	HBD
hardware	HDW
header	HDR
heat	HT
heated	HTD
heater	HTR
heating	HTG
heating, ventilating, and air conditioning	HVAC
height	HGT
hexagon	HEX.
high density overlay	HDO
high point	HPT
highway	HWY
hinge	HNG
hollow-core	HC
hollow metal door	HMD
honeycomb	HNYCMB
horizontal	HOR or HORZ
horsepower	HP
hose bibb	HB
hot air	HA
hot water	HW
humidity	HMD
I	
illuminate	ILLUM
incandescent	INCAND
inch	IN.
inch per second	IPS
inside diameter	ID
install	INSTL
insulation	INS or INSUL
interior	INT
iron	I
J	
jamb	JB or JMB
joint	JT
joist	J
K	
kiln dried	KD
kitchen	K, KT, or KIT.

. . . ABBREVIATIONS . . .

Term	Abbreviation
L	
laminate	LAM
laminated veneer lumber	LVL
landing	LDG
lateral	LATL
lath	LTH
laundry	LAU
laundry tray	LT
lavatory	LAV
leader	L
left hand	LH
length	L, LG, or LGTH
level	LVL
library	LIB
living room	LR
light	LT
light switch	LT SW
limestone	LMS or LS
linen closet	L CL
lining	LN
linoleum	LINO
linoleum floor	LF or LINO FLR
lintel	LNTL
living room	LR
local	LCL
louver	LVR or LV
low point	LP
lumber	LBR
M	
main	MN
makeup	MKUP
manufactured	MFD
marble	MRB or MR
masonry	MSNRY
masonry opening	MO
material	MTL or MATL
maximum	MAX
median	MDN
medicine cabinet	MC
medium	MDM
medium density overlay	MDO
meridian	MER
metal	MET.
metal anchor	MA
metal door	METD
metal flashing	METF
metal threshold	MT
mineral	MNRL
minimum	MIN
mirror	MIR
miscellaneous	MISC
miter	MIT
mixture	MIX.
modular	MOD
molding	MLD or MLDG
mortar	MOR
N	
National Electrical Code®	NEC®
National Electrical Safety Code	NESC
natural grade	NG
negative	(–) or NEG
noncombustible	NCOMBL
north	N
nosing	NOS
not to scale	NTS
O	
obscure glass	OBSC GL
octagon	OCT
on center	OC
one-pole	SP
opening	OPG or OPNG
open web joist	OJ, OW J, or OW JOIST
opposite	OPP
optional	OPT
ordinance	ORD
oriented strand board	OSB
outlet	OUT.
outside diameter	OD
out-to-out	O TO O
overall	OA
overcurrent	OC
overcurrent relay	OCR
overhead	OH or OVHD
P	
paint	PNT
panel	PNL
pantry	PAN.
parallel	PRL
parallel strand lumber	PSL
partition	PTN
passage	PASS.
penny (nails, etc.)	d
perimeter	PERIM
perpendicular	PERP
per square inch	PSI
phase	PH
piping	PP
plaster	PLAS or PL
plastered open	PO
plate	PL
plate glass	PG, PL GL, or PLGL
platform	PLAT
plumbing	PLBG
plywood	PLYWD
point	PT
point of beginning	POB
polyvinyl chloride	PVC
porch	P
pound	LB
power	PWR
power supply	PWR SPLY
precast	PRCST
prefabricated	PFB or PREFAB
prefinished	PFN
property	PROP.
property line	PL
pull switch	PS
pump	PMP
Q	
quadrant	QDRNT
quarry tile	QT
quarry tile base	QTB
quarry tile floor	QTF
quarter	QTR
quarter-round	¼RD
R	
radiator	RAD or RDTR
raised	RSD
random	RDM
range	R
receptacle	RCPT
recessed	REC
rectangle	RECT
redwood	RWD
reference	REF
reference line	REFL
reflected	REFLD
refrigerator	REF or REFR
register	REG or RGTR
reinforce or reinforcing	RE or REINF
reinforced concrete	RC
reinforcing steel	RST
reinforcing steel bar	REBAR
required	REQD
retaining	RETG
revision	REV
revolution per minute	RPM
revolution per second	RPS
right hand	RH
riser	R
road	RD
roof	RF
roof drain	RD
roofing	RFG
room	RM or R
rough	RGH
rough opening	RO or RGH OPNG
rough-sawn	RS
round	RND
rubber	RBR
rubber tile	RBT or R TILE
rustproof	RSTPF
S	
safety	SAF
sanitary	S
S-beam	S
scale	SC
schedule	SCH or SCHED
screen	SCN, SCR, or SCRN
screen door	SCD
screw	SCR
scuttle	S
section	SEC or SECT.
select	SEL

. . . ABBREVIATIONS	
Term	**Abbreviation**
self-cleaning	SLFCLN
self-closing	SELF CL
service	SERV or SVCE
sewer	SEW.
sheathing	SHTH or SHTHG
sheet	SHT or SH
sheeting	SH
sheet metal	SM
shelf and rod	SH&RD
shelving	SH or SHELV
shingle	SHGL
shower	SH
shower and toilet	SH & T
shower drain	SD
shutter	SHTR
sidelight	SI LT
sill cock	SC
single-phase	1PH
single-pole	SP
single-pole double-throw	SPDT
single-pole double throw switch	SPDT SW
single-pole single-throw	SPST
single-pole single-throw switch	SPST SW
single-pole switch	SP SW
single-strength glass	SSG
single-throw	ST
sink	SK or S
skylight	SLT
sliding door	SLD or SL DR
slope	SLP
soffit	SF
soil pipe	SP
soil stack	SSK
solid core	SC
soundproof	SNDPRF
south	S
specific	SP
specification	SPEC
splash block	SB
square	SQ
square feet	SQ FT
square inch	SQ IN.
square yard	SQ YD
stack	STK
stained	STN
stainless steel	SST
stairs	ST
stairway	STWY
standard	STD
steel	ST or STL
steel sash	SS
stone	STN
storage	STO or STG
street	ST or STR
structural	STRL
Structural Clay Products Research Foundation	SCR
structural clay tile	SCT
structural glass	SG
supply	SPLY
survey	SURV
suspended	SUSP
switch	SW or S
T	
telephone	TEL
television	TV
temperature	TEMP
tempered plate glass	TEM PL GL
terra cotta	TC
terazzo	TZ or TER
thermostat	THERMO
thick	THK
threshold	TH
tile base	TB
tile drain	TD
tile floor	TF
timber	TMBR
toilet	T
tongue-and-groove	T & G
township	T
tread	TR or T
typical	TYP
U	
underground	UGND
unexcavated	UNEXC
unfinished	UNFIN or UNF
unit heater	UH
unless otherwise specified	UOS
untreated	UTRTD
utility	U or UTIL
utility room	UR or U RM
V	
vacuum	VAC
valley	VAL
valve	V
variance	VAR
vent	V
vent hole	VH
ventilate	VEN
ventilating equipment	VE
vent pipe	VP
vent stack	VS
vertical	VERT
vestibule	VEST.
vinyl tile	VT or V TILE
vitreous tile	VIT TILR
void	VD
volt	V
voltage	V
voltage drop	VD
volt amp	VA
volume	VOL
W	
wainscot	WSCT, WAIN., or WA
walk in closet	WIC
wall	W
wallboard	WLB
wall receptacle	WR
warm air	WA
washing machine	WM
water	WTR or W
water closet	WC
water heater	WH
water line	WL
water meter	WM
waterproof	WTRPRF
water-resistant	WR
watt	W
weatherproof	WTHPRF or WP
weather-resistant	WR
weather stripping	WS
weep hole	WH
welded wire fabric	WWF
west	W
white pine	WP
wide	W
wide flange	W or WF
window	WDO
wood	WD
wood frame	WF
wrought iron	WI
Y	
yard	YD
yellow pine	YP
Z	
Zone	Z

DECIMAL EQUIVALENTS OF AN INCH							
Fraction	**Decimal**	**Fraction**	**Decimal**	**Fraction**	**Decimal**	**Fraction**	**Decimal**
1/64	0.015625	17/64	0.265625	33/64	0.515625	49/64	0.765625
1/32	0.03125	9/32	0.28125	17/32	0.53125	25/32	0.78125
3/64	0.046875	19/64	0.296875	35/64	0.546875	51/64	0.796875
1/16	0.0625	5/16	0.3125	9/16	0.5625	13/16	0.8125
5/64	0.078125	21/64	0.328125	37/64	0.578125	53/64	0.828125
3/32	0.09375	11/32	0.34375	19/32	0.59375	27/32	0.84375
7/64	0.109375	23/64	0.359375	39/64	0.609375	55/64	0.859375
1/8	0.125	3/8	0.375	5/8	0.625	7/8	0.875
9/64	0.140625	25/64	0.390625	41/64	0.640625	57/64	0.890625
5/32	0.15625	13/32	0.40625	21/32	0.65625	29/32	0.90625
11/64	0.171875	27/64	0.421875	43/64	0.671875	59/64	0.921875
3/16	0.1875	7/16	0.4375	11/16	0.6875	15/16	0.9375
13/64	0.203125	29/64	0.453125	45/64	0.703125	61/64	0.953125
7/32	0.21875	15/32	0.46875	23/32	0.71875	31/32	0.96875
15/64	0.234375	31/64	0.484375	47/64	0.734375	63/64	0.984375
1/4	0.250	1/2	0.500	3/4	0.750	1	1.000

DECIMAL AND METRIC EQUIVALENTS		
Fractions	**Decimal Inches**	**Millimeters**
1/16	0.0625	1.58
1/8	0.125	3.18
3/16	0.1875	4.76
1/4	0.250	6.35
5/16	0.3125	7.97
3/8	0.375	9.52
7/16	0.4375	11.11
1/2	0.500	12.70
9/16	0.5625	14.29
5/8	0.625	15.88
11/16	0.6875	17.46
3/4	0.750	19.05
13/16	0.8125	20.64
7/8	0.875	22.22
1	1.000	25.40

DECIMAL EQUIVALENTS OF A FOOT					
Inches	**Decimal Foot Equivalent**	**Inches**	**Decimal Foot Equivalent**	**Inches**	**Decimal Foot Equivalent**
1/16	0.0052	4 1/16	0.3385	8 1/16	0.6719
1/8	0.0104	4 1/8	0.3438	8 1/8	0.6771
3/16	0.0156	4 3/16	0.3490	8 3/16	0.6823
1/4	0.0208	4 1/4	0.3542	8 1/4	0.6875
5/16	0.0260	4 5/16	0.3594	8 5/16	0.6927
3/8	0.0313	4 3/8	0.3646	8 3/8	0.6979
7/16	0.0365	4 7/16	0.3698	8 7/16	0.7031
1/2	0.0417	4 1/2	0.3750	8 1/2	0.7083
9/16	0.0469	4 9/16	0.3802	8 9/16	0.7135
5/8	0.0521	4 5/8	0.3854	8 5/8	0.7188
11/16	0.0573	4 11/16	0.3906	8 11/16	0.7240
3/4	0.0625	4 3/4	0.3958	8 3/4	0.7292
13/16	0.0677	4 13/16	0.4010	8 13/16	0.7344
7/8	0.0729	4 7/8	0.4063	8 7/8	0.7396
15/16	0.0781	4 15/16	0.4115	8 15/16	0.7448
1	0.0833	5	0.4167	9	0.7500
1 1/16	0.0885	5 1/16	0.4219	9 1/16	0.7552
1 1/8	0.0938	5 1/8	0.4271	9 1/8	0.7604
1 3/16	0.0990	5 3/16	0.4323	9 3/16	0.7656
1 1/4	0.1042	5 1/4	0.4375	9 1/4	0.7708
1 5/16	0.1094	5 5/16	0.4427	9 5/16	0.7760
1 3/8	0.1146	5 3/8	0.4479	9 3/8	0.7813
1 7/16	0.1198	5 7/16	0.4531	9 7/16	0.7865
1 1/2	0.1250	5 1/2	0.4583	9 1/2	0.7917
1 9/16	0.1302	5 9/16	0.4635	9 9/16	0.7969
1 5/8	0.1354	5 5/8	0.4688	9 5/8	0.8021
1 11/16	0.1406	5 11/16	0.4740	9 11/16	0.8073
1 3/4	0.1458	5 3/4	0.4792	9 3/4	0.8125
1 13/16	0.1510	5 13/16	0.4844	9 13/16	0.8177
1 7/8	0.1563	5 7/8	0.4896	9 7/8	0.8229
1 15/16	0.1615	5 15/16	0.4948	9 15/16	0.8281
2	0.1667	6	0.5000	10	0.8333
2 1/16	0.1719	6 1/16	0.5052	10 1/16	0.8385
2 1/8	0.1771	6 1/8	0.5104	10 1/8	0.8438
2 3/16	0.1823	6 3/16	0.5156	10 3/16	0.8490
2 1/4	0.1875	6 1/4	0.5208	10 1/4	0.8542
2 5/16	0.1927	6 5/16	0.5260	10 5/16	0.8594
2 3/8	0.1979	6 3/8	0.5313	10 3/8	0.8646
2 7/16	0.2031	6 7/16	0.5365	10 7/16	0.8698
2 1/2	0.2083	6 1/2	0.5417	10 1/2	0.8750
2 9/16	0.2135	6 9/16	0.5469	10 9/16	0.8802
2 5/8	0.2188	6 5/8	0.5521	10 5/8	0.8854
2 11/16	0.2240	6 11/16	0.5573	10 11/16	0.8906
2 3/4	0.2292	6 3/4	0.5625	10 3/4	0.8958
2 13/16	0.2344	6 13/16	0.5677	10 13/16	0.9010
2 7/8	0.2396	6 7/8	0.5729	10 7/8	0.9063
2 15/16	0.2448	6 15/16	0.5781	10 15/16	0.9115
3	0.2500	7	0.5833	11	0.9167
3 1/16	0.2552	7 1/16	0.5885	11 1/16	0.9219
3 1/8	0.2604	7 1/8	0.5938	11 1/8	0.9271
3 3/16	0.2656	7 3/16	0.5990	11 3/16	0.9323
3 1/4	0.2708	7 1/4	0.6042	11 1/4	0.9375
3 5/16	0.2760	7 5/16	0.6094	11 5/16	0.9427
3 3/8	0.2813	7 3/8	0.6146	11 3/8	0.9479
3 7/16	0.2865	7 7/16	0.6198	11 7/16	0.9531
3 1/2	0.2917	7 1/2	0.6250	11 1/2	0.9583
3 9/16	0.2969	7 9/16	0.6302	11 9/16	0.9635
3 5/8	0.3021	7 5/8	0.6354	11 5/8	0.9688
3 11/16	0.3073	7 11/16	0.6406	11 11/16	0.9740
3 3/4	0.3125	7 3/4	0.6458	11 3/4	0.9792
3 13/16	0.3177	7 13/16	0.6510	11 13/16	0.9844
3 7/8	0.3229	7 7/8	0.6563	11 7/8	0.9896
3 15/16	0.3281	7 15/16	0.6615	11 15/16	0.9948
4	0.3333	7	0.6667	12	1.0000

U.S. CUSTOMARY SYSTEM

		Unit	Abbr	Equivalents
LENGTH 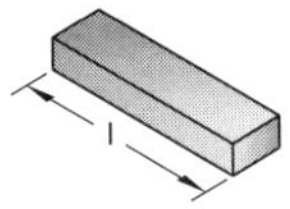		mile	mi	5280′, 320 rd, 1760 yd
		rod	rd	5.50 yd, 16.5′
		yard	yd	3′, 36″
		foot	ft *or* ′	12″, 0.333 yd
		inch	in. *or* ″	0.083′, 0.028 yd
AREA $A = l \times w$ 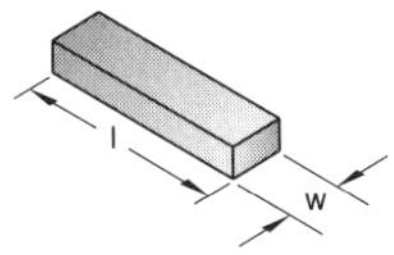		square mile	sq mi *or* mi^2	640 A, 102,400 sq rd
		acre	A	4840 sq yd, 43,560 sq ft
		square rod	sq rd *or* rd^2	30.25 sq yd, 0.00625 A
		square yard	sq yd *or* yd^2	1296 sq in., 9 sq ft
		square foot	sq ft *or* ft^2	144 sq in., 0.111 sq yd
		square inch	sq in. *or* in^2	0.0069 sq ft, 0.00077 sq yd
VOLUME $V = l \times w \times h$ 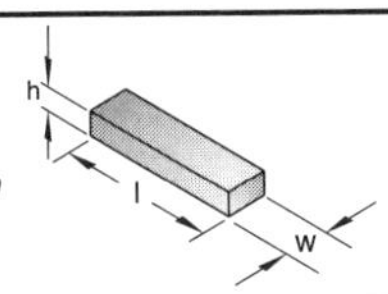		cubic yard	cu yd *or* yd^3	27 cu ft, 46,656 cu in.
		cubic foot	cu ft *or* ft^3	1728 cu in., 0.0370 cu yd
		cubic inch	cu in. *or* in^3	0.00058 cu ft, 0.000021 cu yd
CAPACITY WATER, FUEL, ETC. 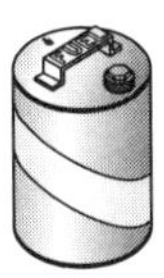	U.S. liquid measure	gallon	gal.	4 qt (231 cu in.)
		quart	qt	2 pt (57.75 cu in.)
		pint	pt	4 gi (28.875 cu in.)
		gill	gi	4 fl oz (7.219 cu in.)
		fluidounce	fl oz	8 fl dr (1.805 cu in.)
		fluidram	fl dr	60 min (0.226 cu in.)
		minim	min	1/6 fl dr (0.003760 cu in.)
VEGETABLES, GRAIN, ETC. 	U.S. dry measure	bushel	bu	4 pk (2150.42 cu in.)
		peck	pk	8 qt (537.605 cu in.)
		quart	qt	2 pt (67.201 cu in.)
		pint	pt	½ qt (33.600 cu in.)
DRUGS 	British imperial liquid and dry measure	bushel	bu	4 pk (2219.36 cu in.)
		peck	pk	2 gal. (554.84 cu in.)
		gallon	gal.	4 qt (277.420 cu in.)
		quart	qt	2 pt (69.355 cu in.)
		pint	pt	4 gi (34.678 cu in.)
		gill	gi	5 fl oz (8.669 cu in.)
		fluidounce	fl oz	8 fl dr (1.7339 cu in.)
		fluidram	fl dr	60 min (0.216734 cu in.)
		minim	min	1/60 fl dr (0.003612 cu in.)
MASS AND WEIGHT COAL, GRAIN, ETC. 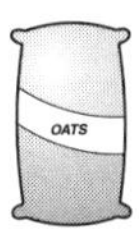	avoirdupois	ton	t	2000 lb
		short ton		2000 lb
		long ton		2240 lb
		pound	lb or #	16 oz, 7000 gr
		ounce	oz	16 dr, 437.5 gr
		dram	dr	27.344 gr, 0.0625 oz
		grain	gr	0.037 dr, 0.002286 oz
GOLD, SILVER, ETC. 	troy	pound	lb	12 oz, 240 dwt, 5760 gr
		ounce	oz	20 dwt, 480 gr
		pennyweight	dwt *or* pwt	24 gr, 0.05 oz
		grain	gr	0.042 dwt, 0.002083 oz
DRUGS 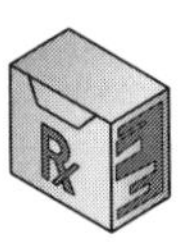	apothecaries'	pound	lb ap	12 oz, 5760 gr
		ounce	oz ap	8 dr ap, 480 gr
		dram	dr ap	3 s ap, 60 gr
		scruple	s ap	20 gr, .333 dr ap
		grain	gr	0.05 s, 0.002083 oz, 0.0166 dr ap

S.I. METRIC SYSTEM			
LENGTH	**Unit**	**Abbreviation**	**Number of Base Units**
	kilometer	km	1000
	hectometer	hm	100
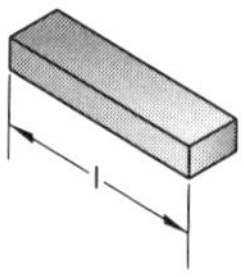	dekameter	dam	10
	meter*	m	1
	decimeter	dm	0.1
	centimeter	cm	0.01
	millimeter	mm	0.001
AREA 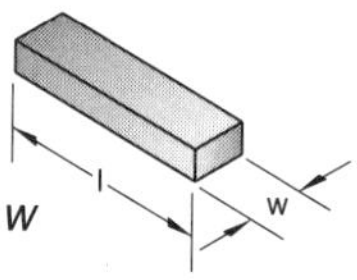	square kilometer	sq km *or* km^2	1,000,000
	hectare	ha	10,000
$A = l \times w$	are	a	100
	square centimeter	sq cm *or* cm^2	0.0001
VOLUME 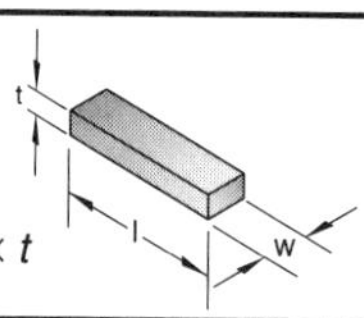	cubic centimeter	cu cm, cm^3, *or* cc	0.000001
	cubic decimeter	dm^3	0.001
$V = l \times w \times t$	cubic meter*	m^3	1
CAPACITY	kiloliter	kl	1000
	hectoliter	hl	100
WATER FUEL, ETC. 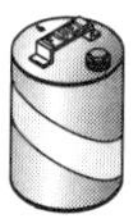	dekaliter	dal	10
	liter*	l	1
VEGETABLES, GRAIN, ETC. 	cubic decimeter	dm^3	1
	deciliter	dl	0.10
DRUGS 	centiliter	cl	0.01
	milliliter	ml	0.001
MASS AND WEIGHT	metric ton	t	1,000,000
COAL GRAIN, ETC. 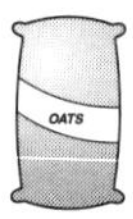	kilogram	kg	1000
	hectogram	hg	100
	dekagram	dag	10
GOLD SILVER, ETC. 	gram*	g	1
	decigram	dg	0.10
DRUGS 	centigram	cg	0.01
	milligram	mg	0.001

* base units

AREA—PLANE FIGURES

$A = l \times w$

where
A = area
l = length
w = width

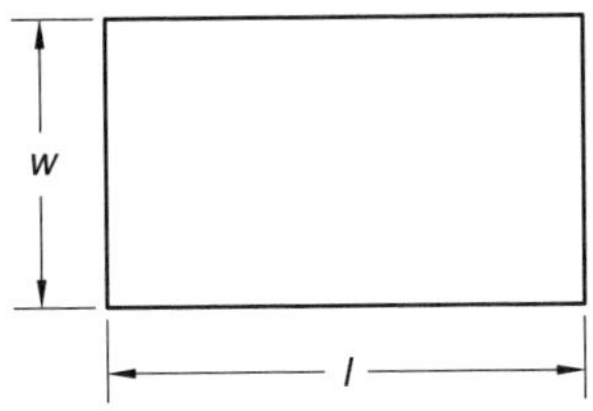

SQUARE OR RECTANGLE

$A = \frac{1}{2} \times b \times h$

where
A = area
b = base
h = height

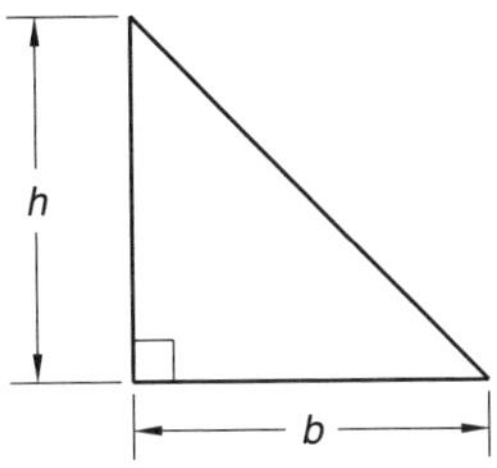

RIGHT TRIANGLE

$A = 0.7854d^2$

where
A = area
0.7854 = constant
d = diameter

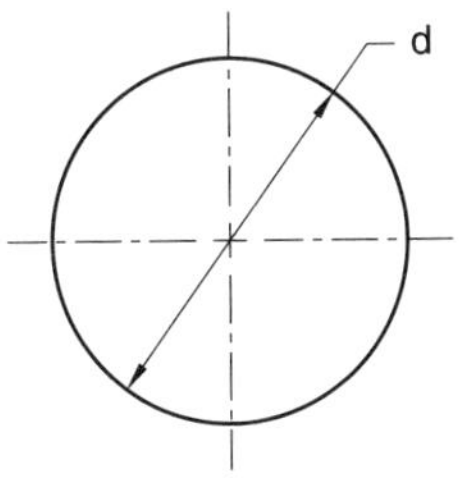

CIRCLE

VOLUME—SOLID FIGURES

$V = l \times w \times h$

where
V = volume
l = length
w = width
h = height

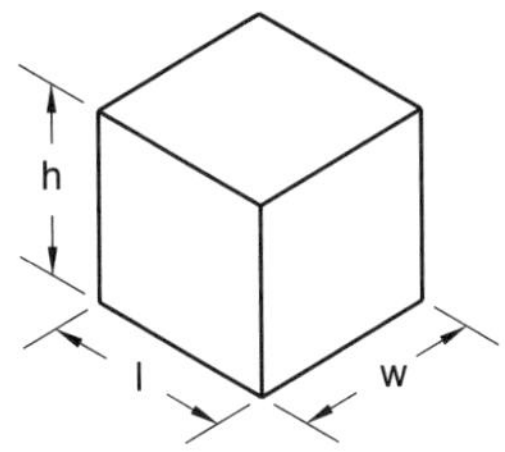

RIGHT RECTANGULAR PRISM

$$V = \frac{h\pi\,[(d_1)^2 + (d_2)^2 + (d_1 \times d_2)]}{12}$$

where
V = volume
h = height
π = 3.1416
d_1 = diameter of top
d_2 = diameter of base

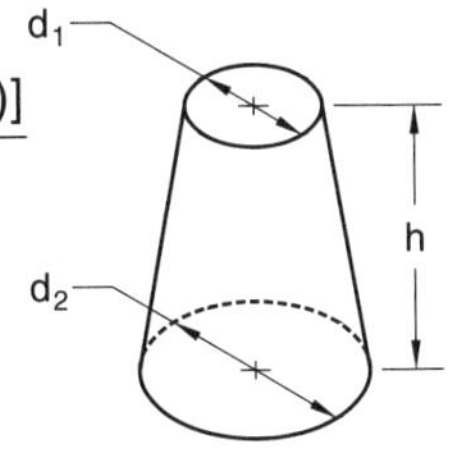

FRUSTUM OF A CONE

$V = 0.7854d^2 \times h$

where
V = volume
0.7854 = constant
d = diameter
h = height

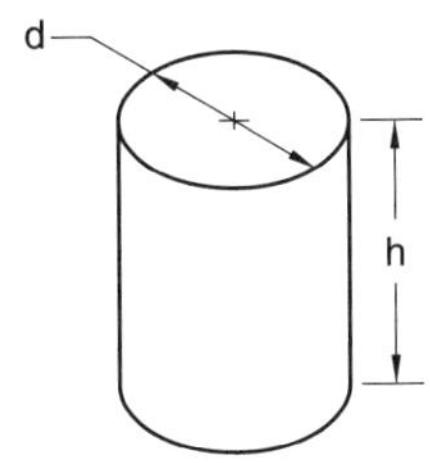

CYLINDER

CYLINDER AND TANK CAPACITIES*																		
Cylinder Diameter†	Cylinder Length‡																	
	1′	5′	6′	7′	8′	9′	10′	11′	12′	13′	14′	15′	16′	17′	18′	20′	22′	24′
1	0.04	0.20	0.24	0.28	0.32	0.36	0.40	0.44	0.48	0.52	0.56	0.60	0.64	0.68	0.72	0.80	0.88	0.96
2	0.16	0.80	0.96	1.12	1.28	1.44	1.60	1.76	1.92	2.08	2.24	2.40	2.56	2.72	2.88	3.20	3.52	3.84
3	0.37	1.84	2.20	2.56	2.92	3.30	3.68	4.04	4.40	4.76	5.12	5.48	5.84	6.22	6.60	7.36	8.08	8.80
4	0.65	3.26	3.92	4.58	5.24	5.88	6.52	7.18	7.84	8.50	9.16	9.82	10.5	11.1	11.8	13.0	14.4	15.7
5	1.02	5.10	6.12	7.14	8.16	9.18	10.2	11.2	12.2	13.3	14.3	15.3	16.3	17.3	18.4	20.4	22.4	24.4
6	1.47	7.34	8.80	10.3	11.8	13.2	14.7	16.1	17.6	19.1	20.6	22.0	23.6	25.0	26.4	29.4	32.2	35.2
7	2.00	10.0	12.0	14.0	16.0	18.0	20.0	22.0	24.0	26.0	28.0	30.0	32.0	34.0	36.0	40.0	44.0	48.0
8	2.61	13.0	15.6	18.2	20.8	23.4	26.0	28.6	31.2	33.8	36.4	39.0	41.6	44.2	46.8	52.0	57.2	62.4
9	3.31	16.5	19.8	23.1	26.4	29.8	33.0	36.4	39.6	43.0	46.2	49.6	52.8	56.2	60.0	66.0	72.4	79.2
10	4.08	20.4	24.4	28.4	32.6	36.8	40.8	44.8	48.8	52.8	56.8	61.0	65.2	69.4	73.6	81.6	89.6	97.6
11	4.94	24.6	29.6	34.6	39.4	44.4	49.2	54.2	59.2	64.2	69.2	74.0	78.8	8.38	88.8	98.4	104	118
12	5.88	29.4	35.2	41.0	46.8	52.8	58.8	64.6	70.4	76.2	82.0	87.8	93.6	99.6	106	118	129	141
13	6.90	34.6	41.6	48.6	55.2	62.2	69.2	76.2	83.2	90.2	97.2	104	110	117	124	138	152	166
14	8.00	40.0	48.0	56.0	64.0	72.0	80.0	88.0	96.0	104	112	120	128	136	144	160	176	192
15	9.18	46.0	55.2	64.4	73.6	82.8	92.0	101	110	120	129	138	147	156	166	184	202	220
16	10.4	52.0	62.4	72.8	83.2	93.6	104	114	125	135	146	156	166	177	187	208	229	250
17	11.8	59.0	70.8	81.6	94.4	106	118	130	142	153	163	177	189	201	212	236	260	283
18	13.2	66.0	79.2	92.4	106	119	132	145	158	172	185	198	211	224	240	264	290	317
19	14.7	73.6	88.4	103	118	132	147	162	177	192	206	221	235	250	265	294	324	354
20	16.3	81.6	98.0	114	130	147	163	180	196	212	229	245	261	277	294	326	359	392
21	18.0	90.0	108	126	144	162	180	198	216	238	252	270	288	306	324	360	396	432
22	19.8	99.0	119	139	158	178	198	218	238	257	277	297	317	337	356	396	436	476
23	21.6	108	130	151	173	194	216	238	259	281	302	324	346	367	389	432	476	518
24	23.5	118	141	165	188	212	235	259	282	306	330	353	376	400	424	470	518	564
25	25.5	128	153	179	204	230	255	281	306	332	358	383	408	434	460	510	562	612
26	27.6	138	166	193	221	248	276	304	331	359	386	414	442	470	496	552	608	662
27	29.7	148	178	208	238	267	297	326	356	386	416	426	476	504	534	594	652	712
28	32.0	160	192	224	256	288	320	352	384	416	448	480	512	544	576	640	704	768
29	34.3	171	206	240	274	309	343	377	412	446	480	514	548	584	618	686	754	824
30	36.7	183	220	257	294	330	367	404	440	476	514	550	588	624	660	734	808	880
32	41.8	209	251	293	334	376	418	460	502	544	586	628	668	710	752	836	920	1004
34	47.2	236	283	330	378	424	472	520	566	614	660	708	756	802	848	944	1040	1132
36	52.9	264	317	370.1	422	476	528	582	634	688	740	792	844	898	952	1056	1164	1268

* in gal.

† in in.

‡ in ft

QUANTITY SHEET

Project: ____________________

Estimator: ____________________

Sheet No. ____________________

Date: ____________________

Checked: ____________________

No.	Description	Dimensions				Unit		Unit		Unit		Unit		Unit

© 2012 American Technical Publishers, Inc. All rights reserved.

ESTIMATE SUMMARY SHEET

Project: ____________________
Estimator: ____________________

Sheet No. ____________________
Date: ____________________
Checked: ____________________

No.	Description	Dimensions			Quantity		Material		Labor		Equipment		Total	
						Unit	Unit Cost	Total	Unit Cost	Total	Unit Cost	Total	Unit Cost	Total
Total														

© 2012 American Technical Publishers, Inc. All rights reserved.

Glossary

A

abutment: A supporting structure at the end of a bridge, arch, or vault.

access door: A door used to enclose an area that houses concealed equipment.

access flooring: A raised floor system supported by short posts and runners.

access road: A temporary road installed by a general contractor to facilitate transportation of materials, equipment, and tradesworkers onto and out of a job site.

addendum: A change to the originally issued construction contract documents.

admixture: A substance other than water, aggregate, and portland cement that is added to concrete to modify its properties.

aggregate: Granular material such as gravel, sand, vermiculite, or perlite that is added to a cement and water mixture to form concrete, mortar, or plaster.

air conditioner: The component in a forced-air system that cools and conditions the air.

air filter: A porous device that removes particles from air.

air handler: A device used to distribute conditioned air to building spaces.

alkyd primer: A thermoplastic or synthetic resin-based primer.

American Standard I-beam: *See* S-shape.

angle of repose: The slope that a material maintains without sliding or caving in.

ashlar: Stone cut to precise dimensions according to the prints.

asphalt: Dark-colored pitch primarily composed of crushed stone and bituminous materials.

assembly: A collection of items needed to complete a particular unit of work.

assembly cost: The price of a number of common construction materials combined into a unit assembly, such as a wall constructed of studs, top and bottom plates, headers, and strap bracing.

astragal: One- or two-piece molding used to cover the opening between the stiles of a pair of doors to provide a weathertight seal.

averaging takeoff: A simplified electrical takeoff method that is useful for a nonspecialist in electrical estimating.

B

backfilling: The replacing of soil around the subsurface structure after construction has been completed.

balloon framing: A framing method in which individual studs extend from the sill plate of the first story to the top plate of the upper story.

baluster: The upright member that extends between the handrail and the treads.

base molding: The molding at the bottom of a wall at the intersection of the wall and floor.

beam: A horizontal structural member that is used to support loads over an opening.

beam-and-column construction: A structural steel construction method consisting of bays of framed structural steel members that are repeated to create large structures.

bent: A structural, interconnected system of timbers contained in a wall.

bid: An offer from a contractor, subcontractor, or supplier to, respectively, perform construction work, provide a construction service, or provide materials or supplies at a stated price.

bituminous roofing: Roofing composed of layers (plies) of asphalt-impregnated felt or fiberglass material that is fastened to the roof deck and mopped with hot tar to create a waterproof surface. Also known as built-up or hot tar roofing.

blocking: Wood fastened between structural members to strengthen the joint, provide structural support, or block the passage of air.

blockout: A frame set between concrete form walls to create a void in the finished concrete structure.

board foot (bf): A unit of measure based on the volume of a piece measuring 1″ thick by 12″ wide by 12″ long (144 cu in).

bond: 1. An insurance agreement for financial loss caused by the act or default of an individual or by some contingency over which the individual has no control. **2.** The arrangement of masonry units in a wall created by overlapping the masonry units one upon another to provide a sturdy structure.

bored pile: A foundation support formed by drilling into the ground to a predetermined depth and diameter and filling the hole with concrete.

boulder: A rock that has been transported by geological action from its formation site to its current location.

box: A metallic or nonmetallic electrical enclosure installed in an electrical system to protect or encase equipment, devices, connections, and pulling and terminating conductors.

brace: A structural piece, either permanent or temporary, designed to resist the weight or pressure of a load.

brick: A masonry unit made of clay or shale that is formed into a rectangular shape while soft and then fired in a kiln.

building information modeling (BIM): A methodology that uses coordinated, consistent formats for electronic design, construction, and product information to optimize planning and construction of a project.

building paper: Felt material saturated with tar to form a waterproof sheet.

built-up roofing: *See* bituminous roofing.

busway: A metal-enclosed distribution system of busbars available in prefabricated sections.

C

cabinet: An enclosure fitted with any combination of shelves, drawers, and doors and is usually used for storage.

cable tray: An open grid rack suspended from structural members to support a series of cables.

CAD and BIM compatibility: The ability of an estimating program to directly transfer quantity takeoff information from CAD-generated drawings or BIM systems.

caisson: A large-diameter cast-in-place concrete pile created by boring a hole and filling it with concrete.

camber: The slight upward curve in a structural member designed to compensate for deflection of the member under load.

carpet: A floor covering composed of natural or synthetic fibers woven through a backing material.

carriage: *See* stringer.

cavity wall: A vertical masonry structure in which the facing and backing wythes are completely separate except for metal ties joining the wythes.

change orders: Any changes to the construction project requested by the owner, contractor, or subcontractor.

channel: Light-gauge metal member used for framing and supporting gypsum board and other wall-finish materials.

circuit breaker: An overcurrent protection device with a mechanism that automatically opens a circuit when an overload condition or short circuit occurs.

clay: Natural mineral material that is compact and brittle when dry but plastic when wet.

clear and grub: The removal of vegetation and other obstacles from a job site in preparation for new construction.

closed stringer: A stringer installed at the meeting of a stairway and wall.

cofferdam: A watertight enclosure used to allow construction or repairs to be performed below the surface of the water.

coil tie: A type of concrete form hardware used in heavy construction in which a bolt screws into the tie and is removed during form removal.

column: A vertical structural member used to support axial compressive loads.

commissioning: The process of reviewing the final performance of building systems to ensure they meet owner and design requirements.

common board: A wood member that has been sawn or milled and has a nominal size from 1 × 4 to 1 × 12.

completion bond: A short-term insurance policy ensuring the owner that a construction project will be fully completed and free from encumbrances and liens when turned over to the owner.

composition shingle: An asphalt or fiberglass shingle coated with a layer of fine mineral gravel.

concrete: A mixture of portland cement, fine and coarse aggregate, and water that solidifies through a chemical reaction called hydration into a hard, strong mass.

concrete masonry unit (CMU): A precast hollow or solid block made of portland cement and fine aggregate, with or without admixtures or pigments.

conductor: A slender rod or wire that is used to control the flow of electrons in an electrical circuit.

conduit: Hollow pipe or tubing that supports and protects electrical conductors.

Construction Specifications Institute (CSI): An organization of individuals and organizations that includes architects, engineers, contractors, specifiers, and suppliers of construction products.

contour line: A dashed (existing grade) or solid (finished grade) line on a site plan used to show elevations of the surface.

contraction joint: *See* control joint.

contractor: An individual or company responsible for performing construction work in accordance with the terms of a contract.

control joint: **1.** A groove made in a vertical or horizontal concrete surface to create a weakened plane and control the location of cracking due to imposed loads. Also known as a contraction joint. **2.** A thin strip of perforated metal or vinyl applied in lath-and-plaster surfaces to relieve stress resulting from expansion and contraction in large ceiling and wall surfaces.

core drilling: The process of making holes in the ground to retrieve soil samples for analysis.

corner bead: A light-gauge, L-shaped galvanized metal or vinyl device used to cover outside corner joints in plaster walls.

corner post: A vertical support member consisting of studs or studs and blocks placed at the outside and inside corners of a framed building.

cost allocation: The designation of overall company equipment costs to particular pieces of equipment.

cost index: A compilation of a number of cost items from various sources in a common reference table.

cost-plus basis: A method of determining profit in which costs are submitted for payment for labor, material, equipment, and overhead, and then an amount or percentage is added automatically for profit.

course: A continuous horizontal layer of masonry units bonded with mortar.

cove molding: The trim material often used to cover the joint between the tread and the stringer and the joint between the tread and the top of the riser.

cradling: The temporary support of existing utility lines in or around an excavated area to properly protect the lines during construction.

crew-based estimate: An estimate in which material quantities, equipment, and labor costs are included in a single calculation based on a specific quantity of construction put in place.

cripple stud: A short wall stud placed between the header and the top plate or between the rough sill and the bottom plate.

curtain wall: A non-load-bearing prefabricated panel suspended on or fastened to structural members.

D

dampproofing: The treatment of a material that makes it moisture-resistant.

database transfer: The direct transfer by an estimating program of stored information from a database into a spreadsheet.

dead load: A permanent, stationary load composed of all construction materials, fixtures, and equipment permanently attached to a structure.

deck: A floor made of light-gauge metal panels (decking).

demolition: The organized destruction of a structure.

demountable partition: A type of room divider that is designed to be disassembled and reassembled as needed to form new room layouts.

depreciation: The accounting practice of reducing the value of equipment by a fixed amount each year.

detail takeoff estimate: An estimate in which each individual construction component is accounted for, or taken off.

digitizer: An electronic input device that converts analog data into digital data.

dimension lumber: Lumber that is precut to a particular size for the construction industry, normally having nominal sizes of 2 × 4 to 4 × 6.

door closer: A device that closes a door and controls the speed and closing action of the door.

driven pile: A steel or wood member that is driven into the ground with a pile-driving rig.

dry film thickness: The thickness of the resultant final mass of a paint application after the VOCs have evaporated.

ductwork: The distribution system for forced-air heating and cooling systems.

E

elastomeric roofing: Roofing made of a pliable synthetic polymer.

electronic estimating method: An estimating method in which quantity takeoff and pricing calculation is accomplished using computer spreadsheets and integrated database systems.

entrance door: The primary means of ingress and egress for pedestrian traffic in a commercial building.

Environmental Protection Agency (EPA): A federal government agency established in 1970 to control and reduce pollution in the areas of air, water, solid waste, pesticides, radiation, and toxic substances.

equipment leasing: The use of equipment based on a contract that stipulates a set amount of time and a set cost for the equipment use.

equipment rental: The use of equipment for a fixed cost per a given unit of time.

estimating: The computation of construction costs of a project.

estimating method: An approach used by an estimator to perform project analysis, quantity takeoff, and pricing in a consistent and organized manner.

estimating practice: A practice used to integrate all parts of the estimating process in a cohesive, consistent, and reliable manner to ensure an accurate final bid.

excavation: Any construction cut, cavity, trench, or depression in the earth's surface formed by earth removal.

exposure: The amount a shingle or shake is visible after installation.

exterior insulation and finish system (EIFS): An exterior finish system composed of a layer of exterior sheathing, insulation board, reinforcing mesh, a base coat of acrylic copolymers, and a textured finish.

F

face brick: Brick made of select clay to impart distinctive, uniform colors and is commonly used for an exposed, finished surface.

final grading: The process of smoothing and sloping a site for paving and landscaping.

finish wood member: A decorative and nonstructural wood component in a structure.

firebrick: Brick manufactured from refractory ceramic materials that resist disintegration at high temperatures.

fire door: A fire-resistant door and assembly, including the frame and hardware, commonly equipped with an automatic door closer.

flashing: **1.** A metal or flexible material embedded in a masonry wall to direct water out of the wall and prevent moisture penetration at exposed wall surfaces. **2.** Material installed at the seam or joint of two building components to inhibit air, water, or fire passage.

fluorescent lamp: A low-pressure discharge lamp in which ionization of mercury vapor transforms ultraviolet energy generated by the discharge into light.

folding door: A door formed of panel sections joined with hinges along the vertical edges of the panels and supported by rollers in a horizontal upper track.

footing: The section of a foundation that supports and distributes structural loads directly to the bearing surface of the soil, piles, or rock below the structure.

forced-air system: A system that uses the circulation of warm or cool air to condition building spaces.

form: A temporary structure or mold used to retain and support concrete while it sets and hardens.

form tie: A metal bar, strap, or wire used to hold concrete forms together at a predetermined distance and resist the outward pressure of fresh concrete.

formwork: The entire system of support for fresh concrete including the forms, hardware, and bracing.

foundation: The primary support for a structure through which all imposed loads are transmitted to the footings or the earth.

framed-tube construction: A structural steel construction method in which the structural steel members form an exterior load-bearing wall that resembles a structural steel tube.

furnace: A self-contained heating unit that includes a blower, burner or burners, a heat exchanger or electric heating elements, and controls.

fuse: An overcurrent protection device with a fusible link that melts and opens a circuit when an overload condition or short circuit occurs.

G

ganged panel form: A large form constructed by joining a series of small panels into a larger unit.

general conditions: The written agreements describing building components and various construction procedures included in the specifications for a construction project.

general contractor: The contractor who has overall responsibility for a construction project.

geotextile material: Any synthetic material in the form of a fabric sheet that stabilizes and retains soil or earth in position on slopes or in other unstable conditions.

girder: A large horizontal structural member that supports loads at isolated points along its length.

girt: A horizontal bracing member placed around the perimeter of a structure.

glass block: Hollow translucent or transparent block made of glass.

glazing: The transparent or translucent material installed in a window opening.

glulam (glued laminated) timber: An engineered lumber product composed of wood laminations (lams) that are bonded together with adhesives.

gooseneck: A curved section of a handrail.

grade beam: A reinforced short, vertical concrete wall placed below the slab at ground level and supported by piles or piers.

grading: **1.** The process of lowering high spots and filling in low spots of earth at a job site. **2.** The classification of various pieces of wood according to their quality and structural integrity.

granite: Extremely hard natural rock consisting of quartz, feldspar, and other minerals produced under intense heat and pressure.

gravel: Pieces of rock that are smaller than boulders and larger than sand.

gridding: The division of a topographic site map of large areas to be excavated or graded into smaller squares or grids.

ground: A wood or metal shape used as an edge for plaster material and a gauge for plaster thickness.

ground-fault circuit interrupter (GFCI) receptacle: A device with a disconnect that interrupts the flow of current when a ground fault exceeding a predetermined value of current occurs.

groundwater: Water present in subsurface material.

grout: A thin, fluid mortar made of portland cement, fine aggregate, lime, and water.

H

handrail: A support member that is grasped by the hand for support when using a stairway.

hardpan: A hard, compacted layer of soil, clay, or gravel.

hardwood: Wood produced from broadleaf, deciduous trees such as ash, birch, maple, oak, and walnut.

hazardous material: A material capable of posing a risk to health, safety, or property.

header: A horizontal framing member placed at the top of a window or door opening.

heavy soil: Soil that can be loosened by picks but is hard to loosen with shovels.

hinge: A pivoting hardware device that joins two surfaces or objects and allows them to swing around a pivot.

hot tar roofing: *See* bituminous roofing.

HP-pile: A steel HP-shape (beam) driven into the ground and is supported by a subsurface layer of earth.

hydronic system: A system that uses water, steam, or another liquid to condition the building spaces.

hydrostatic pressure: Pressure exerted by a fluid at rest.

I

igneous rock: Rock formed from the solidification of molten lava.

incandescent lamp: A lamp that produces light by the flow of current through a tungsten filament inside a gas-filled, sealed glass bulb.

infill material: Material, including sand, gravel, and crushed rock, used to increase the stability of the soil.

insulation: Material used as a barrier to inhibit thermal and sound transmission.

isolation joint: A separation between adjoining sections of concrete used to allow movement of the sections without affecting the adjacent section.

J

job cost analysis: The study of the final costs of building a project as compared to the original estimate.

joint sealant: Material installed between two members to seal the seam between the members.

joist: A horizontal structural member that supports the load of a floor or ceiling.

L

labor unit: The average time it takes an average tradesperson to install a specific type of material.

lagging: Horizontal planks stacked on edge to support the earth on the side of an excavation.

landing: The platform that breaks the stairway flight from one floor to another.

landing newel post: The main post supporting the handrail at a landing.

landscaping: The process of excavating, treating and preparing soil, and installing plants, including shrubs and trees, in specified locations.

latex primer: A waterbased primer that dries when the water evaporates.

lath: The backing fastened to structural members onto which the plaster is applied.

ledger sheet: A grid consisting of rows and columns on a sheet of paper into which item descriptions, locations, code numbers, quantities, and costs are entered.

license: A privilege granted by a state or other government agency to an individual or company to perform work or provide professional services.

lift-slab construction: A method of concrete construction in which concrete slabs are cast on top of each other, lifted into position using jacks, and secured to columns at the desired elevation.

light: A pane of glass or translucent material in a door.

light soil: Soil that is easily shoveled by hand without the aid of machines.

limestone: Sedimentary rock consisting of calcium carbonate.

lintel: A structural member placed over an opening or recess in a wall to support the construction above.

liquidated damages: Penalties assessed against the contractor or subcontractor for failure to complete work within a specified period of time.

live load: The total of all the dynamic loads a structure is capable of supporting including wind loads, human traffic, and any other imposed loads.

lockset: The complete assembly of a bolt, knobs, escutcheon, and all mechanical components for securing a door and providing a means for opening.

long-span construction: A structural steel construction method that consists of fastening large horizontal steel members together and bracing them with angles, channels, and other steel members to create large girders and trusses.

lumber: Wood used for construction that has been sawn and sized.

M

marble: Crystallized limestone.

MasterFormat™: A uniform system of numbers and titles used for organizing information about construction requirements, products, and activities into a standard sequence.

medium-density fiberboard (MDF): An engineered wood panel product manufactured from fine wood fibers mixed with binders and formed with heat and pressure.

medium soil: Soil that can be loosened by picks, shovels, and scrapers.

metal framing: A construction method in which a framework is constructed using light-gauge metal components.

metal-halide lamp: An HID lamp that produces light by an electric discharge through mercury vapor and metal halides in the arc tube.

metal roofing: Roofing made of steel, aluminum, copper, or various metal alloys.

metamorphic rock: Sedimentary or igneous rock that has been changed in composition or texture by extreme heat, pressure, or chemicals.

millwork: Finished wood materials or parts, such as moldings, jambs, and frames, completed in a mill or fabrication facility.

mobilization: The process of moving the required construction equipment and support systems to the current construction site.

moment-resisting construction: A structural steel construction method in which the stability of the steel frame and its wind and seismic resistance depend on a solid connection between the columns and beams.

mortar: A bonding mixture of fine aggregate, cement, and water that is used to fill voids between masonry units, join them together, and reinforce a structure.

movement joint: A continuous vertical or horizontal joint in a masonry wall without mortar or other nonresilient materials that is used to accommodate expansion and contraction of the wall system.

mullion: A vertical dividing member between two window units.

N

negotiated construction work: A project delivery system in which a qualified contractor submits a bid to an owner based on the plans and specifications without a competitive bid process.

nominal size: The size of rough lumber prior to planing.

O

Occupational Safety and Health Administration (OSHA): A federal agency, established under the Occupational Safety and Health Act of 1970, that requires employers to provide a safe environment for their employees.

Office of Federal Contract Compliance Programs (OFCCP): A federal government agency whose mandate is to promote affirmative action and equal employment opportunity on behalf of minorities, women, the physically challenged, and veterans.

open office systems furniture: Furniture that consists of panels that are connected in various configurations to provide working office space such as desktops, storage units, and lighting in a compact, flexible configuration.

open stringer: A stringer that has been cut out to support the treads on the open side of a stairway.

open web joist: A horizontal structural steel member constructed with steel angles that are used as chords, with steel bars or angles extending between the chords at an angle.

operable partition: A panel or partition that is moved by hand or motor along an overhead track.

oriented strand board (OSB): An engineered wood panel product in which wood strands are mechanically oriented and bonded with phenolic resin under heat and pressure.

overdig: The amount of excavation required beyond the dimensions of a structure to provide an area for construction activities.

overhead: Any business expense that is not chargeable to a particular part of a construction project.

overhead door: A door which, when opened, is suspended in a track above the opening.

P

paint: A mixture of minute solid particles (pigment) suspended in a liquid medium that serves as a vehicle.

panelboard: A wall-mounted distribution cabinet containing overcurrent and short-circuit protection devices for lighting, heating, or power circuits.

particleboard: An engineered wood panel product constructed of wood particles and flakes that are bonded together with a synthetic resin.

partition: A vertical divider that separates one area from another.

payline: A measurement used by excavation contractors to compute the amount of required excavation as stated in the specifications.

percentage basis: A method of determining profit in which a percentage is added to the final estimated cost.

performance bond: A short-term insurance agreement guaranteeing that a contractor will execute a construction project in the manner described in the contract documents.

permanent soil stabilization: The process of providing long-term stabilization of the ground at a job site through mechanical compaction or the use of infill materials.

permit: Written permission granted by the governmental authority having jurisdiction over the construction process to perform the work described in the prints and specifications according to local regulations.

pervious concrete pavement: Concrete pavement that is designed to allow surface water to penetrate through the pavement surface and minimize stormwater run-off.

pile: A structural member installed in the ground to provide vertical and/or horizontal support.

pile cap: A large unit of concrete placed on top of a pile or group of piles.

pipe fitting: A device fastened to the end of a pipe to terminate it or connect it to another pipe.

pipe pile: Pile made of steel pipe that is driven into the ground and is supported by friction against the earth.

plan markup: The color coding or marking off of items on the actual plan documents during the quantity takeoff process.

plaster: A mixture of portland cement, water, and sand used as an interior and exterior wall finish.

plastic laminate: Sheet material composed of multiple layers of plastic and resins bonded together under intense heat and pressure.

plate: A horizontal support member at the top and bottom of a framed wall to which studs are attached.

platform framing: A framing method in which single or multistory buildings are built one story at a time, with work proceeding in consecutive layers or platforms.

plug number: A monetary amount used as a temporary placeholder in an estimate during bid development.

plywood: An engineered wood panel product made of wood layers glued and pressed together under intense heat and pressure.

post-tensioning: A method of prestressing concrete in which the reinforcing cables (tendons) are tensioned after the concrete has hardened.

prebid meeting: A conference in which all interested parties in a construction project review the project, ask questions of the architect or owner concerning methods for accomplishing the work, and share information necessary to understand the entire scope of the work.

precast concrete pile: A pile cast with reinforcing steel and is designed to withstand the impact of the pile-driving hammer during installation.

pretensioning: A method of prestressing reinforced concrete in which the steel reinforcing cables (tendons) in the structural member are tensioned before the concrete has hardened.

primer: A coating applied to new and old surfaces to improve the adhesion and final quality of the finish coating.

process piping: Piping installed in commercial and industrial facilities for transporting compressed air, vacuum, gas, or fuel.

profit: The amount of money remaining at the completion of a project in excess of all project and company expenses.

purlin: A horizontal support member that spans across adjacent rafters or between beams, columns, or joists to carry intermediate loads, such as a roof deck or wall panels.

Q

quantity takeoff: The practice of reviewing construction contract documents to determine the quantities of materials to be included in a bid.

R

raceway: An enclosed channel for conductors.

range: A course of any thickness that extends across the face of a wall, but not all courses need to be the same thickness.

receptacle: A device used to connect equipment with a cord and plug into an electrical system.

reflected ceiling plan: A plan view with the viewpoint located beneath the object.

refractory material: A masonry unit that can withstand high temperatures without structural failure.

refractory mortar: Fire-resistant material made with high-temperature cement and carefully selected aggregate that does not expand and deteriorate when heated.

reinforced concrete: Concrete that has increased tensile strength due to tensile members, such as steel rods, bars, or mesh, embedded in the concrete.

reinforcement: Material embedded in or attached to another material to provide additional support or strength.

remediation: The act or process of correcting situations where hazardous materials exist and returning a site to environmentally safe conditions.

retaining wall: A wall constructed to hold back earth.

ribbon board: A supporting ledger applied horizontally across studs to support the end of joists.

riser: The piece forming the vertical face of a step.

roof: The covering for the top exterior surface of a building or structure.

roofing tile: Clay, slate, or lightweight concrete material installed as weather-protective finish on a roof.

roof paver: A flat, precast concrete unit that is set in place on a roof deck to provide a walking surface and to protect waterproof roofing materials.

rough sill: A horizontal framing member placed under window openings to support the window.

rubble: Rough, irregular stone that is used as fill or to provide a rustic appearance.

R value: The unit of measure for resistance to heat flow.

S

safety switch (disconnect): An electrical control device used periodically to remove electrical circuits from their supply source.

sand: Loose granular material consisting of particles that are smaller than gravel but coarser than silt.

sandstone: Sedimentary rock consisting of quartz held together by silica, iron oxide, and/or calcium carbonate.

sanitary sewer: A piping system designed to collect liquid and solid waste from a building and channel it to the appropriate removal or treatment system.

sealant: A liquid or semiliquid material that forms an airtight or waterproof joint.

sealer: A coating applied to a surface to close the pores and prevent penetration of liquids to the surface.

sedimentary rock: Rock formed from sedimentary materials such as sand, silt, and rock and shell fragments.

service: The electrical supply in the form of conductors and equipment that provides electrical power to a building or structure.

shake: A hand-split wood roofing material.

shaking out: The process of unloading steel members and assemblies in a planned manner to minimize any additional handling and moving of pieces during the erection process.

she bolt: A steel rod that threads into the ends of a tie rod.

sheet pile: An interlocking vertical support driven into the ground with a pile-driving rig to form part of a continuous structure.

shelf angle: A steel angle attached to structural backing members to support horizontal courses of brick for high masonry walls.

shingle: A thin piece of wood, asphalt-saturated felt, fiberglass, lightweight concrete, or other material applied to the surface of a roof or wall to provide a waterproof covering.

shoring: A system of wood or metal members used as temporary support for sides of excavations or construction components such as concrete formwork.

sill plate: A wood member, usually a pressure-treated 2 × 6, laid flat and fastened to the top of a foundation wall to provide a nailing base for floor joists or studs.

silt: Fine granular material that has been produced from the disintegration of rock.

slab on grade: A concrete slab that is placed directly on the ground.

slate: A fine-grained metamorphic rock that is easily split into thin sheets.

sliding door: A horizontal-moving door suspended on rollers that travel in a track that is fixed at the top of the opening or that moves on rollers mounted in the bottom of the door.

slope cut: An inclined or sloped wall excavation.

slump test: A test that measures the consistency, or slump, of concrete.

slurry: A mixture of water and fine particles such as cement that improves soil stability.

slurry wall: A wall built around an excavation to retain groundwater.

smoke shelf: A section of a chimney directly over a fireplace and below the flue liner that prevents downdrafts from blowing smoke into the living area.

snap tie: A concrete form tie with ends that are designed to be snapped off after the forms are removed and the concrete is set.

softwood: Wood produced from a conifer (evergreen) tree.

soil amendment: A material added to existing job site soil to increase its compaction and ability to maintain the desired angle of repose or slope at the completion of construction.

soil swell: The volume growth in soil after it is excavated.

spandrel beam: A beam in the perimeter of a structure that spans several columns and typically supports a floor or roof.

specifications: Written supplements to a set of construction documents and drawings that provide additional construction information.

spread footing: A foundation support with a wide base designed to spread a load over a wider area.

spreadsheet: A computer program that uses cells organized into rows and columns to perform various mathematical calculations when formulas and numbers are entered.

squared stone: Rough stone made approximately square either at a quarry or at the job site.

S-shape: A structural steel member with parallel outer flange surfaces and sloping inner flange surfaces that are joined with a perpendicular web. Also known as an American Standard I-beam.

stain: A pigmented wood finish used to seal wood while allowing the wood grain to be exposed.

starting newel post: The main post supporting the handrail at the bottom of a stairway.

storm sewer: A piping system designed to collect stormwater and channel it to a retention pond or other means of removal.

street bond: A bond that guarantees the repair of streets adjacent to a construction project if they are damaged during construction.

stringer: The support for a stairway. Also known as a carriage.

strongback: A vertical support attached to concrete formwork behind the horizontal walers to provide additional strength against form deflection during concrete placement.

structural clay tile: A hollow masonry unit composed of burned clay, shale, or a combination of clay and shale.

structural steel construction: A building process in which a series of horizontal steel beams and trusses and vertical columns are joined to create large structures with open areas.

stud: 1. A vertical steel-framing member in a wall that extends from a bottom track to a top track. **2.** A vertical support member that extends from the bottom to the top plates in a framed wall.

subcontractor: A specialty contractor who works under the direct control of the general contractor.

substrate: The underlying surface of a finish material.

subsurface investigation: The analysis of all materials below the surface of a job site to a predetermined depth.

supply piping: Piping that delivers water from a source to the point of use.

suspended ceiling: A ceiling hung from wires fastened to structural members.

switch: A device that is used to start, stop, or redirect the flow of current in an electrical circuit.

T

temporary soil stabilization: The process of holding existing surface and subsurface materials in place during the construction process.

terrazzo: A mixture of colored stone, marble chips, or other decorative aggregate and a portland cement, modified portland cement, or resinous matrix.

tie: A type of reinforcement used to secure two or more members together.

tie rod: A steel rod that extends between form walls to hold them together after concrete is placed.

tile: A thin building material made of cement, fired clay, glass, plastic, or stone.

tilt-up construction: A method of concrete construction in which concrete members are cast horizontally at a location close to their final position and tilted in place after the forms are stripped.

tilt-up panel: A concrete slab that is precast and lifted into place at the job site.

timber: Heavy lumber that has a 5 × 5 or larger nominal size.

timber framing: A framing method in which large wood members are used to form large open areas.

timber pile: A pile made of tapered, treated timber poles driven into the ground to a predetermined point of refusal.

time and material basis: A method of determining profit in which the owner pays a fixed rate for time spent on a project and pays for all materials.

traditional estimating method: An estimating method in which material quantities are calculated and entered onto ledger sheets for pricing.

transformer: An electric device that uses electromagnetism to change voltage from one level to another.

traprock: Fine-grained, dark-colored igneous rock.

tread: The horizontal surface of a step of a stairway.

trench box: A reinforced wood or metal assembly used to shore the sides of a trench.

truss: A manufactured roof or floor support member with components commonly placed in a triangular arrangement.

U

unit cost pricing: A method where each type of fixture is priced individually.

V

value engineering: A process whereby the owner and contractor agree on changes to the original design documents that will result in a comparable finished construction project with some substitutions and cost savings.

valve: A device that controls the pressure, direction, or rate of fluid flow.

volatile organic compound (VOC): The portion of a paint mixture that evaporates after application.

W

waler: A horizontal member used to align and brace concrete forms or piles.

wall-bearing construction: A structural steel construction method in which horizontal steel beams and joists are supported and reinforced with other construction materials, such as masonry or reinforced concrete.

water-cement ratio: The weight of the water divided by the weight of the cement contained in one cubic yard of concrete.

waterproofing: The treatment of a material that makes it impervious to water.

weave: The pattern or method used to join surface and backing carpet yarns.

well point: A pipe with a perforated point that is driven into the ground to allow for the removal and pumping of groundwater.

wide-flange beam: *See* W-shape.

wire mesh partition: A partition fabricated of wire mesh.

W-shape: A structural steel member with parallel inner and outer flange surfaces that are of constant thickness and are joined with a perpendicular web. Also known as a wide-flange beam.

wythe: A single, continuous vertical masonry wall that is one unit thick.

Z

zero line: A line connecting points on a topographic map where existing and planned elevations are equal.

Index

Numbers in italic refer to figures.

D

E

F

M

N

O

R

S

T

U

V

W

Z

USING THE *CONSTRUCTION ESTIMATING* INTERACTIVE CD-ROM

Before removing the Interactive CD-ROM from the protective sleeve, please note that the book cannot be returned for refund or credit if the CD-ROM sleeve seal is broken.

Windows System Requirements

To use this CD-ROM on a Windows® system, your computer must meet the following minimum system requirements:

- Microsoft® Windows® 7, Windows Vista®, or Windows® XP operating system
- Intel® 1.3 GHz processor (or equivalent)
- 128 MB of available RAM (256 MB recommended)
- 335 MB of available hard disk space
- 1024 × 768 monitor resolution
- CD-ROM drive (or equivalent optical drive)
- Sound output capability and speakers
- Microsoft® Internet Explorer® 6.0 or Firefox® 2.0 web browser
- Active Internet connection required for Internet links

Macintosh System Requirements

To use this CD-ROM on a Macintosh® system, your computer must meet the following minimum system requirements:

- Mac OS X 10.5 (Leopard) or 10.6 (Snow Leopard)
- PowerPC® G4, G5, or Intel® processor
- 128 MB of available RAM (256 MB recommended)
- 335 MB of available hard disk space
- 1024 × 768 monitor resolution
- CD-ROM drive (or equivalent optical drive)
- Sound output capability and speakers
- Apple® Safari® 2.0 web browser or later
- Active Internet connection required for Internet links

Opening Files

Insert the Interactive CD-ROM into the computer CD-ROM drive. Within a few seconds, the home screen will be displayed allowing access to all features of the CD-ROM. Information about the usage of the CD-ROM can be accessed by clicking on Using This Interactive CD-ROM. The Quick Quizzes®, Illustrated Glossary, Flash Cards, Activities, Web links, Getting Started—Sage Timberline Estimating, Tutorial Videos, and ATPeResources.com can be accessed by clicking on the appropriate button on the home screen. Clicking on the ATP web site button (www.go2atp.com) accesses information on related educational products. Unauthorized reproduction of the material on this CD-ROM is strictly prohibited.

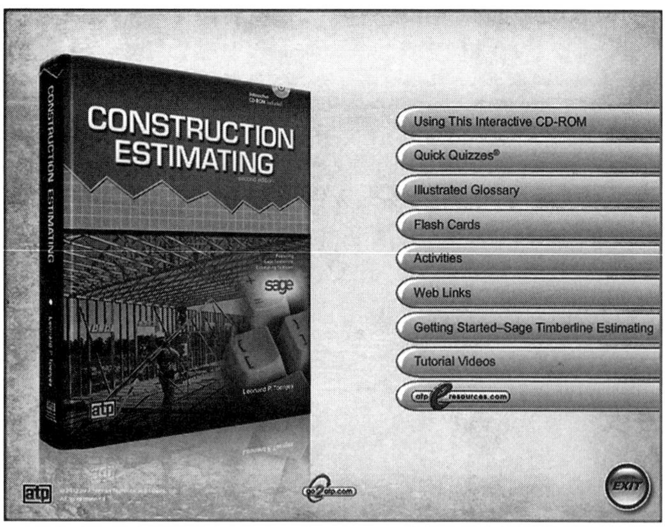

Microsoft, Windows, Windows Vista, PowerPoint, and Internet Explorer are either registered trademarks or trademarks of Microsoft Corporation in the United States and/or other countries. Adobe, Acrobat, and Reader are either registered trademarks of Adobe Systems Incorporated in the United States and/or other countries. Intel is a registered trademark of Intel Corporation in the United States and/or other countries. Firefox is a registered trademark of Mozilla Corporation in the United States and other countries. Apple, Macintosh, and Safari are registered trademarks of Apple, Inc. PowerPC is a registered trademark of International Business Machines Corporation. Quick Quiz, Quick Quizzes, and Master Math are either registered trademarks or trademarks of American Technical Publishers, Inc.